T0183205

Lecture Notes in Computer Science   9676

Commenced Publication in 1973
Founding and Former Series Editors:
Gerhard Goos, Juris Hartmanis, and Jan van Leeuwen

Editorial Board

David Hutchison
  Lancaster University, Lancaster, UK
Takeo Kanade
  Carnegie Mellon University, Pittsburgh, PA, USA
Josef Kittler
  University of Surrey, Guildford, UK
Jon M. Kleinberg
  Cornell University, Ithaca, NY, USA
Friedemann Mattern
  ETH Zurich, Zürich, Switzerland
John C. Mitchell
  Stanford University, Stanford, CA, USA
Moni Naor
  Weizmann Institute of Science, Rehovot, Israel
C. Pandu Rangan
  Indian Institute of Technology, Madras, India
Bernhard Steffen
  TU Dortmund University, Dortmund, Germany
Demetri Terzopoulos
  University of California, Los Angeles, CA, USA
Doug Tygar
  University of California, Berkeley, CA, USA
Gerhard Weikum
  Max Planck Institute for Informatics, Saarbrücken, Germany

More information about this series at http://www.springer.com/series/7407

Claude-Guy Quimper (Ed.)

Integration of AI and OR Techniques in Constraint Programming

13th International Conference, CPAIOR 2016
Banff, AB, Canada, May 29 – June 1, 2016
Proceedings

 Springer

Editor
Claude-Guy Quimper
Université Laval
Quebec, QC
Canada

ISSN 0302-9743   ISSN 1611-3349   (electronic)
Lecture Notes in Computer Science
ISBN 978-3-319-33953-5   ISBN 978-3-319-33954-2   (eBook)
DOI 10.1007/978-3-319-33954-2

Library of Congress Control Number: 2016937948

LNCS Sublibrary: SL1 – Theoretical Computer Science and General Issues

© Springer International Publishing Switzerland 2016
This work is subject to copyright. All rights are reserved by the Publisher, whether the whole or part of the material is concerned, specifically the rights of translation, reprinting, reuse of illustrations, recitation, broadcasting, reproduction on microfilms or in any other physical way, and transmission or information storage and retrieval, electronic adaptation, computer software, or by similar or dissimilar methodology now known or hereafter developed.
The use of general descriptive names, registered names, trademarks, service marks, etc. in this publication does not imply, even in the absence of a specific statement, that such names are exempt from the relevant protective laws and regulations and therefore free for general use.
The publisher, the authors and the editors are safe to assume that the advice and information in this book are believed to be true and accurate at the date of publication. Neither the publisher nor the authors or the editors give a warranty, express or implied, with respect to the material contained herein or for any errors or omissions that may have been made.

Printed on acid-free paper

This Springer imprint is published by Springer Nature
The registered company is Springer International Publishing AG Switzerland

Preface

The 13th International Conference on Integration of Artificial Intelligence and Operations Research Techniques in Constraint Programming, was held in Banff, Canada, May 29 to June 1, 2016. It was co-located with CORS 2016, the conference of the Canadian Operational Research Society.

The aim of the conference is to bring together interested researchers from constraint programming (CP), artificial intelligence (AI), and operations research (OR) to present new techniques or applications in combinatorial optimization and to provide an opportunity for researchers in one area to learn about techniques in the others. A main objective of this conference series is also to give these researchers the opportunity to show how the integration of techniques from different fields can lead to interesting results on large and complex problems. Therefore, papers that actively combine, integrate, or contrast approaches from more than one of the areas were especially solicited. High-quality papers from a single area were also welcome, provided that they are of interest to other communities involved. Application papers showcasing CP/AI/OR techniques on novel and challenging applications or experience reports on such applications were strongly encouraged.

There were 51 papers submitted. Each paper received at least three independent peer reviews. From this process, 33 papers were accepted. Among these accepted papers, four were published in the journal *Constraints*.

The conference included an invited talk given by Pascal Van Hentenryck. The first day of the conference was a Master Class about decomposition methods. Jean-François Cordeau, John Hooker, Christopher Beck, Bernard Gendron, Willem-Jan van Hoeve, and Louis-Martin Rouseau gave a one-hour talk on topics covering the classic and logic-based Benders decomposition, the Lagrangian relaxation in MIP and CP, as well as column generation.

March 2016   Claude-Guy Quimper

Organization

Program Committee

Chris Beck	University of Toronto, Canada
Nicolas Beldiceanu	TASC (CNRS/Inria), Mines Nantes, France
David Bergman	University of Connecticut, USA
Lucas Bordeaux	Microsoft Research, UK
Andre Cire	University of Toronto Scarborough, Canada
Jean-Guillaume Fages	COSLING S.A.S., France
Bernard Gendron	Université de Montréal, Canada
Tias Guns	KU Leuven, Belgium
Emmanuel Hebrard	LAAS, CNRS, France
John Hooker	Carnegie Mellon University, USA
George Katsirelos	INRA, Toulouse, France
Philip Kilby	NICTA and the Australian National University, Australia
Andrea Lodi	École Polytechnique de Montréal, Canada
Michele Lombardi	DISI, University of Bologna, Italy
Laurent Michel	University of Connecticut, USA
Barry O'Sullivan	4C, University College Cork, Ireland
Gilles Pesant	École Polytechnique de Montréal, Canada
Claude-Guy Quimper	Université Laval, Canada
Jean-Charles Regin	University of Nice-Sophia Antipolis/I3S/CNRS, France
Louis-Martin Rousseau	École Polytechnique de Montréal, Canada
Pierre Schaus	UC Louvain, Belgium
Christian Schulte	KTH Royal Institute of Technology, Sweden
Meinolf Sellmann	IBM Research, USA
Paul Shaw	IBM, France
Peter J. Stuckey	University of Melbourne, Australia
Pascal Van Hentenryck	University of Michigan, USA
Willem-Jan Van Hoeve	Carnegie Mellon University, USA
Petr Vilím	IBM, Czech Republic
Mark Wallace	Monash University, Australia
Toby Walsh	NICTA and UNSW, Australia

Additional Reviewers

Bal, Deepak	Carbonnel, Clément
Borghesi, Andrea	Cardonha, Carlos
Borghetti, Alberto	Carlsson, Mats
Bridi, Thomas	Castañeda Lozano, Roberto

den Hertog, Dick
Fontaine, Daniel
Gay, Steven
Harabor, Daniel
Hendel, Gregor
Huguet, Marie-José
Johnson, Greg
Kelly, Richard
Kinable, Joris
Ku, Wen-Yang

Lhomme, Olivier
Malapert, Arnaud
Monaci, Michele
Perez, Guillaume
Salvagnin, Domenico
Siala, Mohamed
Tjandraatmadja, Christian
Tran, Tony T.
Tubertini, Paolo
Yunes, Tallys

Abstracts of Fast Tracked
Journal Papers

Breaking Symmetries in Graph Coloring
Problems with Degree Matrices:
The Ramsey Number R(4, 3, 3) = 30

Michael Codish[1], Michael Frank[1], Avraham Itzhakov[1],
and Alice Miller[2]

[1] Department of Computer Science,
Ben-Gurion University of the Negev, Beersheba, Israel
[2] School of Computing Science, University of Glasgow, Glasgow, Scotland

Ramsey numbers are notoriously hard graph coloring problems. An $(r_1, \ldots, r_k; n)$ Ramsey coloring is a graph coloring in k colors of the complete graph K_n that does not contain a monochromatic complete sub-graph K_{r_i} in color i for each $1 \leq i \leq k$. The set of all such colorings is denoted $\mathcal{R}(r_1, \ldots, r_k; n)$. The Ramsey number $R(r_1, \ldots, r_k)$ is the least $n > 0$ such that no $(r_1, \ldots, r_k; n)$ coloring exists.

The Ramsey number $R(4, 3, 3)$ is often presented as the unknown Ramsey number with the best chance of being found "soon". Yet, its precise value has remained unknown for more than 50 years. This paper presents a methodology based on *abstraction* and *symmetry breaking* that is demonstrated by using it to compute the value $R(4, 3, 3) = 30$.

It was previously known that $30 \leq R(4, 3, 3) \leq 31$ [4]. Kalbfleisch [2] proved in 1966 that $R(4, 3, 3) \geq 30$, Piwakowski [3] proved in 1997 that $R(4, 3, 3) \leq 32$, and one year later Piwakowski and Radziszowski [4] proved that $R(4, 3, 3) \leq 31$. We demonstrate how our methodology applies to computationally prove that $R(4, 3, 3) = 30$. Our approach involves applying an embedding technique to conclude that if a $(4, 3, 3; 30)$ Ramsey coloring exists then it must be $\langle 13, 8, 8 \rangle$ regular. To determine if there exists a $\langle 13, 8, 8 \rangle$ regular $(4, 3, 3; 30)$ Ramsey coloring required first computing the previously unknown set $\mathcal{R}(3, 3, 3; 13)$, which was shown to have size 78,892. To do this we demonstrate that an existing symmetry breaking technique combining SAT solving with *symmetry breaking* [1] works for smaller instances but not for $\mathcal{R}(3, 3, 3; 13)$. Instead we use a new abstraction referred to as *degree matrices*. Having determined $\mathcal{R}(3, 3, 3; 13)$ we then use it within the embedding approach to achieve the major result of this paper: that there is no $(4, 3, 3; 30)$ Ramsey coloring, and so $R(4, 3, 3) = 30$.

Supported by the Israel Science Foundation, grant 82/13. Computational resources provided by an IBM Shared University Award (Israel).

References

1. Codish, M., Miller, A., Prosser, P., Stuckey, P.J.: Breaking symmetries in graph representation. In: Rossi, F. (ed.) Proceedings of the 23rd International Joint Conference on Artificial Intelligence, Beijing, China. IJCAI/AAAI (2013)
2. Kalbfleisch, J.G.: Chromatic graphs and Ramsey's theorem. Ph.D. thesis, University of Waterloo (1966)
3. Piwakowski, K.: On Ramsey number r(4, 3, 3) and triangle-free edge-chromatic graphs in three colors. Discrete Math. **164**(1–3), 243–249 (1997)
4. Piwakowski, K., Radziszowski, S.P.: $30 \leq R(3,3,4) \leq 31$. J. Comb. Math. Comb. Comput. **27**, 135–141 (1998)

Multi-language Evaluation of Exact Solvers in Graphical Model Discrete Optimization (Summary)

Barry Hurley[1], Barry O'Sullivan[1], David Allouche[2],
George Katsirelos[2], Thomas Schiex[2], Matthias Zytnicki[2],
and Simon de Givry[2]

[1] Insight Centre for Data Analytics, University College Cork, Cork, Ireland
{barry.hurley, barry.osullivan}@insight-centre.org
[2] MIAT, UR-875, INRA, 31320 Castanet Tolosan, France
{david.allouche, george.katsirelos, thomas.schiex,
matthias.zytnicki, simon.givry}@toulouse.inra.fr

By representing the constraints and objective function in factorized form, graphical models can concisely define various NP-hard optimization problems. They are, therefore, extensively used in several areas of computer science and artificial intelligence. Graphical models can be deterministic, *e.g.*, Constraint Networks (in Minizinc mzn format) and weighted variants such as Cost Function Networks, aka Weighted Constraint Satisfaction Problems (wcsp), or stochastic, *e.g.*, Bayesian Networks and Markov Random Fields (uai). They optimize a sum *or* product of local functions (constraints being represented as functions with values in $\{0, \infty\}$ or $\{0, 1\}$ *resp.*), defining a joint cost or probability distribution for discrete variables. Simple transformations exist between these two types of models, but also with MaxSAT (wcnf) and linear programming (lp).

We report on a large comparison of exact solvers which are all state-of-the-art for their own target language. These solvers are all evaluated on deterministic and probabilistic graphical models coming from the Probabilistic Inference Challenge 2011, the Computer Vision and Pattern Recognition OpenGM2 benchmark, the Weighted Partial MaxSAT Evaluation 2013, the MaxCSP 2008 Competition, the MiniZinc Challenge 2012 & 2013, and the CFN-Lib, a library of Cost Function Networks.

3026 problems divided into 43 categories	DAOOPT	TOULBAR2	CPLEX	CPLEX$_{tuple}$	MAXHS	MAXHS$_{tuple}$	GECODE
Nb. of problems solved in less than 1 hour	1832	**2433**	1273	1862	1417	1567	202
Borda-score (see Minizinc Chal., norm. by nb. of applicable categories)	2.08 [5]	**4.24 [1]**	3.01 [2]	2.86 [3]	2.66 [4]	1.65 [7]	1.84 [6]

All 3026 instances are made available in five different formats (mzn, wcsp, uai, wcnf, lp) and seven formulations (two encodings for wcnf and lp, including one based

Supported by grants SFI/10/IN.1/I3032, SFI/12/RC/2289, and the GenoToul Bioinfo. platform.

on the so-called local polytope)[1]. The results show that a small number of evaluated solvers are able to perform well on multiple areas. By exploiting the variability and complementarity of solver performances, we show that a portfolio approach based on TOULBAR2[2], MPLP2[3], and CPLEX 12.6, can be very effective, winning the 2014 Uncertainty in Artificial Intelligence (UAI) Evaluation[4,5]. We hope that our collection of benchmarks, available in many formats, will enrich the various competitions in CP, AI, and OR, leading to more robust solvers and new solving strategies.

---

[1] http://genoweb.toulouse.inra.fr/~degivry/evalgm.

[2] http://www.inra.fr/mia/T/toulbar2 (version 0.9.8, parameters -A -V -dee -hbfs).

[3] http://cs.nyu.edu/~dsontag/ (version 2).

[4] http://www.hlt.utdallas.edu/~vgogate/uai14-competition/leaders.html (MAP/Proteus).

[5] https://github.com/9thbit/uai-proteus.

Breaking Symmetries in Graph Search
with Canonizing Sets

Avraham Itzhakov and Michael Codish

Department of Computer Science,
Ben-Gurion University of the Negev, Beersheba, Israel

There are many complex combinatorial problems which involve searching for an undirected graph satisfying given constraints. Such graph search problems are often highly challenging because of the large number of isomorphic representations of their solutions. One common approach to eliminate symmetries is to introduce symmetry breaking constraints [2–4] which rule out isomorphic solutions thus reducing the size of the search space while preserving the set of solutions. A complete symmetry breaking constraint eliminates all symmetries. But, the standard approach to define complete symmetry breaking constraints introduces constraints for each permutation of the graph vertices and is too large to be considered practical. Previous work specifies compact but partial symmetry breaking constraints for graphs [1]. These eliminate some but not all of the symmetries. This paper introduces effective and compact, complete symmetry breaking constraints for small graph search problems with up to 10 vertices. We show that for 10 vertices, instead of considering 10! = 3,628,800 permutations of the vertices, it suffices to consider only 7853 permutations. For small search problems with a larger number of vertices we demonstrate the computation of instance dependent symmetry breaking constraints which are complete. We illustrate the application of complete symmetry breaking constraints to extend two known sequences from the OEIS related to graph enumeration. We also demonstrate the application of a generalization of our approach to fully-interchangeable matrix search problems.

References

1. Codish, M., Miller, A., Prosser, P., Stuckey, P.J.: Breaking symmetries in graph representation. In: Rossi, F. (ed.) Proceedings of IJCAI 2013
2. Crawford, J.M., Ginsberg, M.L., Luks, E.M., Roy, A.: Symmetry-breaking predicates for search problems. In: Aiello, L.C., Doyle, J., Shapiro, S.C. (eds.) Proceedings of KR 1996
3. Shlyakhter, I.: Generating effective symmetry-breaking predicates for search problems. Discrete Applied Math. **155**(12), 1539–1548 (2007)
4. Walsh, T.: General symmetry breaking constraints. In: Benhamou, F. (ed.) Proceedings of CP 2006

Supported by the Israel Science Foundation, grant 182/13. Computational resources provided by an IBM Shared University Award (Israel).

A Branch-and-Price-and-Check Model
for the Vehicle Routing Problem
with Location Congestion

Edward Lam[1,2], and Pascal Van Hentenryck[3]

[1] University of Melbourne, Parkville, VIC 3010, Australia
[2] NICTA, Eveleigh, NSW 2015, Australia
[3] University of Michigan, Ann Arbor, MI 48109-2117, USA

Abstract. The vehicle routing family of problems are combinatorial optimization problems that aim to construct routes for a fleet of vehicles that service requests while minimizing some cost function. Some of these problems may feature additional side constraints, such as time windows that restrict the time at which service of requests can commence. This paper considers a Vehicle Routing Problem with Pickup and Delivery, Time Windows, and Location Congestion (VRPPDTWLC, or VRPLC for short). In the VRPLC, requests are situated at a number of locations. Each location provides cumulative resources that are utilized by vehicles either during service (e.g., forklifts) or for the entirety of their visit (e.g., parking bays). Locations can become congested if insufficient resources are available, upon which vehicles must wait until a resource becomes available before proceeding. Modeling location congestion leads to temporal dependencies between vehicles, and a scheduling substructure not present in conventional vehicle routing problems. Specifically, the VRPLC incorporates both a vehicle routing problem and a resource-constrained project scheduling problem, making it exceptionally challenging from a computational standpoint. The main contribution of this paper is a branch-and-price-and-check (BPC) model that uses a branch-and-price algorithm that solves the underlying vehicle routing problem, and a constraint programming subproblem that checks the feasibility of the resource constraints, and adds combinatorial Benders cuts (or nogoods) to the master problem if any resource constraint is violated. The BPC model is compared to a regular mixed integer programming model and a constraint programming model. The three models are evaluated on instances with up to 300 requests (150 pickup and delivery requests) and both types of resources. Results indicate that the BPC algorithm scales better than both the mixed integer programming and the constraint programming models, optimally solves instances with up to 80 requests in under 10 minutes, and finds high quality solutions to larger problems. The BPC model nicely exploits the strengths of constraint programming for scheduling and branch-and-price for vehicle routing.

Contents

On CNF Encodings of Decision Diagrams

Ignasi Abío[1], Graeme Gange[3], Valentin Mayer-Eichberger[1,2(✉)],
and Peter J. Stuckey[1,3]

[1] NICTA, West Melbourne, Australia
{ignasi.abio,valentin.mayer-eichberger,peter.stuckey}@nicta.com.au
[2] University of New South Wales, Sydney, Australia
[3] University of Melbourne, Melbourne, Australia
gkgange@unimelb.edu.au

Abstract. Decisions diagrams such as Binary Decision Diagrams (BDDs), Multi-valued Decision Diagrams (MDDs) and Negation Normal Forms (NNFs) provide succinct ways of representing Boolean and other finite functions. Hence they provide a powerful tool for modelling complex constraints in discrete satisfaction and optimization problems. Generic propagators for these global constraints exist, but they are complex and hard to implement. An alternative approach to making use of them for solving is to encode them to CNF, using SAT style solving technology to implement them efficiently. This may also have advantages since it is naturally incremental and exposes intermediate literals which may well be useful as search decisions for solving the problem.

In this paper we explore different ways that we can map these constraints to CNF, and the different properties these mappings maintain. Surprisingly the most used encoding of BDDs does not maintain domain consistency in arbitrary BDDs. We also consider the strength of propagation with respect to the intermediate literals. We give experiments which compare the performance of the different encodings.

1 Introduction

Decisions diagrams such as Binary Decision Diagrams (BDDs), Multi Decision Diagrams (MDDs) and Negation Normal Forms (NNFs) provide succinct ways of representing Boolean and other finite functions. Hence they provide a powerful tool for modelling complex constraints in discrete satisfaction and optimization problems.

Constraint programming solvers include generic propagators for propagating constraints represented by BDDs [16], MDDs [8] and NNFs [15], since they are highly flexible, and hence useful in many different models. But these propagators are complex and hard to implement.

An alternative approach to making use of them for solving is to encode them to CNF, using SAT style solving technology to implement them efficiently. If the remainder of the problem is naturally modelled in CNF then this allows a SAT solver to tackle the problem.

© Springer International Publishing Switzerland 2016
C.-G. Quimper (Ed.): CPAIOR 2016, LNCS 9676, pp. 1–17, 2016.
DOI: 10.1007/978-3-319-33954-2_1

A SAT encoding may also be preferable within a CP solver, as it avoids the need for implementing complex propagators, is naturally incremental, and exposes intermediate literals as candidates for search and learning. A good encoding is critical in lazy decomposition approaches [1], where a propagator that participates in many conflicts is replaced by a CNF decomposition during runtime.

In this paper we explore different approaches for encoding decision diagrams to CNF.[1] The contributions of this paper are:

- An investigation of a large design space for encoding decision diagrams.
- We clarify the picture of BDD/MDD/NNF encodings, analyse their propagation strength and correct some misunderstandings in the literature.
- We introduce an encoding of BDDs and MDDs where unit propagation implements propagation completeness.
- Experiments which compare the performance of the different encodings.

2   Preliminaries

2.1   SAT Solving

We denote the Boolean value true by $\top$ and false by $\bot$.

Let $\mathcal{Y} = \{y_1, y_2, \ldots\}$ be a fixed set of propositional *variables*. If $y \in \mathcal{Y}$ then y and $\neg y$ are *positive* and *negative literals*, respectively. The *negation* of a literal l, written $\neg l$, denotes $\neg y$ if l is y, and y if l is $\neg y$. A *clause* is a disjunction of literals $\neg y_1 \vee \cdots \vee \neg y_p \vee y_{p+1} \vee \cdots \vee y_n$, sometimes written as $y_1 \wedge \cdots \wedge y_p \rightarrow y_{p+1} \vee \cdots \vee y_n$. A *CNF formula F* is a conjunction of clauses.

A set of literals A is *contradictory* if $\exists y.\{y, \neg y\} \subset A$. A (partial) *assignment* A is a set of literals which is not contradictory. A literal l is *true* in A if $l \in A$, is *false* in A if $\neg l \in A$, and is *undefined* in A otherwise. An *extension* of an assignment A is an assignment A' where $A' \supset A$. A *complete assignment* is an assignment with no undefined literals. Given a partial assignment A, a *completion* of A is an extension of A which is a complete assignment.

A complete assignment A satisfies formula ϕ if replacing each y in ϕ which is true in A with $\top$ and replacing each y in ϕ which is false in A with $\bot$ gives an expression which evaluates to $\top$. A partial assignment A satisfies formula ϕ, written $A \models \phi$ if every completion of A satisfies ϕ.

Systems that decide whether a CNF formula F has any model are called SAT solvers, and the main inference rule they implement is *unit propagation*: given a CNF F and an assignment A, find a clause in F such that all its literals are false in A except at most one, say l, which is undefined, add l to A and repeat the process until reaching a fix-point. See e.g. [21] for more details.

For some set of clauses C, we shall use $UP_C(A)$ to denote the set of literals inferred by unit propagation on C starting from assignment A. We will omit the C subscript when clear from the context. Note that $UP_C(A)$ may be contradictory, in which case unit propagation has detected unsatisfiability.

---

[1] A longer version of this paper including proofs of all Theorems can be found at http://people.eng.unimelb.edu.au/pstuckey/mddenc.pdf.

2.2   Propositional Encodings

Problems of interest rarely (if ever) begin in CNF form. Boolean formulae ϕ must be first converted into some equisatisfiable conjunction of clauses F_ϕ. The seminal work here is the Tseytin transformation [25], later refined by Plaisted and Greenbaum [22], which introduces a variable for each sub-formula and adds clauses to enforce the semantics of each connective.

While equisatisfiability is sufficient for correctness, the choice of decomposition can have a great impact on solver performance. A major consideration here is *propagation strength* – that is, given some partial assignment A and formula ϕ, what can be said of $UP_{F_\phi}(A)$.

There are a number of properties we may wish of F_ϕ.

- An encoding F_ϕ for a formula ϕ is *correct* if any complete assignment A on $vars(\phi)$ where $A \models \phi$, then A has an extension satisfying F_ϕ, and any complete assignment $A \models \neg\phi$ has no extension satisfying F_ϕ.
- An encoding F_ϕ for a formula ϕ *implements consistency* if for every assignment A over $vars(\phi)$ where $A \models \neg\phi$, then $UP_{F_\phi}(A)$ is contradictory.
- An encoding F_ϕ for a formula ϕ *implements domain consistency* when for each literal l over $vars(\phi)$, if $A \models \phi \to l$ then $l \in UP_{F_\phi}(A)$.
- An encoding F_ϕ for a formula ϕ *implements unit refutation completeness* [26] (also called *SLUR* [19]) when for assignment B over $vars(F_\phi)$ where $B \models \neg F_\phi$, then $UP_{F_\phi}(B)$ is contradictory.
- An encoding F_ϕ for a formula ϕ *implements propagation completeness* [6,19] when for each literal l over $vars(F_\phi)$, $B \models F_\phi \to l$ then $l \in UP_{F_\phi}(B)$.

Another important consideration is the encoding size. In general, smaller encodings are more efficient than larger ones, if both have the same propagation strength.

2.3   At-most-one and Exactly-one Constraints

Given a set of literals $l_1, \ldots, l_n$, the *At-most-one* (AMO) constraint over these literals is defined as $l_1 + l_2 + \ldots + l_n \leq 1$.

There are several ways to encode AMO into SAT [3,7,14]. Here, we consider the ladder encoding. It introduces variables $\{a_i := l_1 \vee \ldots \vee l_i \mid 1 \leq i < n\}$ and clauses $\{a_i \to a_{i+1}, l_i \to a_i, l_{i+1} \to \neg a_i\}$. It is easy to see that this encoding is propagation complete.

Given a set of literals $l_1, \ldots, l_n$, the *Exactly-one* (EO) constraint over these literals is defined as $l_1 + l_2 + \ldots + l_n = 1$. Notice that

$$\mathrm{EO}(\{l_1, \ldots, l_n\}) = \mathrm{AMO}(\{l_1, \ldots, l_n\}) \wedge (l_1 \vee \ldots \vee l_n)$$

This defines a propagation complete encoding for EO given a propagation complete encoding of AMO.

2.4 Direct Encoding for Integer Variables

There are different methods for encoding integer variables into SAT (see for instance [18,27]). In this paper we use the direct encoding.

Let x be an integer variable with domain $[a, b]$. The *direct encoding* introduces Boolean variables $[\![x = i]\!]$ for $a \leq i \leq b$. A variable $[\![x = i]\!]$ is true iff $x = i$. The encoding also introduces the constraint $\mathrm{EO}(\{[\![x = i]\!] \mid a \leq i \leq b\})$.

We will sometimes treat Boolean variables b as integers with domain $[0,1]$.

We will implicitly assume that the direct encoding clauses $\mathrm{EO}(\{[\![x = i]\!] \mid a \leq i \leq b\})$ are part of any encoding of formula using integers x. We also assume all assignments A are closed under unit propagation of these clauses.

We extend the notion of satisfaction to formulae involving integer variables, as follows. A complete assignment A satisfies ϕ if replacing each Boolean variable as before, and each integer variable x_i by j if $[\![x_i = j]\!] \in A$ (since $A \models \mathrm{EO}(\{[\![x_i = j]\!] \mid a \leq j \leq b\})$ there must be exactly one) and evaluating the resulting ground expression gives $\top$. We extend the notation $A \models \phi$ as before.

2.5 Multi-valued Decision Diagrams

A directed acyclic graph $\mathcal{M}$ is called an *ordered Multi-valued Decision Diagram (MDD)* if it satisfies the following properties:

- It has two terminal nodes, namely $\mathcal{T}$ (true) and $\mathcal{F}$ (false).
- Each non-terminal node is labeled by an integer variable $\{x_1, x_2, \cdots, x_n\}$. This variable is called *selector variable*.
- Every node labeled by x_i has the same number of outgoing edges, namely $b_i - a_i + 1$, where $[a_i, b_i]$ is the domain of x_i.
- If an edge connects a node with a selector variable x_i and a node with a selector variable x_j, then $j > i$.

The MDD is *quasi-reduced* if no isomorphic subgraphs exist. It is *reduced* if, moreover, no nodes with only one child exist. A *long edge* is an edge connecting two nodes with selector variables x_i and x_j such that $j > i+1$. In the following we only consider quasi-reduced ordered MDDs without long edges, and we just refer to them as MDDs for simplicity.[2] We refer to [24] for further details about MDDs.

Given an MDD $\mathcal{M}$ we use ρ to refer to its *root node*. Given a node $\nu \in \mathcal{M}$, we write $\mathrm{var}(\nu) = x_j$ when node ν is labelled by x_j. Given an edge $\varepsilon \in \mathcal{M}$, we write $\varepsilon = \mathrm{edge}(\nu, \mu, [\![x_i = j]\!])$ if ε joins the node ν and μ when $x_i = j$.

An MDD represents a formula over integer variables: a MDD node ν with selector x with domain $[a, b]$ and children $\nu_a, \nu_{a+1}, \ldots, \nu_b$ represents the formula ϕ_ν where

$$\phi_\nu \equiv \bigvee_{i \in [a,b]} x = i \wedge \phi_{\nu_i}$$

where ϕ_{ν_i} is the formula represented by node ν_i, and $\phi_{\mathcal{T}} = \top$ and $\phi_{\mathcal{F}} = \bot$.

---

[2] Notice, however, that every result in this paper holds for non-reduced MDDs without long edges, and with some modifications of the rules the results also extend to non-reduced MDDs with long edges.

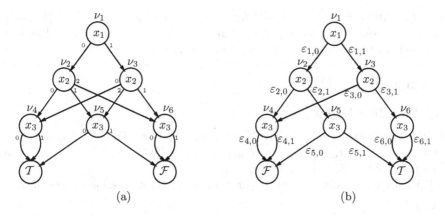

Fig. 1. (a) MDD of $x_2 = 0 \vee (x_3 = 0 \wedge x_2 - x_1 = 1)$ and (b) BDD of $x_2 \wedge (x_1 \vee x_3)$

Example 1. Let us consider the MDD encoding of $x_2 = 0 \vee (x_3 = 0 \wedge x_2 - x_1 = 1)$, with $x_1, x_3 \in \{0, 1\}$ and $x_2 \in \{0, 1, 2\}$, shown in Fig. 1(a). In this case $\rho = \nu_1$, $var(\nu_3) = x_2$, and the rightmost edge from ν_3 is $edge(\nu_3, \nu_6, x_2 = 1)$. $\phi_{\nu_4} \leftrightarrow \top$, $\phi_{\nu_5} \leftrightarrow x_3 = 0$, $\phi_{\nu_6} \leftrightarrow \bot$, and hence $\phi_{\nu_2} \leftrightarrow (x_2 = 0 \wedge \top) \vee (x_2 = 1 \wedge x_3 = 0) \vee (x_2 = 2 \wedge \bot)$ or equivalently $\phi_{\nu_2} \leftrightarrow x_2 = 0 \vee (x_2 = 1 \wedge x_3 = 0)$. □

A *binary decision diagram (BDD)* is an MDD with only Boolean variables. For a BDD $\mathcal{M}$ we can consider a non-terminal node ν as a triple (x, t, f) where there are two outgoing edges $edge(\nu, t, x)$ and $edge(\nu, f, \neg x)$. The BDD node ν represents the formula $\phi_\nu \equiv ITE(x, \phi_t, \phi_f)$ or equivalently $(x \wedge \phi_t) \vee (\neg x \wedge \phi_f)$.

2.6 Negation Normal Form Formulae

A *negation normal form* formula (NNF) is a rooted, directed acyclic graph (DAG) where each leaf node is labeled with x or $\neg x$ and each internal node is labeled with $\wedge$ or $\vee$ and can have arbitrarily many children.

NNFs are a more general form of decision diagram than BDDs, and can be exponentially more compact to represent the same formula [11]. We can use NNFs to express formulae over finite domain integer variables using the direct encoding.

But NNFs in general are too expressive, so usually we require some additional properties, such as:

Decomposable. An NNF $\mathcal{N}$ is *decomposable* if for each conjunction ϕ in $\mathcal{N}$, the conjuncts of ϕ do not share variables. That is, if $\phi_1, \ldots, \phi_n$ are the children of and-node ϕ, then $vars(\phi_i) \cap vars(\phi_j) = \emptyset$ for $i \neq j$.

Deterministic. An NNF $\mathcal{N}$ is *deterministic* if for each disjunction ϕ in $\mathcal{N}$, each two disjuncts of ϕ are logically contradictory. That is, if $\phi_1, \ldots, \phi_n$ are the children of or-node ϕ, then $\phi_i \wedge \phi_j \models \bot$ for $i \neq j$.

Smooth. An NNF $\mathcal{N}$ is *smooth* if for each disjunction ϕ in $\mathcal{N}$, each disjunct of ϕ mentions the same variables. That is, if $\phi_1, \ldots, \phi_n$ are the children of or-node ϕ, then $vars(\phi_i) = vars(\phi_j)$ for $i \neq j$.

3   Encoding MDDs

3.1   Encoding BDDs

The BDD encoding of MiniSat+ [13] is defined as follows: For each non-terminal BDD node $\nu = (x, t, f)$ we generate a Boolean variable ν which represents the truth value of the BDD rooted at ν.

For each non-terminal node $\nu = (x, t, f)$, we generate the following clauses:

B1 $t \wedge x \to \nu$.   B4 $\neg f \wedge \neg x \to \neg \nu$.
B2 $\neg t \wedge x \to \neg \nu$.   B5 $t \wedge f \to \nu$.
B3 $f \wedge \neg x \to \nu$.   B6 $\neg t \wedge \neg f \to \neg \nu$.

Define encoding MiniSAT as B1–B6, together with the terminal and root clauses: $\mathcal{T}$ (the true terminal is true), $\neg \mathcal{F}$ (the false terminal is false) and ρ (the root of the tree must be true).

Note while Een and Sorensen [13] refer to this as a Tseytin encoding, it is not since Tseytin [25] does not include an ITE constructor, so in the Tseytin encoding $ITE(x, t, f)$ needs to be encoded as $(x \wedge t) \vee (\neg x \wedge f)$.

The encoding contains $O(s)$ variables and clauses, where s is the size of the BDD.

Een and Sorensen [13] show that this encoding maintains domain consistency when used to encode (sorted) pseudo-Boolean constraints.

Theorem 1 ([13]). *Unit propagation on the MiniSAT encoding for a BDD for pseudo-Boolean constraint $\sum_{i=1}^{n} c_i x_i \geq d$ maintains domain consistency, assuming the coefficients c_i are in non-increasing order.*   □

This theorem does not hold without the ordering criterion. Consider the BDD encoding $x_1 + 2x_2 + x_3 \geq 3$ (or equivalently $x_2 \wedge (x_1 \vee x_3)$) shown in Fig. 1(b). Any solution of the BDD requires x_2 is $\top$. Unit propagation on the MiniSAT encoding generates $\neg \mathcal{F}, \mathcal{T}, \nu_1, \neg \nu_4, \nu_6$ and nothing else.

Theorem 2. *Unit propagation on the clauses (B2), (B4), (B6), $\neg \mathcal{F}$, ρ for a BDD maintains consistency.*   □

All in all, the encoding is compact (especially if only clauses (B2), (B4), (B6), $\neg \mathcal{F}$ and ρ are used), but the propagation strength is low.

3.2   Encodings MDDs with One Variable Per Node

The first set of encodings for MDDs, used for example in [2], are generalizations of the MiniSat+ encoding. This is natural since they are also used to encode pseudo-Boolean and linear constraints.

For each node ν at level i, with children $\nu_{a_i}, \nu_{a_i+1}, \ldots, \nu_{b_i}$, where the domain of x_i is $[a_i, b_i]$.

M1 $\neg\nu_j \wedge [\![x_i = j]\!] \rightarrow \neg\nu$ (generalizes B2 and B4).
M2 $\nu_j \wedge [\![x_i = j]\!] \rightarrow \nu$ (generalizes B1 and B3).
M3 $\nu_{a_i} \wedge \nu_{a_i+1} \wedge \cdots \wedge \nu_{b_i} \rightarrow \nu$ (weakly generalizes B5).
M4 $\neg\nu_{a_i} \wedge \neg\nu_{a_i+1} \wedge \cdots \wedge \neg\nu_{b_i} \rightarrow \neg\nu$ (weakly generalizes B6).

With these clauses, we can define different encodings:

Minimal: Clauses M1, $\neg\mathcal{F}$, ρ.
GenMiniSAT: Clauses M1–M4, $\mathcal{T}$, $\neg\mathcal{F}$, ρ.

Minimal is very compact, but its propagation strength is low, moreover when the original variables are fixed it does not necessarily fix all the node variables, and hence does not preserve solution counts. GenMiniSAT is the natural generalization of the BDD encoding from [13] to MDDs. Again, it is not the Tseytin encoding [25] of the MDD. Both encodings use $O(s)$ variables and $O(sd)$ clauses, where s is the MDD size and d is the maximum domain size of variables x.

Proposition 1. *Let $A = \{[\![x_i = v_i]\!] \mid 1 \le i \le n\}$ be a complete assignment on variables x satisfying the MDD $\mathcal{M}$. Then, there exists a complete assignment $B \supset A$ over the variables x, ν satisfying clauses GenMiniSAT.* □

Proposition 2. *Let $A = \{[\![x_i = v_i]\!] \mid 1 \le i \le n\}$ be a complete assignment on variables x not satisfying the MDD $\mathcal{M}$, then clauses ρ and M1 propagate $\mathcal{F}$.* □

Corollary 1. *Minimal and GenMiniSAT are correct.* □

These two encodings, however, do not detect inconsistencies:

Example 2. Consider again the MDD of $x_2 = 0 \vee (x_3 = 0 \wedge x_2 - x_1 = 1)$, with $x_1, x_3 \in \{0, 1\}$ and $x_2 \in \{0, 1, 2\}$ shown in Fig. 1(a).
After simplification, GenMiniSAT consists of the following clauses:

$$\neg[\![x_1 = 0]\!] \vee \nu_2, \qquad \neg[\![x_1 = 1]\!] \vee \nu_3, \qquad \nu_2 \vee \nu_3, \qquad \neg[\![x_2 = 0]\!] \vee \nu_2,$$
$$\neg\nu_4 \vee \neg[\![x_2 = 1]\!] \vee \nu_2, \nu_4 \vee \neg[\![x_2 = 1]\!] \vee \neg\nu_2, \neg[\![x_2 = 2]\!] \vee \neg\nu_2 \; \neg[\![x_2 = 0]\!] \vee \nu_3,$$
$$\neg\nu_4 \vee \neg[\![x_2 = 2]\!] \vee \nu_3, \nu_4 \vee \neg[\![x_2 = 2]\!] \vee \neg\nu_3, \neg[\![x_2 = 1]\!] \vee \neg\nu_3 \; \neg[\![x_3 = 0]\!] \vee \nu_4,$$
$$\neg[\![x_3 = 1]\!] \vee \neg\nu_4.$$

Consider the partial assignment $A = \{\neg[\![x_2 = 0]\!], \neg[\![x_3 = 0]\!], [\![x_3 = 1]\!]\}$. It cannot be extended to a complete assignment satisfying the MDD. However, unit propagation does not fail.
The same happens with Minimal, since it is a subset of GenMiniSAT.   □

3.3   Tseytin Encoding of an MDD

In this section we describe an alternative encodings for an MDD, the Tseytin encoding [25]. It detects inconsistencies with respect to the original variables but does not enforce domain consistency.

The Tseytin encoding introduce Boolean variables representing the formula of each edge. Let ν be a node at level i, with outgoing edges $\{\varepsilon_j \mid j \in J\}$. Let $\varepsilon = \text{edge}(\nu, \mu, [\![x_i = j]\!])$ be an edge of $\mathcal{M}$, then the Boolean variable ε encoding the edge represents the formula $[\![x_i = j]\!] \wedge \phi_\mu$.

The clauses of the Tseytin encoding are, for each node ν and edge ε

T1 $\nu \rightarrow \bigvee_j \varepsilon_j$.
T2 $\varepsilon \rightarrow \nu$.
T3 $\varepsilon \rightarrow \mu$.
T4 $\varepsilon \rightarrow [\![x_i = j]\!]$.
T5 $\mu \wedge [\![x_i = j]\!] \rightarrow \varepsilon$.

The Tseytin encoding, Tseytin, consists of clauses T1–T5, $\mathcal{T}$, $\neg\mathcal{F}$ and ρ. Therefore, it consists in $O(sd)$ variables and clauses, where s is the MDD size and d the maximum domain size of variables x.

Proposition 3. *Let $A = \{[\![x_i = v_i]\!] \mid 1 \leq i \leq n\}$ be a complete assignment on variables x satisfying the MDD $\mathcal{M}$. Then, there exists a complete assignment $B \supset A$ over the variables x, ν, ε satisfying clauses Tseytin.* ☐

Proposition 4. *Let A be a partial assignment on variables $\{x_i, x_{i+1}, \ldots, x_n\}$, and let ν be a node of $\mathcal{M}$ at level i. Assume that there is no completion A' of A satisfying the MDD rooted at ν. Then, unit propagation on clauses Tseytin and A enforces $\neg\nu$.* ☐

As a corollary, we can prove:

Theorem 3. *Tseytin is correct; i.e., given a complete assignment of the input variables, this encoding finds an inconsistency if and only if the assignment does not satisfy $\mathcal{M}$. Moreover, it implements consistency.* ☐

However, Tseytin does not preserve domain consistency.

Example 3. Let us consider the BDD of $x_2 \wedge (x_1 \vee x_3)$, shown in Fig. 1(b). Tseytin, once simplified, generates the following clauses:

$$
\begin{array}{llll}
\varepsilon_{1,0} \vee \varepsilon_{1,1}, & \neg\nu_2 \vee x_1 \vee \varepsilon_{1,0}, & \neg\varepsilon_{1,0} \vee \neg x_1, & \neg\varepsilon_{1,0} \vee \nu_2, \\
\neg\nu_3 \vee \neg x_1 \vee \varepsilon_{1,1}, & \neg\varepsilon_{1,1} \vee x_1, & \neg\varepsilon_{1,0} \vee \nu_3, & \neg\nu_2 \vee \varepsilon_{2,1}, \\
\neg\nu_5 \vee \neg x_2 \vee \varepsilon_{2,1}, & \neg\varepsilon_{2,1} \vee \nu_2, & \neg\varepsilon_{2,1} \vee x_2, & \neg\varepsilon_{2,1} \vee \nu_5, \\
\neg\nu_3 \vee \varepsilon_{3,1}, & \neg x_2 \vee \varepsilon_{3,1}, & \neg\varepsilon_{3,1} \vee \nu_3, & \neg\varepsilon_{3,1} \vee x_2, \\
\neg\nu_5 \vee \varepsilon_{5,1}, & \neg x_3 \vee \varepsilon_{5,1}, & \neg\varepsilon_{5,1} \vee \nu_5, & \neg\varepsilon_{5,1} \vee x_3.
\end{array}
$$

Consider the partial assignment $A = \emptyset$. Notice that x_2 is not propagated even though that there is no solution of $\mathcal{M}$ with $\neg x_2$. Clause $x_2 \vee \varepsilon_{2,0} \vee \varepsilon_{3,0}$ would propagate x_2. ☐

Also, Tseytin does not implement unit refutation completeness:

Example 4. Consider the BDD of the constraint $\text{XOR}(x_1, x_2, x_3, x_4)$ shown in Fig. 2. Node ν_2 represents the constraint $\text{XOR}(x_2, x_3, x_4)$, and node ν_3 represents $\neg \text{XOR}(x_2, x_3, x_4)$. It is clear, therefore, that the partial assignment $B = \{\nu_2, \nu_3\}$ cannot be extended to a complete assignment satisfying $\mathcal{M}$. However, Tseytin does not find any conflict. ☐

3.4   Path-Based Encodings

Under the encodings described in Sects. 3.2 and 3.3, the semantics of variables match the Boolean formula they represent – a node/edge variable is true (in a complete assignment) iff the corresponding formula is true.

In this section, we describe a set of *path-based* encodings. Like the Tseytin encoding these introduce one variable per node and per edge, but the interpretation of these variables is different. Under a path-based encoding, ν (or ε) is true iff the path from the root r to $\mathcal{T}$ defined by the selector variables passes through ν (resp. ε).

Unlike the previous encodings, the variables introduced here cannot be re-used if a sub-formula occurs in multiple constraints. However, we shall see that this interpretation allows us to make much stronger inferences.

A related treatment of path-based encodings of the **regular** constraint to CNF can be found in Bacchus work in [4] and by Quimper and Walsh in [23] in context of the **grammar** constraint. Our study provides a complete analysis of such encodings for decision diagrams and introduces a novel encoding with stronger propagation properties.

We generate clauses for each node ν and connecting it to each of its outgoing edge ε_j and each of it incoming edges δ_j, as well as clauses for each edge $\varepsilon = \mathrm{edge}(\nu, \mu, [\![x_i = j]\!])$.

P1 $\nu \wedge [\![x_i = j]\!] \rightarrow \varepsilon_j$.
P2 $\nu \rightarrow \bigvee_j \delta_j$ where $\nu \neq \rho$.
P3 $[\![x_i = j]\!] \rightarrow \bigvee \{\varepsilon' \mid \varepsilon' = \mathrm{edge}(\nu, \mu, [\![x_i = j]\!])$ for some $\nu, \mu \in \mathcal{M}\}$.
P4 $\mathrm{EO}(\{\nu' \in \mathcal{M} \mid \mathrm{Level}(\nu') = i\})$.

Clauses P1 enforce that a node on the path puts its outgoing edge on the path. Clauses P2 require each node on the path (except the root) has an incoming edge. Clauses P3 require that each integer value has an edge that supports it. Clauses P4 require that exactly one node on each level is T.

With these clauses, we can define different encodings:

BasicPath: Clauses P1–P2, T1–T4, $\mathcal{T}$, $\neg \mathcal{F}$, ρ.
NNFPath: BasicPath and clauses P3.
LevelPath: BasicPath and clauses P4.
CompletePath: BasicPath and clauses P3–P4.

All the encodings require $O(sd)$ variables and clauses, where s is the MDD size and d the maximum domain size of variables x.

A complete assignment A over the variables x_i defines a path in $\mathcal{M}$ in the obvious way. This path is denoted by $\nu_1 = \rho, \varepsilon_1, \nu_2, \varepsilon_2, \ldots$ By definition of the MDD, the assignment is compatible with $\mathcal{M}$ if and only if $\nu_{n+1} = \mathcal{T}$.

A complete assignment B over variables x_i, ν, ε is compatible with $\mathcal{M}$ if

- $A := B \cap (\{[\![x_i = j]\!] \mid 1 \leq i \leq n, j \in [a_i, b_i]\}\{\neg[\![x_i = j]\!] \mid 1 \leq i \leq n, j \in [a_i, b_i]\})$
  is compatible with $\mathcal{M}$.

- $\nu \in B$ iff $\nu = \nu_i$ for some i on the path defined by A.
- $\varepsilon \in B$ iff $\varepsilon = \varepsilon_i$ for some i on the path defined by A.

Proposition 5. *Given a complete assignment A on the variables x compatible with $\mathcal{M}$, there exists a complete assignment $B \supset A$ over the variables x, ν, ε satisfying clauses CompletePath.* □

Proposition 6. *Let A be a partial assignment on variables x. Let $UP(A)$ be the set of propagated literals with BasicPath. Let ν be a node of $\mathcal{M}$, and ε be an edge of $\mathcal{M}$. Then:*

- $\neg\nu \in UP(A)$ *if* $A \wedge \nu \models \neg\mathcal{M}$.
- $\neg\varepsilon \in UP(A)$ *if* $A \wedge \varepsilon \models \neg\mathcal{M}$.   □

Let us explain the idea behind the proof. If ν has not been propagated to false, we can create a path from ρ to $\mathcal{T}$ passing through ν, where all the nodes of this path have not been propagated to false. This path will define a completion B satisfying $\mathcal{M}$ with $\nu \in B$.

To build this path, we start from ν. Since $\neg\nu \notin UP(A)$, ν must have a parent that has also not been propagated to false. This node, again, has a parent that has not been propagated to false, etc. That gives a path from ρ to ν. In the same way, ν has a child that has not been propagated to false, and this child has a child that has not been propagated to false, etc. That gives a path from ν to $\mathcal{T}$. Concatenating both paths, we obtain the desired path from ρ to $\mathcal{T}$.

Theorem 4. *BasicPath maintains consistency by unit propagation.*   □

BasicPath, however, does not maintain domain consistency. For that we need clauses P3.

Example 5. Let us consider the BDD of $x_2 \wedge (x_1 \vee x_3)$, shown at Fig. 1(b). BasicPath, once simplified, generates the following clauses:

$$x_1 \vee \varepsilon_{1,0}, \quad \neg x_1 \vee \varepsilon_{1,1}, \ \neg\nu_2 \vee x_2, \quad \neg\nu_3 \vee x_2,$$
$$\neg\nu_5 \vee x_3, \quad \varepsilon_{1,0} \vee \varepsilon_{1,1}, \ \neg\nu_2 \vee \varepsilon_{2,1}, \ \neg\nu_3 \vee \varepsilon_{3,1},$$
$$\neg\nu_5 \vee \varepsilon_{5,1}, \ \neg\nu_2 \vee \varepsilon_{1,0}, \ \neg\nu_3 \vee \varepsilon_{1,1}, \ \neg\nu_5 \vee \varepsilon_{2,1},$$
$$\varepsilon_{3,1} \vee \varepsilon_{5,1} \ \neg\varepsilon_{2,1} \vee \nu_2, \ \neg\varepsilon_{3,1} \vee \nu_3, \ \neg\varepsilon_{5,1} \vee \nu_5,$$
$$\neg\varepsilon_{1,0} \vee \nu_2, \ \neg\varepsilon_{1,1} \vee \nu_3, \ \neg\varepsilon_{2,1} \vee \nu_5, \ \neg\varepsilon_{1,0} \vee \neg x_1,$$
$$\neg\varepsilon_{1,1} \vee x_1, \ \neg\varepsilon_{2,1} \vee x_2, \ \neg\varepsilon_{3,1} \vee x_2, \ \neg\varepsilon_{5,1} \vee x_3.$$

Consider the partial assignment $A = \emptyset$. Then, unit propagation does not propagate x_2 even though that there is no solution of $\mathcal{M}$ with $\neg x_2$. Clause $x_2 \vee \varepsilon_{2,0} \vee \varepsilon_{3,0}$, from P3, would propagate x_2.   □

As Corollary of Proposition 5 and Theorem 4, it follows that

Theorem 5. *Encodings BasicPath, NNFPath, LevelPath and CompletePath are correct; i.e., given a complete assignment of the input variables, these encodings find an inconsistency if and only if the assignment does not satisfy $\mathcal{M}$.*   □

Theorem 6. *NNFPath maintains domain consistency by unit propagation.*   □

NNFPath maintains domain consistency with respect to the original variables. However, since a SAT solver will not differentiate between original variables and auxiliary ones, partial assignments, in general, contain both type of variables. And, without clauses P4, the encodings are not propagation complete:

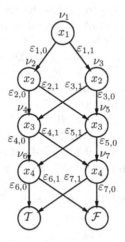

Fig. 2. BDD of $XOR(x_1, x_2, x_3, x_4)$

Example 6. Consider the MDD shown in Fig. 2, representing the constraint $XOR(x_1, x_2, x_3, x_4)$. Consider the partial assignment $B = \{\nu_4, \nu_5\}$. It is clear that B cannot be extended to a complete assignment satisfying $\mathcal{M}$, since no path can contain two nodes on the same level. However, NNFPath does not find any conflict.   □

To maintain consistency with respect to all variables, clauses P4 are needed. In that case, we can generalize the previous results to assignments containing auxiliary variables:

Proposition 7. *Let B be a partial assignment on all the variables. Let $UP(B)$ be the set of propagated literals with LevelPath. Let ν be a node of $\mathcal{M}$, and ε be an edge of $\mathcal{M}$. Then:*

1. $\neg\nu \in UP(B)$ if $B \wedge \nu \models \neg\mathcal{M}$.
2. $\neg\varepsilon \in UP(B)$ if $B \wedge \varepsilon \models \neg\mathcal{M}$.
3. $\nu \in UP(B)$ if $B \wedge \neg\nu \models \neg\mathcal{M}$.
4. $\varepsilon \in UP(B)$ if $B \wedge \neg\varepsilon \models \neg\mathcal{M}$.

Theorem 7. *LevelPath is unit refutation complete.*   □

LevelPath does not maintain domain consistency on all variables, though. Example 5 shows a counterexample. To obtain domain consistency we once more need the clauses P3.

Theorem 8. *CompletePath is propagation complete.*   □

The path based encoding do have one weakness compared to the Tseytin encoding. Since they require only a single path throught the MDD, we cannot allow different MDD constraints that share a sub-MDD to reuse the same encoding, we need a different copy of the encoding for each constraint. This is not the case for Tseytin encdings where the node variable ν just represents the truth value of the sub-formula encoded by the MDD rooted at ν. To our knowledge this restriction is not very significant in the CP context. No such sharing exists in any of our benchmarks. The bulk of nodes in an MDD are in the middle and unlikely to be shared. Moreover, separating MDDs per constraint for translation allows us to use different variable orderings for each MDD and thus reduce the number of nodes required. On the other hand, if substantial sharing of nodes among the different MDDs happens then a Tseytin encoding could be beneficial, since it translates this sharing to the CNF level.

The table below shows the sizes and propagation strength of the different encodings. As before, s is the size of the MDD, d is the maximum domain size of variables x and n is the number of variables x. Notice that usually $n \ll s$.

	Minimal	GMinisat	Tseytin	BasicP	NNFP	LevelP	ComplP
Variables	s	s	$s(d+1)$	$s(d+1)$	$s(d+1)$	$s(d+2)$	$s(d+2)$
Clauses	sd	$s(2d+2)$	$s(4d+1)$	$s(4d+2)$	$s(4d+2)$ $+nd$	$s(4d+5)$	$s(4d+5)$ $+nd$
Consistent	✗	✗	✓	✓	✓	✓	✓
Dom. Consis.	✗	✗	✗	✗	✓	✗	✓
Ref. Compl.	✗	✗	✗	✗	✗	✓	✓
Prop. Compl	✗	✗	✗	✗	✗	✗	✓

4 Encoding NNFs

BDDs are a special case of NNFs and hence NNF encodings provide an alternate approach to encoding BDDs. There is an existing encoding for NNFs given by [20]. When applied correctly to MDDs it results in the NNFPath (hence the name). But care has to be taken in NNF encodings, without the right restrictions on the form of the NNF the encodings are incorrect!

An encoding of an NNF $\mathcal{N}$ to clauses is given by [20]. Each node ν is associated with a literal, also called ν. For leaf nodes the literal is just the label of the node. For non-leaf nodes the literal is a new Boolean variable. The clauses we make use of are

N1 $\nu \rightarrow \nu_1 \vee \cdots \vee \nu_k$ for each $\vee$-node ν with children $\nu_1, \ldots, \nu_k$
N2 $\nu \rightarrow \nu_i, 1 \leq i \leq k$ for each $\wedge$-node ν with children $\nu_1, \ldots, \nu_k$
N3 $\nu \rightarrow p_1 \vee \cdots \vee p_m$ for each node ν with incoming edges from nodes $p_1, \ldots, p_m$.

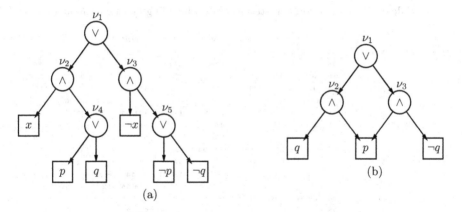

Fig. 3. NNF for formula (a) $(x \wedge (p \vee q)) \vee (\neg x \wedge (\neg p \vee \neg q))$ and (b) $(\neg q \wedge p) \vee (p \wedge q)$

We consider two encodings: BaseNNF Clauses N1–N2 and ρ, and ExtNNF Clauses N1–N3 and ρ as defined in [20].

Theorem 9. *Given an NNF $\mathcal{N}$ then BaseNNF is a correct encoding.*   □

Note that this *correctness* result *does not apply* to ExtNNF unless the NNF is smooth and decomposable. Jung [20] also claim that ExtNNF enforces domain consistency for decomposable NNFs, but this too is incorrect.

Example 7. The NNF shown in Fig. 3(a) is decomposable, deterministic but not smooth (e.g. the two children of node ν_4 do not mention the same variables). The ExtNNF encoding is

$$N1 : \nu_1 \rightarrow \nu_2 \vee \nu_3 \quad \nu_4 \rightarrow p \vee q \quad \nu_5 \rightarrow \neg p \vee \neg q$$
$$N2 : \nu_2 \rightarrow x \quad \nu_2 \rightarrow \nu_4 \quad \nu_3 \rightarrow \neg x \quad \nu_3 \rightarrow \nu_5$$
$$N3 : \nu_2 \rightarrow \nu_1 \quad \nu_3 \rightarrow \nu_1 \quad x \rightarrow \nu_2 \quad \nu_4 \rightarrow \nu_2 \quad \neg x \rightarrow \nu_3$$
$$\nu_5 \rightarrow \nu_3 \quad p \rightarrow \nu_4 \quad q \rightarrow \nu_4 \quad \neg p \rightarrow \nu_5 \quad \neg q \rightarrow \nu_5$$
$$\rho : \nu_1$$

Consider the assignment $A = \{x, \neg q\}$ unit propagation determines ν_1, ν_2, ν_4, $p, \nu_5, \nu_3, \neg x$ and hence a contradiction. This is wrong since there is a model of the NNF $\{x, \neg q, p\}$.   □

Example 8. Consider the smooth, decomposable and deterministic NNF for $(\neg q \wedge p) \vee (p \wedge q)$ shown in Fig. 3(b). Then the clauses of ExtNNF are

$$\rho : \nu_1 \qquad N1 : \qquad \nu_1 \rightarrow \nu_2 \vee \nu_3$$
$$N2 : \nu_2 \rightarrow \neg q \quad \nu_2 \rightarrow p \quad \nu_3 \rightarrow p \qquad \nu_3 \rightarrow q$$
$$N3 : \nu_2 \rightarrow \nu_1 \quad \nu_3 \rightarrow \nu_1 \quad \neg q \rightarrow \nu_2 \quad p \rightarrow \nu_2 \vee \nu_3 \quad q \rightarrow \nu_3$$

Any model of the formula must make p true, but unit propagation on these clauses derives only ν_1. What is missing is information that $\neg p$ does not appear in the NNF. This means p *must hold*!   □

Table 1. Results on nurse rostering, shift scheduling and pentominoes.

Bench	Type	Search	#Inst		Prop	Minimal	GMinisat	Tseytin	BasicP	NNFP	LevelP	ComplP
Nurse	SAT	VSIDS	286	#sol	**282**	88	<u>195</u>	184	150	185	157	187
			78	com	**1.97**	–	<u>5.33</u>	27.81	58.73	14.09	42.51	24.86
			286	all	**23.82**	903.64	<u>395.26</u>	473.05	617.42	457.79	607.16	457.85
		prog	179	#sol	132	143	<u>151</u>	**156**	**156**	108	**156**	104
			80	com	**3.42**	–	<u>6.19</u>	6.61	18.39	54.63	29.96	50.86
			179	all	329.63	284.73	212.5	181.65	**171.95**	516.36	<u>177.19</u>	526.63
	UNSAT	VSIDS	46	#sol	32	29	**46**	27	31	<u>33</u>	32	32
			26	com	42.73	–	**8.09**	229.45	98.31	<u>26.87</u>	71.55	69.26
			46	all	402.57	626.02	**231.34**	631.03	450.35	<u>380.4</u>	413.69	437.38
Shift	OPT	VSIDS	120	#sol	<u>114</u>	85	96	97	**116**	115	110	**116**
			78	com	109.8	–	166.51	161.91	**51.54**	88.07	<u>68.91</u>	117.44
			120	all	<u>213.84</u>	535.59	457.94	444.8	**174.65**	252.11	224.41	276.17
		prog	56	#sol	49	48	**56**	48	<u>55</u>	50	52	48
			48	com	100.11	–	<u>28.64</u>	113.06	**24.52**	74.09	34.02	79.97
			56	all	257.02	240.44	**60.42**	268.34	<u>161.76</u>	232.17	176.28	239.97
Pent	ALL	prog	14	#sol	**14**	<u>12</u>	<u>12</u>	6	<u>12</u>	9	<u>12</u>	6
			6	com	**6.67**	–	<u>8.21</u>	18.27	14.57	16.02	8.8	15.36
			14	all	**279.43**	<u>352.82</u>	505.92	693.54	626.07	653.08	387.67	692.3

To fix Jung's encoding we add the following clauses

N4 $\neg l$ for each literal l for $vars(\mathcal{N})$ which does not appear in $\mathcal{N}$.

We denote by FullNNF Clauses N1–N4 and ρ.

Theorem 10. *Given a smooth decomposable NNF $\mathcal{N}$ then FullNNF is a correct encoding.*

Theorem 11. *Given a smooth decomposable NNF $\mathcal{N}$, then unit propagation on FullNNF enforces domain consistency.* □

It follows that FullNNF is equivalent to NNFPath if applied to MDDs rewritten as NNF. To summarise the results in this section we provide the following table.

	BaseNNF	ExtNNF	FullNNF
Clauses	N1–N2	N1–N3	N1–N4
Correctness	Always	Smooth and Decomposable	Smooth and Decomposable
Domain consistent	✗	✗	✓

5   Experiments

We show results on three benchmarks: nurse rostering, shift scheduling and pentominoes (Nurse, Shift and Pent).[3] The MDD encodings are implemented as eager translations of MDDs within the LCG solver Chuffed [9,10] and compared with a native MDD propagator with learning [17]. We use SAT branching heuristics

---

[3] Benchmarks are available from http://people.eng.unimelb.edu.au/pstuckey/mddenc. tar.gz.

(VSIDS) and the programmed search as specified in the models (prog). We omit instances not solved by any solver using that search. For each model we show: (#sol) the number of instances solved (SAT and UNSAT for Nurse, to optimality for Shift, all solutions for Pent); (com) the mean solving time in seconds for all benchmarks solved by all solvers (except Minimal); and (all) the mean solving time of all benchmarks using timeout (1200 s) for unsolved instances. The results on the encoding Minimal are omitted for *com* and for Pent since it does not preserve solution counting. Best results are in bold, and second best are underlined (Table 1).

In case of satisfiable instances of Nurse the results show that encodings do not compete with the native propagator. This is not surprising as the search quickly finds the solutions without being disturbed by the complete CNF model generated by the eager encodings. For the UNSAT instances decompositions and their intermediate literals show their strength and beat the propagator. GenMiniSAT shows best performance for these UNSAT instances with VSIDS. The encodings also have an advantage over the propagator when programmed search is used, but it is unclear which one dominates.

For Shift the results show that when using activity based search and branching takes place on auxiliary variables, the path based approaches are generally superior.

The main advantage of the native propagator is that its explanations are built in a more deterministic fashion and hence tend to be more reusable. Furthermore, since the propagator only generates a fraction of the variables of the eager encoding, the search is less likely get trapped in an unfruitful search space using VSIDS. The difference in results on SAT and UNSAT instances of Nurse clearly indicate that a combination of the propagator and a lazy encoding as in [1] would be a strong approach.

6   Conclusion and Future Work

This paper resulted from discussions that uncovered our own misunderstanding of the strength of decision diagram encodings. We were surprised to discover that the usual BDD encoding is not domain consistent. In this paper we seek to remove this confusion, and demonstrate a wealth of different encoding possibilities, with different properties.

The experimental results show that there is unlikely to be one single best encoding for MDDs, and hence an important direction of future work is to determine when each encoding is best. Possibly a portfolio approach varying over encodings of each constraint is a fruitful and pragmatic technique to solve hard problems in practice.

Another interesting direction of future work is to determine a propagation complete encoding for NNFs. It appears the result may require restricting to Sentential Decision Diagrams [12] a form of NNF with a uniform V-tree.

The literature on CNF encodings focuses on consistencies wrt. primary variables of the constraint, whereas we have shown that consistency on auxiliary variables are worthwhile to look at. Our work concentrated on translations of decision diagrams and we would like to extend this research to other constraints like

linear and **sequence**. State-of-the-art CNF encodings of **cardinality** are the next candidate for this investigation.

In case of theoretical results, an interesting direction is to establish lower bounds on the size of encodings implementing certain consistencies for concrete constraints. The strong relationship between CNF encodings and monotone circuits established in [5, 19] demonstrates a powerful tool for this purpose.

Acknowledgement. NICTA is funded by the Australian Government as represented by the Department of Broadband, Communications and the Digital Economy and the Australian Research Council through the ICT Centre of Excellence program.

References

1. Abío, I., Nieuwenhuis, R., Oliveras, A., Rodríguez-Carbonell, E., Stuckey, P.J.: To encode or to propagate? The best choice for each constraint in SAT. In: Schulte, C. (ed.) CP 2013. LNCS, vol. 8124, pp. 97–106. Springer, Heidelberg (2013)
2. Abío, I., Stuckey, P.J.: Encoding linear constraints into SAT. In: O'Sullivan, B. (ed.) CP 2014. LNCS, vol. 8656, pp. 75–91. Springer, Heidelberg (2014)
3. Ansótegui, C., Manyà, F.: Mapping problems with finite-domain variables into problems with boolean variables. In: SAT 2004 - The Seventh International Conference on Theory and Applications of Satisfiability Testing, Vancouver, BC, Canada, Online Proceedings, 10–13 May 2004
4. Bacchus, F.: GAC via unit propagation. In: Bessière, C. (ed.) CP 2007. LNCS, vol. 4741, pp. 133–147. Springer, Heidelberg (2007)
5. Bessiere, C., Katsirelos, G., Narodytska, N., Walsh, T.: Circuit complexity and decompositions of global constraints. In: Proceedings of the 21st International Joint Conference on Artificial Intelligence, IJCAI 2009, Pasadena, California, USA, 11–17 July 2009, pp. 412–418 (2009)
6. Bordeaux, L., Marques-Silva, J.: Knowledge compilation with empowerment. In: Bieliková, M., Friedrich, G., Gottlob, G., Katzenbeisser, S., Turán, G. (eds.) SOFSEM 2012. LNCS, vol. 7147, pp. 612–624. Springer, Heidelberg (2012)
7. Chen, J.: A new SAT encoding of the at-most-one constraint. In: Proceedings of the Tenth International Workshop of Constraint Modelling and Reformulation (2010)
8. Cheng, K.C.K., Yap, R.H.C.: Maintaining generalized arc consistency on Ad Hoc r-Ary constraints. In: Stuckey, P.J. (ed.) CP 2008. LNCS, vol. 5202, pp. 509–523. Springer, Heidelberg (2008)
9. Chu, G.: Chuffed. https://github.com/geoffchu/chuffed
10. Chu, G.G.: Improving combinatorial optimization. Ph.D. thesis, The University of Melbourne (2011)
11. Darwiche, A.: On the tractable counting of theory models and its application to truth maintenance and belief revision. J. Appl. Non-class. Logics **11**(1–2), 11–34 (2001)
12. Darwiche, A.: SDD: a new canonical representation of propositional knowledge bases. In: IJCAI, pp. 819–826 (2011)
13. Eén, N., Sörensson, N.: Translating pseudo-boolean constraints into SAT. JSAT **2**(1–4), 1–26 (2006)
14. Frisch, A.M., Peugniez, T.J., Doggett, A.J., Nightingale, P.: Solving non-boolean satisfiability problems with stochastic local search: a comparison of encodings. J. Autom. Reason. **35**(1–3), 143–179 (2005)

15. Gange, G., Stuckey, P.J., Van Hentenryck, P.: Explaining propagators for edge-valued decision diagrams. In: Schulte, C. (ed.) CP 2013. LNCS, vol. 8124, pp. 340–355. Springer, Heidelberg (2013)
16. Gange, G., Stuckey, P.J., Lagoon, V.: Fast set bounds propagation using a BDD-SAT hybrid. J. Artif. Intell. Res. (JAIR) **38**, 307–338 (2010)
17. Gange, G., Stuckey, P.J., Szymanek, R.: MDD propagators with explanation. Constraints **16**(4), 407–429 (2011)
18. Gent, I.P.: Arc consistency in SAT. In: Proceedings of the 15th Eureopean Conference on Artificial Intelligence, ECAI 2002, Lyon, France, pp. 121–125, July 2002
19. Gwynne, M., Kullmann, O.: A framework for good SAT translations, with applications to CNF representations of XOR constraints. CoRR abs/1406.7398 (2014). http://arxiv.org/abs/1406.7398
20. Jung, J.C., Barahona, P., Katsirelos, G., Walsh, T.: Two encodings of DNNF theories. In: ECAI 2008 Workshop on Inference Methods Based on Graphical Structures of Knowledge (2008)
21. Nieuwenhuis, R., Oliveras, A., Tinelli, C.: Solving SAT and SAT modulo theories: from an abstract davis-putnam-logemann-loveland procedure to DPLL(T). J. ACM **53**(6), 937–977 (2006)
22. Plaisted, D.A., Greenbaum, S.: A structure-preserving clause form translation. J. Symb. Comput. **2**(3), 293–304 (1986)
23. Quimper, C.-G., Walsh, T.: Decomposing global grammar constraints. In: Bessière, C. (ed.) CP 2007. LNCS, vol. 4741, pp. 590–604. Springer, Heidelberg (2007)
24. Srinivasan, A., Ham, T., Malik, S., Brayton, R.: Algorithms for discrete function manipulation. In: 1990 IEEE International Conference on Computer-Aided Design, ICCAD-90, Digest of Technical Papers, pp. 92–95 (1990)
25. Tseytin, G.: On the complexity of derivation in propositional calculus. Stud. Constr. Math. Math. Logic Part **2**, 115–125 (1968)
26. del Val, A.: Tractable databases: how to make propositional unit resolution complete through compilation. In: Proceedings of the 4th International Conference on Principles of Knowledge Representation and Reasoning (KR 1994), Bonn, Germany, 24–27 May 1994, pp. 551–561 (1994)
27. Walsh, T.: SAT v CSP. In: Dechter, R. (ed.) CP 2000. LNCS, vol. 1894, pp. 441–456. Springer, Heidelberg (2000)

Time-Series Constraints: Improvements and Application in CP and MIP Contexts

Ekaterina Arafailova[1], Nicolas Beldiceanu[1], Rémi Douence[1], Pierre Flener[2],
María Andreína Francisco Rodríguez[2(✉)], Justin Pearson[2],
and Helmut Simonis[3]

[1] TASC/ASCOLA (CNRS/INRIA), Mines Nantes, 44307 Nantes, France
{Ekaterina.Arafailova,Nicolas.Beldiceanu,Remi.Douence}@mines-nantes.fr
[2] Department of Information Technology,
Uppsala University, 751 05 Uppsala, Sweden
{Pierre.Flener,Maria.Andreina.Francisco,Justin.Pearson}@it.uu.se
[3] Insight Centre for Data Analytics, University College Cork, Cork, Ireland
Helmut.Simonis@insight-centre.org

Abstract. A checker for a constraint on a variable sequence can often be compactly specified by an automaton, possibly with accumulators, that consumes the sequence of values taken by the variables; such an automaton can also be used to decompose its specified constraint into a conjunction of logical constraints. The inference achieved by this decomposition in a CP solver can be boosted by automatically generated implied constraints on the accumulators, provided the latter are updated in the automaton transitions by linear expressions. Automata with non-linear accumulator updates can be automatically synthesised for a large family of time-series constraints. In this paper, we describe and evaluate extensions to those techniques. First, we improve the automaton synthesis to generate automata with fewer accumulators. Second, we decompose a constraint specified by an automaton with accumulators into a conjunction of linear inequalities, for use by a MIP solver. Third, we generalise the implied constraint generation to cover the entire family of time-series constraints. The newly synthesised automata for time-series constraints outperform the old ones, for both the CP and MIP decompositions, and the generated implied constraints boost the inference, again for both the CP and MIP decompositions. We evaluate CP and MIP solvers on a prototypical application modelled using time-series constraints.

1 Context and Motivation

Frameworks are given in [4,14] for specifying a constraint on a sequence of variables in a high-level way by means of a finite automaton, possibly augmented with accumulators in the framework of [4]. An automaton can be seen as a checker for ground instances of the specified constraint. For example, in a nonogram puzzle, a row constrained to contain two stretches of black cells, of lengths 4 and 3 in this order, separated by at least one white cell but preceded and

© Springer International Publishing Switzerland 2016
C.-G. Quimper (Ed.): CPAIOR 2016, LNCS 9676, pp. 18–34, 2016.
DOI: 10.1007/978-3-319-33954-2_2

followed by any amounts of white cells, can be checked by an automaton equivalent to the regular expression $w^* b^4 w^+ b^3 w^*$, where the row is represented by a sequence of variables whose domain value 'w' stands for white and 'b' for black. Accumulators enable the specification of a constraint γ on a variable sequence X by an automaton whose size does not depend on the length of X: accumulators are initialised at the start state and are updated through the transitions; upon acceptance, the accumulators are linked to another variable of γ via an arithmetic constraint. For example, one could constrain the number of white cells between the two black stretches in the nonogram constraint above to be at most half the length of the row.

The framework of [14] lifts an automaton without accumulators into a propagator for the specified constraint; it maintains domain consistency in polynomial time. The more general framework of [4] lifts an automaton, possibly with accumulators, into a decomposition of the specified constraint in terms of constraints with existing propagators; in the presence of accumulators, this decomposition does not maintain domain consistency in general [2]. Encoding the potential accumulator values in the states of the automaton may lead to an exponentially large automaton. In this paper, we focus on automata with accumulators.

The propagation achieved by the automaton decomposition of [4] in a CP solver can be boosted by invariants, seen as implied constraints, on the accumulators. If the latter are updated in the automaton transitions by linear expressions on the accumulators — such as increments and decrements by constant amounts (as in $c := c + 1$) or by other accumulators (as in $c := c + r$), or resets (as in $c := 0$) — then such implied constraints can be automatically generated [11].

Automata with non-linear accumulator updates can be automatically synthesised for a large family of structural time-series constraints [3]. A time series is here a sequence of integers, corresponding to measurements taken over a time interval. Time series are common in many application areas, such as the power output of electric power stations over multiple days, or environmental data (temperature, humidity, CO_2 level) in buildings. Time series are constrained by physical or organisational limits, which restrict the evolution of the series.

After a summary of the background material in Sect. 2, the **contributions** and **impact** of this paper are as follows:

- We improve the automated automaton synthesis of [3] so as to synthesise automata with fewer accumulators and simpler accumulator updates, using fewer 'min' and 'max' operators, say (Sect. 3).
- We decompose a constraint specified by an automaton *with* accumulators into a linear-sized conjunction of linear inequalities, for use by a mixed-integer programming (MIP) solver (Sect. 4).
- We generalise the implied constraint generation of [11] so as to cover the entire family of time-series constraints of [3] and to rank the generated implied constraints by decreasing propagation strength, thereby easing the human selection of which implied constraints actually to use (Sect. 5).
- We show that the newly synthesised automata for time-series constraints outperform the automata of [3], for both the CP and MIP decompositions, and

that the newly generated implied constraints boost the inference, again for both the CP and MIP decompositions (Sect. 6).

– We evaluate CP and MIP solvers on a prototypical application modelled with the help of time-series constraints (Sect. 7).

2   Specifying (Time-Series) Constraints Using Automata

We showed in [3] that many constraints $\gamma(N, \langle X_0, \ldots, X_{n-1}\rangle)$ on an unknown time series $\langle X_0, \ldots, X_{n-1}\rangle$ of given length n can be specified as a triple $\langle p, f, g\rangle$, where p is a regular expression over the alphabet $\{<, =, >\}$ and is called the *pattern*; $f \in \{\text{max, min, one, range, surface, width}\}$ is called the *feature*; and $g \in \{\text{Max, Min, Sum}\}$ is called the *aggregator*. The semantics is that integer variable N is required to be the aggregation, computed using g, of the list of features f of all maximal words matching p within the sequence $\langle S_0, \ldots, S_{n-2}\rangle$ of variables, called the *signature sequence*, which is linked to the time series via the *signature constraints* $(X_i < X_{i+1} \Leftrightarrow S_i = \text{`<'}) \wedge (X_i = X_{i+1} \Leftrightarrow S_i = \text{`='}) \wedge (X_i > X_{i+1} \Leftrightarrow S_i = \text{`>'})$ for all $i \in [0, n-2]$. A list of 23 patterns was identified, giving 266 constraints. We now introduce our running example.

Example 1. The   MaxWidthStrictlyDecreasingSequence(N, X) constraint, requiring N to be the maximum width of the maximal strictly decreasing sequences within the time series X, is specified by the pattern $>^+$, the feature `width`, and the aggregator `Max`. The time series $\langle 4, 4, 3, 2, 2, 6, 3, 5\rangle$ contains two maximal strictly decreasing sequences, namely $4 > 3 > 2$ and $6 > 3$, of widths 3 and 2, so their maximum width is $N = 3$. The following figure shows how to check   MaxWidthStrictlyDecreasingSequence$(3, \langle 4, 4, 3, 2, 2, 6, 3, 5\rangle)$ by (I) building the signature sequence by comparing adjacent time-series values; (II) finding all maximal words matching the regular expression $>^+$; (III) computing the feature `width` of each such strictly decreasing sequence; and (IV) aggregating the feature values using the `Max` aggregator:

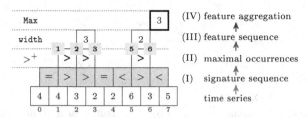

An *automaton* with a memory of $m \geq 0$ integer accumulators [4] is a tuple $\langle Q, \Sigma, \delta, q_0, I, A, \alpha\rangle$, where Q is the set of *states*, Σ the *alphabet*, $\delta:(Q \times \mathbb{Z}^m) \times \Sigma \to Q \times \mathbb{Z}^m$ the *transition function*, $q_0 \in Q$ the *start state*, I the m-tuple of *initial values* of the accumulators, $A \subseteq Q$ the set of *accepting states*, and $\alpha : \mathbb{Z}^m \to \mathbb{Z}$ the *acceptance function*, transforming the memory of an accepting state into an integer. If the left-to-right consumption of the symbols of a word w in Σ^* transits from q_0 to some accepting state and the m-tuple C of current accumulator values, then the automaton *returns* the value $\alpha(C)$, else it *fails*.

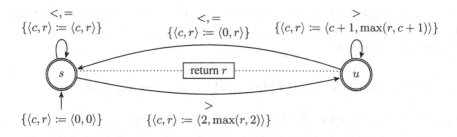

Fig. 1. Automaton for MaxWidthStrictlyDecreasingSequence

Example 2. A ground instance of the constraint of Example 1 holds if and only if its value of N is returned by the automaton in Fig. 1 after consuming the signature sequence linked to its time series X. The automaton uses $m = 2$ accumulators: at any moment, accumulator c has the length of the *current* strictly decreasing sequence, while r has the length of the *longest* strictly decreasing sequence seen so far. The state set Q is $\{s, u\}$: at s the current sequence is *not* strictly decreasing, and at u the current sequence *is* strictly decreasing. The start state $q_0 = s$ is indicated by an arc coming from nowhere, annotated within braces by the initialisation to zero of both c and r, hence $I = \langle 0, 0 \rangle$. The alphabet Σ is $\{<, =, >\}$. The arc from s to u depicts the transition of δ from s to u upon consuming symbol $>$, and is annotated within braces by accumulator updates: r is updated to its maximum with 2, and c is set to 2. All states are accepting, hence $A = Q$. The acceptance function α transforms a memory $\langle c, r \rangle$ into r at both states, and is given in a box linked to s and u by dotted lines.   □

An automaton can be seen as a constraint *checker*. The framework of [14] lifts an automaton with $m = 0$ accumulators into a CP *propagator* for the specified constraint; it maintains domain consistency in time polynomial in the automaton size and sequence length. The more general framework of [4] lifts an automaton with $m \geq 0$ accumulators into a CP *decomposition* of the specified constraint in terms of constraints with existing CP propagators; when $m \geq 1$, this decomposition does not maintain domain consistency in general [2]. Encoding the potential accumulator values in the states of the automaton, so as to get an automaton with $m = 0$ accumulators, may lead to a large automaton.

In this paper, we focus on automata with $m \geq 1$ accumulators, motivated [4] by the wish to specify a constraint on a sequence X by an automaton whose size does not depend on the length of X; this is the case for the automaton in Fig. 1. In Sect. 3, we improve our synthesiser [3] of automata from $\langle p, f, g \rangle$ specifications of time-series constraints, so that it automatically synthesises automata with fewer accumulators and simpler accumulator updates, namely linear accumulator updates rather than updates involving the min and max operators. In Sect. 4, we lift an automaton with $m \geq 1$ accumulators into a MIP decomposition of linear inequalities. In Sect. 5, we boost the inference achieved for the CP and MIP decompositions by generalising our generator [11] of constraints implied by an automaton, so that it covers the entire family of time-series constraints of this section and [3]. Those sections are orthogonal and any subset thereof can be read in any sequence.

3   Simplification of Synthesised Time-Series Automata

In [3] we synthesise automatically an automaton from a triple $\langle p, f, g \rangle$ specifying a time-series constraint. The synthesis relies on a declarative encoding of procedural knowledge into what we call *decoration tables* [3]. Each pattern is specified by a transducer [6,15] obeying wellformedness conditions. The decoration tables are parametrised by features and aggregators, and define substitution rules on the transducers that allow an automaton with $m = 3$ accumulators to be synthesised. The future work in [3] included simplifying the synthesised automata, as they often have more accumulators and more complex accumulator updates than manually designed ones: this may slow down the checker and weaken CP or MIP decompositions of the constraint specified by the synthesised automaton.

In this paper, we largely overcome this bottleneck. Rather than designing a procedural minimisation algorithm for automata with accumulators, we have again opted for capturing such procedural knowledge in a *declarative* and thus more easily reusable way: it suffices to specialise the decoration tables of [3] for some combinations of algebraic properties of pattern-feature-aggregator triples.

First, we recall the concept of pattern e-occurrence from [3], capturing where a feature value is extracted from the time series.

Definition 1. *Given a pattern p; a sequence $X_0, \ldots, X_{n-1}$; its signature sequence $S_0, \ldots, S_{n-2}$; and a non-empty subsequence $S_i, S_{i+1}, \ldots, S_j$ forming a maximal word that matches p, with $0 \leq i \leq j \leq n-2$; the e-occurrence of that maximal word is the interval $[\ell, u]$ of corresponding indices within $X_0, \ldots, X_{n-1}$.*

In Example 1, the sequence $X = \langle 4, 4, 3, 2, 2, 6, 3, 5 \rangle$ gives the signature sequence $S = \langle =, >, >, =, <, >, < \rangle$, which contains two maximal words matching the pattern $>^+$ of strictly decreasing sequences, namely $\langle S_1, S_2 \rangle = \langle >, > \rangle$ and $\langle S_5 \rangle = \langle > \rangle$, corresponding to the strictly decreasing sequences $\langle X_1, X_2, X_3 \rangle = \langle 4, 3, 2 \rangle$ and $\langle X_5, X_6 \rangle = \langle 6, 3 \rangle$, hence the e-occurrences are $[1, 3]$ and $[5, 6]$. A pattern occurrence $\langle S_i, \ldots, S_j \rangle$ within the signature sequence has the e-occurrence $[i, j+1]$ for this constraint, but it could be $[i+1, j]$ for other constraints [3].

All synthesised automata in [3] have the accumulators c, d, and r, which respectively denote the feature value of the *current* pattern e-occurrence (such as accumulator c in Fig. 1); the feature value of a *potential* part of a pattern e-occurrence (no such accumulator is needed in Fig. 1, and achieving this is the purpose of this section); and the aggregated *result* value for the feature values of the pattern e-occurrences already encountered (such as accumulator r in Fig. 1). Figure 2B and C gives the functions used to compute the feature and aggregation values. If the pattern, feature, and aggregator satisfy some properties, then either it is enough to perform the accumulator update only on one specific

Simplification	Percentage
aggregate once	28.9 %
immediate aggreg.	45.9 %
other properties	11.6 %
unchanged automata [3]	13.6 %

(A)

Aggregator g	default$_{g,f}$
Max	min$_f$
Min	max$_f$
Sum	0

(B)

Feature f	id$_f$	min$_f$	max$_f$	ϕ_f	δ^i_f
one	1	1	1	1	1
width	0	0	n	+	1
surface	0	$-\infty$	$+\infty$	+	X_i
max	$-\infty$	$-\infty$	$+\infty$	max	X_i
min	$+\infty$	$-\infty$	$+\infty$	min	X_i
range	0	0	$+\infty$	n/a	X_i

(C)

Fig. 2. (A) Percentage, among the 266 time-series constraints, of automata that can be simplified using the discovered properties. (C) Features: their identity, minimum, and maximum values; the functions ϕ_f and δ^i_f are used to compute recursively the feature value v_u of a sequence $\langle X_\ell, \ldots, X_u \rangle$ by $v_\ell = \phi_f(\text{id}_f, \delta^\ell_f)$ and $v_i = \phi_f(v_{i-1}, \delta^i_f)$ for $i > \ell$; note that δ^i_f provides the contribution of X_i to the value of feature f; (B) Aggregators and their default values.

transition of the automaton, as in Definition 3, or it is possible to start aggregating immediately upon finding an e-occurrence, as in Definition 4. To state these properties, we need another concept.

Definition 2. *A transition from state q to state q' in an automaton is called a 'found' transition if it is the only transition on some path from the initial state q_0 to q' that modifies the accumulator c.*

For example, the transition from the start state s to state u in Fig. 1 is a 'found' transition, as it sets c to 2.

Definition 3. *Given a time-series constraint γ on feature f, an e-occurrence $[\ell, u]$ of its pattern such that X_s triggers a 'found' transition of its automaton, with $s \in [\ell, u]$, we say that γ is an aggregate-once constraint if δ^s_f equals* $\phi_f(\phi_f(\ldots \phi_f(\text{id}_f, \delta^\ell_f), \ldots, \delta^{u-1}_f), \delta^u_f)$, *where ϕ_f and δ^i_f are as in Fig. 2B.*

For aggregate-once constraints the feature value of an e-occurrence depends only on the value of δ^s_f, hence we need only one counter for aggregating.

For example, any constraint with feature $f = \text{one}$, i.e., any constraint counting the number of occurrences of a pattern, is an aggregate-once constraint, because for any e-occurrence $[\ell, u]$ and any $i, i+1 \in [\ell, u]$ we have $\phi_f(\delta^i_f, \delta^{i+1}_f) = \delta^\ell_f = \delta^{\ell+1}_f = \cdots = \delta^u_f = 1$. Also, consider any constraint with feature $f = \text{max}$ and pattern '$< (< \mid =) * (> \mid =)* >$', which means there is a strict increase followed by a non-strictly increasing subsequence, possibly a plateau, and then a non-strictly decreasing subsequence, followed by a strict decrease. The maximal value δ_f^s of an e-occurrence $[\ell, u]$ of that pattern is found already when we traverse the 'found' transition for $s \in [\ell, u]$, which is the first transition on signature symbol '>': there is no need then to consider other elements of the e-occurrence because the rest of the pattern is a non-strictly decreasing sequence, so we can aggregate once we know δ_f^s. Formally, such a constraint is an aggregate-once constraint, because for any e-occurrence $[\ell, u]$ we have that $\phi_f(\phi_f(\ldots \phi_f(\mathrm{id}_f, \delta_f^\ell), \ldots, \delta_f^{u-1}), \delta_f^u) = \max(\mathrm{id}_f, \delta_f^\ell, \ldots, \delta_f^u) = \max(\mathrm{id}_f, X_f^\ell, \ldots, X_f^u) = X_s = \delta_f^s$, where X_s triggers a 'found' transition of the automaton, with $s \in [\ell, u]$.

The second kind of time-series constraints, in Definition 4 below, is characterised by a combination of feature and pattern properties for which we can start aggregating a current feature value into the result accumulator r as soon as when we find out that we are within a pattern e-occurrence, i.e., without waiting for the end of that pattern e-occurrence. To understand how a synthesised automaton works, we define the following functions, parametrised by entries from Fig. 2B and C, representing the updates of the accumulators c and r:

- $F_f \;\; : \mathbb{Z} \times \mathbb{Z} \to \mathbb{Z} \times \mathbb{Z} \qquad (c_i, r_i) \mapsto (\phi_f(c_i, \delta_f^i), \; r_i)$
- $G'_{f,g} : \mathbb{Z} \times \mathbb{Z} \to \mathbb{Z} \times \mathbb{Z} \qquad (c_i, r_i) \mapsto (\mathrm{id}_f, \; g(r_i, \phi_f(c_i, \delta_f^i)))$
- $G''_{f,g} : \mathbb{Z} \times \mathbb{Z} \to \mathbb{Z} \times \mathbb{Z} \qquad (c_i, r_i) \mapsto (\phi_f(c_i, \delta_f^i), \; g(r_i, \phi_f(c_i, \delta_f^i)))$

When a synthesised automaton from [3] computes the value of feature f for an e-occurrence $[\ell, u]$ and aggregates it into the result accumulator r, the new value of r is computed by first applying $u - \ell$ times the function F_f and then applying the function $G'_{f,g}$. However it is often possible to aggregate this feature value into r without waiting for the end of the e-occurrence. There are two such situations: either (a) before aggregating, we must evolve the feature value of the e-occurrence in accumulator c; or (b) we need not evolve this feature value in c, but after each aggregation c is reset to the id_f value from Fig. 2B. We apply $u - \ell$ times the function $G''_{f,g}$ or $G'_{f,g}$ for the situations (a) and (b) respectively. Finally $G'_{f,g}$ is applied once for both (a) and (b), since we do not have to keep in accumulator c the feature value when we are at the end of the e-occurrence. The old [3] order of accumulator updates corresponds to $G'_{f,g} \circ F_f \circ \cdots \circ F_f$, called order (1), while the new order of updates corresponds to either $G'_{f,g} \circ G'_{f,g} \circ \cdots \circ G'_{f,g}$, called order (2), or $G'_{f,g} \circ G''_{f,g} \circ \cdots \circ G''_{f,g}$, called order (3).

Definition 4. *A time-series constraint is an* immediate-aggregation constraint *if for any e-occurrence the use of order* (1) *has the same result as using either order* (2) *or order* (3).

Due to the immediate-aggregation property, we do not have to distinguish the potential and current parts anymore. In [3], updating r is done after the end of an e-occurrence, taking into account the current feature value in c. However, we need not aggregate after the end of an e-occurrence, as the update of r should happen when we are sure that the current element X_i belongs to the e-occurrence, so we can use c for keeping both the potential and current parts.

	before	update 1	$\cdots$	update $u - \ell$	update $u - \ell + 1$
order (1)					
c update		$c_\ell = c_{\ell-1} + 1$	$\cdots$	$c_{u-1} = c_{u-2} + 1$	$c_u = 0$
r update		$r_\ell = r_{\ell-1}$	$\cdots$	$r_{u-1} = r_{u-2}$	$r_u = \max(r_{u-1}, c_{u-1} + 1)$
(c, r)	$(0, r_{\ell-1})$	$(1, r_{\ell-1})$	$\cdots$	$(u - \ell, r_{\ell-1})$	$(0, \max(r_{\ell-1}, u - \ell + 1))$
order (3)					
c update		$c_\ell = c_{\ell-1} + 1$	$\cdots$	$c_{u-1} = c_{u-2} + 1$	$c_u = 0$
r update		$r_\ell = \max(r_{\ell-1}, c_{\ell-1} + 1)$	$\cdots$	$r_{u-1} = \max(r_{u-2}, c_{u-2} + 1)$	$r_u = \max(r_{u-1}, c_{u-1} + 1)$
(c, r)	$(0, r_{\ell-1})$	$(1, \max(r_{\ell-1}, 1))$	$\cdots$	$(u - \ell, \max(r_{\ell-1}, u - \ell))$	$(0, \max(r_{\ell-1}, u - \ell + 1))$

Fig. 3. MaxWidthStrictlyDecreasingSequence immediately aggregates

For example, the MaxWidthStrictlyDecreasingSequence constraint is an immediate-aggregation constraint. This is illustrated in Fig. 3, where c_i and r_i respectively denote the values of accumulators c and r after consuming X_i: we consider an e-occurrence $[\ell, u]$ and apply the two orders (1) and (3); after the last update, the value of the accumulator r coincides for both orders. The column 'before' contains the value of the accumulators just before the e-occurrence $[\ell, u]$. The simplified automaton for this constraint is given in Fig. 1.

The percentage of constraints for which we can simplify the automata using the different types of simplifications is given in Fig. 2A.

4   MIP Decomposition of Automaton-Based Constraints

Consider a constraint $\gamma(N, \langle X_0, \ldots, X_{n-1} \rangle)$ and signature constraints linking its n variables X_j to $n + 1 - w$ signature variables S_i, each S_i being functionally determined by a linear relation on w consecutive X_j variables. For ease of notation, we here assume $w = 2$: each S_i is linked to X_i and X_{i+1}, as for the time-series constraints in Sect. 2. (Other frequent scenarios are $w = 1$, where each S_i is linked to X_i only, and the absence of signature constraints, in which case one would assume $S_i = X_i$ are the signature constraints, also with $w = 1$).

Assume a ground instance of $\gamma(N, \langle X_0, \ldots, X_{n-1} \rangle)$ holds iff an automaton $\mathcal{A}$ with $m \geq 1$ accumulators a_j that are updated by linear expressions ϕ, possibly using the 'max' and 'min' operators, returns the value of its variable N, called the *result variable*, after consuming the values of its signature variables $S_0, \ldots, S_{n-2}$.

Following [1], we decompose γ for a MIP solver by formulating *logical* constraints that model the triggering of transitions in $\mathcal{A}$ (Sect. 4.1) and *linearising* those constraints (Sect. 4.2). For $m = 0$, there is the flow-based MIP decomposition of [8]. For $m = 1$ accumulator that is only updated through increments by positive integers, there is the column-generation approach of [9].

4.1   Logical Constraints

Beside the integer variables $X_0, \ldots, X_{n-1}$ and N of γ, to model the behaviour of $\mathcal{A} = \langle Q, \Sigma, \delta, q_0, I, A, \alpha \rangle$ on the signature variables $S_0, \ldots, S_{n-2}$ over Σ, the key idea is to represent the states visited by $\mathcal{A}$ using *state variables* $Q_0, \ldots, Q_{n-1}$ over Q: each Q_i denotes the state reached *after* consuming S_{i-1}, with $Q_0 = q_0$.

Also, we need *transition variables* $T_0, \ldots, T_{n-2}$ over the set $T = Q \times \Sigma$ of constants denoting all the transitions of the total function δ: each T_i denotes the $(i+1)^{st}$ triggered transition of $\mathcal{A}$, that is *while* consuming S_i.

Last, we need *accumulator variables* $A_{i,j}$ for $i \in [0, n-1]$ and $j \in [1, m]$: each integer $A_{i,j}$ denotes the value of accumulator a_j *after* the ith transition of $\mathcal{A}$, that is *after* consuming S_{i-1}; each $A_{0,j}$ is given in the tuple I of initial values.

The *signature constraints* functionally determine each signature variable S_i from a linear relation on X_i and X_{i+1}. For example, the signature constraints for time-series constraints are given at the beginning of Sect. 2.

The *transition constraints* encode the transitions of δ as follows:

$$Q_0 = q_0$$
$$Q_i = q \wedge S_i = \sigma \Rightarrow Q_{i+1} = \delta(q, \sigma) \wedge T_i = \langle q, \sigma \rangle, \ \forall i \in [0, n-2], \ \forall q \in Q, \ \forall \sigma \in \Sigma$$

For example, a representative transition constraint for the automaton of Fig. 1 is: $Q_i = s \wedge S_i = \, '<' \Rightarrow Q_{i+1} = s \wedge T_i = \langle s, < \rangle, \ \forall i \in [0, n-2]$.

The *accumulator constraints* are of three kinds: the values of the accumulator variables $A_{0,j}$ before any transitions are found in the m-tuple I of initial values; there is an implication constraint for each transition of δ with its accumulator updates; and the values of the accumulator variables $A_{n-1,j}$ after all transitions are linked to the result variable N according to the acceptance function α. If $A \subsetneq Q$, then we have to pose the additional constraint $Q_{n-1} \in A$.

For example, the accumulator constraints for the automaton in Fig. 1 are as follows, using the accumulator variables C_i and L_i for denoting the successive values of the accumulators c and ℓ respectively: the constraints $L_0 = 0$ and $C_0 = 0$ correspond to the pair $I = \langle 0, 0 \rangle$ of initial values; the constraint $N = L_{n-1}$ stems from the acceptance function; further:

$$
\begin{aligned}
T_i = t &\Rightarrow C_{i+1} = C_i, & \forall t \in \{\langle s, < \rangle, \langle s, = \rangle\}, \ \forall i \in [0, n-2] \\
T_i = \langle s, > \rangle &\Rightarrow C_{i+1} = 2, & \forall i \in [0, n-2] \\
T_i = t &\Rightarrow C_{i+1} = 0, & \forall t \in \{\langle u, < \rangle, \langle u, = \rangle\}, \ \forall i \in [0, n-2] \\
T_i = \langle u, > \rangle &\Rightarrow C_{i+1} = C_i + 1, & \forall i \in [0, n-2] \\
T_i = t &\Rightarrow L_{i+1} = L_i, & \forall t \in \{\langle s, < \rangle, \langle s, = \rangle, \langle u, < \rangle, \langle u, = \rangle\}, \ \forall i \in [0, n-2] \\
T_i = \langle s, > \rangle &\Rightarrow L_{i+1} = \max(L_i, 2), & \forall i \in [0, n-2] \\
T_i = \langle u, > \rangle &\Rightarrow L_{i+1} = \max(L_i, C_i + 1), & \forall i \in [0, n-2]
\end{aligned}
$$

For n variables X_i and m accumulators, there are $n-1$ signature variables, n state variables, $n-1$ transition variables, and mn accumulator variables, hence $\Theta(n)$ variables in total, since m is a constant. Since $\mathcal{A}$ has a constant size, each variable occurs in a constant number of constraints, so there are $\Theta(n)$ constraints.

4.2   Linearising the Logical Constraints

To obtain a linear model, we linearise each group of logical constraints.

For each variable S_i over Σ, we introduce 0-1 variables S_i^σ, with 1 denoting truth and 0 denoting falsity, hence the semantics $S_i^\sigma = 1 \Leftrightarrow S_i = \sigma$ for all $i \in [0, n-2]$ and $\sigma \in \Sigma$. This requires that exactly one of the S_i^σ takes value 1:

$$\sum_{\sigma \in \Sigma} S_i^\sigma = 1, \ \forall i \in [0, n-2] \tag{1}$$

We replace each atom $S_i = \sigma$ by the Boolean S_i^σ in each logical constraint.

We perform the same operation for the Q_i and T_i variables with respect to their domains, getting variables Q_i^q and T_i^t for all $q \in Q$ and $t \in T$. If $A \subsetneq Q$, then we additionally require $Q_{n-1}^q = 0$ for all $q \in Q \backslash A$.

To linearise the transition constraints, which are now implications where both sides are conjunctions of Boolean variables, we use the technique of [17, pages 172–177].

The accumulator constraints have the general logical form

$$T_i = t \Rightarrow A_{i+1,j} = \phi, \text{ with } i \in [0, n-2], \ j \in [1, m], \text{ and } t \in T$$

where ϕ is here a linear expression, possibly using the 'max' and 'min' operators, that mentions variables $A_{i,j}$ denoting accumulator values *before* the considered i^{th} transition. We linearise such an implication as follows:

$$A_{i+1,j} - \phi \leq M_j \cdot (1 - T_i^t), \text{ with } i \in [0, n-2], \ j \in [1, m], \text{ and } t \in T$$
$$A_{i+1,j} - \phi \geq M_j \cdot (T_i^t - 1), \text{ with } i \in [0, n-2], \ j \in [1, m], \text{ and } t \in T$$

where constant M_j, chosen with respect to the function ϕ, is such that the constraints above always hold. Computation of M_j may also require calculation of the values serving as plus and minus infinities. For example, for a time-series constraint specified by a triple $\langle p, f, g \rangle$, we have that each M_j depends on the extrema of feature f. If ϕ uses the 'max' and 'min' operators, then we first linearise it using the technique of [10, pages 4–5], introducing a constant number of new variables.

We linearise the signature constraints by using the following technique, explained on the example of time-series constraints, where the minimum difference between two consecutive integer variables X_i is 1. We rewrite the signature constraint $X_i < X_{i+1} \Leftrightarrow S_i = {}'{<}'$ as two linear inequalities enforcing $S_i^< = 1$ if $X_i < X_{i+1}$, and $S_i^< = 0$ otherwise:

$$\frac{X_{i+1} - X_i}{M_i'} \leq S_i^< \leq \frac{X_{i+1} - X_i}{M_i'} + \frac{2M_i' - 1}{2M_i'}, \ \forall i \in [0, n-2]$$

where constant M_i' is $\max\limits_{v \in \text{dom}(X_i), \ w \in \text{dom}(X_{i+1})} |w - v| + 1$, for all $i \in [0, n-2]$, assuming $\text{dom}(Y)$ denotes the domain of variable Y. The linearisation of $X_i > X_{i+1} \Leftrightarrow S_i = {}'{>}'$ is symmetric. The linearisation of $X_i = X_{i+1} \Leftrightarrow S_i = {}'{=}'$ is $S_i^< = 0 \wedge S_i^> = 0$, since the instance $S_i^< + S_i^= + S_i^> = 1$ of (1) implies $S_i^= = 1$.

For n variables X_i and m accumulators, there are $(n-1) \cdot |\Sigma|$ signature variables, $n \cdot |Q|$ state variables, $(n-1) \cdot |Q| \cdot |\Sigma|$ transition variables, and mn accumulator variables. Linearising any of the $(n-1) \cdot |Q| \cdot |\Sigma|$ accumulator constraints requires a constant number of new variables, if any. So we still have $\Theta(n)$ variables in total, since m, $|Q|$, and $|\Sigma|$ are constants; for the time-series constraints, we have $|Q| \leq 4$ for 240 of the 266 automata and $|Q| \leq 13$ otherwise, $m \leq 3$ upon the improvements in Sect. 3, and $|\Sigma| = 3$. Since each variable occurs in a constant number of constraints, there still are $\Theta(n)$ constraints.

5 Improved Generation of Implied Constraints

Given an automaton $\mathcal{A}$ with $m \geq 1$ accumulators a_j, our tool ImpGen [11] generates *invariants* of the form $\alpha_1 a_1 + \cdots + \alpha_m a_m + \gamma \geq 0$: these inequalities hold at every state of $\mathcal{A}$ for any symbols consumed so far. Let variable $A_{i,j}$ denote the value of accumulator a_j after $\mathcal{A}$ has consumed the first i symbols of a sequence of n symbols: these variables appear in the CP decomposition [4], for a sequence of n variables S_i, of the constraint specified by $\mathcal{A}$. This decomposition in general does not achieve domain consistency when $m \geq 1$ [2]: achieving it is NP-hard for such a constraint in general [5]. Each invariant translates into $n + 1$ constraints of the form $\alpha_1 A_{i,1} + \cdots + \alpha_m A_{i,m} + \gamma \geq 0$, for all $0 \leq i \leq n$. We showed in [11] that these constraints are implied by the mentioned CP decomposition, and that the implied constraints translating a suitable selection of invariants improve the propagation strength and speed of that decomposition. The generation of implied constraints is specific to an automaton, but neither to a constrained sequence of variables S_i nor to its length n, and can thus be done offline.

ImpGen handles automata where each accumulator update is a linear expression on accumulators. This includes increments and decrements by constant amounts (as in $c := c + 1$) or other accumulators (as in $c := c + \ell$), resets (as in $c := 0$), etc. This excludes updates via the 'max' and 'min' operators, for instance: ImpGen handles only 64 of the 266 time-series constraints in Sect. 2.

Towards handling all the time-series constraints, we need to extend ImpGen to handle also conditional accumulator updates of the form $c := $ if ρ then ϕ else ψ, where ρ is a linear (in)equality and ϕ, ψ are linear expressions on accumulators: following an idea in [16], we extend the encoding of automaton transitions by allowing preconditions to be expressed. ImpGen now automatically first rewrites accumulator updates containing the binary 'min', 'max', or 'abs' operators into conditional updates. For example, the accumulator update on the arc from s to t in Fig. 1 is rewritten as $\langle c, \ell \rangle := \langle 2, $ if $\ell > 2$ then ℓ else $2 \rangle$.

Finally, we extend ImpGen to *rank* the implied constraints by decreasing propagation strength when added to the CP decomposition: this is done based on a series of random instances. This enables *automated* selection via a top-k rule for a user-chosen parameter k, as opposed to the previous *manual* selection among a *set* of implied constraints. For example, the top three implied constraints generated from the automaton in Fig. 1 are $L_i \geq L_{i-1}$, $L_i \geq L_{i-2}$, and $L_i + L_{i-1} \geq 2 \cdot L_{i-2}$, where L_i denotes the value of accumulator ℓ after consuming the first i symbols. The new tool is available online.[1]

Intuitively, the implied constraints generated by ImpGen can improve inference also for the MIP decomposition of Sect. 4 because they are generated directly from an automaton and are not necessarily linear combinations of the linear inequalities in that decomposition [13]. Our experiments in the next section confirm that implied constraints that improve the propagation of the CP decomposition can also improve the inference of the MIP decomposition.

---

[1] http://www.it.uu.se/research/group/astra/software/impGen.zip.

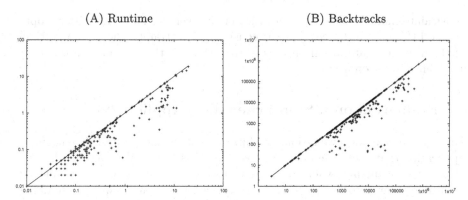

Fig. 4. Time in seconds (left) and backtracks (right) to maximise the result variable for random instances under SICStus Prolog 4.3.2 on a 2011 MacBook Pro 2.2 GHz quad-core Intel Core i7-950 machine with 6MB cache and 16 GB memory. The x-axis is for the new automata and the y-axis is for the old automata: points below the diagonal represent good results for the new automata.

6 Benchmark on CP and MIP Solvers

To evaluate the CP and MIP decompositions of the time-series constraints, we compared their old automata [3] against the new automata of Sect. 3, and the new automata with and without implied constraints generated as in Sect. 5.

To compare the old automata against the new automata for CP, we generated instances for all the 266 time-series constraints over time series of length 15 over the domain $\{1, 2, 3\}$. Note that a domain of size 3 is large enough to allow all patterns to occur and to focus the propagation effort on the transition constraints and accumulator constraints but not on the signature constraints. We maximised the result variable, and used a timeout of 100 s. As can be seen in Fig. 4, the decompositions of the new automata are almost always faster (actually 1.6 times faster on average) and always have fewer backtracks (actually 25 % fewer backtracks on average) than those of the old automata.

To compare the new automata with and without implied constraints both for CP and MIP, we generated 40 instances for each constraint used in Sect. 7 below over time series of length 100 and random sub-intervals of $[0, 1000]$ as domains. We maximised the result variable, and used a timeout of 300 s.

Using SICStus Prolog [7], we chose a static search strategy, assigning the variables X_i by increasing index and trying values from smallest to largest. This means that the first solution found is the same with and without implied constraints, and that the times and backtrack counts are directly comparable. The decompositions of the new automata are always faster in the presence of the top two implied constraints, namely 3.33 times faster on average, and always have fewer backtracks, by up to 5 orders of magnitude. In particular, all instances of half the constraints are now solved in less than 1 s instead of timing out.

Using the Gurobi 6.5 [12] MIP solver, the decompositions of the new automata are almost always faster in the presence of the top two implied constraints, namely also 3.33 times faster on average, and can solve to optimality 14 % more instances. For the considered constraints, the decompositions of the new automata are always faster than those of the old automata, namely 1.63 times faster on average.

7   Evaluation on a Staff Scheduling Application

For a more realistic evaluation, we introduce a prototypical staff scheduling application that uses a number of time-series constraints. We consider the case of a service company, where demand varies over time, and has to be met at each time point. In order to provide the service level required, we have to define a manpower resource profile over time. Resource cost may vary over time, i.e., employees may be paid different rates at different times. If we could hire and fire personnel arbitrarily, we could follow the demand curve exactly, but this is not allowed, as business processes, employment rules, and union contracts limit how quickly we can change the number of persons employed. We are therefore required to sometimes employ more people than strictly necessary. Note that we are not dealing with a shift rostering problem, where the demand must be covered by people working different shift patterns. In the current problem we are only interested in the total manpower curve, over a long-term horizon.

The overall problem is to cover the given resource demand over time, while minimising overall resource cost, and at the same time satisfying the given time-series constraints.

7.1   Notation, Constants and Variables

In our benchmark, we use a time resolution of one week over a one year horizon, i.e. we consider $n = 52$ time points. The integer variables X_i describe the scheduled resource level at time i. These variables form a single time-series $X_1, \ldots, X_n$, all constraints are expressed over this time-series or over one of its sub-sequences. The symbols d_i define the given, fixed demand at each time point i. The symbols c_i define the cost of a resource unit at time point i. For each constraint we also introduce an integer variable which represents the aggregated feature value for the constraint. The lower or upper domain bound of these variables will be constrained.

7.2   Objective Function

The objective is to minimise the total cost of the schedule, i.e.

$$obj^* = \min \sum_{i=1}^{n} X_i c_i$$

The overhead $obj^* - \sum_{i=1}^{n} d_i c_i$ is the increase in cost due to the working rules. We can use the overhead also to evaluate the potential cost/savings due to adding/removing a specific working rule. Another lower bound is the sum of the lower domain bounds after initial propagation: we use this to compute the finite-domain optimality gap in our evaluation.

7.3   Constraints

There are two types of constraints, one concerning the demand profile, and the other a set of time-series constraints. At each timepoint, the resources provided must exceed the required demand $X_i \geq d_i$.

The constraints on the time series are given in natural language form below, we also note the constraints used, following the naming scheme in [3].

1. The manpower profile can have at most two peaks. This is expressed with a NBPEAK constraint with a parameter variable with an upper bound of two.
2. The manpower profile can have at most two valleys. This is handled by the NBVALLEY constraint.
3. The maximal manpower level at any peak of employment is 250. The numbers employed at the start or end of the planning period can be higher. The MAXMAXPEAK constraint handles this condition.
4. We can hire at most 5 persons in one week. This limit is caused by the induction training required. The induction covers safety training, where spaces in each course are limited. We use the MAXRANGEINCREASING constraint to model this condition.
5. We can fire at most 7 persons in one week (expressed with a MAXRANGEDE-CREASING constraint).
6. We can only have at most four consecutive increases of personnel in the planning period. This is expressed by the MAXWIDTHSTRICTLYINCREAS-INGSEQUENCE constraint, considering that four consecutive increases lead to a pattern of width five.
7. We can only have at most six consecutive decreases of personnel numbers in the planning period (using MAXWIDTHSTRICTLYDECREASINGSEQUENCE from Example 1).
8. If we reach a peak in the employment, the profile has to stay constant for at least 10 weeks. Otherwise, we will be violating a "hire and fire" union rule. This is handled by a MINWIDTHPLATEAU constraint.
9. If we fire a person, we can not hire another person for four weeks. Instead, we should keep on employing the person (MINWIDTHPLAIN).
10. We are not allowed to fire persons in the two weeks before Christmas (expressed with a NBDECREASING constraint on a sub-sequence).
11. In every month, we can have at most 20 new hires. This is due to limitations of the human resources department. For this we use one SUMRANGEIN-CREASING constraint for each month.
12. The difference between the highest and lowest peak should not be more than 30. We already have a MAXMAXPEAK constraint to constrain the level of the highest peak. A MINMAXPEAK constrains the height of the lowest peak, an inequality between the parameters limits the difference to at most 30.

Manually Generated Redundant Constraints. In order to find solutions more easily, we initially manually defined some redundant constraints controlling the domain envelope. Constraint (4) can be approximated by inequalities $X_{i+1} \leq X_i + c$ with a constant c equal to five (this is also generated by ImpGen), while constraints (4) and (6) imply inequalities of the form $X_{i+p+1} \leq X_i + pc$, as any sequence of $p + 1$ intervals can contain at most $p = 4$ increases. These constraints are currently out of the scope of ImpGen because they are linear only at the instance level.

7.4   Search Routine and Experimental Setup

In order to evaluate the impact of different implementations of the constraints, we choose a static search strategy, assigning the X_i variables by increasing index, and enumerating values from smallest to largest. This means that the first solution found is the same for all CP models used, and the times and backtrack counts are directly comparable.

We create random sample problem instances that follow a common structure. There are demand peaks in Spring and Autumn, and reduced demand during Summer and Winter. The minimal difference between peaks and valleys is controlled by a parameter P, which we vary from 10 to 40 in steps of 5. For each parameter value, we generate 100 instances.

We compare different implementations of the time-series constraints, together with manually or automatically generated implied constraints, using the solvers described in Sect. 6, on the hardware introduced in Fig. 4. On their own, the time-series constraints perform quite poorly. Both the old and the new automata definitions only solve instances for the easiest instance set (P=10), finding solutions for 12, respectively 16, of the 100 problems. Adding either manually defined constraints or the top two implied constraints as described in Sect. 5 to the new automata allow us to find solutions for all problem instances for all parameter values. Using the old automata with the manually defined constraints solves 90, 70, 45, 36, 31, 35, and 32 out of 100 instances for parameter values 10 to 40.

For the combinations of automata and implied constraints that solve all instances we compare backtracks and solution times for the CP model in Table 1, which also shows the average and maximal optimality gap for both the CP and MIP models. Note that the finite-domain solver typically only finds a first solution, and cannot prove optimality within the timeout period. We report results for finding that first solution. At the moment, the MIP solver, even when using the implied constraints and with a timeout of 300 s, only finds optimal solutions for some of the problem instances (column Opt), and performs worse than the CP model for some instances.

We can see that both automatically and manually generated implied constraints are important, and that their combination significantly reduces the search space explored. On average, the best CP solutions found are within 4 % of the lower bound, but for some instances the gap is as large as 17 %. The average MIP optimality gap is smaller, but the worst cases are even higher, and do not occur for the same instances as for the CP model.

Table 1. Backtracks, Execution Times, Solution Quality

	new+implied				new+manual				new+impl.+man.				optimality gap				
	back		time		back		time		back		time		cp		mip		
p	avg	max	avg	max	avg	max	avg	max	avg	max	avg	max	avg	max	avg	max	opt
10	20	55	0.08	0.10	478	2168	0.37	1.41	12	35	0.09	0.12	2.86	8.45	1.75	7.97	14
15	80	730	0.11	0.34	548	2144	0.47	1.59	18	42	0.09	0.12	3.27	11.25	1.82	7.22	13
20	200	990	0.17	0.63	496	3921	0.49	4.07	18	43	0.09	0.12	3.42	9.67	2.28	18.77	27
25	1034	17719	0.60	9.30	766	6119	0.73	5.30	35	448	0.10	0.33	3.20	10.54	2.15	17.25	24
30	1001	17726	0.68	13.01	789	6452	0.80	5.85	34	452	0.10	0.35	3.20	8.02	2.04	6.34	26
35	1247	17726	0.86	15.17	824	6621	0.85	6.96	36	460	0.10	0.40	3.38	8.25	2.03	6.21	28
40	1992	25986	1.23	15.44	962	7369	1.02	5.80	37	468	0.10	0.39	3.51	17.32	1.97	10.47	18

8 Conclusion

Within the context of automaton-specified constraints in general, and time-series constraints in particular, the theoretical contributions of this paper have been shown to improve significantly both CP and MIP models. We hope our work motivates the quest for other general results that have a positive impact on different solving technologies, such as CP, MIP, local search, and SAT.

Acknowledgements. We thank Michel Minoux for his feedback on the integer linear programming decomposition in Sect. 4. We thank Mats Carlsson for his useful input during the early discussions of this paper. We also thank the anonymous referees for their helpful comments. The first and second authors are partially supported by the Gaspard-Monge programme. The authors at Mines Nantes are supported by project GRACeFUL, which has received funding from the European Union's Horizon 2020 research and innovation programme under grant agreement № 640954. The authors at Uppsala University are supported by grants 2011-6133 and 2012-4908 of the Swedish Research Council (VR). The last author was supported by Science Foundation Ireland under Grant Number SFI/10/IN.1/I3032. The INSIGHT Centre for Data Analytics is supported by Science Foundation Ireland under Grant Number SFI/12/RC/2289.

References

1. Arafailova, E.: Reformulation of automata for time series constraints as linear programs. Master's thesis, Université de Nantes, France (2015)
2. Beldiceanu, N., Carlsson, M., Debruyne, R., Petit, T.: Reformulation of global constraints based on constraints checkers. Constraints **10**(4), 339–362 (2005)
3. Beldiceanu, N., Carlsson, M., Douence, R., Simonis, H.: Using finite transducers for describing and synthesising structural time-series constraints. Constraints **21**(1), 22–40 (2016). http://dx.doi.org/10.1007/s10601-015-9200-3
4. Beldiceanu, N., Carlsson, M., Petit, T.: Deriving filtering algorithms from constraint checkers. In: Wallace, M. (ed.) CP 2004. LNCS, vol. 3258, pp. 107–122. Springer, Heidelberg (2004)
5. Beldiceanu, N., Flener, P., Pearson, J., Van Hentenryck, P.: Propagating regular counting constraints. In: AAAI 2014, pp. 2616–2622. AAAI Press (2014)
6. Berstel, J.: Transductions and Context-Free Languages. Teubner, Berlin (1979)

7. Carlsson, M., Ottosson, G., Carlson, B.: An open-ended finite domain constraint solver. In: Hartel, P.H., Kuchen, H. (eds.) PLILP 1997. LNCS, vol. 1292, pp. 191–206. Springer, Heidelberg (1997)
8. Côté, M.-C., Gendron, B., Rousseau, L.-M.: Modeling the regular constraint with integer programming. In: Van Hentenryck, P., Wolsey, L.A. (eds.) CPAIOR 2007. LNCS, vol. 4510, pp. 29–43. Springer, Heidelberg (2007)
9. Demassey, S., Pesant, G., Rousseau, L.M.: A Cost-Regular based hybrid column generation approach. Constraints **11**(4), 315–333 (2006)
10. FICO: MIP formulations and linearizations. Fair Isaac Corporation, June 2009. http://www.fico.com/en/node/8140?file=5125
11. Francisco Rodríguez, M.A., Flener, P., Pearson, J.: Implied constraints for automaton constraints. In: GCAI 2015. EasyChair Epic Series in Computing (forthcoming), preprint at http://www.it.uu.se/research/group/astra/publications
12. Gurobi Optimization, Inc.: Gurobi optimizer reference manual (2015). http://www.gurobi.com
13. Minoux, M.: Personal communication, July 2015
14. Pesant, G.: A regular language membership constraint for finite sequences of variables. In: Wallace, M. (ed.) CP 2004. LNCS, vol. 3258, pp. 482–495. Springer, Heidelberg (2004)
15. Sakarovitch, J.: Elements of Language Theory. Cambridge University Press (2009)
16. Sankaranarayanan, S., Sipma, H.B., Manna, Z.: Constraint-based linear-relations analysis. In: Giacobazzi, R. (ed.) SAS 2004. LNCS, vol. 3148, pp. 53–68. Springer, Heidelberg (2004)
17. Williams, H.P.: Model Building in Mathematical Programming. Wiley, New York (2015)

Finding a Collection of MUSes Incrementally

Fahiem Bacchus[1] and George Katsirelos[2(✉)]

[1] Department of Computer Science, University of Toronto, Toronto, ON, Canada
fbacchus@cs.toronto.edu
[2] MIAT, INRA, Toulouse, France
george.katsirelos@toulouse.inra.fr

Abstract. Minimal Unsatisfiable Sets (*MUSes*) are useful in a number
of applications. However, in general there are many different *MUSes*,
and each application might have different preferences over these *MUSes*.
Typical MUSER systems produce a single *MUS* without much control over
which *MUS* is generated. In this paper we describe an algorithm that can
efficiently compute a collection of *MUSes*, thus presenting an application
with a range of choices. Our algorithm improves over previous methods
for finding multiple *MUSes* by computing its *MUSes* incrementally. This
allows it to generate multiple *MUSes* more efficiently; making it more
feasible to supply applications with a collection of *MUSes* rather than
just one.

1 Introduction

When given an unsatisfiable CNF $\mathcal{F}$, SAT solvers can return a core, i.e., a
subset of $\mathcal{F}$ that remains unsatisfiable. Many applications, e.g., program type
debugging, circuit diagnosis, and production configuration [6], need cores in their
processing. In many cases these applications can be made much more effective
if supplied with minimal unsatisfiable sets (*MUSes*), which are cores that are
minimal under set inclusion. That is, no proper subset of a *MUS* is unsatisfiable.

This makes the problem of efficiently extracting a *MUS* an important and
well studied problem, see [5,6,9,13,18,20,21] for a more extensive list. In fact,
the problem of finding a minimal set of constraints sufficient to make a problem
unsolvable is important in other areas as well. For example in operations research
it is often useful to find irreducible inconsistent subsystems (IISes) of linear
programs and integer linear programs [8,24], and in CP a minimal unsatisfiable
set of constraints [12].

In various applications the preference for *MUSes* over arbitrary cores goes
further, and some *MUSes* might be preferred to others. Most algorithms for
computing *MUSes*, however, return an arbitrary *MUS*. There has been some
work on the problem of computing specific types of *MUSes*. In [19] the problem of computing lexicographic preferred *MUSes* is addressed. Furthermore, the
problem of computing the smallest *MUS* has been addressed in [10,11,15]. However, algorithms for extracting specific *MUSes*, especially those for extracting
the smallest *MUS*, can be considerably less efficient than state-of-the-art *MUS*
extraction algorithms returning an arbitrary *MUS*.

© Springer International Publishing Switzerland 2016
C.-G. Quimper (Ed.): CPAIOR 2016, LNCS 9676, pp. 35–44, 2016.
DOI: 10.1007/978-3-319-33954-2_3

In this paper we address this issue by trying to quickly return a collection of *MUSes*, rather than trying to compute a specific type of *MUS*. The application can then choose its best *MUS* from that collection. So, e.g., although our approach cannot guarantee returning the smallest *MUS*, the application can choose the smallest *MUS* from among the collection returned. This approach is advantageous when algorithms for computing the most preferred *MUS* are too costly (e.g., computing the smallest *MUS*), or when there is no known algorithm for computing the most preferred *MUS* (e.g., the application's preference criteria is not lexicographic).

We accomplish this task through an extension of a recent *MUS* algorithm [3]. The advantage of our algorithm is that it can exploit information computed when finding previous *MUSes* to speed up finding future *MUSes*. Hence, it can find multiple *MUSes* more efficiently. This algorithm has the drawback, however, that it cannot keep on finding more *MUSes* when given more time: it computes a set of *MUSes* of indeterminate size and then stops. Adopting the power set exploration idea of [14] we address this drawback, presenting a method that can eventually compute all *MUSes* while still enumerating them at a reasonable rate. We show that our algorithms improve on the state of the art.

2   Background

Let $\mathcal{F}$ be an unsatisfiable set of clauses.

Definition 1 (MUS). A **Minimal Unsatisfiable Set** (*MUS*) of $\mathcal{F}$ is a unsatisfiable subset $M \subseteq \mathcal{F}$ that is minimal w.r.t. set inclusion. That is, M is *unsat* but no proper subset is.

Definition 2 (MSS). A **Maximal Satisfiable Subset** (*MSS*) of $\mathcal{F}$ is a satisfiable subset $S \subseteq \mathcal{F}$ that is maximal w.r.t set inclusion.

Definition 3 (MCS). A correction subset of $\mathcal{F}$ is a subset of $\mathcal{F}$ whose complement in $\mathcal{F}$ is *sat*. A **Minimal Correction Subset** (*MCS*) of $\mathcal{F}$ is a correction subset that is minimal w.r.t. set inclusion.

Note that if C is an *MCS* of $\mathcal{F}$ then its complement $\mathcal{F} \setminus C$ is an *MSS* of $\mathcal{F}$.

Definition 4. A clause $c \in \mathcal{F}$ is said to be **critical** for $\mathcal{F}$ (also known as a *transition* clause [7]) when $\mathcal{F}$ is *unsat* and $\mathcal{F} - \{c\}$ is *sat*.

Intuitively, a *MUS* is an unsatisfiable set that cannot be reduced without causing it to become satisfiable; a *MSS* is a satisfiable set that cannot be added to without causing it to become unsatisfiable; and an *MCS* is a minimal set of removals from $\mathcal{F}$ that causes $\mathcal{F}$ to become satisfiable.

A critical clause for $\mathcal{F}$ is one whose removal from $\mathcal{F}$ causes $\mathcal{F}$ to become satisfiable. If c is critical for $\mathcal{F}$ then (a) c must be contained in every *MUS* of $\mathcal{F}$ and (b) $\{c\}$ is an *MCS* of $\mathcal{F}$. Furthermore, M is a *MUS* if and only if every $c \in M$ is critical for M. Note that a clause c that is critical for a set S is not necessarily critical for a superset $S' \supset S$. In particular, S' might contain other *MUSes* that do not contain c.

Duality. A hitting set H of a collection of sets C is a set that has a non empty intersection with each set in C: $\forall C \in C.H \cap C \neq \emptyset$. A hitting set H of C is minimal (or irreducible) if no subset of H is a hitting set of C.

Let $AllMuses(\mathcal{F})$ ($AllMcses(\mathcal{F})$) be the set containing all $MUSes$ ($MCSes$) $\mathcal{F}$. There is a well known hitting set duality between AllMuses and AllMcses [22]. Specifically, $M \in AllMuses(\mathcal{F})$ iff M is a minimal hitting set of $AllMcses(\mathcal{F})$, and dually, $C \in AllMcses(\mathcal{F})$ iff C is a minimal hitting set of $AllMuses(\mathcal{F})$. The duality also holds for non-minimal sets, e.g., any correction set hits all unsatisfiable subsets. It is useful to point out that if $\mathcal{F}' \subseteq \mathcal{F}$, then $AllMuses(\mathcal{F}') \subseteq AllMuses(\mathcal{F})$. Hence, if f is critical for $\mathcal{F}$ it is critical for all unsatisfiable subsets of $\mathcal{F}$. An MCS C' of $\mathcal{F}' \subset \mathcal{F}$, on the other hand, is not necessarily an MCS of $\mathcal{F}$, however C' can always be extended to an MCS C of $\mathcal{F}$ [3].

3 Enumerating *MUSes*

To the best of our knowledge the current state-of-the-art algorithm for the problem of quickly computing a collection of $MUSes$ is the MARCO system originally developed in [14] and later improved in [16]. MARCO$^+$ (the new optimized version of MARCO [16]) was compared with previous approaches [4,17] and shown to be superior at this task. Therefore we confine our attention in this paper to comparing with the MARCO$^+$ approach.

Algorithm 1 shows the algorithm used by MARCO$^+$. MARCO$^+$ uses the technique of representing subsets of $\mathcal{F}$, the input set of clauses, with a CNF, *ClsSets*. *ClsSets* contains a variable s_i for each clause $c_i \in \mathcal{F}$. Every satisfying solution of *ClsSets* specifies a subset of $\mathcal{F}$: the set of clauses c_i corresponding to true s_i in the satisfying solution. Initially, *ClsSets* contains no clauses, and thus initially its set of satisfying solutions corresponds to $\mathcal{F}$'s powerset.

MARCO$^+$ uses *ClsSets* to keep track of which subsets of $\mathcal{F}$ have already been tested so that each MUS it enumerates is distinct. When *ClsSets* becomes *unsat* all subsets of $\mathcal{F}$ have been tested and all $MUSes$ have been enumerated. Otherwise, the truth assignment π (line 4) provides a subset S of unknown status.

MARCO$^+$ forces the sat solver to assign variables to true in each decision. Hence, if S is *sat* it is guaranteed to be an MSS (see [2] or [23] for a simple proof). S and all of its subsets are thus now known to be *sat* so they can be blocked in *ClsSets*. This means that all future solutions of *ClsSets* must have a non-empty intersection with $\mathcal{F} \setminus S$, i.e., they must hit the complement of S, a (minimal) correction set. The update of *ClsSets* is accomplished with the subroutine call **hitCorrectionSet**$(\mathcal{F} \setminus S)$ (line 6) which returns a clause asserting that some s_i corresponding to a clause in $F \setminus S$ must be true.

Otherwise S is *unsat* and it contains at least one MUS. MARCO$^+$ then invokes a MUS finding algorithm to find one of S's $MUSes$. In addition, MARCO$^+$ informs the MUS algorithm of all singleton $MCSes$ it has found. The computed MUS M has to include the union of these singleton $MCSes$ as it must hit every MCS.

M and all of its supersets are known to be *unsat* and are blocked in *ClsSets* by a clause computed by **blockSuperSets**(M) asserting that some s_i corresponding

---

Algorithm 1. MARCO$^+$ *MUS* enumeration algorithm

Input: $\mathcal{F}$ an **unsatisfiable** set of clauses
Output: All *MUSes* of $\mathcal{F}$, output as they are computed

1 $ClsSets \leftarrow \emptyset$ $\triangleleft$ Initially, $ClsSets$ admits all subsets of $\mathcal{F}$ as solutions.
2 **while** *true* **do**
   // If C is *sat*, **SatSolve**(C, π) returns true and puts truth assignment in π
3   **if SatSolve**$(ClsSets, \pi)$ **then**
4       $S \leftarrow \{c_i \in \mathcal{F} \mid \pi[s_i] = true\}$ $\triangleleft$ All decisions set to *true* so S is maximal
5   **if SatSolve**(S, π) **then**
6          $ClsSets \leftarrow ClsSets \cup \textbf{hitCorrectionSet}(\mathcal{F} \setminus S)$ $\triangleleft$ $\mathcal{F} \setminus S$ is a *MCS*
7   **else**
8          $M \leftarrow \textbf{findMUS}(S, \{\text{all singleton } MCSes\})$
9   output(MUS)
10         $ClsSets \leftarrow ClsSets \cup \textbf{blockSuperSets}(M)$
11   **else return**

---

to $c_i \in M$ must be false [14]. After all subsets of $\mathcal{F}$ have been identified as being *sat* or *unsat* (detected by $ClsSets$ becoming *unsat*), the algorithm returns.

One advantage of MARCO$^+$ is that it can utilize any *MUS* algorithm. Thus once it has identified a subset of $\mathcal{F}$ to be *unsat* it can enumerate a new *MUS* as efficiently as finding a single *MUS*. Another advantage is that it will continue to enumerate *MUSes* until it has enumerated them all. On the negative side, each new *MUS* is computed with an entirely separate computation. This *MUS* computation only knows about the prior singleton *MCSes* but does not otherwise share much information with prior *MUS* computations (beyond some learnt clauses).

4   A New Algorithm for Enumerating *MUSes*

Algorithm 2 shows our new algorithm for generating multiple *MUSes* from a formula. The grayed out lines will be used when multiple initial calls are made to the algorithm, they will be discussed in the next section. For now it can be noted that these lines have no effect if $ClsSets$ is initially an empty set of clauses.

The algorithm is a modification of the recently proposed state-of-the-art *MUS* algorithm MCS-MUS [3]. It extends MCS-MUS by performing a backtracking search over a tree in which the branch points correspond to the different ways the *MUSes* to be output can hit a just computed *MCS*.

The algorithm maintains a current formula $F' \subseteq F$, such that F' is *unsat*, partitioned into a set of clauses known to be critical for F', *crits*, and a set of clauses of unknown status, *unkn*. It starts by identifying an *MCS*, *cs*, of *crits* $\cup$ *unkn*, such that $cs \subseteq unkn$, using a slight modification of existing *MCS* algorithms [3]. If no such *MCS* exists, then *crits* is unsatisfiable and since all of its clauses are critical, it is a *MUS*. This *MUS* is reported and backtrack occurs. If *cs* does exist, it creates a choice point. By duality we know that every *MUS* must hit *cs*, and by minimality of *cs* we know that for every clause $c \in cs$ there is a *MUS* whose intersection with *cs* is only c. Hence, we select a clause $c \in cs$

Algorithm 2. MCS-MUS-BT ($unkn, crits, ClsSets$): Output a collection of $MUSes$ of $unkn \cup crits$ using MCS duality. To find some $MUSes$ of $\mathcal{F}$ the initial call MCS-MUS-BT ($\mathcal{F}, \{\}, ClsSets = \emptyset$) is used.

Input: $unkn$ a set of clauses of unknown status such that $unkn \cup crits$ is $unsat$
Input: $crits$ a set of clauses critical for $unkn \cup crits$
Input: $ClsSets$ a CNF representing subsets of the input formula of unknown status.
Output: Some $MUSes$ of $unkn \cup crits$, output as computed
Output: $ClsSets$ is modified.

1   $crits \leftarrow crits \cup \{c_i \mid s_i \in UP(ClsSets \cup \{(\neg s_j) \mid c_j \notin crits \cup unkn\})\}$

2   $unkn \leftarrow unkn \setminus crits$

3   $(cs, \pi) \leftarrow$ **findMCS**($crits, unkn$) $\triangleleft$ *Find cs, an MCS contained in $unkn$.*

4   **if** $cs = null$ **then**

5   output($crits$) $\triangleleft$ *$crits$ is a MUS of $crits \cup unkn$*

6       $ClsSets \leftarrow ClsSets \cup$ **blockSuperSets**($crits$)

7   **return**

8   **else**

9       $ClsSets \leftarrow ClsSets \cup$ **hitCorrectionSet**($\{c \mid \pi \not\models c\}$) $\triangleleft$ *Correction set of $\mathcal{F}$*

10      $unkn \leftarrow unkn \setminus cs$

11   **for** $c \in cs$ **do**

12          $crits' \leftarrow crits \cup \{c\}$

13          $unkn' \leftarrow$ **refineClauseSet**($crits', unkn$)

14          $C \leftarrow$ **recursiveModelRotation**($c, crits, unkn, \pi$)

15   MCS-MUS-BT ($unkn' \setminus C, crits' \cup C$)

---

to mark as critical (line 12) removing the rest from $unkn$ (line 10). This ensures that all $MUSes$ enumerated in the recursive call contain c and hence hit cs.

Before the recursive call, we can use two standard techniques that are critical for performance, *clause set refinement* [21] and *recursive model rotation* [7].

Theorem 1. *All sets output by* MCS-MUS-BT *are $MUSes$ of its input $\mathcal{F} = unkn \cup crits$. Furthermore, if $\mathcal{F}$ unsatisfiable a least one MUS will be output. Finally, if only one MUS is output, then $\mathcal{F}$ contains only one MUS.*

We omit the straightforward proof to save space. Although the theorem shows that MCS-MUS-BT will generate at least one MUS (as efficiently as the state-of-the-art MCS-MUS algorithm), the number of $MUSes$ it will generate is indeterminate, as this depends on the $MCSes$ it happens to generate. Furthermore, it cannot, in general, generate all $MUSes$. Intuitively, by removing cs from $unkn$ at line 10, we block it from generating any MUS M with $|M \cap cs| > 1$.

The main advantage of this algorithm is that it shares computational effort among many $MUSes$. Namely, after the first MUS is generated, computation for the second MUS starts with at least one (potentially many) known MCS, and may also have several clauses in $crits$ and a smaller set of clauses in $unkn$. Hence, it can more efficiently generate several $MUSes$.

4.1   Enumerating All *MUSes*

While MCS-MUS-BT may be able to generate a sufficiently large collection of *MUSes*, the unpredictability of the size of this collection might be unsuitable in some cases. In such cases we may of course fall back to MARCO⁺, giving up the advantages of MCS-MUS-BT.

Another option is to embed MCS-MUS-BT in MARCO⁺. It is straightforward to modify Algorithm 1 so that it uses MCS-MUS-BT instead of **findMUS** and blocks all *MUSes* discovered during one call. However, without modifying MARCO⁺ this allows only limited information to flow between MARCO⁺ and MCS-MUS-BT. In particular, sharing information beyond singleton correction sets is not supported.

A third option then is deeper integration of MCS-MUS-BT into a MARCO-like algorithm. We show this in Algorithm 3, which is based on the MCS-MUS-ALL algorithm of [3]. The outline of MCS-MUS-ALL-BT is broadly similar to that of MARCO⁺. Like MARCO⁺ it uses a CNF *ClsSets* to represent subsets of $\mathcal{F}$ with unknown status and uses the same **hitCorrectionSet** and **blockSuperSets** procedures to block *MSSes* and *MUSes*, respectively. When *ClsSets* becomes unsatisfiable all *MUSes* have been enumerated (line 3). Each solution π of *ClsSets* yields a set S of unknown status, which is then tested for satisfiability.

If it is satisfiable, S is guaranteed to be an *MSS* since we require the solver to assign variables to true in each decision as in MARCO⁺. We can then block S and all of its subsets by forcing *ClsSets* to hit its complement with **hitCorrectionSet**.

If S is unsatisfiable, then it is given to MCS-MUS-BT to extract some of its *MUSes*. We generalize MARCO⁺, however, by providing all previously discovered correction sets to MCS-MUS-BT, not just the singleton *MCSes*. These correction sets can be exploited to discover new critical clauses. In particular, all previously discovered correction sets result in clauses being added to *ClsSets* by **hitCorrectionSet**. We can use unit propagation (line 1 of Algorithm 2) to determine if the clauses currently excluded from the *MUSes* being enumerated ($\mathcal{F} \setminus (crits \cup unkn)$) make some prior correction set cs a singleton (of course all correction sets that are already singleton will also be found, so this method obtains at least as much information as MARCO⁺). If so then all *MUSes* of the current subset $crits \cup unkn$ must include that single remaining clause $c \in cs$ since all *MUSes* must hit cs; i.e., c is critical for $crits \cup unkn$.

Thus our algorithm has two advantages over using MCS-MUS-BT in the MARCO⁺ framework. First, individual calls to MCS-MUS-BT may produce *MUSes* more quickly because our generalization of MARCO⁺'s technique of exploiting singleton *MCSes* (at line 1) can detect more critical clauses, either initially or as *unkn* shrinks. Second, the multiple correction sets that can be discovered within MCS-MUS-BT are all added to *ClsSets*. Hence, their complementary satisfiable sets will not appear as possible solutions to *ClsSets* in the main loop of Algorithm 3. This can reduce the time spent processing satisfiable sets.

Algorithm 3. MCS-MUS-ALL-BT($\mathcal{F}$): Enumerate all *MUSes* of $\mathcal{F}$.

Input: $\mathcal{F}$ an **unsatisfiable** set of clauses
Output: All *MUSes* of $\mathcal{F}$, output as there are computed
1   $ClsSets \leftarrow \emptyset$ $\lhd$ *Initially, ClsSets admits all subsets of $\mathcal{F}$ as solutions.*
2   **while** *true* **do**
3   **if** *not SatSolve(ClsSets,π)* **then return**; $\lhd$ All *MUSes* enumerated
4      $S \leftarrow \{c_i \mid c_i \in \mathcal{F} \wedge \pi \models s_i\}$ $\lhd$ *All decisions set to true so S is maximal*
5   **if** *SatSolve(S,π)* **then**
6          $ClsSets \leftarrow ClsSets \cup \textbf{hitCorrectionSet}(\mathcal{F} \setminus S)$ $\lhd$ *S is an MSS*
7   **else** MCS-MUS-BT ($\mathcal{F}$, *crits*, *unkn*, *ClsSets*)

---

5   Empirical Results

In this section we evaluate our algorithms which we implemented in C++ on top of MiniSAT. We used the benchmark set of [1] containing 300 problems. We used a cluster of 48-core 2.3 GHz Opteron 6176 nodes with 378 GB RAM available.

First we tested MCS-MUS-BT (Algorithm 2) against the Marco$^+$ system [16]. MCS-MUS-BT can only generate some *MUSes*, while Marco$^+$ can potentially generate all. So in the scatter plot (a) of Fig. 1 we plotted for each instance the time each approach took to produce the first k *MUSes*, where k is the minimum of the number of *MUSes* produced by the two approaches on that instance when run with a 3600 s timeout. In the plot, points above the 45° line are where MCS-MUS-BT is better than Marco$^+$. The data shows that MCS-MUS-BT outperforms Marco$^+$ on most instances.

We also tested how many *MUSes* are typically produced by MCS-MUS-BT. When run on the 300 instances it yielded no *MUSes* on 20 instances (in 3600 s), 1 on 111 instances, 2–10 on 29 instances, and more than 10 on 140 instances. On 6 instances it yielded over 10,000 *MUSes*. So we see that MCS-MUS-BT often yielded a reasonable number of *MUSes*, but in some cases perhaps not enough.

To go beyond MCS-MUS-BT, potentially generating all *MUSes*, we used two variations of our complete algorithms. The first we call Marco-Many. This is MCS-MUS-BT integrated into an implementation of the Marco$^+$ algorithm, with MCS-MUS-BT called when a *MUS* is to be computed and returning multiple *MUSes*. The second variation is MCS-MUS-ALL-BT, from the previous section. We also compare these against Marco$^+$ [1] and our previous *MUS* enumeration algorithm MCS-MUS-ALL [3].

Figure 1(b) compares MCS-MUS-ALL-BT with Marco$^+$. Here we plotted for each instance the number of *MUSes* produced by each approach within a 3600 s timeout. Points above the line represent instances where MCS-MUS-ALL-BT generated more *MUSes* than Marco$^+$. The picture here is not completely clear. However, overall MCS-MUS-ALL-BT showed better performance: it generated more *MUSes* in 170 cases, an equal number in 48 cases, and fewer in 82 cases. Furthermore, notice that as we move up the x and y axis the instances

---

[1] Version 1.1, downloaded from https://sun.iwu.edu/~mliffito/marco/.

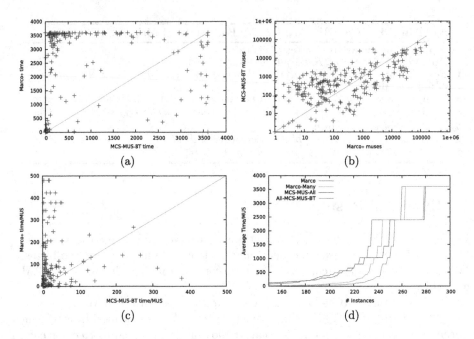

Fig. 1. (a) Time for MARCO⁺ to generate as many *MUSes* as MCS-MUS-BT (b): number of *MUSes* MCS-MUS-ALL-BT against MARCO⁺ (logscale). (c) Average time/*MUS* MCS-MUS-ALL-BT against MARCO⁺. (d) Cactus plot of Average time/*MUS* of all solvers.

become easier, i.e., many more MUSes can be generated per second in these instances. The instances in which MARCO⁺ outperformed MCS-MUS-ALL-BT tend to be towards the upper right of the plot.

Besides the number of instances we are also interested in the rate at which *MUSes* are generated. For each instance we calculated the average time needed to generate a *MUS* by MCS-MUS-ALL-BT and MARCO⁺. Figure 1(c) shows a scatter plot of these points. The cactus plot of Fig. 1(d) elaborates on this data showing the other algorithms as well.

In scatter plot (c) the axes have been inverted so that once again points above the line represent instances in which MCS-MUS-ALL-BT is better than MARCO⁺. We zoomed this plot into the range [0,500] s per *MUS* as most of the data was clustered into this region. These instances show a convincing win for MCS-MUS-ALL-BT. The plot excludes 100 instances. Of these, 43 instances could not be plotted as one or both algorithms produced zero *MUSes*: on 18 both produced zero *MUSes*; on 22 MCS-MUS-ALL-BT generated a *MUS* but MARCO⁺ did not; on 3 the inverse happened. The other 57 instances were excluded because of the plot range. Among them 3 were below the line, 23 above the line and 31 on the line. Of these excluded instances the most extreme win for MARCO⁺ was an instance where MARCO⁺ generated 3 *MUSes* and MCS-MUS-ALL-BT only 1; and the most extreme win for MCS-MUS-ALL-BT was an instance where MARCO$^+$ generated only 1 *MUS* and MCS-MUS-ALL-BT generated 800. We see that with few exceptions, the average time to generate a *MUS* with MCS-MUS-ALL-BT is smaller. This is confirmed by the cactus plot (d), where we see that the average time to generate a *MUS* by MCS-MUS-ALL-BT remains well below that of other algorithms. The corresponding lines only meet for the hardest instances, where all methods generate one or no *MUSes*. The cactus plot also confirms that simply integrating MCS-MUS-BT into a MARCO-like algorithm (i.e., MARCO-MANY) is not sufficient. Additionally, we see that the MCS-MUS-ALL-BT provides a good improvement over the previous MCS-MUS-ALL.

References

1. MUS track of the sat competition (2011). http://www.maxsat.udl.cat
2. Bacchus, F., Davies, J., Tsimpoukelli, M., Katsirelos, G.: Relaxation search: a simple way of managing optional clauses. In: Proceedings of the AAAI National Conference (AAAI), pp. 835–841 (2014)
3. Bacchus, F., Katsirelos, G.: Using minimal correction sets to more efficiently compute minimal unsatisfiable sets. In: Kroening, D., Păsăreanu, C.S. (eds.) CAV 2015. LNCS, vol. 9207, pp. 70–86. Springer, Heidelberg (2015)
4. Bailey, J., Stuckey, P.J.: Discovery of minimal unsatisfiable subsets of constraints using hitting set dualization. In: Hermenegildo, M.V., Cabeza, D. (eds.) PADL 2004. LNCS, vol. 3350, pp. 174–186. Springer, Heidelberg (2005)
5. Belov, A., Heule, M.J.H., Marques-Silva, J.: MUS extraction using clausal proofs. In: Sinz, C., Egly, U. (eds.) SAT 2014. LNCS, vol. 8561, pp. 48–57. Springer, Heidelberg (2014)
6. Belov, A., Lynce, I., Marques-Silva, J.: Towards efficient MUS extraction. AI Commun. **25**(2), 97–116 (2012)
7. Belov, A., Marques-Silva, J.: Accelerating MUS extraction with recursive model rotation. In: Formal Methods in Computer-Aided Design (FMCAD), pp. 37–40 (2011)
8. Chinneck, J.W.: Feasibility and Infeasibility in Optimization: Algorithms and Computational Methods. International Series in Operations Research and Management Sciences, vol. 118. Springer, USA (2008). ISBN 978-0387749310
9. Grégoire, É., Mazure, B., Piette, C.: On approaches to explaining infeasibility of sets of boolean clauses. In: International Conference on Tools with Artificial Intelligence (ICTAI), pp. 74–83 (2008)
10. Ignatiev, A., Janota, M., Marques-Silva, J.: Quantified maximum satisfiability: a core-guided approach. In: Järvisalo, M., Van Gelder, A. (eds.) SAT 2013. LNCS, vol. 7962, pp. 250–266. Springer, Heidelberg (2013)
11. Ignatiev, A., Previti, A., Liffiton, M., Marques-Silva, J.: Smallest MUS extraction with minimal hitting set dualization. In: Pesant, G. (ed.) CP 2015. LNCS, vol. 9255, pp. 173–182. Springer, Heidelberg (2015)
12. Junker, U.: QUICKXPLAIN: preferred explanations and relaxations for over-constrained problems. In: Proceedings of the Nineteenth National Conference on Artificial Intelligence, Sixteenth Conference on Innovative Applications of Artificial Intelligence, San Jose, California, USA, 25–29 July 2004, pp. 167–172 (2004)

13. Lagniez, J.-M., Biere, A.: Factoring out assumptions to speed up MUS extraction. In: Järvisalo, M., Van Gelder, A. (eds.) SAT 2013. LNCS, vol. 7962, pp. 276–292. Springer, Heidelberg (2013)

14. Liffiton, M.H., Malik, A.: Enumerating infeasibility: finding multiple muses quickly. In: Gomes, C., Sellmann, M. (eds.) CPAIOR 2013. LNCS, vol. 7874, pp. 160–175. Springer, Heidelberg (2013)

15. Liffiton, M.H., Mneimneh, M.N., Lynce, I., Andraus, Z.S., Marques-Silva, J., Sakallah, K.A.: A branch and bound algorithm for extracting smallest minimal unsatisfiable subformulas. Constraints 14(4), 415–442 (2009)

16. Liffiton, M.H., Previti, A., Malik, A., Marques-Silva, J.: Fast, flexible MUS enumeration. Constraints 21(2), 223–250 (2015)

17. Liffiton, M.H., Sakallah, K.A.: Algorithms for computing minimal unsatisfiable subsets of constraints. J. Autom. Reason. 40(1), 1–33 (2008)

18. Marques-Silva, J., Janota, M., Belov, A.: Minimal sets over monotone predicates in boolean formulae. In: Sharygina, N., Veith, H. (eds.) CAV 2013. LNCS, vol. 8044, pp. 592–607. Springer, Heidelberg (2013)

19. Marques-Silva, J., Previti, A.: On computing preferred MUSes and MCSes. In: Sinz, C., Egly, U. (eds.) SAT 2014. LNCS, vol. 8561, pp. 58–74. Springer, Heidelberg (2014)

20. Nadel, A., Ryvchin, V., Strichman, O.: Efficient MUS extraction with resolution. In: Formal Methods in Computer-Aided Design (FMCAD), pp. 197–200 (2013)

21. Nadel, A., Ryvchin, V., Strichman, O.: Accelerated deletion-based extraction of minimal unsatisfiable cores. J. Satisfiability Boolean Model. Comput. (JSAT) 9, 27–51 (2014)

22. Reiter, R.: A theory of diagnosis from first principles. Artif. Intell. 32(1), 57–95 (1987)

23. Di Rosa, E., Giunchiglia, E.: Combining approaches for solving satisfiability problems with qualitative preferences. AI Commun. 26(4), 395–408 (2013)

24. van Loon, J.: Irreducibly inconsistent systems of linear equations. Eur. J. Oper. Res. 8(3), 283–288 (1981)

Decomposition Based on Decision Diagrams

David Bergman[1]([⊠]) and Andre A. Cire[2]

[1] Department of Operations and Information Management,
University of Connecticut, Mansfield, USA
david.bergman@business.uconn.edu
[2] Department of Management, University of Toronto Scarborough,
Toronto, Canada
acire@utsc.utoronto.ca

Abstract. In recent years, decision diagrams (DDs) have proven useful
for solving a variety of optimization problems, often closing long-standing
instances from classical benchmarks. This success is primarily driven by
a DDs ability to capture structure. This paper exploits this characteristic
and proposes a novel solution method which decomposes a problem into
highly-structured portions, where the solution set of each portion can
be compactly represented using a DD. This technique is applied to a
special case of the independent set problem and to unconstrained binary
quadratic programming. Preliminary computational results suggest that
the proposed decomposition approach can improve upon both standard
integer programming models and a single DD approach.

1 Introduction

A decision diagram (DD) is a graphical data structure originally introduced to
compactly represent Boolean functions [1], with several applications in circuit
design and formal verification [14,18]. In recent years, DDs have also been applied
to encode the solution set of discrete optimization problems, serving for a variety
purposes such as cut generation in mixed-integer linear programming [5], to
enhance propagation in constraint programming [3], and in novel general branch-
and-bound procedures for combinatorial optimization problems [8].

In the context of optimization, the successful applications of DDs are ample.
Examples include solving long-standing open benchmark instances of the maximum cut problem [8] and of variants of the traveling salesman problem [15].
Moreover, DDs have also been incorporated into state-of-the-art integer programming and constraint programming technology to substantially improve optimality gaps [9–11] and solution times [16] on a number of applications.

One of the key reasons for the successful application of DDs lies in the fact
that they are particularly well-suited to capture complete inference for certain
problem structures. For example, the size of a DD encoding the feasible solutions
of a set covering problem can be bounded by a function of the *bandwidth* of the
constraint matrix [11,17]. As a result, DD-based methods perform well when the
bandwidth of the matrix is small, but tend to lose effectiveness for larger band-
widths. Analogously, DDs provide strong optimization bounds for the maximum

© Springer International Publishing Switzerland 2016
C.-G. Quimper (Ed.): CPAIOR 2016, LNCS 9676, pp. 45–54, 2016.
DOI: 10.1007/978-3-319-33954-2_4

clique problem when the underlying graph is dense [9], and can improve solution times in scheduling problem by orders of magnitude depending on the structure of the precedence graph [15].

In this paper we exploit this inherent characteristic of DDs and propose the notion of *decision diagram decomposition*. The idea is to decompose an optimization problem into distinct subproblems, each capturing some complicating problem structure for which the associated DD is provably small in size. Once equipped with a valid decomposition, the original discrete problem then reduces to finding a path in each DD that mutually agree on the assignment of the decision variables, which can be solved in many ways.

In particular, we propose a methodology in which each DD is assigned a network flow relaxation, thereby transforming the representation from a discrete structure into a mixed-integer linear programming (MILP) model. The network flow relaxations are combined through linking constraints, stated either generally or as problem specific constraints, so as to harness the power of MILP solvers.

The contributions of this work are hence threefold. First, we provide a decomposition approach which captures problem structure in a novel systematic way. Distinct from existing methods, the decomposition of the instance can be based not only on the constraints, but also on the objective function, or on both. Second, the resulting MILP from our methodology effectively yields new *extended formulations* [4] in a generic way, i.e. MILP models in a higher dimension that can be stronger than other existing models. Finally, our methodology can also be used to improve the robustness of current DD-based methods in constraint programming and operations research, since one can use different DDs to represent different substructures as opposed to a single DD for all the problem.

The paper is organized as follows. Section 2 introduces BDDs, and Sect. 3 explicitly expresses the decomposition framework. Next, Sect. 4 provides the MILP formulation for linking the distinct BDDs, and preliminary study cases are presented in Sect. 5 on the maximum independent set problem and the binary quadratic programming problem. A conclusion is provided in Sect. 6.

2   Decision Diagrams for Binary Optimization

For the purposes of this paper, we focus on solving *binary optimization problems* (BOPs), which are of the form $\max\{f(x) : x \in S, x \in \mathbb{B}^n\}$, where n is the number of variables, f is any function mapping binary vectors into the set of real numbers, and S is an arbitrarily defined constraint set.

A BDD $B = (U, A)$ is a directed acyclic graph with nodes U and arcs A. The nodes are partitioned into n_B layers $L_1, \ldots, L_{n_B}$, and each node $u \in U$ is in layer $\ell(u) \in \{1, 2, \ldots, n_B\}$; thus, $L_i = \{u \mid \ell(u) = i\}$. In particular, $L_1 = \{\mathbf{r}\}$ and $L_{n_B} = \{\mathbf{t}\}$, where $\mathbf{r}$ and $\mathbf{t}$ are referred to as the *root* node and *terminal* node, respectively. Each arc $a \in A$ has an *arc-weight* $w(a) \in \mathbb{R}$ and an *arc-domain* $d(u) \in \{0, 1\}$, where a is a *0-arc* when $d(a) = 0$ and an *1-arc* otherwise. An arc $a = (h(a), t(a))$ has *head* $h(a)$ and *tail* $t(a)$, with $\ell(t(a)) - \ell(h(a)) = 1$. Each node $u \in U \backslash \{t\}$ is the head of at most one 0-arc and at most one 1-arc.

We use BDDs is to represent solutions (or partial solutions) to BOPs as $r-t$ paths, where the length of the path corresponds to objective function values through the total weight of the $\mathbf{r}-\mathbf{t}$ path. Let $\mathcal{P}(B)$ be the set of arc-specified $r-t$ paths in B. Each layer, with the exception of the terminal layer, is associated with a variable through the injection $\sigma^B : \{1,\ldots,n_B-1\} \rightarrow \{1,\ldots,n\}$. For a path $p = (a_1,\ldots,a_{n_B-1}) \in \mathcal{P}(B)$, the arc-domains along p yield a partial solution $x(p) = (x_{\sigma^B(1)}, x_{\sigma^B(2)}, \ldots, x_{\sigma^B(n_B-1)}) = (d(a_1), d(a_2), d(a_{n_B-1}))$. Let $\mathcal{X}(p) = \{x \in \mathbb{B}^n : x_{\sigma^B(i)} = x_i(p), i = 1,\ldots,n_B-1\}$ be the set of possible *completions* of partial solutions, and define the *solutions* of B as $\mathrm{Sol}(B) = \bigcup_{p \in \mathcal{P}(B)} \mathcal{X}(p)$.

Let the *weight* $w(p)$ of path p be $\sum_{i=1}^{n_B-1} q(a_i)$. An *exact BDD* B for a BOP is one in which (1) $\mathrm{Sol}(B)$ coincides with the set of feasible solutions to the BOP and (2) for each path p and each solution $x \in \mathcal{X}(p)$, $f(x) = w(p)$. Such a BDD encodes all feasible solutions along with their objective function values, so that a longest path, which can be computed in linear time in $|U|$, corresponds to an optimal solution and its length corresponds to the optimal solution value.

To illustrate, consider the BOP $\mathcal{P}'$ defined by max $f(x) = \sum_{i=1}^{5} x_i$ subject to $x_1 + x_2 \leq 1, x_1 + x_3 \leq 1, x_2 + x_3 \leq 1, x_2 + x_4 \leq 1, x_3 + x_4 \leq 1$, and $x_4 + x_5 \leq 1$. The optimal solution value is 2. Figure 1(a) depicts an exact BDD for this BOP. The variables corresponding to each layer appear on the left of the BDD and dashed/solid arcs correspond to setting variables on that layer to 0/1.

3   Decomposition Based on Binary Decision Diagrams

A *decomposition based on binary decision diagrams* (DBDD) for a BOP is a collection of BDDs $B_1, \ldots, B_m$ with the following two properties: (1) $\bigcap_{k=1}^{m} \mathrm{Sol}(B_k)$ coincides with the feasible set of the BOP, and (2) for any feasible solution x, the set of paths p_k in BDD B_k for which $x \in \mathcal{X}(p_k), k = 1,\ldots,m$, satisfy $f(x) = w(p_1) + \cdots + w(p_k)$. These two conditions enforce that every feasible solution to the BOP corresponds to some path in each of the BDDs, that every vector in $\mathbb{B}^n$ which is infeasible does not correspond to any path in at least one of the BDDs, and that the sum of the weights of these paths coincide with the objective function value of that solution. In particular, the latter condition

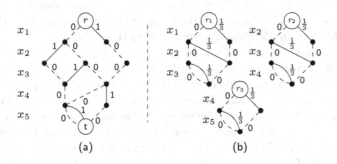

(a)   (b)

Fig. 1. (a) Exact BDD and (b) Exact DBDD for example BOP.

can be enforced by setting the weights of the 1-arc in layer k of each BDD as $f(x_k)/m(x_k)$, where $m(x_k)$ is the number of BBDs which include variable x_k. Finally, note that a single BDD is a DBDD.

Figure 1(b) depicts a DBDD for P': Any solution is a collection of 3 paths, one in each BDD, for which the variable assignments coincide. The optimal solution $(1, 0, 0, 1, 0)$ corresponds to the solid-dashed-dashed/dashed-dashed-solid/dashed-solid path through the BDD with roots r_1, r_2, r_3, respectively.

4   Solving a DBDD

Given a BDD for a BOP, finding the optimal solution reduces to a longest path computation which, since a BDD is a directed-acyclic graph, can be identified in $\mathcal{O}(|U|)$. The same is not true for a DBDD — even determining whether or not $\bigcap_{k=1}^{m} \mathrm{Sol}(B_k)$ is empty is NP-hard. To establish this, consider the *set packing problem*, which asks for a binary vector x that maximizes a linear function and that satisfies $Ax \leq 1$, where A is a 0/1 matrix. The exact BDD for each constraint has width of 2 [10], the collection of exact BDD for each individual constraint is a DBDD, and finding a common solution would solve the set packing problem.

In order to solve the underlying DBDD optimization problem, we formulate an MILP model based on interpreting the longest path of each BDD as a network flow. Let $B = (U, A)$ be a BDD and consider the MILP model $\texttt{netflow}(B)$:

$$
\texttt{netflow}(B) = \max_{\substack{x \in \{0,1\}^n \\ y \in [0,1]^{|A|}}} \left\{ \sum_{a \in A} w(a) y_a : \sum_{a \in A : h(a) = \mathbf{r}} y_a = 1, \sum_{a \in A : t(a) = \mathbf{t}} y_a = 1, \right.
$$

$$
\left. \sum_{a \in A : t(a) = u} y_a - \sum_{a : h(a) = u} y_a = 0 \; \forall u \in U, \; x_{\sigma^B(i)} = \sum_{a \in A : \ell(a) = i, d(a) = 1} y(a) \; \forall i \right\}
$$

Each arc $a \in A$ has a variable y_a and the constraints are the typical network flow constraints which enforce that each feasible solution must correspond to a $\mathbf{r} - \mathbf{t}$ path, with an additional set of constraints that relate the flow variables with the BOP variables. These constraints enforce that the BOP variable $x_{\sigma^B(i)} = 1$ if and only if the arc of the path on layer i is a 1-arc. Note that a similar formulation can be created for other forms of BDDs, including *zero-compressed* BDDs [9], which will be used in the computational results that follow.

It is well known that $\texttt{netflow}(B)$ is an integral polytope so that relaxing the integrality constraints results in a linear programming (LP) for which the corner points are integral. Moreover, if the BDD is exact, the projection onto the x-variables is a convex-hull relaxation of the feasible set of the BOP [5].

Hence, a DBDD $\{B_k\}_{k=1}^m$ can be formulated as an MILP model that combine the polytopes $\texttt{netflow}(B_k)$, which are naturally linked through the variables x. In particular, the feasible solutions to $\texttt{netflow}(B_k)$ is $\mathrm{Sol}(B_k)$, and therefore the conjunction of the polytopes (represented by the combined model) yields an extended formulation for the BOP. One of the key advantages is that each B_k

may capture some combinatorial structure that is non-trivial to encode as linear constraints, thus yielding better MILP models. In order to make the approach scalable, each BDD in the DBDD should be as small as possible.

5 Study Cases

We examine two problem classes as preliminary study cases which exemplify how problems can be decomposed into structured subproblems with limited-size BDDs. All experiments ran on an Intel(R) Core(TM) i7-4770 CPU @ 3.40 GHz processor, 32 GB RAM, using IBM ILOG CPLEX 12.6 (one thread).

Independent Sets on Social Networks. The application that motivated the design of DBDDs was finding *independent sets* in social network graphs. Let $G = (V, E)$ be an undirected graph with vertices $V = \{1, \ldots, n\}$ and edge set $E \subseteq V \times V$. An independent set I is a subset of V such that no two vertices in I are adjacent to each other, i.e. $i, j \in I$ if $(i, j) \notin E$. The *maximum independent set problem* asks for an independent set I with the largest cardinality. This problem is typically written as the BOP $\max \left\{ \sum_{i \in V} x_i : x_i + x_j \leq 1 \ \forall (i, j) \in E, x \in \mathbb{B}^n \right\}$.

Fig. 2. Relaxed caveman graph. Picture from Judd et al. [20].

In graphs representing social networks, independent sets play a key role in identifying interpersonal relations [6], in game theoretical models for the provision of goods [13], and as a measure of fairness [20] and diversity [19]. Here we focus on a particular type of social network graph denoted by *relaxed caveman* graphs. A caveman graph represents large groups of mutually adjacent connections with sporadic links to other groups, as depicted in Fig. 2. Independent sets in such graphs are used, e.g., in procedures for detection of communities [2]. The decomposition in this case consists of building the exact BDD for each denser component (the "caves"), and then using single edge inequalities to enforce that the endpoints of the redirected edges cannot both be in an independent set. The BDD for a single clique has width 2 [9,17] and, upon removal of one edge, the width of each layer grows by at most a factor of 2. Therefore, if k edges are removed, the size of the exact BDD is at most 2^k, although in general will be much smaller. Since there are typically few interconnections in this class of graphs, the BDDs will be of practical size.

For our experiments, we constructed random graphs specified by a triple (c, s, p): First, c cliques of size s are generated, and then the endpoints of the edges in a clique are randomly assigned to a vertex in another clique with probability p. We considered $c = 10$, $s \in \{15, 20, \ldots, 40\}$, and $p \in \{0.10, 0.15, \ldots, 0.50\}$. Three instances per triple (c, s, p) were generated, yielding in total 175 instances. We compared three approaches: a clique-based MILP formulation for the problem [9], the MILP generated from the DBDD, and *Cliquer* [21], a specialized method for the problem. The clique-based MILP method exploited the clique structure imposed by the caves. A time limit of 30 min was set in all cases.

Figure 3(a) depicts the number of solved instances over time, and shows that the DBDD MILP is more robust then all other methods combined. Figure 3(b) compares times between the DBDD and Cliquer, where the size of a point indicates the number of vertices in each cave (s) and the color gradation indicates the density. In particular, DBDD performs better for any instance that can be solved in more than 10 s by both methods, and it is particularly more effective for larger problems with a higher density between caves.

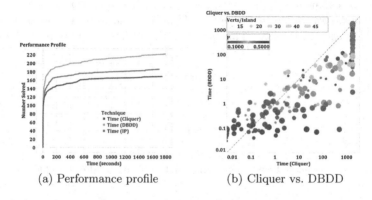

(a) Performance profile   (b) Cliquer vs. DBDD

Fig. 3. Results on relaxed caveman graphs.

Unconstrained Binary Quadratic Programming (UBQP). We present a case where the *objective function* is decomposed as opposed to the constraint set. A instance of UBQP is specified by an objective function matrix $Q \in \mathcal{S}^n$, where $\mathcal{S}^n$ is the set of symmetric $n \times n$ matrices (i.e., $\forall i, j \in \{1, \ldots, n\}, q_{i,j} = q_{j,i}$). The UBQP is defined as $\max\{x^T Q x : x \in \mathbb{B}^n\}$. For notation purposes, let $x|k$ be the *partial solution* of $x \in \mathbb{B}^n$ on only the first k indices: $x|k = (x_1, \ldots, x_k)$.

An exact BDD B for BOP can be compiled with a top-down approach based on a dynamic programming (DP) formulation for the problem, as presented in Bergman et al. [7,8]. DP models in our context are defined by a state $s(.)$ and a value function $v(.)$, which will be encoded as nodes of the BDD and as length of arcs, respectively. For the UBQP with $Q = \{q_{i,j}\}$, the state is defined as $s(x'|k+1)_j = 0$ if $j \leq k+1$, and $s(x'|k)_j + 2 \cdot q_{k+1,j} \cdot x'_{k+1}$, if $j > k+1$. The value function is written as $v(x'|k + 1) = v(x'|k) + x'_{k+1} \cdot s(x'|k)_{k+1}$. Intuitively, the value in the jth coordinate of the state represents the marginal effect of setting $x_j = 1$ given the values assigned to $x_1, \ldots, x_k$. We omit the proof for brevity, but note that they are similar to the DP model of the maximum cut problem [8]. Figure 4 depicts an exact BDD for an objective function matrix Q'.

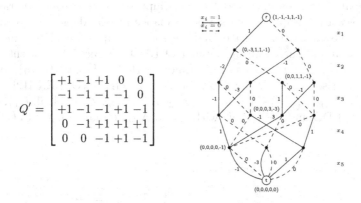

$$Q' = \begin{bmatrix} +1 & -1 & +1 & 0 & 0 \\ -1 & -1 & -1 & -1 & 0 \\ +1 & -1 & -1 & +1 & -1 \\ 0 & -1 & +1 & +1 & +1 \\ 0 & 0 & -1 & +1 & -1 \end{bmatrix}$$

Fig. 4. Objective function matrix for an UBQP instance with an exact BDD. States are indicated on nodes in which the path corresponding to solution $(1, 0, 1, 1, 0)$ traverses.

Given a graph $G = (V, E)$, a *path decomposition* of G is a sequence of subsets $V_i \subseteq V$ for which (1) $\forall e \in E, \exists V_i$ s.t. $e \subseteq V_i$ and (2) $\forall i \leq j \leq k, V_i \cap V_k \subseteq V_j$. The *width* of a path decomposition is one less that the maximum cardinality set, $\max_i |X_i| - 1$. The *path-width*, $pw(G)$, of a graph G is the minimum width over all path decomposition. For a symmetric $n \times n$ matrix Q, let $G(Q) = (V, E')$ be the graph with $E' = \{(i, j) \in E : q_{i,j} \neq 0\}$. We state the following result, which follows a similar proof as in previous works bounding the size of DDs [17].

Theorem 1. *Let Q be an $n \times n$ symmetric matrix. There exists an exact BDD for the UBQP defined on Q for which $w(B) \leq 2^{pw(G(Q))-1}$.*

Consider Q' in Fig. 4. A path decomposition for $G(Q')$ consists of vertex sets $\{1, 2, 3\}, \{2, 3, 4\}, \{3, 4, 5\}$, proving that there exists a BDD with width bounded by $2^{3-1} = 4$. The BDD in Fig. 4 is such a witness.

Theorem 1 provides a bound on the size of the width of BDDs based on the pathwidth of $G(Q)$. This indicates that for matrices whose corresponding graphs have a limited pathwidth, a DD-based approach will work well. However, if Q does not possess such a characteristic, the DD can grow exponentially large. We exploit this idea to decompose the matrix Q as follows: fix $p, 1 \leq p \leq n$ as the desired pathwidth for each element of the decomposition, and let $m = \lceil n/p \rceil$ be the number of elements created. Define Q^k by, for $i \leq j$, $q_{i,j}^k = q_{i,j}$ if $k(p-1) \leq i < kp$ and $j \geq i$, and $q_{i,j}^k = 0$ otherwise, with the elements in indices with $i > j$ defined so that q^k is symmetric (i.e., $q_{i,j}^k = q_{j,i}^k$). Each Q^k satisfies that $pw(G(Q^k)) \leq p$, thereby limiting the size of the exact BDD for each Q^k.

In order to test the effectiveness of the approach, random symmetric matrices were generated, with $n = 40$ and fixed Q to have bandwidth $L \in \{10, 20, 30, 40\}$, with 10 instances per configuration. Each q_{ij} took a value uniformly at random from $\{-5, -4, \ldots, 4, 5\}$ if $i \neq j$, and the diagonal entries were set as $q_{i,i} = -1 \cdot \sum_{j \neq i} q_{i,j}$ if $|i - j| \leq L$, and 0 otherwise.

Our goal is to show how DBDDs can improve upon integer programming methodology. Figure 5(a) presents a performance profile, depicting the effect of varying p, the pathwidth of each BDD in the decomposition. Each line corresponds to a p and the resulting number of BDDs. Figure 5(b) presents a plot depicting the average time to solve instances with $L \in \{10, 12, 14, 16\}$. We provide a comparison of the IP linearization implemented in CPLEX [12], namely $\min\{\sum_{i,j} q_{i,j} y_{i,j} : x_i + x_j - 1 \leq y_{i,j}; x_i, x_j \geq y_{i,j}; x_i, y_{i,j} \in \{0, 1\}, \text{all } i, j\}$, with a single exact BDD and the DBDD with $p = 4$. A time limit was set to 600 s.

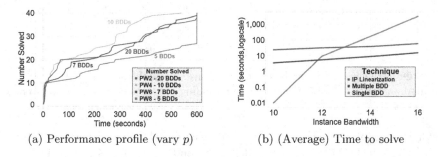

(a) Performance profile (vary p)   (b) (Average) Time to solve

Fig. 5. UBQP Results.

Figure 5 shows how a single exact BDD approach is impractical because of the exponential growth in its size as L grows. Both IP and the DBDD approaches scale much better as L grows, with the latter finding the optimal solutions slightly faster. As the difficulty in the instances grows, the DBDD solution time grows at essentially the same pace as IP, which also elucidates the promise of investigated DBDDs in instances which lack structure. We remark that CPLEX is capable of solving quadratic models directly, which has a better performance than the methods presented here. Nonetheless, it transforms Q into a positive semi-definite matrix, yielding a different model that we do not have access to.

6   Conclusion

This paper presents a novel decomposition-based optimization approach to binary optimization problems. The proposed technique combines the strength of binary decision diagrams, which are particularly successful at representing portions of problems with structure, with the strength of integer programming technology. Computational experiments indicate promise in the approach for solving independent set and binary quadratic programming problems.

Acknowledgements. This research was supported in part by the Natural Sciences and Engineering Research Council of Canada (NSERC), Discovery Grant.

References

1. Akers, S.B.: Binary decision diagrams. IEEE Trans. Comput. **C–27**(6), 509–516 (1978)
2. Aldecoa, R., Marín, I.: Surprise maximization reveals the community structure of complex networks. Scientific Reports 3(1060) (2013)
3. Andersen, H.R., Hadzic, T., Hooker, J.N., Tiedemann, P.: A constraint store based on multivalued decision diagrams. In: Bessière, C. (ed.) CP 2007. LNCS, vol. 4741, pp. 118–132. Springer, Heidelberg (2007)
4. Balas, E.: Projection, lifting and extended formulation in integer and combinatorial optimization. Ann. Oper. Res. **140**(1), 125–161 (2005). http://dx.doi.org/10.1007/s10479-005-3969-1
5. Behle, M.: Binary Decision Diagrams and Integer Programming. Ph.D. thesis, Max-Planck Institute for Computer Science (2007)
6. Belik, I.: The analysis of split graphs in social networks based on the k-cardinality assignment problem. NHH Dept. of Business and Management Science Discussion Paper No. 2014/8, (February 2014). http://dx.doi.org/10.2139/ssrn.2405010
7. Bergman, D.: New Techniques for Discrete Optimization. Ph.D. thesis, Tepper School of Business, Carnegie Mellon University (2013)
8. Bergman, D., Cire, A.A., van Hoeve, W.-J.: Discrete optimization with decision diagrams. INFORMS J. Comput. (2015, to appear)
9. Bergman, D., Cire, A.A., van Hoeve, W.-J., Hooker, J.N.: Optimization bounds from binary decision diagrams. INFORMS J. Comput. **26**(2), 253–268 (2014)
10. Bergman, D., Cire, A., van Hoeve, W.J., Yunes, T.: Bdd-based heuristics for binary optimization. J. Heuristics **20**(2), 211–234 (2014)
11. Bergman, D., van Hoeve, W.-J., Hooker, J.N.: Manipulating MDD relaxations for combinatorial optimization. In: Achterberg, T., Beck, J.C. (eds.) CPAIOR 2011. LNCS, vol. 6697, pp. 20–35. Springer, Heidelberg (2011)
12. Bliek, C., Bonami, P., Lodi, A.: Solving mixed-integer quadratic programming problems with IBM-CPLEX: a progress report. In: Proceedings of the Twenty-Sixth RAMP Symposium (2001)
13. Bramoull, Y., Kranton, R.: Public goods in networks. J. Econ. Theor. **135**(1), 478–494 (2007). http://www.sciencedirect.com/science/article/pii/S0022053106001220
14. Bryant, R.E.: Graph-based algorithms for boolean function manipulation. IEEE Trans. Comput. **C–35**, 677–691 (1986)
15. Cire, A.A., van Hoeve, W.J.: Multivalued decision diagrams for sequencing problems. Oper. Res. **61**(6), 1411–1428 (2013)
16. Bergman, D., Cire, A.A., van Hoeve, W.-J.: MDD propagation for sequence constraints. J. Artif. Intell. Res. **50**, 697–722 (2014)
17. Haus, U.U., Michini, C.: Representations of all solutions of boolean programming problems. In: ISAIM (2014)
18. Hu, A.J.: Techniques for efficient formal verification using binary decision diagrams. Thesis CS-TR-95-1561, Stanford University, Department of Computer Science, December 1995
19. Johnson, M., Paulusma, D., van Leeuwen, E.J.: Algorithms to measure diversity and clustering in social networks through dot product graphs. In: Cai, L., Cheng, S.-W., Lam, T.-W. (eds.) Algorithms and Computation. LNCS, vol. 8283, pp. 130–140. Springer, Heidelberg (2013). http://dx.doi.org/10.1007/978-3-642-45030-3_13

20. Judd, S., Kearns, M., Vorobeychik, Y.: Behavioral conflict and fairness in social networks. In: Chen, N., Elkind, E., Koutsoupias, E. (eds.) Internet and Network Economics. LNCS, vol. 7090, pp. 242–253. Springer, Heidelberg (2011). http://dx.doi.org/10.1007/978-3-642-25510-6_21

21. Östergård, P.R.: A fast algorithm for the maximum clique problem. Discrete Appl. Math. **120**(13), 197–207 (2002). http://www.sciencedirect.com/science/article/pii/S0166218X01002906. Special issue devoted to the 6th Twente Workshop on Graphs and Combinatorial Optimization

Logic-Based Decomposition Methods for the Travelling Purchaser Problem

Kyle E.C. Booth[(✉)], Tony T. Tran, and J. Christopher Beck

Department of Mechanical and Industrial Engineering,
University of Toronto, Toronto, ON M5S 3G8, Canada
{kbooth,tran,jcb}@mie.utoronto.ca

Abstract. We present novel branch-and-check and logic-based Benders decomposition techniques for the Travelling Purchaser Problem, an important optimization problem with applications in vehicle routing, logistics, and warehouse management. Our master problem determines a set of markets and directed travel arcs that satisfy product purchase constraints with relaxed travel costs. Our subproblem identifies subtours within this master assignment and produces a set of generalized subtour elimination cuts. We show that the proposed technique demonstrates strong performance on the asymmetric problem variants, finding optimal solutions to previously unsolved instances, while performing competitively on a number of symmetric problem classes. Furthermore, our model is implemented unchanged for the four problem variants whereas other state-of-the-art approaches are variant-specific.

1 Introduction

Given a set of markets, each with a set of available products, the Travelling Purchaser Problem (TPP) aims to determine a simple cycle among a subset of markets that minimizes the sum of travel cost and purchase cost for the set of products required by the traveller. The problem is $\mathcal{NP}$-Hard [23] and generalizes both the Travelling Salesman Problem (TSP) [10] and the Uncapacitated Facility Location Problem (UFLP) [9].

Our primary contribution is the development of an exact decomposition model to solve the TPP. The decomposition uses mixed-integer programming for the master problem and a straightforward subtour identification algorithm to generate cuts. The method is simple to implement in commercially available solvers, and does not require sophisticated separation procedures, nor an in-depth polytope analysis [19,24]. To our knowledge, there is no other single approach that has been used without modification to efficiently solve the uncapacitated, capacitated, asymmetric, and symmetric problem variants. Our approach achieves strong performance on both the capacitated and uncapacitated asymmetric instances while remaining competitive on symmetric problems.

As far as we are aware, this work is the first application of branch-and-check and logic-based Benders decomposition (LBBD) for the TPP, though an LBBD-inspired heuristic approach has been investigated [6] and served as an initial inspiration for this work.

© Springer International Publishing Switzerland 2016
C.-G. Quimper (Ed.): CPAIOR 2016, LNCS 9676, pp. 55–64, 2016.
DOI: 10.1007/978-3-319-33954-2_5

2   Background

In this section we define the TPP and review existing relevant work, focusing on exact algorithms rather than heuristic approaches (e.g., [5,12,13]).

2.1   Problem Definition

Following Laporte et al. [19], consider a set of markets $M := \{v_0, v_1, ..., v_n\}$, where v_0 is a depot, and a set of available products $K := \{p_1, ..., p_m\}$. The demand for product p_k, the quantity of product that must be purchased, is d_k and the price of p_k at v_i is b_{ki}. Each product, p_k, can be purchased at a subset of the markets, M_k, and the quantity of p_k available at v_i is q_{ki}. We define $M^* := \{v_0\} \cup \{v_i \in M : \exists\, p_k \in K$ such that $\sum_{v_j \in M_k \setminus \{v_i\}} q_{kj} < d_k\}$ as the set of required markets. The travel cost between markets v_i and v_j is c_{ij}. We must find a simple cycle among a subset of markets such that all product demand is satisfied and the sum of the travel and product purchase costs are minimized.

We present a mixed-integer programming (MIP) model in Fig. 1. The model is based on Laporte et al. [19] but uses the lifted Miller-Tucker-Zemlin (MTZ) subtour elimination formulation [8,21]. The decision variables are:

$z_i := 1$ if market v_i is visited and 0 otherwise

$x_{ij} := 1$ if market v_i is visited directly before v_j and 0 otherwise

$y_{ki} :=$ the purchased quantity of p_k at market v_i

$u_i :=$ a positive variable used for MTZ subtour elimination [21]

Equation (1) minimizes the sum of travel and product purchase costs. Constraints (2) and (3) represent the degree constraints for each market. Constraint (4) ensures that the demand for each product is satisfied while Constraint (5) ensures that quantity of a product purchased at a market is contingent on both the decision to visit that market and the product quantity available. Constraint (6) represents the lifted MTZ [8,21] formulation for subtour elimination. Constraint (8) ensures that z_i is set to a value of 1 if market v_i is a required market.

2.2   Problem Variants

The majority of TPP variants addressed in the literature fall along two dimensions: capacitated vs. uncapacitated, and symmetric vs. asymmetric. A problem is uncapacitated if each market sells enough of its products to satisfy the traveller's demand for those products (i.e., $q_{ki} \geq d_k, \forall v_i \in M_k, p_k \in K$) and, therefore, a given product is always satisfied by a single market. In capacitated problems, the quantity of a product at a market may or may not completely satisfy the demand (i.e., $0 < q_{ki} \leq d_k, \forall v_i \in M_k$) and so the traveller may have to visit several markets to satisfy the demand for a single product. A problem is symmetric if $c_{ij} = c_{ji}$ holds, and asymmetric if it does not. These dimensions combine to form four problem variations: uncapacitated-symmetric (U-STPP), uncapacitated-asymmetric (U-ATPP), capacitated-symmetric (C-STPP), and capacitated-asymmetric (C-ATPP).

$$\min \sum_{v_i \in M} \sum_{v_j \in M, v_j \neq v_i} c_{ij} x_{ij} + \sum_{p_k \in K} \sum_{v_i \in M_k} b_{ki} y_{ki} \tag{1}$$

$$\text{s.t.} \quad \sum_{v_i \in M, v_i \neq v_j} x_{ij} = z_j \qquad \forall \, v_j \in M, \tag{2}$$

$$\sum_{v_j \in M, v_j \neq v_i} x_{ij} = z_i \qquad \forall \, v_i \in M, \tag{3}$$

$$\sum_{v_i \in M_k} y_{ki} = d_k \qquad \forall \, p_k \in K, \tag{4}$$

$$y_{ki} \leq z_i q_{ki} \qquad \forall \, p_k \in K; v_i \in M_k, \tag{5}$$

$$u_i - u_j + (|M| - 1)x_{ij} + (|M| - 3)x_{ji} \leq |M| - 2 \quad \forall \, v_i, v_j \geq v_1; v_i \neq v_j, \tag{6}$$

$$1 \leq u_i \leq |M| - 1 \qquad \forall \, v_i \geq v_1, \tag{7}$$

$$z_i = 1 \qquad \forall \, v_i \in M^*, \tag{8}$$

$$z_i \in \{0, 1\} \qquad \forall \, v_i \in M \setminus M^*, \tag{9}$$

$$x_{ij} \in \{0, 1\} \qquad \forall \, v_i, v_j \in M; v_i \neq v_j, \tag{10}$$

$$0 \leq y_{ki} \leq d_k \qquad \forall \, p_k \in K; v_i \in M_k. \tag{11}$$

Fig. 1. A MIP Model for the Travelling Purchaser Problem based on Laporte et al. [19] with lifted MTZ subtour elimination constraints [8, 21].

The MIP model in Fig. 1 is valid for all four problem variants as the differences are embodied in the instance data: the symmetricity difference is due to the travel cost matrix data and the uncapacitated variants simply assign $d_k = 1, \forall p_k \in K$ and $q_{ki} = 1, \forall p_k \in K; v_i \in M_k$.

2.3 Related Work

The first exact approach to the TPP was a lexicographic search algorithm capable of solving the U-STPP and U-ATPP instances with $|M| = 12$ and $|K| = 10$ (12×10) [23]. Singh et al. [25] proposed a branch-and-bound method for the U-STPP and U-ATPP that utilized the relaxation of UFLP constraints to generate lower bounds. The approach solved asymmetric instances of size 25×50 and symmetric instances of size 25×30. Laporte et al. [19] proposed the first capacitated formulation of the symmetric TPP and developed a branch-and-cut approach for the U-STPP and C-STPP. This method solved instances of size 200×200 and remains a state of the art for symmetric variants. Riera et al. [24] developed a state-of-the-art branch-and-cut approach for the U-ATPP and C-ATPP, solving instances of 200×200. More recently, Cambazard et al. [7] developed a constraint programming approach for the U-STPP, solving instances of size 250×200, often out-performing Laporte et al.

3   LBBD and Branch-and-Check for the TPP

We investigate both logic-based Benders decomposition (LBBD) and branch-and-check (B&C), with the notion that the TPP can benefit from the delayed enforcement of certain constraints. The decomposition structure is the same for both approaches, the difference concerning whether the sub-problem is solved at optimal or feasible solutions to the master problem. LBBD [15] uses logic-based subproblems to produce valid Benders cuts [11] for the master problem. In LBBD, the master problem is solved to optimality and the solution to this relaxed problem is utilized to solve the subproblem(s) and generate cuts. The master problem is then re-solved, and this iterative process continues until the solution to the master problem, with all generated cuts, is valid with respect to the subproblems, and thus is a globally optimal solution. B&C [26] is a variation of LBBD where the subproblem(s) are solved whenever a feasible solution is found during the branch-and-bound search of the master problem. Problems with more difficult master problems, as compared to the subproblems, are more suited for branch-and-check, whereas difficult subproblems comparatively favor LBBD [4].

Assignment Master Problem. In the decomposition proposed, the master problem is a relaxation of the full MIP model (Fig. 1) through omission of the subtour elimination constraints and associated variables: that is, the removal of Constraints (6) and (7). A solution to the master problem consists of an integer set of assigned markets, z_i, and directed travel arcs, x_{ij}, that satisfy product purchase requirements while allowing subtours. It is natural, therefore, for the subproblem to identify subtours and eliminate them through cut generation.

Subtour Identification Subproblem. Our approach consists of identifying these subtours, evaluating their candidacy as globally feasible solutions, and producing generalized subtour elimination [16] cuts when appropriate.

Due to Constraints (2) and (3), the master solution consists of a set of one or more disjoint tours of the selected markets. Since our master assignments are integer, a trivial depth first search is sufficient to identify the unique set of subtours, S^h, in the hth master problem solution where each $s_\ell^h \in S^h$ consists of a set of markets in a subtour.

For each subtour we first assess whether it is, by itself, a feasible solution to the global problem; that is, does s_ℓ^h satisfy all product purchase requirements and include the depot, $v_0 \in s_\ell^h$. While such subtours will not exist for LBBD, due to the optimality of the master problem solution, for B&C at most one such subtour may exist per iteration.[1] If such a subtour, $\hat{s}^h$, exists, we remove it from S^h and use it as a new global incumbent solution. At the same time, for each subtour $s_\ell^h \in \bar{S}^h := S^h \setminus \{\hat{s}^h\}$, we introduce a generalized subtour elimination cut defined as follows:

$$\sum_{v_i \in s_\ell^h} \sum_{v_j \in s_\ell^h, v_j \neq v_i} x_{ij} \leq \sum_{v_i \in s_\ell^h} z_i + \psi_{s_\ell^h} - 1 \qquad \forall s_\ell^h \in \bar{S}^h, \tag{12}$$

$$\psi_{s_\ell^h} + z_i \leq 1 \qquad \forall v_i \in s_\ell^h; s_\ell^h \in \bar{S}^h, \tag{13}$$

$$\psi_{s_\ell^h} \in \{0,1\} \qquad \forall s_\ell^h \in \bar{S}^h. \tag{14}$$

---

[1] The limit of one per iteration is due to the depot inclusion condition.

The left hand side of Cut (12) is the number of chosen arcs in the (complete) sub-graph induced by the markets in subtour s_ℓ^h. The right hand side defines an upper bound for this value, prohibiting the creation of a subtour of any permutation of the markets as well as removing some, but not all, subtours among subsets of the markets. The subset prohibitions are achieved through the use of the $\sum_{v_i \in s_\ell^h} z_i$ term instead of simply the cardinality of the markets, $|s_\ell^h|$.

The cut adds of a new auxiliary variable, $\psi_{s_\ell^h}$, for each $s_\ell^h \in \bar{S}^h$ which, due to (12), is constrained to the value 1 if none of the markets in the subtour are chosen in a subsequent iteration (i.e., if $\sum_{v_i \in s_\ell^h} z_i = 0$) and, due to (13), is constrained to 0 otherwise. Functionally, $\psi_{s_\ell^h}$ ensures the global validity of the cut. Without its inclusion, if none of the markets $v_i \in s_\ell^h$ were chosen in a subsequent iteration, Cut (12) would reduce to $0 \le 0 - 1$, removing a potentially globally optimal solution.

The validity of the cut is easily seen. Each cut eliminates at least one subtour from the master solution space and as the removed subtour does not itself constitute a globally feasible solution, no such solutions are removed. Convergence to optimality is then based on the finite (though large) number of subtours.

This cut has a similar purpose to the one proposed for the Orienteering Problem [14,16], though is different through the use of variable generation. Initial experimentation suggests that our cut performs more effectively, in general, than the one proposed in Laporte et al. [16], though this is an area we intend to explore more thoroughly. We note that an equivalent generalized connectivity cut [16] can be used as well, using the cut-set of directed arcs.

In traditional cut-based approaches for cycle problems, subtour constraints and integrality requirements are relaxed and violated inequalities are separated based on fractional solutions of the resulting linear program (LP). When the LP is solved, a max-flow (or min-cut) problem is solved to identify and separate violated tour constraints [22]. Additional cutting planes are available, most notably the class of comb inequalities [2]. This standard approach relies heavily on the performance of the LP solver, requiring active management of model size due to the large number of valid inequalities introduced via the various separation procedures. Conversely, subtour elimination based on integer assignments has been applied to the TSP [18,20], but not to our knowledge the TPP.

4 Computational Results

In this section we present benchmark results of the proposed LBBD and B&C formulations on the four main variants of the TPP.

4.1 Benchmark Problems

The instance set we use is well-established in the literature [19,24] and consists of 745 instances across six problems classes. The capacitated instances introduce a parameter λ that dictates how traveller demand, d_k, relates to the available quantity of a product, q_{ki}: $d_k := \lceil \lambda \max_{v_i \in M_k}(q_{ki}) + (1 - \lambda)\sum_{v_i \in M_k} q_{ki} \rceil$ [19]. Each instance set consists of five instances for each combination of $|M|$, $|K|$ and λ considered.

Class 1. U-STPP [19] where $|M| = 33$, $|K| \in \{50, 100, 150, 200, 250\}$, for a total of $\mathcal{N} = 25$ problem instances.
Class 2 and 3. U-STPP [19] where $|M| \in \{50, 100, 150, 200\}$, $|K| \in \{50, 100, 150, 200\}$, for a total of $\mathcal{N} = 80$ problem instances in each.
Class 4. C-STPP [19] where $|M| \in \{50, 100, 150, 200\}$, $|K| \in \{50, 100, 150, 200\}$, and $\lambda \in \{0.5, 0.7, 0.9, 0.99\}$, for a total of $\mathcal{N} = 280$ problem instances.
Class 1A. U-ATPP [24] where $|M| \in \{50, 100, 150, 200\}$, $|K| \in \{50, 100, 150, 200\}$, with a total of $\mathcal{N} = 80$ problem instances.
Class 2A. C-ATPP [24] where $|M| \in \{50, 100\}$, $|K| \in \{50, 100, 150, 200\}$, and $\lambda \in \{0.5, 0.8, 0.9, 0.95, 0.99\}$ with a total of $\mathcal{N} = 200$ instances.

4.2 Experimental Details

We implement our methods with the CPLEX 12.6.2 mixed-integer programming solver in C++. For B&C, we utilize lazy constraints to trigger subproblem solving and cut generation whenever a feasible master solution is found. As CPLEX does not directly support variable generation within branch-and-bound, we pre-allocate a number of $\psi_{s_\ell^h}$ variables which the cuts then make use of. This situation is not ideal, as it leads to a master problem with variables that may never be utilized. Better B&C performance is likely with true variable generation.[2] For the LBBD approach, since each master iteration is solved anew, we do not require in-search variable generation and do not suffer the same B&C limitations.

We compare to published results from the aforementioned state-of-the-art methods. As the four problem variants have never been approached in one study, we adapt our run-times to be appropriate for different experimental designs in the literature. We use a run-time limit of 3,600 s for symmetric experiments, whereas previous papers use longer run-times (7,200 and 18,000 s). For asymmetric instances, we use a run-time limit of 7,200 s to match the limits in the existing papers.

Our experiments use a Xeon 3.5 GHz processor machine with 16 GB of RAM running OS X Yosemite. Laporte et al. and Riera et al. utilize much older Pentium 500 MHz and AMD 1.33 GHz machines, respectively, running Linux with CPLEX 6.0. The Cambazard paper uses a Xeon 2.66 GHz processor machine with 16 GB of RAM running Linux 2.6.25 x64 and customized CP software. Due to the mix of software and hardware, the results should be interpreted with care.

4.3 Results

Table 1 presents the results for the asymmetric variants (Classes 1A and 2A) where the branch-and-cut approaches of Riera et al. [24] represent the state of

---

[2] Experiments using dynamic variable creation in SCIP [1] show a relative improvement in B&C compared to LBBD though both CPLEX implementations are faster.

the art. Two methods are presented in Riera et al.: Riera$_{B\&CUT}$ that uses a customized branch-and-cut algorithm with sophisticated separation procedures, and Riera$_{TRANS}$, that transforms the ATPP into its symmetric counterpart, and then uses the branch-and-cut of Laporte et al. [19].

The results for asymmetric instances with our approach show speed-up factors of 3 to 70 compared to the previous state of the art, including solving a number of previously unsolved instances. We suspect the reason for this superior performance is rooted within our master problem relaxation, namely the market assignment relaxation for the TPP. As demonstrated by Balas et al. [3], the assignment problem (AP) relaxation for the TSP is very strong when the cost matrix, c_{ij}, is generated randomly based on a uniform distribution, and thus asymmetric, resulting in near-optimal solutions. Since the instances for Class 1A and 2A are generated randomly based on a uniform distribution, it would appear this property holds for the TPP as well. Again, hardware and software differences make this comparison less clear cut, though we believe they do provide supporting evidence for our techniques as strong contenders for the new state of the art on asymmetric TPP problems.

Table 2 presents the results on the symmetric instances (Classes 1–4), compared to the state-of-the-art approaches. Laporte$_{B\&CUT}$ [19] utilizes branch-and-cut with valid inequalities to strengthen the linear relaxation with sophisticated separation techniques. The Cambazard [7] approach, Cambazard$_{CP-PM}$, uses a constraint programming (CP) model with a p-median constraint, originally intended for solving TPP problems with a bounded number of visits.

Our proposed decomposition techniques are competitive with the existing state of the art on Classes 1 and 2 while falling short on instances of Class 3 and 4 for larger values of $|M|$. For the symmetric case, our master assignment relaxation is much weaker [3] for the underlying cycle problem, which tends to contain many more subtours of size 2 [17] than the asymmetric counterparts, resulting in excessive computation time for their elimination.

Table 1. Asymmetric results vs. the state of the art. $\mathcal{C}$ is the problem class and $\mathcal{N}$ the number of instances for each value of $|M|$. #F indicates number of instances that were not proved optimal in 7,200 second limit. Run-times are arithmetic mean CPU values.

Asymmetric problems (U-ATPP, C-ATPP)										
Problem			B&C		LBBD		Riera$_{B\&CUT}$		Riera$_{TRANS}$	
$\mathcal{C}$	$\mathcal{N}$	$|M|$	Avg (s)	#F	Avg (s)	#F	Avg (s)	#F	Avg (s)	#F
1A	20	50	1.5	0	2.3	0	14.0	0	33.8	0
	20	100	24.8	0	67.7	0	1,083.0	0	1,076.9	2
	20	150	166.6	0	485.2	1	1,697.6	4	2,779.1	4
	20	200	731.9	1	1,147.1	2	3,331.9	11	3,634.2	11
2A	100	50	1.7	0	2.8	0	100.5	0	152.4	0
	100	100	34.9	0	86.7	0	2,557.9	24	2,397.0	36

Table 2. Symmetric results vs. the state of the art. Notation is identical to Table 1 except with a 3,600 second run-time limit. '−' indicates the method was not attempted.

Symmetric problems (U-STPP, C-STPP)										
Problem			B&C		LBBD		Laporte$_{B\&CUT}$		Cambazard$_{CP-PM}$	
C	$\mathcal{N}$	$|M|$	Avg (s)	#F	Avg (s)	#F	Avg (s)	#F	Avg (s)	#F
1	25	33	6.0	0	85.5	0	28.0	0	−	−
2	20	50	0.9	0	1.1	0	1.9	0	−	−
	20	100	9.6	0	6.9	0	12.2	0	−	−
	20	150	62.4	0	81.5	0	140.4	0	−	−
	20	200	221.1	0	280.0	0	298.5	1	−	−
3	20	50	11.8	0	270.5	0	23.9	0	7.1	0
	20	100	1,872.5	8	2,908.7	15	299.8	0	528.6	2
	20	150	3,244.6	17	3,513.8	19	1,725.1	1	821.94	2
	20	200	3,210.1	16	3,158.5	17	2,621.0	11	1,383.7	7
4	80	50	9.2	0	113.5	0	23.1	0	−	−
	80	100	677.7	8	1,174.2	19	402.3	2	−	−
	80	150	1,603.1	24	1,594.4	31	1,281.2	16	−	−
	40	200	3,020.6	29	2,084.9	20	2,414.6	19	−	−

5   Conclusions

We presented strategies for solving the Travelling Purchaser Problem with branch-and-check and logic-based Benders decomposition. We utilize a MIP model for the master problem, determining a set of markets that satisfy product purchase costs while relaxing the tour requirement. Our subproblem produces generalized subtour elimination cuts with variable generation.

Numerical results indicate strong performance on the asymmetric problem variants (both capacitated and uncapacitated) with order of magnitude speed-ups observed, albeit with differing hardware and software. On the symmetric instances, the performance was weaker, achieving about the same performance as the existing state of the art (on older hardware) on some problem classes but not achieving the same performance on others.

Notably, our model is applicable without modification across all four of the primary variants of the TPP while the current state-of-the-art techniques are variant-specific, exploiting sophisticated valid inequalities, separation schemes, and polytope analyses.

We plan to investigate algorithm extensions including primal solution heuristics, alternate cuts (e.g., [27]), as well as to perform a deeper analysis into the impact of symmetricity in order to improve performance on symmetric instances. We believe that the methods presented in this paper can be applied to more complex TPP-variants, as well as other routing problems.

References

1. Achterberg, T.: SCIP: solving constraint integer programs. Math. Program. Comput. **1**(1), 1–41 (2009)
2. Applegate, D., Bixby, R., Cook, W., Chvátal, V.: On the solution of traveling salesman problems. Rheinische Friedrich-Wilhelms-Universität Bonn (1998)
3. Balas, E., Toth, P.: Branch and bound methods for the traveling salesman problem. Technical report MSRR-488, DTIC Document (1983)
4. Beck, J.C.: Checking-up on branch-and-check. In: Cohen, D. (ed.) CP 2010. LNCS, vol. 6308, pp. 84–98. Springer, Heidelberg (2010)
5. Bontoux, B., Feillet, D.: Ant colony optimization for the traveling purchaser problem. Comput. Oper. Res. **35**(2), 628–637 (2008)
6. Burt, C.N., Lipovetzky, N., Pearce, A.R., Stuckey, P.J.: Approximate unidirectional benders decomposition. In: Proceedings of PlanSOpt-15 Workshop on Planning, Search and Optimization AAAI-15 (2015)
7. Cambazard, H., Penz, B.: A constraint programming approach for the traveling purchaser problem. In: Milano, M. (ed.) CP 2012. LNCS, vol. 7514, pp. 735–749. Springer, Heidelberg (2012)
8. Desrochers, M., Laporte, G.: Improvements and extensions to the Miller-Tucker-Zemlin subtour elimination constraints. Oper. Res. Lett. **10**(1), 27–36 (1991)
9. Erlenkotter, D.: A dual-based procedure for uncapacitated facility location. Oper. Res. **26**(6), 992–1009 (1978)
10. Flood, M.M.: The traveling-salesman problem. Oper. Res. **4**(1), 61–75 (1956)
11. Geoffrion, A.M.: Generalized benders decomposition. J. Optim. Theor. Appl. **10**(4), 237–260 (1972)
12. Goerler, A., Schulte, F., Voß, S.: An application of late acceptance hill-climbing to the traveling purchaser problem. In: Pacino, D., Voß, S., Jensen, R.M. (eds.) ICCL 2013. LNCS, vol. 8197, pp. 173–183. Springer, Heidelberg (2013)
13. Goldbarg, M.C., Bagi, L.B., Goldbarg, E.F.G.: Transgenetic algorithm for the traveling purchaser problem. Eur. J. Oper. Res. **199**(1), 36–45 (2009)
14. Golden, B.L., Levy, L., Vohra, R.: The orienteering problem. Naval Res. Logistics (NRL) **34**(3), 307–318 (1987)
15. Hooker, J.N., Ottosson, G.: Logic-based benders decomposition. Math. Program. **96**(1), 33–60 (2003)
16. Laporte, G.: Generalized subtour elimination constraints and connectivity constraints. J. Oper. Res. Soc. **37**, 509–514 (1986)
17. Laporte, G.: The traveling salesman problem: an overview of exact and approximate algorithms. Eur. J. Oper. Res. **59**(2), 231–247 (1992)
18. Laporte, G., Nobert, Y.: A cutting planes algorithm for the m-salesmen problem. J. Oper. Res. Soc. **31**, 1017–1023 (1980)
19. Laporte, G., Riera-Ledesma, J., Salazar-González, J.-J.: A branch-and-cut algorithm for the undirected traveling purchaser problem. Oper. Res. **51**(6), 940–951 (2003)
20. Miliotis, P.: Integer programming approaches to the travelling salesman problem. Math. Program. **10**(1), 367–378 (1976)
21. Miller, C.E., Tucker, A.W., Zemlin, R.A.: Integer programming formulation of traveling salesman problems. J. ACM (JACM) **7**(4), 326–329 (1960)
22. Padberg, M., Rinaldi, G.: A branch-and-cut algorithm for the resolution of large-scale symmetric traveling salesman problems. SIAM Rev. **33**(1), 60–100 (1991)
23. Ramesh, T.: Traveling purchaser problem. Opsearch **18**(1–3), 78–91 (1981)

24. Riera-Ledesma, J., Salazar-González, J.-J.: Solving the asymmetric traveling purchaser problem. Ann. Oper. Res. **144**(1), 83–97 (2006)
25. Singh, K.N., van Oudheusden, D.L.: A branch and bound algorithm for the traveling purchaser problem. Eur. J. Oper. Res. **97**(3), 571–579 (1997)
26. Thorsteinsson, E.S.: Branch-and-check: a hybrid framework integrating mixed integer programming and constraint logic programming. In: Walsh, T. (ed.) CP 2001. LNCS, vol. 2239, pp. 16–30. Springer, Heidelberg (2001)
27. Tran, T.T., Araujo, A., Beck, J.C.: Decomposition methods for the parallel machine scheduling problem with setups. INFORMS J. Comput. **28**(1), 83–95 (2016)

Lagrangian Decomposition
via Sub-problem Search

Geoffrey Chu[(✉)], Graeme Gange, and Peter J. Stuckey

National ICT Australia, Victoria Laboratory, Department of Computing
and Information Systems, University of Melbourne, Melbourne, Australia
{geoffrey.chu,gkgange,pstuckey}@unimelb.edu.au

Abstract. One of the critical issues that affect the efficiency of branch
and bound algorithms in Constraint Programming is how strong a bound
on the objective function can be inferred at each search node. The
stronger the bound that can be inferred, the earlier failed subtrees can
be detected, leading to an exponentially smaller search tree. Normal CP
solvers are only capable of inferring a bound on the objective function via
propagating the problem constraints. Unfortunately, for many problem
classes, this does not yield a very strong bound. Recently, Lagrangian
decomposition methods have been adapted and applied to Constraint
Programming in order to yield stronger bounds on the objective function. While these methods yield some success, they are somewhat limited
in the types of problems they can be effectively applied to. In particular,
the set of constraints has to be divided into subsets such that each subset can be solved efficiently via a specialized propagator, e.g., consists
of a knapsack problem, or a cost-MDD problem. For many more practical problem classes, such a division of constraints is simply not possible
and thus those methods cannot be applied. In this paper, we propose a
Lagrangian decomposition method where the sub-problems are solved via
search rather than through a specialized propagator. This has the benefit
that the method can be applied to a much wider range of problems. We
present experiments to show the effectiveness of our method.

1 Introduction

Constraint Programming (CP) approaches are state-of-the-art for solving many
combinatorial optimization problems using a branch and bound approach. But
a critical issue that effects the efficiency of branch and bound algorithms in
CP is how strong a bound on the objective function can be inferred at each
search node. The stronger the bound that can be inferred, the earlier failed
subtrees can be detected, leading to an exponentially smaller search tree. Normal
CP solvers are only capable of inferring a bound on the objective function via
propagating the problem constraints. Unfortunately, for many problem classes,
this does not yield a very strong bound. Indeed for this reason Mixed Integer
Programming (MIP) approaches are preferable to CP for solving many forms of
combinatorial optimization problem – they have strong bounds derived from the
linear programming relaxation of the problem.

© Springer International Publishing Switzerland 2016
C.-G. Quimper (Ed.): CPAIOR 2016, LNCS 9676, pp. 65–80, 2016.
DOI: 10.1007/978-3-319-33954-2_6

Recently, Lagrangian decomposition methods have been adapted and applied to Constraint Programming in order to yield stronger bounds on the objective function [1,2]. Lagrangian decomposition allows us to break an optimization problem down into parts that act independently, analogous to the way that CP solvers treat different constraints. For Lagrangian decomposition each of these parts is a constrained optimization problem, and together they generate bounds on the objective, which can be much stronger than simply propagating the objective constraint. The use of Lagrangian decomposition in CP is an exciting development, exactly because it gives us scope for the same powerful heterogeneous approach to constraint satisfaction used in CP, through separate communicating propagators, to be used for constraint optimization, through separate communicating optimizers.

While the introduction of Lagrangian decomposition to CP is an important development, current methods are quite limited in the types of problems they can be applied to. In particular, the set of constraints have to be divided into subsets such that each subset can be solved efficiently via a specialized propagator/optimizer. Examples considered so far restrict the sub-problems to be either knapsack problems [2], or problems specified by a cost-MDD constraint [1] (although this can theoretically express any COP).

For many problem classes, such a division of constraints is simply not possible and thus those methods cannot be applied. In this paper, we propose a Lagrangian relaxation method where the sub-problems are solved via search rather than through a specialized propagator. This has the benefit that the method can be applied to a much wider range of problems. We present experiments to show the effectiveness of our method.

The contributions of this paper are:

- A generic approach to Lagrangian decomposition, applicable to any problem with a linear objective.
- A meta-search based approach to solving Lagrangian decomposed sub-problems, in order to improve bounds on the objective.
- Experiments showing that the search based approach to Lagrangian decomposition can be highly effective.

2 Background and Definitions

2.1 Constraint Programming with Lazy Clause Generation

Let $\mathcal{V}$ be a set of (integer) variables (we will treat Boolean variables as 0–1 integers).

A *valuation*, θ, is a mapping of variables to values, denoted $\{x_1 \mapsto d_1, \ldots, x_n \mapsto d_n\}$. Define $vars(\theta) = \{x_1, \ldots, x_n\}$. We can apply a valuation to a variable $\theta(x_i)$ to return the value d_i, and extend application of valuations θ to arbitrary expressions involving $vars(\theta)$ in the obvious way.

A *primitive constraint*, c, is a set of valuations over a set of variables $vars(c)$. A valuation θ is a *solution* of c if $\{x \mapsto \theta(x) \mid x \in vars(c)\} \in c$. A *constraint* C is a conjunction of primitive constraints, which we often treat as a set. A valuation

θ is a solution of constraint C if it is a solution for each $c \in C$. We write $C_1 \models C_2$ if every solution of C_1 is a solution of c_2. We extend this notation to valuations, writing $\theta \models C$ if $\bigwedge_{i=1}^{n} x_i = d_i \models C$ where $\theta = \{x_1 \mapsto d_1, \dots, x_n \mapsto d_n\}$.

A *literal* is a unary constraint (we can restrict to the forms $x = d, x \neq d, x \geq d, x \leq d$), or *false*. A *domain* D is a conjunction of literals over $vars(D)$. D is a *false domain* if it has no solutions. We use notation $D(x) = \{\theta(x) \mid \theta \text{ is a solution of } D\}$. We use *range notation* $[l..u] = \{d \mid l \leq d \leq u\}$. We can map a valuation θ to a domain $D_0 = \bigwedge_{x \in vars(D)} x = \theta(x)$.

A *propagator* $p(c)$ for constraint c is an inference algorithm, it maps a domain D to a conjunction of literals $p(c)(D)$, where $D \wedge c \models p(c)(D)$. We assume each propagator is *checking*, that is if $\forall x \in vars(c).|D(x)| = 1$ then $p(c)(D) = \emptyset$ if θ_D is a solution of c and $\{false\}$ otherwise. A propagation solver $prop(P, D)$ applied to a set of propagators P and a domain D repeatedly applies the propagators $p \in P$ until $p(D') = \emptyset$ for $p \in P$, and returns D'.

A *constraint satisfaction problem (CSP)* is a constraint C, often broken into a domain constraint and the remainder $C \Leftrightarrow D \wedge C'$. A *constraint optimization problem (COP)* is of the form $z = \min\{e \mid C\}$, where e is an expression to be minimized and C is a constraint.

In *lazy clause generation (LCG)* solvers [3] propagators are also required to return explanations for each new consequence $l \in p(c)(D)$, that is an explanation clause $e \equiv l_1 \wedge \cdots l_n \rightarrow l$ where $\forall 1 \leq i \leq n, D \models l_i$ and $c \models e$. LCG solvers, like SAT solvers, create an implication graph, where every new consequence is attached to a reason. On failure this is used to create a *nogood* by repeatedly replacing literals in the explanation of failure until only one literal that became true after the last decision remains. This nogood is guaranteed to generate new propagation information. See [3] for more details.

2.2   Lagrangian Decomposition

Lagrangian decomposition is a well understood application of Lagrangian relaxation in order to decompose an optimization problem into parts. Consider an optimization problem of the form $z = \min\{cx \mid C_1(x) \wedge C_2(x)\}$ where z is the objective value, c are the coefficients and x the decisions of the linear objective, and $C_1(x)$ and $C_2(x)$ are arbitrary constraints, then we can provide a lower bound on the objective using

$$\begin{aligned}
z = \min\{cx \mid C_1(x) \wedge C_2(x)\} &= \min\{cx \mid C_1(x) \wedge C_2(y) \wedge x = y\} \\
&= \min\{cx + \lambda(x - y) \mid C_1(x) \wedge C_2(y) \wedge x = y\} \\
&\geq \min\{cx + \lambda(x - y) \mid C_1(x) \wedge C_2(y)\} \\
&= \min\{(c + \lambda)x \mid C_1(x)\} + \min\{-\lambda y \mid C_2(y)\}
\end{aligned}$$

The problem is decomposed by duplicating the variables and adding a Lagrange multiplier penalty λ to try to force the duplicate variables to be the same.

The above reasoning shows how to break a problem into two parts, the approach straightforwardly generalizes into $n + 1$ parts by creating n copies of the variables and n sets of equations relating the copied variables with the original variables.

The correctness of the lower bound holds regardless of the values of λ, but we can get stronger bounds by solving the Lagrangian dual to obtain the best values for λ. In the CP space, since the constraints $C_i(x)$ are arbitrary the usual approach to do this is the subgradient method [4].

In CP many integer variables represent different choices, and the order of the integers is irrelevant, hence applying a Lagrangian penalty like $x_i - y_i$ makes no sense since if it is non-zero it simply represents that two different choices are made of (original) variable x_i. Hence Lagrangian decomposition approaches for CP usually break such integer variables into separate 0–1 variables representing which choice is taken. Given $D_{init}(x_i) = [l..u]$ we replace x_i by 0–1 variables $x_i^j, l \leq j \leq u$ where $x_i^j = 1 \leftrightarrow x_i = j$. We then replace $x_i = y_i$ by the conjunction $\bigwedge_{l \leq j \leq u} x_i^j = y_i^j$. The key advantage for Lagrangian decomposition is that we have separate Lagrange multipliers λ for each such equation.

The CP based Lagrangian decomposition approaches, solve the original problem $z = \min\{cx \mid C_1(x) \wedge C_2(x)\}$ by effectively solving the problem $\min\{z \mid C_1(x) \wedge C_2(x) \wedge z = cx \wedge z \geq z_1 + z_2 \wedge z_1 = \min\{(c + \lambda)x \mid C_1(x)\}\} \wedge z_2 = \min\{-\lambda y \mid C_2(y)\}\}$. Each of the constraints in the master problem are represented by propagators. This requires a propagation algorithm for the each of optimization sub-problem constraints. This is a distinct restriction on the approaches. Some practical approaches to building these propagation algorithms are:

- Restrict to a well understood problem: e.g. $z = \min\{-\lambda y \mid dy \leq d_0\}$ is an instance of the knapsack problem, for which many algorithms are known, and indeed quick approximation algorithms are available.
- Encode the problem using an existing global: e.g. $z = \min\{-\lambda y \mid C_2(y)\}$ can be represented as a cost-MDD constraint, where the MDD encodes $C_2(y)$. As long as the MDD constraint is not too large then we can use the global cost-MDD propagation algorithms [5,6].

In the end the difficulty of creating efficient propagators for the optimization sub-problem can severely limit the applicability of Lagrangian decomposition to CP.

3   Objective Splitting Lagrangian Decomposition

The existing approaches to Lagrangian decomposition decompose the problem in a constraint centric way, splitting up the constraints into disjoint subsets and assigning the corresponding part of the objective to each subset. We advocate a decomposition based on breaking up the objective function directly and assigning the corresponding parts of the constraints to each part of the objective. Unlike normal Lagrangian decomposition where each constraint can only belong to one sub-problem, we project the original constraints onto each sub-problem, meaning that each original constraint could have a projection in more than one sub-problem.

3.1   Problem Decomposition

We consider a similar Lagrange decomposition based on splitting on the objective for $z = \min\{cx + du \mid C(x, u, v)\}$. We consider three classes of variables: x

and u appear in the objective and v are auxiliary variables; and split the constraints $C(x, u, v)$ into three categories: $C_1(x, v)$ are constraints only involving x and auxiliaries, $C_2(u, v)$ are constraints only involving u and auxiliaries and $C_0(x, u, v)$ are the remaining constraints. In practice if we have auxiliaries v only related to x we can add them to x, treating them as having coefficient 0 in the objective, similarly for auxiliaries only related to u. This reduces the number of Lagrange multipliers required.

The objective splitting decomposition is based on the following reasoning:

$$
\begin{aligned}
z &= \min\{cx + du \mid C(x, u, v)\} \\
&= \min\{cx + du \mid C_0(x, u, v) \wedge C_1(x, v) \wedge C_2(u, v)\} \\
&= \min\{cx + du \mid C_1(x, v) \wedge C_{02}(x, u, v) \wedge C_2(u, v) \wedge C_{01}(x, u, v)\} \\
&= \min\{cx + du \mid C_1(x, v) \wedge C_{02}(x, u, v) \wedge C_2(u, v') \wedge C_{01}(x, u, v') \wedge v = v'\} \\
&= \min\{cx + du + \lambda(v - v') \mid C_1(x, v) \wedge C_{02}(x, u, v) \wedge C_2(u, v') \wedge C_{01}(x, u, v') \wedge v = v'\} \\
&\geq \min\{cx + du + \lambda(v - v') \mid C_1(x, v) \wedge (\exists u. C_{02}(x, u, v)) \wedge C_2(u, v') \wedge (\exists x. C_{01}(x, u, v'))\} \\
&= \min\{cx + \lambda v \mid C_1(x, v) \wedge \exists u. C_{02}(x, u, v)\} + \min\{du - \lambda v' \mid C_2(u, v') \wedge \exists x. C_{01}(x, u, v')\} \\
&= \min\{cx + \lambda v \mid C_1(x, v) \wedge \exists u. C_{02}(x, u, v)\} + \min\{du - \lambda v \mid C_2(u, v) \wedge \exists x. C_{01}(x, u, v)\} \\
&= \min\{cx + \lambda v \mid \exists u. C_1(x, v) \wedge C_{02}(x, u, v)\} + \min\{du - \lambda v \mid \exists x. C_2(u, v) \wedge C_{01}(x, u, v))\} \\
&= \min\{cx + \lambda v \mid \exists u. C(x, u, v)\} + \min\{du - \lambda v \mid \exists x. C(x, u, v)\} \\
&\geq \min\{cx + \lambda v \mid \bar{\exists} u. C(x, u, v)\} + \min\{du - \lambda v \mid \bar{\exists} x. C(x, u, v)\}
\end{aligned}
$$

where $C_{02}(x, u, v) \Leftrightarrow C_0(x, u, v) \wedge C_2(u, v)$, $C_{01}(x, u, v) \Leftrightarrow C_0(x, u, v) \wedge C_1(x, v)$, and the *quasi projection*, defined later, $\bar{\exists} y. C$ is a formula such that $\exists y. C \Rightarrow \bar{\exists} y. C$. The two weakening steps hold since weakening the constraints can only reduce the minimum value.

The resulting CP optimization problem is $\min\{z \mid C(x, u, v) \wedge z = cx + du \wedge z \geq z_1 + z_2 \wedge z_1 = \min\{cx + \lambda v \mid \bar{\exists} u. C(x, u, v)\} \wedge z_2 = \min\{du - \lambda v \mid \bar{\exists} x. C(x, u, v)\}\}$. Notice that all sub-problems use the same variables, and all constraints are present (in a quasi projected form) in every sub-problem.

We can generalize this to separating into m components

$$
\begin{aligned}
z &= \min\{c_1 x_1 + \cdots c_m x_m \mid C(x, v)\} \\
&\geq \min\{c_1 x_1 + (\lambda_2 + \cdots \lambda_m) v \mid \bar{\exists} x_2 ... x_m. C(x, v)\} \\
&\quad + \min\{c_2 x_2 - \lambda_2 v \mid \bar{\exists} x_1 x_3 ... x_m. C(x, v)\} + \cdots \\
&\quad + \min\{c_m x_m - \lambda_m v \mid \bar{\exists} x_1 ... x_{m-1}. C(x, v)\}
\end{aligned}
$$

Now the resulting problem appears far more complex than the original problem, since we have m sub-problems that appear to be (almost) copies of the original. The advantage that arises is that we have weakened the constraints in the sub-problem and still get correct bounds. Of course if we weaken them too much the bounds will be useless.

The objective based decomposition makes use of existential quantification to allow us to separate constraints that involve objective variables from different classes. Since projection is impractical to compute we weaken the projection. While the logic holds for an arbitrary weakening we will use a certain form we call *quasi projection*.

Given a constraint C with $vars(C) = \mathcal{V}$ a *quasi projection* of C onto variables V, written $\bar{\exists}\{\mathcal{V} - V\}.C$ or $\bar{\exists}_{-V}.C$, is the set of solutions θ over variables V such that $prop(\{p(c) \mid c \in C\}, D_\theta)$ does not return a false domain. We call V the *local variables* of the qausi projection.

Proposition 1. $\exists W.C \Rightarrow \bar{\exists} W.C.$

Proof. By definition each solution σ of $\exists W.C$ is such that there exists θ solution of C where $\sigma = \{x \mapsto \theta(x) \mid x \in \mathcal{V} - W\}$. The call $prop(\{p(c) \mid c \in C\}, D_\sigma)$ cannot return a false domain since this would eliminate the solution θ erroneously. Consider the propagator that did this, i.e. $\theta \in D'$ and $D'' = p(D')$ where $\theta \notin D''$. Now $\theta \models D \wedge c$ but $\theta \not\models p(D)$ which contradicts the definition of a propagator. Hence $\sigma \in \bar{\exists} W.C.$ □

Example 1. Consider the constraint $C \equiv x < y \wedge y < z \wedge y \mod 3 = 0 \wedge x \in 0..4 \wedge y \in 0..4 \wedge z \in 0..4$. Assuming bounds propagators for the inequalities and a mod propagator that only wakes when y is fixed, the quasi projection onto $\{x, z\}$ is $\{\{x \mapsto 0, z \mapsto 3\}, \{x \mapsto 0, z \mapsto 4\}, \{x \mapsto 1, z \mapsto 4\}, \{x \mapsto 2, z \mapsto 4\}\}$. Note how $\{x \mapsto 0, z \mapsto 2\}$ causes failure since propagation fixes y to 1 where the mod constraint then fails. The actual projection eliminates the first solution. If the propagator for $y \mod 3 = 0$ was stronger, changing the domain of y to $\{3\}$ then the quasi projection would return the projection. □

Example 2. Consider a nurse rostering problem. We have n nurses working for m days and on each day we must choose a shift type in S for each nurse (including a day off). The model has complex restrictions on the sequence of shifts that each nurse can undertake, typically encoded by a `regular` constraint using some finite automata FA, and vectors of upper u and lower l limits on the number of nurses assigned to each shift type on a day, typically encoded by a global cardinality constraint. Finally each nurse i has a preferred shift p_{ij} for each day j, and the aim is to maximize the number of preferences that are met by the schedule. Let x_{ij} represent the shift type chosen for nurse i on day j, then the model is

$$z = \text{minimise} - \sum_{j=1}^{m} \sum_{i=1}^{n} (x_{ij} = p_{ij})$$
$$\text{subject to } \texttt{gcc\_low\_up}([x_{ij} | i \in 1..n], S, l, u), \ j \in 1..m$$
$$\texttt{regular}([x_{ij} | j \in 1..m], FA), \quad i \in 1..n$$

We decompose the objective into days, arriving at the following m subproblems $P(j)$, of the form $y_j = \min\{-\sum_{i=1}^{n}(x_{ij} = p_{ij}) \mid \bar{\exists}_{-V_j}.C\}$ where $V_j = \{x_{ij} \mid i \in 1..m\}$ and C is all the constraints. Note that the constraint $\texttt{gcc\_low\_up}([x_{ij} | i \in 1..n], S, l, u)$ will certainly be satisfied by any solution of the quasi-projection since none of its variables are projected away. In this problem since there are no auxiliary variables we need no Lagrange multipliers. □

Example 3. Consider the problem of max density still life, building a $2m \times 2m$ square which is stable under the Conway's Game of Life rules, and has the maximum number of live cells. The best model [7] for this minimizes wastage (wasted opportunities for placing live cells) which can be computed from each

3×3 subsquare. Let $x_{ij}, 1 \le i, j \le 2m$ be the 01 decisions for each cell: 1 is live, 0 is dead. Let $w_{ij}, 2 \le i, j \le 2m - 1$ be the wastage for the 3×3 subsquare centered at (i, j). A (simplified for ease of exposition) model for the problem is

$$z = \text{minimise} \sum_{i=2}^{2m-1} \sum_{j=2}^{2m-1} w_{ij}$$

$$\text{subject to } \texttt{table}([\begin{matrix} x_{i-1,j-1}, & x_{i-1,j}, & x_{i,j+1}, \\ x_{i,j-1}, & x_{i,j}, & x_{i,j+1}, & w_{ij}], T), i, j \in 2..2m - 1 \\ x_{i+1,j-1}, & x_{i+1,j}, & x_{i+1,j+1}, \end{matrix}$$

where T is a table relating 3×3 patterns to their wastage. We will use $wastage_{ij}$ as shorthand for the table constraint. We consider a decomposition of the objective into 4 quadrants $z = \text{minimise} \sum_{i=2}^{m} \sum_{j=2}^{m} w_{ij} + \sum_{i=2}^{m} \sum_{j=m+1}^{2m-1} w_{ij} + \sum_{i=m+1}^{2m-1} \sum_{j=2}^{m} w_{ij} + \sum_{i=m+1}^{2m-1} \sum_{j=m+1}^{2m-1} w_{ij}$ The auxiliary x variables for columns and rows m and $m + 1$ are shared by sub-problems and need Lagrange multipliers, the remaining x variables only appear in one sub-problem and do not. The top left quadrant sub-problem has objective

$$z = \text{minimise} \begin{aligned} & \sum_{i=2}^{m} \sum_{j=2}^{m} w_{ij} \\ & + \sum_{i=2}^{m} \lambda_{i,m} x_{i,m} + \sum_{j=2}^{m} \lambda_{m,j} x_{m,j} \\ & - \sum_{i=2}^{m} \lambda_{i,m+1} x_{i,m+1} - \sum_{j=2}^{m} \lambda_{m+1,j} x_{m+1,j} - \lambda_{m+1,m+1} x_{m+1,m+1} \end{aligned}$$

The quasi projection (quasi)eliminates all variables not in top left quadrant except those included in the last line of the objective. All of the $wastage$ constraints for the top left quadrant will be guaranteed to be solved since none of their variables are projected out, hence the bound will understand the effect of their interaction on the objective. □

How to split the objective expression into parts remains a question for all Lagrangian decomposition methods. In many problem classes, the partitioning is somewhat natural. It is often the case that problems have a certain amount of locality, where there are certain groups of variables which are strongly related to each other, but weakly related to other variables. We propose to partition the objective function into groups of closely related terms.

Example 4. Example 2 shows how we can meaningful split the nurse scheduling objective into individual days. This make sense for improving the lower bound since the nursed preferred shifts will contradict the gcc constraint, and we will get a much better idea of how many it is possible to simultaneously satisfy. Another possibility is to split it into groups of consecutive days, since these are more tightly related by the regular constraint, so the sub-problems then learn about the interaction of regular and gcc. Alternatively, we could imagine splitting it into individual nurses, thus capturing the effect of the regular on the objective. □

Our objective based decomposition differs from the constraint based decomposition of earlier methods [1,2], and has both advantages and disadvantages. Some points of interest are as follows:

– When a variable in the objective function appears in two or more sub-problems, it is not clear which sub-problem this objective term should be assigned to in order to maximize the effectiveness of the Lagrangian decomposition. In the constraint based decomposition, the constraint split does not completely tell us how to split the objective terms, and indeed we often have to make a second set of decisions as to how to split the objective terms. This decision is handled in a somewhat ad-hoc manner in [1, 2]. Sometimes one sub-problem gets the term, sometimes it is split into two or more parts. It is hard to understand what sort of assignment/split gives the best bound in general. In the objective based decomposition method however, this question does not arise, as the objective split has already fully decided which objective term belongs to each sub-problem, lowering the total amount of decisions that need to be made. Further, we have a good general policy for splitting the objective function, which is to split the objective terms into groups of strongly related terms.

– The constraint based decompositions allow objective terms to be split between sub-problems whereas our proposed objective based decomposition does not. This may be an advantage of the constraint based decomposition as there may be problem classes where splitting objective terms gives a better bound than we can if we are not allowed to split them. On the other hand, our objective based decomposition approach allows constraints to be split (projected) onto multiple sub-problems, whereas in the constraint based decomposition method, each constraint can only appear in one sub-problem. This may be an advantage of the objective based decomposition as many COPs have global constraints that may include all, or many of the variables defining the problem. If we place such a global constraint in only one sub-problem, then all the other sub-problems are substantially weakened. By projecting the global constraint into all of the sub-problems, all of them can get the pruning benefit of the relevant part of that global constraint.

– The constraint based decomposition approach can handle non-linear objective functions by for example assigning the entire objective function to one subproblem. This is not possible in general for the objective based decomposition. However, the objective based decomposition could also potentially handle certain forms of non-linear objective functions in a different way. For example, a $min(x_1, \ldots, x_n)$ objective function could be split so that each of the x_i is the objective function for one subproblem.

3.2 Solving the Sub-problems

The main difference between our approach and the recent approaches is that we do not require the sub-problems to be of a special form which can be solved via a specialized propagator. Instead, we are going to solve them via standard CP search. This means that our approach can be applied to virtually any CP problem with a linear objective function, rather than only to those which so happen to decompose into sub-problems of specialized forms.

```
bandb(D, V, P, z, S)
    u := max D(z)
    θ := ⊥
    repeat
        best := θ
        θ := search(D, V, P, z, {p(z ≤ u)}, S)
        u := θ(z) - 1
    until θ = ⊥
    return best


search(D, V, P, z, Q, S)
    D := propagate(D, V, P, Q, S)
    if (∃x ∈ V.D(x) = ∅) return ⊥
    if (∀x ∈ V.|D(x)| = 1)
        let θ = {x ↦ d_x | x ∈ V, D(x) = {d_x}}
        return θ
    else
        {c_1, ..., c_m} := branch(D, V)
        for i ∈ 1..m
            θ := search(D, V, P ∪ Q, z, {p(c_i)}, S)
            if (θ ≠ ⊥) return θ
        return ⊥
```

```
propagate(D, V, P, Q, S)
    P := P ∪ Q
    repeat
        while (∀x ∈ V.D(x) ≠ ∅ ∧ ∃p' ∈ Q)
            Q := Q - {p'}
            D' := D ∧ p'(D)
            Q := Q ∪ new(P, D, D')
            D := D'
        if (∃s ∈ S.Θ(s) ⊭ D)
            D' := subbound(s, D, P)
            Q := Q ∪ new(P, D, D')
            D := D'
    until Q = ∅
    return D


subbound(s, D, P)
    let s ≡ z = min([o|∃̄_{-V_s}.C ∧ S_s])
    θ := bandb(D, V_s, P, o, S_s)
    D := D ∧ z ≥ θ(o)
    Θ(s) := θ
    return D
```

Fig. 1. Pseudo-code for evaluating LD COPs.

Our approach is as follows. We decide on a splitting of the objective and create the Lagrangian optimization sub-problems. Note that since these problems also have a linear objective we can apply the splitting *recursively* constructing a nested Lagrangian decomposition.

First, we add a new variable z_j representing the objective value of each of those Lagrangian decomposed optimization sub-problems. We add a constraint $z \geq \sum_{i=1}^{m} z_j$ to relate the original objective to these variables. Finally we add the optimization sub-problems defining the z_j. Note that we do not create multiple copies of variables when they belong to multiple sub-problems. Instead, we only require the original copy. In addition, even if a constraint appears in multiple sub-problems, we only need to post one copy of that constraint in the CP solver. This is important because it increases the reusability of nogoods.

The solving of the Lagrangian decomposed COP $z = \min\{c.x \mid C\}$ is as follows. We begin by calling $\mathsf{bandb}(D_{init}, \mathcal{V}, \{p(c) \mid c \in P \cup \{z \geq \sum_{j=1}^{m} z_j\}\}, z, S)$ (shown in Fig. 1) where $s_j \in S$ is a Lagrangian decomposed optimization problem of the form $s_j \equiv z_j = \min\{o_j \mid \bar{\exists}_{-V_j}.C\}$.

Notice that each optimization sub-problem is of the same form as the original problem, with a different linear objective and some variables quasi projected. Hence we can apply Lagrangian decomposition on the sub-problems, nesting new Lagrangian decomposed problems within them. The algorithms in Fig. 1 handle arbitrary depth of nesting of optimization sub-problems.

Branch and bound search calls search to search for a solution, repeatedly, and then adds constraints to search for better solutions, returning the best solution when it is proved optimal. The search routine is almost standard except: it passes around the sub-problem constraints S, it terminates when all the local variables V are fixed (as opposed to all variables in the problem $\mathcal{V}$), and the branching

```
subbound(s, D, P)
    let s ≡ z = min([o|∃₋Vₛ.C ∧ Sₛ])
    l := min D(o)
    while (θ ≡ search(D, Vₛ, P, o, {p(o ≤ l)}, Sₛ) = ⊥)
        l := l + 1
    D := D ∧ z ≥ l
    Θ(s) := θ
    return D
```

Fig. 2. Subsearch using destructive lower bounding search.

decisions returned by branch are restricted to only involve local variables V. This implements quasi projection.

The propagation routine propagate is standard, except that it wakes up a sub-problem s, when its incumbent optimal solution $\Theta(s)$ is no longer a support for the lower bound since it is incompatible with the current domain, using subbound to calculate a new lower bound. Initially the incumbent solution $\Theta(s)$ for each sub-problem s is set to $\bot$. We assume $\Theta(s)$ is a backtracking global variable.

The subbound procedure finds the optimal solution θ to the optimization sub-problem $s \equiv z = \min\{o \mid \exists_{-V}, C \wedge S_s\}$ where S_s are optimization sub-problems local to S. It uses branch and bound search to minimize o. Crucially the variables of interest are limited to the objective variables for this sub-problem V. Note that the variable reduction is critical for solving the sub-problem more efficiently, since we only look for "solutions" where each local variable is fixed (and the propagators do not detect failure). It sets the lower bound of the sub-problem variable z to that value, as well as storing θ as the incumbent solution.

Alternatively we can use destructive lower bounding search to raise the lower bound of the sub-problem. Unlike normal branch and bound where we iteratively find better and better solutions, in destructive bounding, we start with the tightest bound on the objective function and repeatedly loosen that bound until we find a solution. Destructive bounding is more suitable than normal branch and bound for re-solving a sub-problem when the previous incumbent solution has become invalid. It will immediately try to find a replacement solution which is at least as good as the previous one, and if that fails, it will be able to strengthen the bound proved and then try to find a solution which is one unit worse, etc. This is generally better than re-solving the sub-problem from scratch via normal branch and bound. Destructive bounding is described in Fig. 2. Note that we can break the **while** loop at any time and still get a correct lower bound, although the there will be no incumbent solution in this case. This may be useful in cases where we want to put a time limit on solving the sub-problems. In practice we use normal branch and bound for the first solve of a sub-problem, and use destructive bounding for all re-solves.

3.3   Nogood Learning

Our search-based method integrates seamlessly with nogood learning [3]. In nogood learning, each propagation has to have an explanation clause which explains why that propagation is valid given the current domain. If we want to use nogood learning, then when we derive a bound on z_j via the sub-search on $z_j = \min\{o_j \mid \bar{\exists}_{-V_j}.C\}$, we have to be able to generate a clause which explains the bound on z_j given the current domain. Fortunately, this occurs naturally without any need to modify the nogood learning solver. When we enter a sub-search, any domain changes made by the master search will act as "assumptions" in the sub-search. Any literals representing those initial domain conditions which are relevant to failures in the sub-search will be kept in the nogoods derived during that sub-search. At the end of the proof of optimality phase of the sub-search, we will end up deriving a nogood of form: $l_1 \wedge \ldots \wedge l_n \wedge o_j < \theta(o_j) \rightarrow false$ where l_i are conditions on the variables V which are sufficient to force that bound on o_j. We can translate this to an explanation for the lower bound $l_1 \wedge \ldots \wedge l_n \rightarrow z_j \geq \theta(o_j)$.

Another benefit of nogood learning is that many of the things learned during one sub-search are encapsulated in nogoods and can be reused in subsequent sub-searches of the same sub-problem. Thus we do not have to re-solve those sub-problems from scratch each time, but rather, much of the failed subspace is still encapsulated in the nogoods and can be immediately pruned.

Example 5. Consider an instance of the nurse rostering problem of Example 2 with 8 nurses, requiring at least 3 on day shift (d) and at least 2 on night shift (n), where shift regulations require: no day shift immediately after a night shift, no more than 3 days shifts in a row, no more than 2 night shifts in a row, and no more than one dayoff (o) in a row. Consider a sub-problem instance, for a day j where all nurses request a day shift. Running the sub-problem at the root will discover that $z_j \geq -6$ since at most 6 nurses can get a day shift. When the branch and bound code searches with $o_j \leq -7$ the search fails with explanation $o_j \leq -7 \rightarrow false$, since this makes use only of globally true information. The resulting explanation of the bound is simply $z_j \geq -6$.

Now consider waking the sub-problem when on the previous day $j - 1$ we have assigned the first four nurses to night shift, and the last 4 to day shift. Then only the last 4 nurses can be assigned to a day shift on day j. Branch and bound fails when we add $o_j \leq -5$ with explanation $x_{1j-1} = n \wedge x_{2j-1} = n \wedge x_{3j-1} = n \wedge x_{4j-1} = n \wedge o_j \leq -5 \rightarrow false$. Generating the explanation for $z_j \geq -4$ as $x_{1j-1} = n \wedge x_{2j-1} = n \wedge x_{3j-1} = n \wedge x_{4j-1} = n \rightarrow z_j \geq -4$.   □

3.4   Lagrangian Multipliers

In order to take maximum advantage of Lagrangian multipliers for CP, we differentiate between two different types of integer variables; bounds type integer variables, and value type integer variables. Bounds-type variables are those which are mainly involved in bounds type constraints like linear inequalities. Whereas value-type variables are those which are mainly involved in value type constraints like `alldifferent`, `table` or `regular` constraints. For the latter class of variable we break them into separate 0–1 variables representing each possible value.

We could update Lagrangian multipliers at each call to search using the subgradient method. However, on the problem classes we tried, it appears that updating the Lagrangian multipliers at each node is usually not worth it. Instead we simply calculate the Lagrange multipliers at the root node and use the same multipliers throughout the computation, like [1]. This has an advantage for nogood learning, because the Lagrange multipliers are globally fixed we do not need to include any assumptions about them in explanations for propagation. Using the subgradient method [4] at the root, we update the Lagrangian multipliers for a fixed number of iterations or until the bound derived no longer improves.

3.5 Lazier Bounding

Re-solving the optimization sub-problems does not always yield a better bound on z_j. If the bound does not improve, then all the effort done in the sub-search is wasted. Thus we want to try to only resolve the sub-problems when we have a good chance of improving its bound. The pseudo-code naively re-solves a sub-problem each time its incumbent solution becomes inconsistent with the current domain of the master search, as there is a chance that the bound may be improved. However, this may be too costly.

We propose the following dynamic policy for determining whether to perform the sub-search. For each sub-problem s_j, we have an activation chance p_j which determines whether to re-solve the sub-problem when the incumbent solution becomes inconsistent with the master search. p_j starts at 1. Each time re-solving s_j yields a better bound, we increase p_j by α, capped at 1. Each time re-solving s_j does not yield a better bound, we decrease p_j by β, capped at 0.1. Some reasonable values for α and β are 0.1 and 0.05, and varying these values somewhat did not appear to make much difference. The main idea is that if re-solving the sub-problem often does not yield anything, then p_j will eventually decrease and we will rarely re-solve that sub-problem again, lowering the overhead of the method.

We propose another policy for reducing the overhead of the method. When the master search is searching on the variables of a particular sub-problem, the incumbent solution of that sub-problem will become invalid at almost every decision. If we follow the normal policy of re-solving a sub-problem whenever its incumbent solution becomes invalid, then we will end up re-solving a sub-problem at almost every node in the master search tree. This is clearly very expensive. It is also often redundant work, because since the master search is searching on the variables of that sub-problem, the bound of that sub-problem will quickly be fixed by the master search anyway and worth trying to strengthen that bound through sub-search. Thus we modify our policy so that if the master search has just made a decision on variable v, then any sub-problem involving v will not be woken up for sub-search at that search node.

4 Experiments

In this section, we describe a few problems as well as how we partitioned their objective function.

Nurse Scheduling Problem. This problem is described in Example 2. In the instance we use there are 4 possible shifts including a "day off" shift, and requirements on the number of nurses for each shift and minimum and maximum requirements on the number of holidays per time span. We partition the objective function into partial sums each representing one day, as described in Example 2.

Maximum Density Still Life Problem. This problem is described in Example 3, although there we only give a simplified model of the optimization problem, ignoring edge effects, and require the size n is even $(2m)$. We partition the objective function by chopping the $n \times n$ region into 9 equal sized square chunks (rather than the 4 of Example 3).

Concert Hall Scheduling Problem. In this problem, we have k concert halls and a bunch of orders. Each order hires a hall from a certain start time to a certain end time and gives a certain profit. The problem is to pick the subset of orders to satisfy such that we maximize the profit. Clearly, orders which are close together in terms of their time are more closely related then orders which are far apart in terms of their time. Thus we can quite naturally partition the objective function according to time. We partition the objective function by dividing the time span into 4 equal sized chunks and putting an order in a chunk based on their starting time.

Talent Scheduling Problem. In this problem, we have some actors and some scenes. Each scene requires a subset of the actors. Each actor has a cost. The scenes are shot in a certain order. Each actor has to be on-scene from the first scene that they are in until the last scene that they are in is finished. For each day they are on-scene, they have to be paid their corresponding cost. The problem is to find the order of the scenes which minimizes the total cost of the actors. Again, terms which represent costs close in time are more closely related together than those far in time. Thus we partition the objective function into 4 equal sized chunks according to time.

Resource Constrained Project Scheduling Problem with Tardiness. In this problem, we have some tasks, each of which requires a certain amount of resources on each of the machines. Each machine has a maximum resource capacity. There are some precedences between the tasks. Each task has a due date. For each unit of time past the due date the task is finished by, there is a penalty. The problem is to find the schedule with the least penalty. Tasks which are closer together in terms of their time are more closely related. Thus we partition the objective function into 4 equal sized chunks based on time.

Sweatshop Scheduling Problem. In this problem, we have rows of benches, each with some machines. Each machine is assigned a type of garment to make. Each type of garment will cost a different amount of power. Each bench and each bench column has limitations on the total amount of power used. Each person in a row has to work on a different garment. There are also global constraints on the minimum and maximum amount of each garment made. The problem is to find the assignment of garments which maximizes the profit. Clearly, terms belonging to the same bench/row/column are more closely related to each other than to terms belong to different bench/row/columns. We partition the objective function based on rows.

Traveling Salesman Problem. In this problem, we have a number of cities. We have a salesman which must tour all the cities and visit each one exactly once. The problem is to minimise the total amount of distance traveled. Cities which are geographically closer together are more closely related to each other. Thus we divide the objective function by partitioning the cities geographically into 9 square chunks.

The experiments were performed on Xeon Pro 2.4 GHz processors using the state of the art LCG solver CHUFFED. We use 20 instances of each problem class, except for Still Life where we use just one.[1] We compare running the above problems with the new sub-problem search-based Lagrangian decomposition (LD via search), with the decomposition but no Lagrangian multipliers (D via search), and without any decomposition (Normal). We also try to compare against the cost-MDD based Lagrangian decomposition method [1] (Bergman et al.). Unfortunately, it is not very clear how that method can be applied to these particular problem classes, as most of them do not decompose into a set of MDD's. Thus we only compare against that method on the Nurse Scheduling problem, as the constraints in that problem can easily be modeled as MDD's. In all the new methods, we use the lazier bounding as described in Sect. 3.5. We use constructive brand and bound for the first solve of each subproblem and destructive bounding for all subsequent re-solves. We update the Lagrangian multipliers at the root for 100 iterations using the subgradient method or until the bound no longer improves. We use a time out of 10 min. The results are shown in Table 1 and Fig. 3.

Table 1. Comparison between using and not using the search-based Lagrangian decomposition method.

Problem	Normal		LD via search		D via search		Bergman et al.	
	Fails	*Time*	*Fails*	*Time*	*Fails*	*Time*	*Fails*	*Time*
Nurse scheduling	618537	56.32	54325	**7.23**	54325	**7.23**	48630	13.02
Still life	478182	57.12	**23415**	3.45	24218	**3.37**	—	—
Concert hall	389799	35.14	**45231**	**5.27**	158962	12.56	—	—
Talent scheduling	**1535814**	**215.1**	2589576	417.82	2620582	428.07	—	—
RCPSP with tardiness	**234758**	**68.54**	1834758	487.29	1834758	487.29	—	—
Sweatshop scheduling	682934	24.87	**124562**	**5.27**	124562	**5.27**	—	—
Traveling salesman	256375	94.56	**185239**	**70.82**	185239	**70.82**	—	—

Table 1 shows the total number of fails and time spent on the benchmarks. These numbers include the fails and time spent in the master search and the sub-

---

[1] Available from `people.unimelb.edu.au/pstuckey/lgadec`.

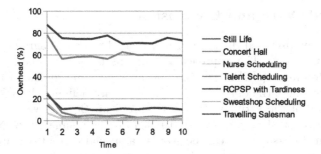

Fig. 3. Overhead of sub-searches per time partition

problem searches. In all of these benchmarks, the Lagrangian decomposition can give a better bound than that found through normal propagation. However, it is not always worth the overhead. We get significantly stronger bounds for the Nurse Scheduling, Still Life, Concert Hall and Sweatshop Scheduling problems, but relatively weak improvements to the bound for Talent Scheduling, RCPSP with tardiness and the Traveling Salesman Problem. It can be seen that when the bound we derive is significantly better, the overhead of the Lagrangian decomposition method is often worth it. Whereas when the improvement in bound is small, the reduction in the search space of the master search may well be swamped out by the overhead of the sub-searches. For example, in RCPSP with tardiness, the sub-searches occur frequently but the improvement in bound is too insignificant to be worth it. The Lagrangian multipliers are only useful for the Concert Hall problem. For the rest they either have no use because there are no shared variables between sub-problems, or their effect is statistically insignificant. We suspect that this is because Lagrangian multipliers do not work well when sub-problems are connected via value type integer variables, whereas they work far better when sub-problems are connected via bounds type integer variables. The approach of [1] requires less search, since instead of solving the sub-problems via search, the sub-problems are solved by the global propagator instead, which does not contribute to the node count. However, this costs more than the gain in run time compared to our approach.

Figure 3 shows the time overhead of the sub-searches as a percentage of the overall search when the search time is split evenly into 10 parts. It can be seen that for some problem classes, much more time is spent on solving the sub-problems near the start of the search than in the rest of the search. There are several reasons. The first is that the first solve via branch and bound is often expensive, whereas subsequent re-solves using destructive bounding are often very quick. The second is that the learned nogoods which describe which boundary conditions force certain bounds for the sub-problem may often immediately propagate to avoid some redundant re-solving of sub-problems, making re-solving sub-problems quicker later in the search.

5 Related Work and Conclusion

We have already discussed the most closely related work on Lagrangian Decomposition for CP [1,2]. The approach we present is closely related to Nested Constraint Programs (NCPs) [8], the optimization sub-problems can be seen as nested CPs where the domain of the variables are defining the sub-problem, for a new copy of the variables in the sub-problem. Because we use the same variables and constraints we can avoid much of the complexity of NCP. Similarly Russian doll search [9] can be seen as a special case of our approach to Lagrangian Decomposition, where there is exactly one optimization sub-problem per level, and no recomputation of the optimization sub-problems.

Lagrangian Decomposition is an exciting development for CP, allowing the same heterogeneous approach to satisfaction to be extended to optimization. In this paper we show how to create a very general scheme for Lagrangian Decomposition using sub-problem search, which, together with learning provides a powerful method for tackling optimization problems that can be meaningfully decomposed. We have shown that the new method can provide significant speedups on some realistic problem classes.

Acknowledgments. NICTA is funded by the Australian Government through the Department of Communications and the Australian Research Council through the ICT Centre of Excellence Program. This work was partially supported by Asian Office of Aerospace Research and Development grant 15-4016.

References

1. Bergman, D., Cire, A.A., van Hoeve, W.-J.: Improved Constraint Propagation via Lagrangian Decomposition. In: Pesant, G. (ed.) CP 2015. LNCS, vol. 9255, pp. 30–38. Springer, Heidelberg (2015)
2. Hà, M.H., Quimper, C.-G., Rousseau, L.-M.: General bounding mechanism for constraint programs. In: Pesant, G. (ed.) CP 2015. LNCS, vol. 9255, pp. 158–172. Springer, Heidelberg (2015)
3. Ohrimenko, O., Stuckey, P., Codish, M.: Propagation via lazy clause generation. Constraints **14**(3), 357–391 (2009)
4. Shor, N.: Minimization Methods for Non-differentiable Functions. Springer, Heidelberg (1985)
5. Demassey, S., Pesant, G., Rousseau, L.: A cost-regular based hybrid column generation approach. Constraints **11**(4), 315–333 (2006)
6. Gange, G., Stuckey, P.J., Van Hentenryck, P.: Explaining propagators for edge-valued decision diagrams. In: Schulte, C. (ed.) CP 2013. LNCS, vol. 8124, pp. 340–355. Springer, Heidelberg (2013)
7. Chu, G., Stuckey, P.: A complete solution to the maximum density still life problem. Artif. Intell. **184–185**, 1–16 (2012)
8. Chu, G., Stuckey, P.J.: Nested constraint programs. In: O'Sullivan, B. (ed.) CP 2014. LNCS, vol. 8656, pp. 240–255. Springer, Heidelberg (2014)
9. Verfaillie, G., Lemaître, M., Schiex, T.: Russian doll search for solving constraint optimization problems. In: Proceedings of the Thirteenth National Conference on Artificial Intelligence and Eighth Innovative Applications of Artificial Intelligence Conference, AAAI Press/The MIT Press, pp. 181–187 (1996)

Non-linear Optimization of Business Models in the Electricity Market

Allegra De Filippo[(✉)], Michele Lombardi, and Michela Milano

DISI, University of Bologna, Viale del Risorgimento 2, 40136 Bologna, Italy
{allegra.defilippo,michele.lombardi2,michela.milano}@unibo.it

Abstract. Demand Response mechanisms and load control in the electricity market represent an important area of research at international level, and the market liberalization is opening new perspectives. This calls for the development of methodologies and tools that energy providers can use to define specific business models. In this work we develop an optimization model to provide recommendations on time-of-use based prices to providers, taking into account some key factors of the customer and market behavior. We have tested our model on data from the Italian energy market, merging statistical census and population information. The main advantage of the model is that it provides a tool for sensitivity analysis, namely for understanding the impact of economical and behavioral parameters on the consumption profiles.

Keywords: Non-linear optimization · Demand response · Tariff optimization · Business model definition · Customer behavior modeling

1 Introduction

Global energy consumption is expected to grow by 37 % within 2040, with a consequent increase of polluting emissions[1] (International Energy Agency) thus negatively impacting the environment and the quality of life. Therefore we need to adopt energy efficiency measures that on one hand lead to *lower* electricity consumption, and on the other hand to *better* electricity consumption via demand shifting.

Demand shifting can provide a number of advantages to the energy system:

– Load management can improve *system security* by allowing a demand reduction in emergency situations.
– In periods of peak loads even a limited reduction in demand can lead to *significant reductions in electricity prices* on the market.
– If users receive information about prices, energy consumption is more closely related to the energy cost, thus increasing *market efficiency*: the demand is moved from periods of high load (typically associated with high prices) to periods of low load.

---

[1] See http://www.iea.org/textbase/npsum/weo2014sum.pdf.

© Springer International Publishing Switzerland 2016
C.-G. Quimper (Ed.): CPAIOR 2016, LNCS 9676, pp. 81–97, 2016.
DOI: 10.1007/978-3-319-33954-2_7

– Load management can limit the need for expensive and polluting power generators, leading to *better environmental conditions*.

Potential benefits and implementation schemes for demand response mechanisms are well documented in international literature: one implementation approach in particular consists in defining economically and environmentally sustainable energy pricing schemes.

The objective of this paper is the development of methodologies and tools for energy providers to manage the sector via the definition of new business models. We developed a mathematical optimization model to provide support for political and economic decisions. This model allows to compare alternative scenarios in terms of user behavior and consequent demand shift.

The model can be employed in a decision support system for utilities, helping them to shape pricing schemes, taking into account on one hand the economic sustainability, and on the other hand the customer flexibility and response to these prices. The model could also be used by policy makers to shape incentive mechanisms, and to evaluate energy policies, and define policy goals.

We have employed our approach to obtain a simplified, approximate model of the Italian energy market. By solving the problem under different scenarios, it is possible to identify trends and assess how the characteristics of the market and the customers affect the consumption profiles.

The rest of the paper is organized as follows. Section 3 describes the proposed optimization model, which takes into account the behavior of multiple customers and of one energy provider. Section 4 presents how we applied our model to the Italian energy market, and provides results for two interesting scenarios. Concluding remarks are in Sect. 5.

2   Related Work

Demand Response can be defined as the occurrence of deviations from the usual consumption pattern in response to stimuli, such as dynamic prices, incentives for load reductions, tax exemptions, or subsidies. An overview of demand response schemes can be found in [3,4,12]. Probably the most widespread demand response mechanism in practice is given by Time of Use (ToU) based tariffs, where the price of electricity is dynamic and follows a weekly pattern.

Optimization approaches to define dynamic prices have been proposed in [1,2,8]. All such works focus in the definition of day-ahead prices for a period of 24 h and for a single customer (or a single group of homogeneous customers). Works [1,2] take into account also other incentive schemes, and rely on an elastic model proposed in [10] to model the demand-response behavior. In this paper, we have adapted such model to ToU based prices over an year-long period.

Only a few papers in related contexts have considered multiple customers or intermediate actors: for example, [15] developed an investment model of renewable energy (solar parks) through crowdfunding. The authors of [15] devise a game theory approach that takes into account interactions between crowd-funders, the owner of the solar park, and a power company that buys renewable energy generated from the solar farm itself.

Consumption and cost awareness has an important role for the effectiveness of demand response schemes. The paper [14] describes a system architecture for monitoring the electricity consumption and displaying consumption profiles to increase awareness. Works [7,11] study how customers respond to price changes, and which price indicators are more relevant on this respect.

The authors of [7] try also to account for mis-perception of energy consumption, which are further analyzed in [5]. The latter work in particular attempts to design a model for the relashionship between real and perceived consumption via regression techniques (i.e. function fitting). The conclusion is that customers tend to slightly overestimate low-energy activities and significantly underestimate activities with high energy consumption.

In this complex and heterogeneous context, our work aims to create a comprehensive non linear optimization model that can be easily customized to different electricity market conditions. We collect in a single model the main parameters relevant to the design of sustainable energy tariffs for demand response.

3   Model Description

The main contribution of this paper is a Non-linear Optimization Model that can be used to simulate the behavior of some key actors in the electricity market and obtain recommendations. For example, our model can be employed to obtain suggestions on Time of Use (ToU) prices for new tariffs, to identify possible points of equilibrium, or to investigate the effect of changes in the customer behavior on the power consumption profiles.

The main actors considered in our model are *a single energy provider* and *multiple groups of homogeneous customers*. Our model consist of several components that take into account: (1) the existence of multiple tariffs with ToU based prices; (2) the demand-response behavior of customers; (3) some cognitive aspects of the customer behavior, in particular their ability to correctly estimate their consumption, and their risk aversion when switching to a new tariff; (4) the relation between the global consumption profile and the wholesale price of electricity. We consider an year-long time horizon. To the best of our knowledge, this is the first approach that tries to take into account multiple tariffs, and cognitive aspects of the customer behavior.

Each model component is presented here separately. The most natural application of our model is obtaining ToU price recommendation for new tariffs, and therefore all components will be presented from this perspective. In the concluding remarks (i.e. Sect. 5) we discuss how our approach can be modified to achieve different goals.

3.1   Set of Tariffs

We consider a set T of ToU based tariffs τ_i, defined over a common price band scheme. The scheme specifies a set P of pre-defined price bands π_j: the exact configuration (e.g. start, end) of each price band is left unspecified, but we assume that for each band π_j the total number of hours $|\pi_j|$ over a period of one year is available.

Each ToU based tariff is defined by a price value $p_{i,j}$ for each band. We assume that a subset T_f of the tariffs is *fixed*, i.e. it cannot be altered by the energy provider. For each $t_i \in T_f$, the prices are constant values. The remaining tariffs are *variable*, and their prices are the main decision variables in our model. We assume that the prices take values over a bounded range, i.e. $p_{i,j} \in [\underline{p}_{i,j}, \overline{p}_{i,j}]$. As long as the bounds are large enough, this assumption is sufficiently general to handle practical scenarios. Finally, we assume that a subset T_o of tariffs is *owned* by the energy provider, i.e. the provider earns profit (and pays the cost) for the electricity consumed under such tariffs. All variable tariffs are owned.

This setup is sufficient to handle a number of interesting cases. Variable tariffs are those for which we wish to obtain price recommendations. Tariffs that are both fixed and owned represent pre-existing contracts that cannot be altered. Tariffs that are fixed and not owned are those offered by competitor providers. To the best of our knowledge, our approach is the first to provide support for multiple tariffs and for modeling the existence of competitors.

3.2   Tariff Choice and Customer Risk Aversion

We take into account the behavior of a set C of homogeneous groups of customers, often referred to as "customer classes" in the remainder of the paper. Each customer class $\kappa_k \in C$ is associated to an original tariff $\tau(\kappa_k)$, which must be fixed (i.e. $\tau(\kappa_k) \in T_f$). Moreover, we assume that the approximate number of customers in each class (let this be $|\kappa_k|$) is known.

Customers may switch to a new tariff, based on its economical benefits. In particular, we assume that each customer in a group can switch tariff with a probability that depends on the obtained savings. Formally, let $c_{k,i}$ be the electricity cost for customer class κ_k under tariff τ_i. The $c_{k,i}$ term is a constant if τ_i is fixed, and a decision variable if τ_i is variable: the computation of such cost values will be discussed in the forthcoming Sect. 3.4. The savings for class κ_k under tariff τ_i are given by:

$$s_{k,i} = \max(0, c_{k,\tau(\kappa_k)} - c_{k,i}) \tag{1}$$

Equation (1) does not take into account the fact that staying with the current tariff is in practice more convenient and less risky than switching. Technically, we say that the customers are likely to exhibit a certain degree of *risk aversion*. We take this into account by adjusting Eq. (1) as follows:

$$s_{k,i} = \begin{cases} \rho_k\, c_{k,\tau(\kappa_k)} & \text{if } \tau_i = \tau(\kappa_k) \\ \max(0, c_{k,\tau(\kappa_k)} - c_{k,i}) & \text{otherwise} \end{cases} \tag{2}$$

where $0 < \rho_k < 1$ is a risk aversion coefficient. In practice, *staying with the current tariffs is considered equivalent to saving a factor ρ_k of the current cost*: this provides an intuitive approach to define the value of ρ_k based on questionnaires or existing data.

As we mentioned, we model the tariff switching as a stochastic process. In particular, we assume that all tariff choices for a given customer class are independent and identically distributed. Formally, we introduce a set of stochastic binary variables $Y_{k,i}$ that are equal to 1 if a customer in class κ_k adopts tariff τ_i. The variable has a discrete probability distribution, given by:

$$P(Y_{k,i} = 1) = \frac{s_{k,i}}{\sum_{\tau_i \in T} s_{k,i}} \qquad P(Y_{k,i} = 0) = 1 - P(Y_{k,i} = 1) \qquad (3)$$

i.e. the probability is proportional to the savings from Eq. (2). The number of customers of class κ_k that adopt tariff τ_i can be obtained by summing $Y_{k,i}$ for $|\kappa_k|$ times. The expected value of this expression is given by:

$$E\left[\sum_{h=0}^{|\kappa_k|-1} Y_{k,i}\right] = |\kappa_k| E[T_{k,i}] = |\kappa_k| \frac{s_{k,i}}{\sum_{\tau_i \in T} s_{k,i}} \qquad (4)$$

Therefore, on average the customers in each class spread over the available tariffs proportionally to the value of $s_{k,i}$. We use this information to define the *tariff selection (and risk aversion) component of our model*, which is given by:

$$y_{k,i} = \frac{s_{k,i}}{\sum_{\tau_i \in T} s_{k,i}} \qquad \forall \kappa_k \in C, \tau_i \in T \qquad (5)$$

$$\text{Equation (2)} \qquad \forall \kappa_k \in C, \tau_i \in T \qquad (6)$$

$$y_{k,i} \in [0,1] \qquad \forall \kappa_k \in C, \tau_i \in T \qquad (7)$$

$$c_{k,i} \in \mathbb{R}^+ \qquad \forall \kappa_k \in C, \tau_i \in T \qquad (8)$$

The $y_{k,i}$ variables represent the fraction of customers of class κ_k that adopt tariff τ_i. Due to the presence of "max" operators in Eq. (2), this model component is non-smooth. The "max" operators can be linearized in a standard fashion by using additional binary variables and big-Ms. In our experimentation, however, we employ the modeling system GAMS and let the software take care of the linearization.

3.3   Demand Response Behavior

We assume that customers can shift their consumption depending on the energy prices, i.e. they are capable of a demand response behavior.

Many demand response programs (including ToU based prices) have been considered in the literature and a few mathematical models have been provided. We have developed a variant of one of the most widely employed approaches [10], which is based on a *cross-elasticity matrix*. Essentially, the approach uses a linear transformation to map variation of prices to variations of demand:

$$\tilde{d} = \epsilon \tilde{p} \tag{9}$$

where $\tilde{d}$ is a vector of demand variations over multiple time periods, $\tilde{p}$ is a vector of (normalized) price variations, and ϵ is called cross-elasticity matrix. The original approach from [10] and employed in [1,2,6] is designed for time periods of homogeneous duration and day-ahead prices.

We adapted the model to ToU based tariffs with price bands of non-uniform duration, over an year-long time period. The main idea is simply to introduce variables to represent the variation in the yearly electricity demand of an individual customer, for each price band π_j. Since we consider multiple customer classes and tariffs in our model, we need separate variables $\tilde{d}_j^{(k,i)}$ for each class κ_k and tariff τ_i. The demand variation is connected to the tariff prices by:

$$\tilde{d}_j^{(k,i)} = \sum_{\pi_h \in P} \hat{d}_h^{(k)} \epsilon_{j,h} \, \tilde{p}_h^{(k,i)} \tag{10}$$

where the $\tilde{p}_h^{(k,i)}$ variables represent normalized price variations. The term $\hat{d}_h^{(k)}$ is a problem parameter, representing the original demand for an individual customer of class κ_k, in price band π_h.

The terms on the diagonal of ϵ are always non-positive and are called self-elasticity coefficients. The other terms are always non-negative. For normalizing the price variations, we use the average price under the original tariff, i.e.:

$$\tilde{p}_j^{(k,i)} = \frac{p_{i,j} - p_{\tau(\kappa_k),j}}{\sum_{\pi_h \in P} p_{\tau(\kappa_k),h}} \tag{11}$$

Our choice is based on insights from works [7,11], which show how customers tend to reason in terms of average prices.

Having weighted the contributions by $\hat{d}_j^{(k)}$ and normalized the prices provides us with a way to intuitively interpret the $\epsilon_{j,h}$ coefficients. In particular, if the price in band π_h roughly doubles (i.e. the normalized variation is 1), then:

– The demand in band π_h (the same band) *decreases* by a factor $|\epsilon_{h,h}|$ of the original demand (we recall that self-elasticity coefficients are non-positive).
– A factor $\epsilon_{j,h}$ of the original demand of band π_h *shifts* to band π_j

Intuitively, the self-elasticity coefficients describe how the demand within each band depends on the prices. The other terms in the matrix represent how the demand shifts *from* price bands (columns to rows) and *to* price bands (rows to columns). The sum of the coefficients on each column corresponds to the net increase/decrease of consumption when the normalized prices increase/decrease: we refer to such quantity as *loss factor*. If the loss factor is zero, changing the prices may alter the distribution of the demand between the price bands, but not its total value. Figure 1 reports an example of cross-elasticity matrix, with

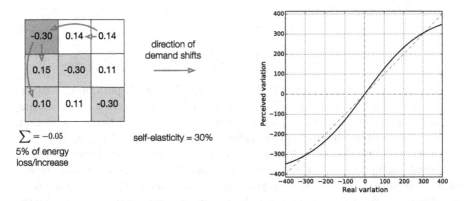

Fig. 1. (left) Interpretation of the cross-elasticity matrix. (right) perception model

a visual depiction of the demand flows. Overall, the demand-response component of our model is given by:

Equation (10)	$\forall \kappa_k \in C, \tau_i \in T, \pi_j \in P$	(12)
Equation (11)	$\forall \kappa_k \in C, \tau_i \in T, \pi_j \in P$	(13)
$\tilde{d}_j^{(k,i)}, \tilde{p}_j^{(k,i)} \in \mathbb{R}$	$\forall \kappa_k \in C, \tau_i \in T, \pi_j \in P.$	(14)

3.4   Energy Demand and Perception Accuracy

There is a growing awareness that a correct perception of the electricity consumption may be a key factor to enable energy savings and make demand response schemes more effective: this is shown by the increasing diffusion of smart-meters and energy monitoring systems in general. However, only a limited number of works have tried to characterize the dynamics of consumer perception: a few papers (e.g. [7,11]) have focused on perceived prices, and even fewer (e.g. [5]) on the accuracy of the consumption estimates.

In particular, the authors of [5] propose to relate perceived and real consumption via a polynomial model in logarithmic scale. The model is calibrated over the estimates provided by a group of users for the consumption of some electric appliances. The authors conclude that people tend to slighly over-estimate low consumption value and considerably under-estimate large values.

In this paper, we take into account the perception accuracy in the demand response behavior. The main idea is to view the $\tilde{d}_j^{(k,i)}$ variables from Sect. 3.3 as *perceived variations*. We then introduce a second set of variables $\tilde{r}_j^{(k,i)}$ to represent the corresponding *real variations*. The two sets of variables are related by a custom model based on results from [5]. In particular, our model is based on a sigmoid function in the form:

$$y = sig(x, \alpha, \beta_\kappa) = \alpha \left(\frac{2}{1 + e^{-2\frac{1}{\alpha \beta_k} x}} - 1 \right) \tag{15}$$

The α parameter determines the scale of the sigmoid, the β_k parameter controls the growth of the sigmoid, which is specific to each customer class κ_k. Assuming that y is a perceived variation, that x is a real variation, and that $\beta_k > 0$, then the sigmoid exhibits the qualitative behavior reported in [5]: for values close to 0, y is a overestimation of x; for values close to α, y is an understimation of x. The break-even point depends on the value of β_k. If $\beta_k < 1$ there is a bias toward overestimation (i.e. for a wide range of values, the perceived variation is larger than the real one), if $\beta_k > 1$ there is a bias toward underestimation (i.e. for a wide range of values, the perceived variation is smaller than the real one). Figure 1 (right) shows this behavior in the sigmoid function for $\alpha = 400$, $\beta_k = 0.75$.

We use our sigmoid function to relate the average perceived and real *power* variations. Those are obtained by dividing the variation variables $\tilde{d}_j^{(k,i)}$ and $\tilde{r}_j^{(k,i)}$ (which represent *energy* values) by the total number of hours in the price bands. Therefore, we obtain:

$$\frac{1}{|\pi_j|} \tilde{d}_j^{(k,i)} = sig \left(\frac{1}{|\pi_k|} \tilde{r}_j^{(k,i)}, \alpha, \beta_k \right) \tag{16}$$

We can then use the real variation variable to compute the total electricity demand for each individual customer of class κ_k, in each price band, and for each tariff. Formally, we have:

$$d_j^{(k,i)} = \hat{d}_j^{(k,i)} + \tilde{r}_j^{(k,i)} \tag{17}$$

where the $d_j^{(k,i)}$ represents the total demand. For the original tariff, i.e. $\tau_i = \tau(\kappa_k)$, the total demand will be the same as the original demand, i.e. $d_j^{(k,i)} = \hat{d}_j^{(k,i)}$. The demand variables $d_j^{(k,i)}$ can be used to compute the cost of energy for each individual customer under each tariff, i.e. the value of the $c_{k,i}$ variables from Sect. 3.2. This is given by:

$$c_{k,i} = \sum_{\pi_j \in P} p_{i,j} \, d_j^{(k,i)} \tag{18}$$

Overall, the perception and demand component of our model is given by:

Equation (16)	$\forall \kappa_k \in C, \tau_i \in T, \pi_j \in P$	(19)
Equation (17)	$\forall \kappa_k \in C, \tau_i \in T, \pi_j \in P$	(20)
Equation (18)	$\forall \kappa_k \in C, \tau_i \in T, \pi_j \in P$	(21)
$\tilde{r}_j^{(k,i)} \in \mathbb{R}$	$\forall \kappa_k \in C, \tau_i \in T, \pi_j \in P$	(22)
$d_j^{(k,i)} \in \mathbb{R}^+$	$\forall \kappa_k \in C, \tau_i \in T, \pi_j \in P.$	(23)

3.5   Wholesale Energy Price

The price of the electricity on the wholesale market (i.e. the electricity cost for the provider) depends on the national energy demand, with larger demand values leading to larger prices. This allows a provider to increase profit by exploiting the demand response behavior and *reduce the wholesale energy price*. In ToU based tariffs, demand shifts are obtained by lowering prices: hence, the dependency of the wholesale electricity on the demand provides opportunities for win-win situations, where both the provider and the customers obtain some gain.

Many works (e.g. [4]) assume the wholesale price-demand curve to be super linear, based on the idea that more expensive power generators are activated when the demand is large. We have tested this conjecture for the Italian energy market, which is organized by a (government controlled) corporation (GME) that issues a reference electricity price and keeps track of the national consumption on a hourly basis. By checking this data we have observed a weak and linear, rather than super-linear, correlation (see Fig. 2).

We have therefore decided to use a linear relation to estimate the wholesale energy price in our model. Formally, we have:

$$w_j = \mu_1 \left(\sum_{\kappa_k \in C} demand(\kappa_k, \pi_j) + b_j \right) + \mu_0 \tag{24}$$

where $demand(\kappa_k, \pi_j)$ is the total demand of the customers of class κ_k in price band π_j. This is given by:

$$demand(\kappa_k, \pi_j) = \sum_{\tau_i \in T} |\kappa_k| \, y_{k,i} \, d_j^{(k,i)} \tag{25}$$

The product $|\kappa_k| \, y_{k,i}$ is the number of customers of class κ_k that adopt tariff τ_i. The b_j term in Eq. (24) represents a baseline consumption, which cannot be altered by adjusting the variable tariffs. This is useful, for example, to model the consumption of *industrial* customers, when the goal is to design tariffs for *residential* customers. The μ_1 and μ_0 terms are the coefficients of the linear relation. Overall, the wholesale price component of our model consists of:

Equation (24)	$\forall \pi_j \in P$	(26)
Equation (25)	$\forall \kappa_k \in C, \pi_j \in P$	(27)
$w_j \in \mathbb{R}^+$	$\forall \pi_j \in P.$	(28)

3.6   Provider Profit and Problem Objective

The most natural problem objective for designing a new tariff is to maximize the provider profit. This is also the natural problem objective if we want to employ

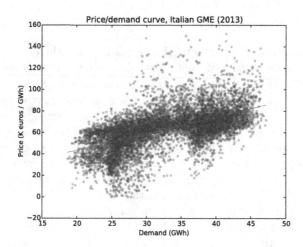

Fig. 2. Wholesale electricity price over demand, Italian market (2013)

our model just to estimate/simulate the behavior of a provider in a given market configuration. In detail, the objective function of our models is:

$$\max z = \sum_{\kappa_k \in C} \sum_{\pi_j \in P} profit(\kappa_k, \pi_j) \tag{29}$$

i.e. the sum of the profit for each customer class and price band, with:

$$profit(\kappa_k, \pi_j) = \sum_{\tau_i \in T_o} |\kappa_k| \, y_{k,i} \, (p_{i,j} - w_j - o_j) \, d_j^{(k,i)} \tag{30}$$

where $p_{i,j}$ is the tariff price for tariff τ_i and w_j is the wholesale energy price. The o_j term represents an overhead value, which captures indirect costs due to (e.g.) energy distribution services or taxes: this is often a very significantly part of the energy costs. The summation in Eq. (30) is performed only on the tariffs owned by the provider.

4   A Case Study on the Italian Residential Market

As a case study, we have used our approach to define a simplified model of the Italian residential energy market. ToU based tariffs in Italy are defined over three standard price bands, roughly corresponding to office hours, evenings-and-saturdays, nights-and-sundays. The total amount of hours for the three bands is $|\pi_0| = 2,860$, $|\pi_1| = 2,132$, and $|\pi_2| = 3,768$. The coefficients for the wholesale demand-price relation have been obtained based on data from the national energy market management corporation[2] (GME): in particular, we have $\mu_1 = {\sim}1.39$ K€/GW and $\mu_0 = 0.013$ K€.

---

[2] See http://www.mercatoelettrico.org/En/Default.aspx.

We focus on residential energy consumption, which makes $\sim 22\%$ of the national energy demand. We consider 5 customer classes corresponding to families with varying number of members, namely 1 (single persons), 2 (couples), 3, 4, and 5 or more. The number of customers per class and their total consumption comes from public data from the Italian national institute for statistics[3] (Istat).

The consumption distribution over the price bands has been estimated based on public data, with larger families having flatter profiles. The risk aversion coefficient ρ has been estimated and it is equal to 0.95 for all customers. For the perception model we have $\alpha = 400$ and $\beta_k = 0.75$ for all classes: intuitively, this means that an *average* consumption of 400 W on a price band is considered large, and that customers tend to over-estimate demand variations (i.e. the real variation is typically smaller than expected).

The coefficients of the cross-elasticity matrix have been defined based on intuitive considerations: in particular, families with fewer members are assumed to be more flexible and more prone to change their net consumption in case of price changes, i.e. to have more significant loss factor. Conversely, larger families are assumed to be less flexible in terms of electricity demand. Clearly, having real data we could derive precise cross-elasticity coefficients for the matrix. All the customer parameters used in our case study are summarized in Table 1.

The baseline consumption has been obtained by subtracting the residential consumption from the national consumption reported in the GME data. In particular, we have (approximately) $b_0 = 75,300$ GWh, $b_1 = 57,800$ GWh, and $b_2 = 56,700$ GWh. The non-residential consumption has a peak in π_0, while residential consumption is more relevant in π_1 and π_2. The overhead electricity costs are assumed to be 250 K€/GWh for all price bands.

We consider two simplified market scenarios: in the *first scenario*, we assume that all customers start with a fixed tariff offered by a *competitor utility*, with a price of 360 K€/GWh (i.e. 0.360 €/KWh) for each band. There is a single variable tariff with $\underline{p}_{i,j} = 72$ and $\overline{p}_{i,j} = 720$ K€/GWh (i.e. from 1/10 to twice the price of the fixed tariff). In the *second scenario* the situation is identical, except that the initial tariff is now *owned by the provider*.

In both cases, our approach tries to define a new tariff that is *beneficial for both the provider* (because the profit is the problem objective) *and the customers* (so that they switch tariff). In the second scenario, such a new tariff exists if a win-win situation is possible, i.e. if a market equilibrium exists even in absence of competition. For solving our model, we used SCIP via the GAMS modeling system on the Neos server for optimization[4], with a time limit of 300 s.

Most of the discussion is devoted to scenario 1, since for scenario 2 we currently have only *a negative, but relevant, result*. We discuss here the results on scenario 2 and we report in the next section results on scenario 1. It seems that no equilibrium is possible under reasonable assignments of all the parameters in our model. This result is consistent with other analyses of the Italian market independently performed by ENEL [13], the main Italian electricity company. The main

---

[3] Available at http://dati-censimentopopolazione.istat.it.

[4] Available at http://www.neos-server.org/neos/.

Table 1. Values of all customer related parameters

#	$|\kappa_k|$ (millions)	$\hat{d}_0^{(k)}$ (GWh)	$\hat{d}_1^{(k)}$ (GWh)	$\hat{d}_2^{(k)}$ (GWh)	Self-elas.	Loss factor
1	~7.6	~1,600	~2,800	~3,600	−35 %	−21 %
2	~6.6	~3,800	~4,600	~6,900	−30 %	−16 %
3	~4.6	~3,500	~2,900	~5,300	−25 %	−11 %
4	~3.9	~4,200	~4,700	~3,800	−20 %	−6 %
>5	~1.5	~1,800	~1,800	~1,600	−15 %	−1 %

reason for this lack of equilibrium points seems to be the linear correlation between wholesale energy prices and demand, which offers limited opportunities to exploit demand shifting. In this situation, competition (which is captured by scenario 1) is the most reliable approach to yield benefits for the customers.

4.1 Results of the Experimentation (for Scenario 1)

Solving the first scenario with our model leads to an optimal tariff with prices (approximately) equal to 155/437/406 K€/GWh. Perhaps counter-intuitively, this has the effect of shifting some demand in π_0 (office hours), which is the less loaded band for residential consumption. The corresponding demand values, pre- and post- the introduction of the new tariff, are reported in Fig. 3 (left).

The amount of the shift is less significant than one may expect, for two reasons: first, the users tend to overestimate the demand variations due to the perception bias. Second, not all the users adopt the new tariff. The fraction of users in each class that make the switch is reported in Fig. 3 (right): the figure shows that the new tariff is more beneficial for larger families, while single persons (i.e. customer 1) tend to stick with the competitor tariff. The corresponding savings (w.r.t. the initial cost) go from a negligible 0.2 % for customer 1 to an 11 % for customer 5. The estimated provider profit is (around) 1.19 M€.

Since several of the parameters of our model have been estimated based on intuition, it is reasonable (and very interesting) to wonder what the effect of changing such parameters would be. As a first attempt in this direction, we have tried to *increase the elasticity of all customers* by multiplying all terms of the ϵ matrix by a factor of 1.5. This modification increases the rate of demand shifts, and it causes a proportional growth of the loss factor (i.e. the sum of the coefficients on each column): as a result, more significant changes of the net energy consumption are likely to occur.

The prices for the optimal tariff in this modified scenario are 447/194/390 K€/GWh. These prices favor an increase in the consumption for band π_1, which is evident in Fig. 4 (left). The fraction of customers in each class that make the switch is reported in Fig. 4 (left) and the corresponding savings (w.r.t. the initial cost) are 3 %/5 %/2 %/8 %/7 %. It is interesting to observe that the new prices strike a very different trade-off compared to the initial setup: in particular, the savings are more evenly spread among customer classes, and attracting customers

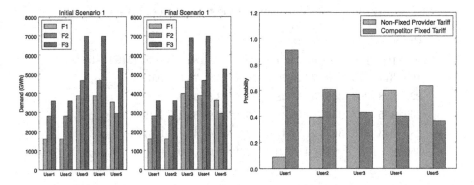

Fig. 3. (left) Demand values for scenario 1 before and after the introduction of the new tariff. (right) Fraction of customers switching to the new tariff, for each class.

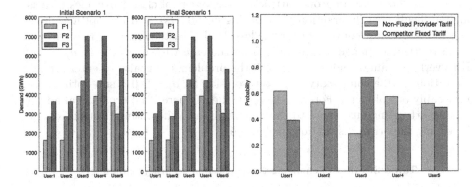

Fig. 4. (left) Demand values for scenario 1, with increased elasticity. (right) Fraction of customers adopting the new tariff, for each class.

of class 1 (single persons) has now become economically appealing. The provider profit is now larger, with a value of (around) 1.3 M€. Such general improvements are possible since the increased elasticity enables a reduction of the wholesale electricity price via demand shifting. In particular, the wholesale price in each a band is 36.8/36.9/21.1 K€/GWh, down from 44.4/49.0/28.9 K€/GWh in the original setup.

The value of the overhead o_j is a major component of the electricity cost for the provider and may have a strong impact on the profit margin. We have investigated the effect of *changing the value of the fixed overhead costs* from 250 K€/GWh down to 100 and up to 350. The results of this evaluation are reported in Table 2, which shows the overhead value (in K€/GWh), the prices of the optimized new tariff (in K€/GWh), the percentage of switching customers for each class, and the provider profit. We also report the status of the problem at the time limit.

Table 2. Effect of changing the fixed overhead costs

Prices				% switch						
o_j	p_0	p_1	p_2	κ_0	κ_1	κ_2	κ_3	κ_4	profit (M€)	status
100	371	273	238	76.2	76.7	76.8	75.9	76.5	10.64	opt
220	328	300	360	58.6	58.4	57.3	63.8	64.2	2.32	opt
250	155	437	436	8.9	39.3	56.8	59.9	63.4	1.19	opt
280	343	328	362	41.8	41.6	40.3	47.6	48.1	0.611	feas
350	–	–	–	0.0	0.0	0.0	0.0	0.0	0.0	opt

When the overhead is low, the profit margin is very high and it is possible to lower the prices to that point that the new tariff is competitive for all customer classes. As the overhead grows, attracting customers of class 1 becomes increasingly difficult (see the fraction of switching customers), and the optimal prices become those most convenient for the less flexible customers: the price for band π_0 in particular changes significanly from the case with $o_j = 100$ to $o_j = 250$. For overhead values slightly above 250 the problem seems also to enter a phase transition, with a complexity peak: our solution for $o_j = 280$ is sub-optimal and noticeably out of trend. For very large overhead, there is no economical advantage in attracting customers to the new tariff, and the optimal choice consists in leaving the original situation unchanged.

Table 3. Effect of changing the perception accuracy parameters

Prices					% switch						
o_j	β_k	p_0	p_1	p_2	κ_0	κ_1	κ_2	κ_3	κ_4	profit (M€)	status
250	0.75	155	437	436	8.9	39.3	56.8	59.9	63.4	1.19	opt
250	1.0	–	–	–	0.0	0.0	0.0	0.0	0.0	0.0	opt
250	1.25	–	–	–	0.0	0.0	0.0	0.0	0.0	0.0	opt
220	0.75	328	300	360	58.6	58.4	57.3	63.8	64.2	2.32	opt
220	1.0	–	–	–	0.0	0.0	0.0	0.0	0.0	0.0	opt
220	1.25	–	–	–	0.0	0.0	0.0	0.0	0.0	0.0	opt
100	0.75	371	273	238	76.2	76.7	76.8	75.9	76.5	10.64	opt
100	1.0	320	260	343	70.9	70.8	69.9	74.4	74.8	6.64	feas
100	1.25	72	211	705	77.4	74.9	62.4	74.9	72.4	6.87	opt

Finally, we have investigated the effect of *changing the perception accuracy parameters*. In particular, besides the initial setup with $\beta_k = 0.75$, we have performed experiments with $\beta_k = 1.0$ (i.e. a quite accurate perception, at the relevant range), and $\beta_k = 1.25$ (i.e. bias toward underestimated variations). The results are reported in Table 3. In both cases, our approach was no longer able to find an economically favorable tariff for the provider, i.e. the optimal decision was to leave all customers to the competitor. Apparently, customers that are less prone to over-estimation are more difficult to manage for an energy provider, either because they are better capable of exploiting dynamic prices (due to accurate perception) or they are likely to cause wholesale price increases (due to underestimated variations). We have performed a limited test of this conjecture by repeating our experiments with lower overhead costs, i.e. 220 and 100 K€/GWh. The results for $o_j = 220$ are similar to those for $o_j = 250$. For $o_j = 100$, tariffs that are beneficial for both the provider and the customers can be found even for $\beta_k = 1.0$ and $\beta_k = 1.25$, although their profit is considerably lower than the $\beta_k = 0.75$ case.

5 Concluding Remarks

In this work we have devised a non-linear optimization model for optimizing ToU based tariffs for electricity. The model is rather comprehensive and takes into account: (1) the presence of multiple tariffs and (most importantly) competitors, (2) the demand response behavior of the customers, (3) the effect of demand shifts on the wholesale energy price, and finally (4) cognitive aspects of the customers, in particular their risk aversion and the accuracy of their consumption estimates. The problem objective is to maximize the provider profit.

Our model can be employed directly by an energy utility to obtain tariff recommendations, or it can be used to assess the behavior of a provider and multiple customers in real or hypothetical market scenarios. This second, indirect, way of using our model may allow a policy maker to evaluate the effectiveness of energy policies via what-if analysis, or to prioritize policy goals (e.g. identify the most important parameters that should be changed).

In principle, it should be relatively easy to modify our model for at least two more relevant use cases. First, the model could be used to support the definition of tailored tariffs for public buildings, or large industrial customers: in this case, each κ_k would represent an individual customer rather than a class, the effect of demand shifts on the wholesale price should be disregarded, and finally the tariff choice should be deterministic rather than stochastic. This deterministic choice behavior could be modeled by formulating the KKT optimality conditions for a tariff selection subproblem.

A second alternative use case consists in employing the model to obtain recommendations about changes in the customer behavior parameters (e.g. elasticity, perception accuracy): this could be achieved by assuming that all tariffs are fixed, by turning the parameters for which we want recommendations into decision variables, and finally by adjusting the objective function. This alternative use case may be particularly appealing for policy makers.

We are aware that some components in our model could be improved. In particular, our perception model could be made more flexible and should be validated on real data. The elastic model that we employ for demand shifting could be augmented to take into account limits to the maximal acceptable variation, or to take into account the comfort loss/gain due changes in the net energy consumption. Obtaining a better characterization of the elasticity of the residential consumption may require to develop a feedback system to keep the customers informed about their energy usage, as proposed in [9]. We are also considering the possibility to split the yearly consumption into multiple months, which would allow one to model different types of tariffs and to (partially) take into account weather conditions. We are confident that we could provide support for a wider range of tariff schemes (e.g. discounted rates). We are actively working on some of these topics and some new results may be released in the coming months.

Acknowledgments. This work is partially supported by the EU FP7 project DAREED (g.a. 609082). We thank the anonymous reviewers for their comments.

References

1. Aalami, H.A., Moghaddam, M.P., Yousefi, G.R.: Modeling and prioritizing demand response programs in power markets. Electr. Power Syst. Res. **80**(4), 426–435 (2010)
2. Aalami, H.A., Moghaddam, M.P., Yousefi, G.R.: Demand response modeling considering interruptible/curtailable loads and capacity market programs. Appl. Energy **87**(1), 243–250 (2010)
3. Albadi, M.H., El-Saadany, E.F.: Demand response in electricity markets: an overview. In: Proceedings of IEEE Power Engineering Society General Meeting, pp. 1–5. IEEE (2007)
4. Albadi, M.H., El-Saadany, E.F.: A summary of demand response in electricity markets. Electr. Power Syst. Res. **78**(11), 1989–1996 (2008)
5. Attari, S.Z., DeKay, M.L., Davidson, C.I., De Bruin, W.B.: Public perceptions of energy consumption and savings. Proc. Nat. Acad. Sci. **107**(37), 16054–16059 (2010)
6. Baboli, P.T., Eghbal, M., Moghaddam, M.P., Aalami, H.: Customer behavior based demand response model. In: 2012 IEEE Proceedings of Power and Energy Society General Meeting, pp. 1–7. IEEE (2012)
7. Borenstein, S.: To what electricity price do consumers respond? residential demand elasticity under increasing-block pricing. Preliminary Draft, 30 April 2009
8. Conejo, A.J., Morales, J.M., Baringo, L.: Real-time demand response model. IEEE Trans. Smart Grid **1**(3), 236–242 (2010)
9. Jessoe, K., Rapson, D.: Knowledge is (less) power: experimental evidence from residential energy use. Working Paper 18344, National Bureau of Economic Research, August 2012. http://www.nber.org/papers/w18344
10. Kirschen, D.S., Strbac, G., Cumperayot, P., de Paiva Mendes, D.: Factoring the elasticity of demand in electricity prices. IEEE Trans. Power Syst. **15**(2), 612–617 (2000)
11. Koichiro, I.: Do consumers respond to marginal or average price? evidence from nonlinear electricity pricing. Am. Econ. Rev. **104**(2), 537–563 (2014)
12. Palensky, P., Dietrich, D.: Demand side management: demand response. Intell. Energ. Syst. Smart Loads **7**(3), 381–388 (2011)
13. Scalari, S.: Personal Communication (2015)

14. Tanaka, R., Schmidt, M., Ahlund, C., Takamatsu, Y.: An energy awareness study in a smart city lessons learned. In: IEEE Ninth International Conference on Proceedings of Intelligent Sensors, Sensor Networks and Information Processing (ISSNIP), pp. 1–4. IEEE (2014)
15. Zheng, R., Xu, Y., Chakraborty, N., Sycara, K.: Crowdfunding investment for renewable energy. In: Proceedings of the International Conference on Autonomous Agents and Multiagent Systems, pp. 1751–1752. International Foundation for Autonomous Agents and Multiagent Systems (2015)

Weighted Spanning Tree Constraint with Explanations

Diego de Uña[1]([⊠]), Graeme Gange[1], Peter Schachte[1], and Peter J. Stuckey[1,2]

[1] Department of Computing and Information Systems,
The University of Melbourne, Melbourne, Australia
d.deunagomez@student.unimelb.edu.au,
{gkgange,schachte,pstuckey}@unimelb.edu.au
[2] Victoria Laboratory, National ICT Australia, Melbourne, Australia

Abstract. Minimum Spanning Trees (MSTs) are ubiquitous in optimization problems in networks. Even though fast algorithms exist to solve the MST problem, real world applications are usually subject to constraints that do not let us apply such methods directly. In these cases we confront a version of the MST called the "Weighted Spanning Tree" (WST) in which we look for a spanning tree in a graph that satisfies other side constraints and is of minimum cost. In this paper we implement this constraint using a lower bound and learning to accelerate the search and thus reduce the solving time. We show that having this propagator is tremendously beneficial for solvers and we show the benefits of learning.

1 Introduction

Given a connected weighted graph $G = (N, E)$, the Minimum Spanning Tree (MST) T of G is a connected acyclic sub-graph of G that contains all the nodes in N and is of minimum weight. Finding the MST of a graph can be done using Kruskal's algorithm (among others) which is $\mathcal{O}(|N|log(|E|))$. Nevertheless, many interesting variants of the MST are NP-hard. In these variants, there are side constraints that make these algorithms unusable.

Some examples where side constraints make the MST problem NP-hard are the capacitated MST [5,17], the degree-constrained MST [14], the min-degree MST [2], the constrained MST [19], or the diameter-constrained MST [1]. These and other variants can be found in the real world. For instance, cable layout for offshore wind farms [13] combines the capacitated MST, the degree-constrained MST and an extra constraint disallowing cable crossing.

In Constraint Programming (CP), the Weighted Spanning Tree (WST) constraint is defined as follows: given a graph $G = (N, E)$ and a weight function ws that maps every edge $e \in E$ to an integer called the *weight* of e, find a tree T that is a subgraph of G, spans all nodes in N and is of cost at most w (w being an integer variable). The decisions made by this constraint are Boolean variables c_e representing for each edge $e \in E$ whether it is chosen to be part of T or not. Let $B = \{c_e | e \in E\}$, we write the constraint $wst(N, E, ws, B, w)$. Because this is

© Springer International Publishing Switzerland 2016
C.-G. Quimper (Ed.): CPAIOR 2016, LNCS 9676, pp. 98–107, 2016.
DOI: 10.1007/978-3-319-33954-2_8

a constraint, it can be used in combination with other constraints and therefore applied to the above stated optimization problems.

The first appearance of this constraint was in [9] (called "Not-too heavy" spanning tree). Their work was followed up in [21] with a simpler algorithm for propagation maintaining the same strength for the propagator. They re-named the constraint "WST", which is the term that we will use. Here the propagation was proved to be arc-consistent. Later, in [22], the *ccTree* data-structure was improved to decrease the complexity of their algorithms. Similar work was done in [8], although this constraint forced to solution to be a *minimum* spanning tree. The contribution of these papers are the filtering algorithms they provide, but no implementation or experiments are reported. Nonetheless, in Constraint Programming, constant factors in the complexity are crucial and the asymptotic complexity of their algorithms gives only partial information on performance. Also, no previous work explored the use of explanations in this useful constraint.

In this paper we present our implementation of the WST constraint in the CP solver CHUFFED [7]. We use learning [16] to accelerate the search. We show that the explanations on this global constraint are tremendously beneficial in practice. We compare our implementation to the one available in the CHOCO3 CP solver [18] and show the benefits of learning.

We illustrate the use of this constraint on the Diameter-Constrained MST (DCMST) problem, because it has been recently addressed in Constraint Programming by [15] and has a large number of applications in wireless network routing [3], telecommunications [26], distributed mutual exclusion in computer networks [20] and data compression [4]. For this problem there has been work on both approximation and exact algorithms. In approximations, [11] presented an approach using Variable Neighbourhood Search, followed by another heuristic approach [12]. For exact solutions [23] presented a Mixed Integer Programming formulation of the problem that was later improved in [10]. The latest exact algorithm was presented by [15] using CP and it outperforms all other approaches known to the authors. Our approach to the DCMST is also CP, so it is only comparable to the last one. Nevertheless, the solver they used is not the same as ours, and thus comparisons (especially in time) should be considered with care.

Section 2 briefly introduces Lazy Clause Generation. Section 3 describes our algorithms and implementation of the WST constraint, including the computation of explanations. Section 4 summarizes our experimental results on the DCMST.

2   Lazy Clause Generation

Lazy Clause Generation (LCG, [16]) is a technique by which CP solvers can learn from what they have explored in the search space. Constraints can be transformed into a number of clauses over Booleans, and this is typically how SAT solvers work. The idea of LCG is to make propagators generate these clauses on the fly when they propagate. These clauses capture the reason for propagation and thus we say they "explain" propagation. These *explanations* are then given to the solver that uses them as a way of remembering what propagators inferred. This way, we run a relatively expensive algorithm once, do an inference and remember it for the rest of the solving process. On the other hand, propagators might need to do some extra computation to compute the explanations.

3   WST Propagator with Explanations

In our problem, the decision variables are the edges: which edges form the tree and which edges do not. We say that an edge is *mandatory* if it has been set to be part of the tree. We call *forbidden* the edges that have been set to not be part of the tree. Other edges are *undecided*. Let M be the set of mandatory and F the set of forbidden edges.

Here we present our WST propagator with explanations. We first introduce a novel lower bound with explanations followed by a propagation rule (which was already introduced in [21]).

We define a *substitute edge* of an edge e in a spanning tree $T = (N, T_E)$ as any edge e_s such that $(N, T_E \backslash \{e\} \cup \{e_s\})$ is a spanning tree. Also, following the definition of [21], given a tree T and a non-tree edge $e = (i, j)$, let e' be the edge of maximum cost in the path from i to j in T. Then e' is called the *support* of e.

3.1   Lower Bound with Explanations

Assume we are looking for a solution of cost w lower than K. When we branch (i.e. we make a decision) we can compute a lower bound of the problem that will tell us if a better solution can exist in this branch. If that's not the case, we can stop the search. This is known as branch-and-bound.

The most accurate lower bound for w in the WST propagator is naturally the MST of the graph given the decisions so far. That is, the tree $T^* = (N, E^*)$ of minimum weight W_{T^*} such that $M \subseteq E^*$ and $E^* \cap F = \emptyset$.

It is easy to see that applying Kruskal's algorithm where the edges in M have been pre-added and the edges in F are not used yields T^*.

Now, if $W_{T^*} \geq K$ then no solution of cost lower than K exists in the current search space, and we can cut the search. A trivially correct explanation is $\bigwedge_{e \in F} \neg c_e \wedge \bigwedge_{e \in M} c_e \Rightarrow W_{T^*} \geq K$, but it is possible to build a better explanation.

Let F_c be the set of forbidden edges e_F such that $T^* \cup \{e_F\}$ forms a cycle where e_F is not the most expensive edge. Let M_S be the set of edges $e \in M$ having some substitute e' such that $ws(e') < ws(e)$. Let S_S be a mapping $M_S \mapsto E$ from each edge in M_S to the substitute of minimum weight for that edge. We then select a subset $M_H \subseteq M_S$ such that the inequality $\sum_{e \in M_H} (ws[S_S[e]] - ws[e]) + W_{T^*} \geq K$ holds. Note that multiple such sets M_H may exist.

A better explanation is given by Theorem 1.

Theorem 1. *A correct explanation for the failure of* $wst(N, E, ws, B, w)$ *is:*

$$\bigwedge_{e \in F_c} \neg c_e \wedge \bigwedge_{e \in M \backslash M_H} c_e \Rightarrow W_{T^*} \geq K$$

Proof. **Forbidden Edges:** Clearly, $F_c \subseteq F$. Let $e = (u, v) \in F \backslash F_c$. By definition of F_c, e is the most expensive edge in the cycle formed by $T^* \cup \{e\}$. Because the queue in Kruskal's algorithm is sorted in increasing order, the path P between u and v in T^* was already built before considering e. Therefore, whether e is forbidden or not does not affect the cost of P and consequently does not affect W_{T^*} and the explanation $\bigwedge_{e \in F_c} \neg c_e \wedge \bigwedge_{e \in M} c_e \Rightarrow W_{T^*} \geq K$ holds.

Mandatory Edges: By construction, M_H is a set of edges that, when removed and substituted by the best possible edge available, the cost of the tree is still higher than K. Therefore, the edges in M_H do not need to be in the explanation for it to hold. $\qquad \square$

Note that because several sets M_H may exist, different explanations can be computed. Evaluating which explanation is better than another is highly dependent on the instance of the problem. We ran different tests and could not determine a way of choosing M_H that dominated others in all cases. In our final implementation we start by putting the cheapest edges in M_H.

The algorithm to detect failure and compute the explanation is Algorithm 1. To construct explanations, we use the Rerooted-Union-Find data structure described in [25]. This is a modification of a classic Union-Find that allows the user to retrieve paths between nodes. Lines 7 to 9 pre-add all the mandatory edges. Lines 10 to 20 follow the classic Kruskal's algorithm with some modifications. Lines 12 and 17 add to the explanation any forbidden edge that should have been used. Lines 14 and 15 compute the cheapest substitute for each mandatory edges (if any). Once the tree T^* is computed, we build M_H in lines 22 and 23, leaving all the other mandatory edges (that have substitutes) in the explanation in line 25. The final explanation for $W_{T^*} \geq K$ is the set X. The complexity of the algorithm is $\mathcal{O}(|E|(|N| + log(|E|)))$.

The same explanations can be used for failure if $cost > K$:

$$\bigwedge_{e \in F_c} \neg c_e \wedge \bigwedge_{e \in M \backslash M_H} c_e \wedge [\![w < K]\!] \Rightarrow \mathit{false}$$

In the example of Fig. 1, we are looking for a solution of cost less than $K = 27$. W_{T^*} is 31, so we must fail. When we consider edge e_1, the fact that it is mandatory causes no trouble, as there is no other substitute to this edge that would connect h. When we consider e_4, we must skip it because it is forbidden, which means that we will use a more expensive edge to reach c (here e_8). When considering e_6, e and g are already connected by a path containing the mandatory edges e_7, e_9 and e_{10} and the undecided edge e_3. Therefore e_6 is the substitute of all of them. We later compute that: $W_{T^*} - ws[e_7] + ws[e_6] = 30 > K$, then $30 - ws[e_9] + ws[e_6] = 28 > K$ and lastly $28 - ws[e_{10}] + ws[e_6] = 21 < K$. Therefore, the explanation will be $\neg c_{e_4} \wedge c_{e_{10}} \wedge [\![w < K]\!] \Rightarrow \mathit{false}$.

3.2 Propagation Rule with Explanations

We use the propagation rule exposed in Proposition 3 of [21], that is: given the best possible tree T^* and an upper bound for the solution K such that $W_{T^*} < K$,

Algorithm 1. Computing the lower bound with explanation.

```
 1: procedure MANDATORY_KRUSKAL(G = (N, E), M, F, K)
 2:     Q ← sort(E)
 3:     uf ← RerootedUF()
 4:     c ← 0,  cost ← 0
 5:     X ← ∅, sub ← array(|E|, nil)
 6:     for all e = (u, v) ∈ M do                              ▷ Pre-add mandatory edges
 7:         uf.unite(u, v)
 8:         c ← c + 1; cost ← cost + ws[e]
 9:     for all e = (u, v) ∈ Q do                              ▷ (in order)
10:         if ¬uf.connected(u, v) ∧ e ∈ F then
11:             X ← X ∪ {¬c_e}                                 ▷ Should add e, but it is forbidden
12:         else if uf.connected(u, v) then
13:             for all ep ∈ uf.path(u, v) do
14:                 sub[ep] = min_w(sub[ep], e)
15:                 if ws[ep] > ws[e] ∧ e ∈ F then
16:                     X ← X ∪ {¬c_e}                         ▷ e would be cheaper
17:         else if c < |N| − 1 ∧ ¬uf.connected(u, v) then
18:             uf.unite(u, v)
19:             c ← c + 1; cost ← cost + ws[e]
20:     for all e ∈ M ∧ sub[e] ≠ nil do
21:         if cost − ws[e] + ws[sub[e]] ≥ K then
22:             cost ← cost − ws[e] + ws[sub[e]]              ▷ e ∈ M_H
23:         else if ws[sub[e]] ≠ ws[e] then
24:             X ← X ∪ {c_e}                                  ▷ e ∉ M_H
25:     return [X ⇒ [w ≥ cost]]
```

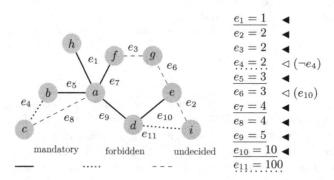

Fig. 1. Example of a graph during solving. The weights are indicated on the right. Symbol '◀' indicates an edge that was used in the solution, whereas '◁' indicates edges that should have been used (accompanied by the explanation).

for any non-tree edge e^* of support e', e^* can be part of the solution if and only if $W_{T^*} - ws[e'] + ws[e^*] < K$. That is, e^* is a valid substitute of e'. If this is not the case, we must remove e^* from the possible edges since using it would increase the weight of T^* above the upper bound K. It is easy to see that the previous explanation applies as well in this case. Let $M'_H = ((M \backslash M_H) \backslash \{e'\}) \cup \{e^*\}$.

$$\bigwedge_{e \in F_c} \neg c_e \wedge \bigwedge_{e \in M'_H} c_e \wedge [\![w < K]\!] \Rightarrow false$$

$$\Leftrightarrow \bigwedge_{e \in F_c} \neg c_e \wedge \bigwedge_{e \in M'_H \backslash e^*} c_e \wedge [\![w < K]\!] \Rightarrow \neg e^*$$

We execute this rule after the previously described algorithm in the case where no failure is detected.

4 The Diameter Constrained Minimum Spanning Tree

The DCMST is formally defined as follows: given a graph $G = (N_G, E_G)$ find a sub-graph $T = (N_T, E_T)$ of G such that T is a tree, $N_T = N_G$ and the longest distance between any two nodes in T is at most D (called the *diameter* of T). The distance between two nodes u and v is the number of nodes in the path from u to v.

4.1 Modeling DCMST

This problem is separated in two cases whether D is even or odd. If D is even, then there exists a node r that is the root of T and the height of all the other nodes has to be at most $\lfloor D/2 \rfloor$. If D is odd, there exists an edge $e = (a, b)$ that acts as the root of the tree (e is therefore in the tree), meaning that the height of a and b is zero and all the other nodes must have at most height $\lfloor D/2 \rfloor$. Notice that r, a and b are not given in the input: these are variables.

We used the same model as [15] with the only addition of our propagator. The matrix *adj* gives for each node the set of neighbour nodes. For the DCMST-specific constraints, we use an array of heights of nodes h, and an array of parenthood of nodes p. Two variables a and b are the end-nodes of the edge that acts as root in the odd case, or are both the root in the even case (in that case, $a = b$). The model is *minimize*(w) such that:

$$wst(N, E, ws, B, w) \tag{1}$$
$$D \bmod 2 = 0 \Leftrightarrow a = b \tag{2}$$
$$(h[a] = 0 \wedge p[a] = b) \wedge (h[b] = 0 \wedge p[b] = a) \tag{3}$$
$$\forall n \in N \backslash \{a, b\}, \; h[n] = h[p[n]] + 1 \tag{4}$$
$$\forall n \in N \backslash \{a, b\}, \; p[n] \in adj[n] \tag{5}$$
$$\forall e = (u, v) \in E, \; c_e \Leftrightarrow p[u] = v \vee p[v] = u \tag{6}$$

Constraint 2 states that in the even case a and b are the same node (the root r). Constraint 3 forces a and b to be at height 0 and be each others parents. Constraint 4 makes every node (other than the root(s)) be one level below its parent. Constraint 5 forces each node to chose a parent that is adjacent to it. Constraint 6 links the edge variables of the graph with the parenthood relations.

Although our main intention is to compare the improvement that the WST propagator and explanations bring to the solver, we also compare our work to [15] (we name their results "NRS") as their results are the state of the art in DCMST as far as we can tell. They used a Pentium 4, 2.8 GHz and 2 GB of RAM.

For better comparison, we implemented the exact same search strategy they describe in their paper ([15] Sect. 3, Fig. 2). First, for each node n we compute the sum s_n of the shortest paths from n to any other node. Then, we associate to each pair of nodes (a, b) the minimum of s_a and s_b, noted $s_{(a,b)} = min(s_a, s_b)$. The search is as follows. Start by taking each pair of nodes (a, b) in increasing order of $s_{(a,b)}$. Then for each possible value of the height (from 1 to $\lfloor D/2 \rfloor$), remove that value from the domain of all the nodes (when possible) taking the nodes in decreasing order of the shortest path to either a or b. Here "shortest path" is in weight of the edges.

They use a dominance rule in the search, which we converted into a dominance-breaking constraint [6] in our model, for ease of implementation: $\forall \{e_1 = (u, v), e_2 = (u, y)\} \in E^2$, $ws[e_1] < ws[e_2] \wedge h[v] \leq h[y] \Rightarrow p[u] \neq y$. This states that if it is cheaper to connect u to v than to y and the height of v is lower than the height of y, we can connect u to v with a lower cost. This is because if using e_2 does not violate the diameter constraint, neither does e_1.

4.2 Experimental Results

We run our experiments on a Linux 3.16 Intel® Core™ i7-4770 CPU @ 3.40 GHz, 15.6 GB of RAM. We used 5 min as the time limit. The results from NRS are extracted from their paper where they used the solver IBM OPL. Benchmarks can be found in [24]. We give different versions in CHUFFED: NOPROP uses learning but does not use our propagator, NOEXPL uses our propagator without learning, EXPL uses the propagator with explanations from Sect. 4, and NAIVEEXPL uses our propagator with naive explanations (i.e. all fixed elements in the graph are in the explanation). All use the same strategy.

As we can see in Table 1, the use of the propagator is absolutely beneficial. The total time is improved by 48.02 % when using the propagator without explanations against no propagator. Furthermore, our version with explanation (EXPL) is 90.5 % faster than the version without explanations (NOEXPL) and 95.1 % faster than the version with no propagator at all. Also, our total time is 36.6 % shorter than NRS. Most of the tests with NOPROP and CHOCO3 got to the optimal solution, but timed-out when proving optimality. This also illustrates the need for this propagator.

In CP, the number of nodes represents the size of the search space explored before proving optimality. Here we see an obvious dominance of EXPL as it almost always has less nodes than other versions. It also has an improvement on the total number of nodes for all benchmarks of 99.4 % over NOPROP and 99.0 % over NOEXPL. Additionally, it has an improvement of 96 % over NRS.

Table 1. Comparison in time (seconds) and nodes for the DCMST models.

Instance			NoProp		NoExpl		NaiveExpl		Expl		Choco3		NRS (IBM OPL)	
$|N|$	$|E|$	D	Nodes	Time	Nodes	Time	Nodes	Time	Nodes	Time	Nodes	Time	Nodes	Time
15	105	4	6825	0.45	784	0.23	448	0.2	**447**	0.19	6256	1.92	1044	**0.08**
15	105	5	35322	2.28	1921	0.39	1003	0.39	**1001**	0.38	301269	44.36	2850	**0.22**
15	105	6	133259	10.31	5997	0.63	2235	0.48	**2101**	0.45	160445	37.06	6960	**0.28**
15	105	7	258317	22.91	5873	0.54	2312	0.41	**2221**	0.41	2182510	300	8240	**0.38**
15	105	9	493166	39.80	6049	0.47	1968	0.21	**1731**	**0.19**	2623006	300	11743	0.47
15	105	10	550536	40.93	24259	1.95	2872	0.39	**2831**	**0.29**	898948	156.63	11830	0.41
20	190	4	192965	20.39	2651	1.86	**1261**	1.70	1266	1.69	200651	82.51	3143	**0.20**
20	190	5	1869837	186.76	9387	4.85	4452	4.56	**4432**	4.48	2064050	300	18283	**1.06**
20	190	6	2585912	300	49673	14.97	9018	4.55	**8462**	4.08	862115	300	35383	**2.03**
20	190	7	2661381	300	16690	4.67	**4252**	2.04	4288	1.97	1857850	300	19142	**0.97**
20	190	9	2433234	300	157236	34.43	6336	1.71	**5972**	**1.60**	1738525	300	119906	5.01
20	190	10	2628419	300	315618	52.61	9050	3.96	**8645**	**3.56**	1067170	300	151969	6.08
25	300	4	1898689	300	20202	58.53	**6166**	17.10	6217	17.1	592738	300	28842	**1.48**
25	300	5	2415919	300	86662	93.57	32787	80.62	**26547**	68.88	1553235	300	37608	**2.83**
25	300	6	2262702	300	402861	300	16150	17.66	**15147**	**15.99**	847691	300	534222	39.14
25	300	7	2045173	300	449104	210.13	76272	87.72	**61098**	63.68	1448142	300	812957	**56.06**
25	300	9	1929801	300	462886	300	21195	18.66	**19724**	**17.43**	1270399	300	2655810	114.14
25	300	10	1961836	300	620555	261.71	**21453**	11.86	21565	**11.50**	586552	300	1126130	55.47
20	50	4	14548	0.48	1219	**0.05**	558	0.05	558	**0.05**	4489	0.61	**389**	**0.05**
20	50	5	55748	2.58	307392	10.87	2258	0.26	**2227**	0.24	426762	40.91	3611	**0.17**
20	50	6	52217	2.34	68384	0.75	1574	0.10	**1475**	**0.08**	41892	5.28	2678	0.13
20	50	7	66676	3.45	25043	0.68	1381	0.12	**1238**	**0.09**	1389117	133	1975	0.14
20	50	9	274583	16.59	14016	0.33	**1117**	0.06	1261	**0.06**	3820792	300	13040	0.45
20	50	10	310688	18.93	**410**	**0.01**	564	0.03	564	0.02	329333	42.87	17937	0.64
40	100	4	3426079	300	45199	6.27	**13766**	4.41	13901	**4.30**	1180714	300	130480	5.44
40	100	5	3261615	300	9596291	300	36496	14.22	**26970**	9.69	3196955	300	161961	**7.31**
40	100	6	4836734	300	8161773	300	10708	2.59	**5037**	**0.84**	1851687	300	91022	4.72
40	100	7	4709441	300	5979528	300	38153	11.03	**18504**	**4.49**	2989047	300	778669	34.38
40	100	9	3646022	300	4468371	300	88837	25.12	**36572**	**7.08**	2873734	300	769161	40.16
Total			47017644	4868.2	31306034	2530.50	414642	312.21	**302002**	**240.81**	38365804	6245.15	7556985	379.84

The comparison between Expl and NaiveExpl shows that computing our explanations is worthwhile. The NaiveExpl uses the same algorithms as described throughout this paper, only the explanations contain all the fixed c_n and c_e. This makes the explanations more strict and thus less reusable. As we would expect, this makes the explanations much longer: the average length in the explanations for NaiveExpl is 128.88 literals, whereas the length of our explanations is 73.18 literals in average. We see the consequences of this in the Table 1: naive explanations most often slow down the solving step. The version Expl is 22.9 % faster and has 27.2 % less nodes.

We observe that our propagator dominates specially when the diameter is big. This is because in that case, the lower bound is more accurate as it violates fewer diameter constraints. When the diameter is small, Algorithm 1 is not aware of it and just computes an MST thus rapidly violating the diameter constraints.

5   Conclusion

In this paper we have given an efficient algorithm to compute explanations for the lower bound for the WST constraint, and we have implemented an already existing propagation rule in our solver. Our major contribution is the computation of explanations that, as we can see in the experiments, are absolutely beneficial to solve large instances of optimization problems on spanning tree.

Acknowledgement. Diego de Uña thanks "la Caixa" Foundation for partially funding his Ph.D. studies at The University of Melbourne.

References

1. Achuthan, N., Caccetta, L., Caccetta, P., Geelen, J.: Algorithms for the minimum weight spanning tree with bounded diameter problem. Optim. Tech. Appl. **1**(2), 297–304 (1992)
2. Akgün, İ., Tansel, B.Ç.: Min-degree constrained minimum spanning tree problem: new formulation via miller-tucker-zemlin constraints. Comput. Oper. Res. **37**(1), 72–82 (2010)
3. Bala, K., Petropoulos, K., Stern, T.E.: Multicasting in a linear lightwave network. In: Proceedings of the Twelfth Annual Joint Conference of the IEEE Computer and Communications Societies, Networking: Foundation for the Future, IEEE, INFOCOM 1993, pp. 1350–1358. IEEE (1993)
4. Bookstein, A., Klein, S.T.: Compression of correlated bit-vectors. Inf. Syst. **16**(4), 387–400 (1991)
5. Chandy, K.M., Lo, T.: The capacitated minimum spanning tree. Networks **3**(2), 173–181 (1973)
6. Chu, G., Stuckey, P.J.: Dominance breaking constraints. Constraints **20**(2), 155–182 (2015)
7. Chu, G.G.: Improving combinatorial optimization. Ph.D. thesis, The University of Melbourne (2011)
8. Dooms, G., Katriel, I.: The *minimum spanning tree* constraint. In: Benhamou, F. (ed.) CP 2006. LNCS, vol. 4204, pp. 152–166. Springer, Heidelberg (2006)
9. Dooms, G., Katriel, I.: The "not-too-heavy spanning tree" constraint. In: Van Hentenryck, P., Wolsey, L.A. (eds.) CPAIOR 2007. LNCS, vol. 4510, pp. 59–70. Springer, Heidelberg (2007)
10. Gruber, M., Raidl, G.R.: A new 0–1 ILP approach for the bounded diameter minimum spanning tree problem. In: Gouveia, L., Mourão, C. (eds.) Proceedings of the 2nd International Network Optimization Conference 2005, Lisbon, Portugal, vol. 1, pp. 178–185 (2005). https://www.ac.tuwien.ac.at/files/pub/gruber-05.pdf
11. Gruber, M., Raidl, G.R.: Variable neighborhood search for the bounded diameter minimum spanning tree problem. In: Hansen, P., Mladenovi, N., Pérez, J.A.M., Batista, B.M., Moreno-Vega, J.M. (eds.) Proceedings of the 18th MiniEuro Conference on Variable Neighborhood Search, Tenerife, Spain (2005). https://www.ac.tuwien.ac.at/files/pub/gruber-05a.pdf
12. Gruber, M., Raidl, G.R.: (Meta-)Heuristic separation of jump cuts in a branch & cut approach for the bounded diameter minimum spanning tree problem. In: Maniezzo, V., Stützle, T., Voß, S. (eds.) Hybridizing Metaheuristics and Mathematical Programming. Annals of Information Systems, vol. 10, pp. 209–229. Springer, Heidelberg (2010)
13. Klein, A., Haugland, D., Bauer, J., Mommer, M.: An integer programming model for branching cable layouts in offshore wind farms. In: An Le Thi, H., Dinh, T.P., Nguyen, N.T. (eds.) Modelling, Computation and Optimization in Information Systems and Management Sciences, pp. 27–36. Springer, Switzerland (2015)
14. Narula, S.C., Ho, C.A.: Degree-constrained minimum spanning tree. Comput. Oper. Res. **7**(4), 239–249 (1980)

15. Noronha, T.F., Ribeiro, C.C., Santos, A.C.: Solving diameter-constrained minimum spanning tree problems by constraint programming. Int. Trans. Oper. Res. **17**(5), 653–665 (2010)
16. Ohrimenko, O., Stuckey, P., Codish, M.: Propagation via lazy clause generation. Constraints **14**(3), 357–391 (2009). http://dx.doi.org/10.1007/s10601-008-9064-x
17. Papadimitriou, C.H., Vazirani, U.V.: On two geometric problems related to the travelling salesman problem. J. Algorithms **5**(2), 231–246 (1984)
18. Prud'homme, C., Fages, J.G., Lorca, X.: Choco3 Documentation. TASC, INRI-ARennes, LINA CNRS UMR 6241, COSLING S.A.S (2014). http://www.choco-solver.org
19. Ravi, R., Goemans, M.: The constrained minimum spanning tree problem. Algorithm Theory SWAT 1996, pp. 66–75 (1996)
20. Raymond, K.: A tree-based algorithm for distributed mutual exclusion. ACM Trans. Comput. Syst. (TOCS) **7**(1), 61–77 (1989)
21. Régin, J.-C.: Simpler and incremental consistency checking and arc consistency filtering algorithms for the weighted spanning tree constraint. In: Trick, M.A. (ed.) CPAIOR 2008. LNCS, vol. 5015, pp. 233–247. Springer, Heidelberg (2008)
22. Régin, J.-C., Rousseau, L.-M., Rueher, M., van Hoeve, W.-J.: The weighted spanning tree constraint revisited. In: Lodi, A., Milano, M., Toth, P. (eds.) CPAIOR 2010. LNCS, vol. 6140, pp. 287–291. Springer, Heidelberg (2010)
23. dos Santos, A.C., Lucena, A., Ribeiro, C.C.: Solving diameter constrained minimum spanning tree problems in dense graphs. In: Ribeiro, C.C., Martins, S.L. (eds.) WEA 2004. LNCS, vol. 3059, pp. 458–467. Springer, Heidelberg (2004)
24. de Uña, D.: Weighted spanning tree benchmarks (2015). http://people.eng.unimelb.edu.au/pstuckey/wst/
25. de Uña, D., Gange, G., Schachte, P., Stuckey, P.J.: Steiner tree problems with side constraints using constraint programming. In: Proceedings of the Thertieth AAAI Conference on Artificial Intelligence. AAAI Press (2016)
26. Wang, S., Lang, S.: A tree-based distributed algorithm for the k-entry critical section problem. In: International Conference on Parallel and Distributed Systems, 1994, pp. 592–597. IEEE (1994)

Forward-Checking Filtering for Nested Cardinality Constraints: Application to an Energy Cost-Aware Production Planning Problem for Tissue Manufacturing

Cyrille Dejemeppe[1]([✉]), Olivier Devolder[2], Victor Lecomte[1], and Pierre Schaus[1]

[1] ICTEAM, UCLouvain, Louvain-la-neuve, Belgium
{cyrille.dejemeppe,pierre.schaus}@uclouvain.be,
victor.lecomte@student.uclouvain.be
[2] N-SIDE, Louvain-la-neuve, Belgium
ode@n-side.com

Abstract. Response to electricity price fluctuations becomes increasingly important for industries with high energy demands. Consumer tissue manufacturing (toilet paper, kitchen rolls, facial tissues) is such an industry. Its production process is flexible enough to leverage partial planning reorganization allowing to reduce electricity consumption. The idea is to shift the production of the tissues (rolls) requiring more energy when electricity prices (forecasts) are lower. As production plans are subject to many constraints, not every reorganization is possible. An important constraint is the order book that translates into hard production deadlines. A Constraint Programming (CP) model to enforce the due dates can be encoded with p Global Cardinality Constraints (GCC); one for each of the p prefixes of the production variable array. This decomposition into separate GCC's hinders propagation and should rather be modeled using the global nested_gcc constraint introduced by Zanarini and Pesant. Unfortunately it is well known that the GAC propagation does not always pay off in practice for cardinality constraints when compared to lighter Forward-Checking (FWC) algorithms. We introduce a preprocessing step to tighten the cardinality bounds of the GCC's potentially strengthening the pruning of the individual FWC filterings. We further improve the FWC propagation procedure with a global algorithm reducing the amortized computation cost to $\mathcal{O}(log(p))$ instead of $\mathcal{O}(p)$. We describe an energy cost-aware CP model for tissue manufacturing production planning including the nested_gcc. Our experiments on real historical data illustrates the scalability of the approach using a Large Neighborhood Search (LNS).

1 Introduction

The share of renewable energy production, such as wind or solar power is growing fast in several countries of the EU [19]. While the production of nuclear and fossil energy tends to be stable, renewable energy production is highly dependent

© Springer International Publishing Switzerland 2016
C.-G. Quimper (Ed.): CPAIOR 2016, LNCS 9676, pp. 108–124, 2016.
DOI: 10.1007/978-3-319-33954-2_9

of both climatic conditions and time of the day considered. Renewable resources add a huge variability on the offer and demand, and thus on the price of electricity. As an example, Fig. 1 shows the historical electricity prices in Europe on March 3rd, 2014. In this example, the electricity prices fluctuate with a multiplicative factor higher than 3.5. Performing activities requiring more energy when electricity price is low represents both an economical and ecological advantage (the energy produced is not "wasted").

In [14], Simonis and Hadzic propose a cumulative constraint that links the energy consumption of activities with evolving electricity prices. We believe this kind of energy-aware optimization will become increasingly present in the industries with order-driven production planning that can be easily split into different steps. It generally offers enough flexibility to reduce the energy costs by scheduling tasks requiring more energy when the electricity price is lower. This paper addresses the problem of energy-efficient scheduling in consumer tissue production planning. Consumer tissue production planning offers several levers of flexibility, allowing to drastically reduce the energy costs for a given set of orders. Indeed, the paper machine receiving paper pulp as input and producing paper rolls consumes an amount of energy that depends on the tissue properties (quality, density of fibers, thickness, etc.). Therefore, our CP model attempts to schedule the production of paper rolls requiring more energy when electricity price forecasts are lower. The order book limits the flexibility and is modeled using a nested_gcc [20]. A flow based GAC filtering for this constraint is proposed in [20]. This paper introduces a light filtering algorithm particularly well suited to tackle large instances in a Large Neighborhood Search (LNS) requiring fast restarts. A preprocessing step to tighten the initial cardinality bounds allows to obtain additional pruning compared to a naive decomposition with Forward Checking (FWC) GCCs. Furthermore, we propose a general refined FWC propagation procedure allowing to reduce its amortized time complexity from $\mathcal{O}(p)$ with the decomposition into multiple GCCs to $\mathcal{O}(\log(p))$.

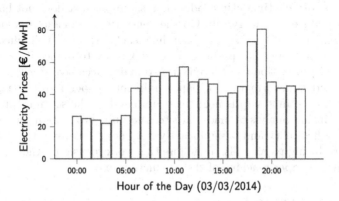

Fig. 1. Historical evolution of electricity prices on the EU market on March 3rd, 2014.

In Sect. 2, we describe the consumer tissue manufacturing problem. Then, in Sect. 3, we propose a CP model to solve this industrial problem. Section 4 describes the preprocessing step to obtain tighter cardinality bounds as well as our own FWC propagation algorithm for the nested_gcc. Finally, Sect. 5 explains the results obtained on real historical data with our model.

2   Paper Production Planning

An important industrial site in Belgium manufactures hygienic paper (toilet paper and facial tissues are examples of refined paper they produce). Paper rolls are produced before being converted into different products (e.g. toilet papers or kitchen rolls). The production is a two step process: paper roll production, then conversion of paper rolls into final products. In Fig. 2, we give a schematic overview of the different steps in the production of paper on the industrial site considered. The energy consumption can vary up to 15% depending on the type of roll produced. Therefore the company is looking for the less expensive production planning given the electricity price forecasts.

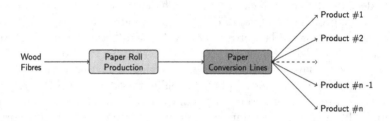

Fig. 2. Production steps in paper industry.

The potential savings depends on the flexibility of the production site. For example, a factory continuously producing a same product does not have much potential to reduce its energy bill. On the contrary, a manufacturer producing many different products on a production line, each requiring a significantly different amount of energy has probably more flexibility to reduce its energy bill. The paper roll production can be split in two main successive steps: paper pulp production and transformation of paper pulp into paper rolls. The potential energy gain on the first step is negligible compared to the second step. Indeed, as reported during our visits made on site, the pulp production part does not contain much flexibility and is significantly less energy-intensive than the paper machines producing paper rolls. Therefore this work focuses on the roll production part on the paper machine of the production line.

2.1   Paper Machine Scheduling

The paper machine transforms paper pulp into paper rolls. This consists in a continuous process in which the paper pulp is spread out on a treadmill passing through several presses and in front of a succession of heating devices or ventilation systems in order to dry the pulp and obtain a sheet of paper that will then be rolled up to form paper rolls.

As this process is continuous, the biggest factor that can impact the consumption of electricity is the kind of paper that has to go through the process. Indeed, depending on the type of paper pulp on the treadmill, the treadmill speed, the temperature of the heater, the speed of ventilation systems and other parameters will vary. The flexibility comes from the possible permutations of paper types according to electricity prices.

A new calibration (of the treadmill and the other components) is necessary for any change of paper type on the machine. This calibration is time consuming and the quality of the paper cannot be ensured during a transition between two different paper types. A minimal duration between any change of paper type is imposed in order to reduce their frequency. The duration for calibration and the loss of paper quality incurred depends on the transition of paper types. Some transitions are more desirable than others. A transition cost can thus be associated for every transition type (i.e. every pair of paper types that will be produced successively).

3   A Planning Model for Paper Roll Production

In this section, we describe a production planning model to represent the transformation of paper pulp into paper rolls. The constraints of this problem are:

- For every demand of paper rolls of a given type at a specified due date, a larger or equivalent amount of paper rolls of the same type has to be produced before the respective due date.
- When a paper type is produced, it has to be produced for a minimum duration before another paper type can be produced.

The objective quantities should be optimized:

- The total energy cost of the production planning has to be minimized.
- The cost (and thus also the number) of transitions between different successive paper types has to be minimized.

A formal definition of the problem is given next. Item indices i, j are ranging on the set $\{1, \ldots, I\}$. Time index t is ranging over $\{1, \ldots, T\}$ where T is the horizon of the planning at an hour basis (since electricity price is changing every hour). The deadline indices are a subset of the time indices: $\{l_1, \ldots, l_L\} \subseteq \{1, \ldots, T\}$.

$$\text{minimize} \quad w_1 \sum_{i,t} (p_t \cdot c_i \cdot x_{t,i}) + w_2 \sum_{i,j,t} (s_{i,j} \cdot y_{i,j,t}) \tag{1}$$

$$\text{subject to} \quad y_{i,j,t} + 1 \geq x_{t,i} + x_{t+1,j} \qquad \forall i, j, t \tag{2}$$

$$\sum_i x_{t,i} = 1 \qquad \forall t \tag{3}$$

$$\text{lower}_l^i \leq \sum_{t=1}^{l} x_{t,i} \leq \text{upper}_l^i \qquad \forall i, l \qquad (4)$$

$$\texttt{contiguous sequence length} \geq k \qquad (5)$$

$$x_{t,i} \in \{0,1\} \qquad \forall i, t \qquad (6)$$

$$y_{i,j,t} \in \{0,1\} \qquad \forall i, j, t \qquad (7)$$

The variable $x_{t,i}$ is a binary variable equal to 1 if paper type i is produced at period t. The variable $y_{i,j,t}$ is true only if there is a transition from paper type i to paper type j occurring at time t. Equation (1) is the objective function composed of two terms weighted by w_1 and w_2. The first term is the energy cost with p_t the price of electricity at time t and c_i is the energy consumption per period for paper type i. The second term is a penalty to pay for the transitions with $s_{i,j}$ the cost associated to the transition between paper type i and paper type j. Equation (2) ensures that $y_{i,j,t} = 1$ only if a product of type i is scheduled at time t and a product of type j at time $t + 1$. Equation (3) ensures that only a single product is scheduled at any time. The constraints at Eq. (4) enforce that the number of products of type i scheduled during the first $l \in \{l_1, \ldots, l_L\}$ periods is within the interval $[\text{lower}_l^i, \text{upper}_l^i]$. The constraint (5) is more difficult to express concisely in a mathematical form. It asks that contiguous sequences of successive variables of a same type should be of length at least k.

CP Model. The problem described above is modeled into Constraint Programming (CP). For every hour t of the planning, a variable x_t with domain $\{1, \ldots, I\}$ is introduced: the paper type to be produced at the hour t. We can compute the electricity consumption c_{x_t} at time t with element constraints[1]. The electricity price to pay is then $\sum_t c_{x_t} \cdot p_t$. The transition cost $s_{x_t,x_{t+1}}$ at every time-point t is also computed with element constraints. The overall transition costs is $\sum_t s_{x_t,x_{t+1}}$. The order book constraint of Eq. (4) can be enforced with a Global Cardinality Constraint (GCC) [11] at every deadline $l \in L$. However, the pruning of this formulation can be improved by using nested_gcc [20]. An efficient FWC algorithm for this constraint is introduced in Sect. 4. The constraint (5) asks that contiguous sequences of a same paper type should be at least of length k. This can easily be expressed in CP with a stretch [4] or a regular [9] constraint. In Fig. 3a, we see a schedule where the constraint is satisfied for $k = 4$ (i.e. there is no succession of rectangles of the same color with length inferior to 4). On the other hand, Fig. 3b shows an unfeasible schedule for the same set of paper types produced since there are two successions of 2 periods where the paper type is blue. The two objectives, minimization of electricity costs and minimization of transition costs between paper types, are aggregated in a sum that is minimized.

---

[1] The element constraint [18] allows to access the value of an array where the index is a variable.

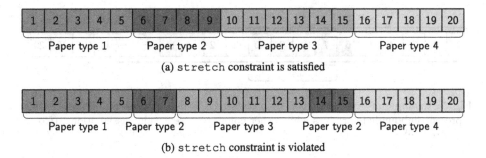

(a) stretch constraint is satisfied

(b) stretch constraint is violated

Fig. 3. Example of feasible and unfeasible schedule for stretch constraint. (Color figure online)

4   A Nested GCC Forward Consistent Propagator

The Global Cardinality Constraint (GCC) [11] on a vector of variables restricts the number of occurrences for each values to be within a specified interval. On our problem, the book order constrains the production on the first $l \in \{l_1, \ldots, l_L\}$ variables to contain at least lower_l^i times the value i. As an example, let us consider a schedule with 20 periods. A first deadline could impose that we produce at least 4 paper rolls of type 1 for period 11 and at least 6 paper rolls of type 1 for period 18. A feasible schedule for this example is shown in Fig. 4.

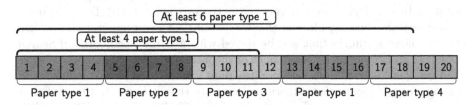

Fig. 4. Feasible schedules with nested GCC (2 deadlines on the paper type 1: at least 4 units at the end of period 11 and at least 6 units at the end of period 18).

Similarly, stock constraints impose a maximum number of times upper_l^i a value i can appear in the first l variables. As deadlines and stock constraints are nested on overlapping variable sets, we are in the special case of a GCC: the nested_gcc [20]. More formally

$$\texttt{nested\_gcc}([x_1, \ldots, x_n], [\text{lower}_{l_1}^1, \ldots, \text{lower}_{l_L}^I], \ldots, [\text{upper}_{l_1}^1, \ldots, \text{upper}_{l_L}^I])$$

enforces the following constraints

$$\text{lower}_{l_k}^i \leq \sum_{t=1}^{l_k} (x_t = i) \leq \text{upper}_{l_k}^i \quad \forall i \in \{1, \ldots, I\}, k \in \{1, \ldots, L\}.$$

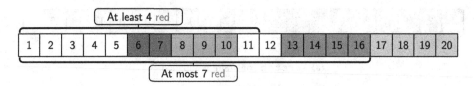

Fig. 5. Example of pruning missed by a classic GCC decomposition. The white cells represent unbounded variables while other colors represent value assignments. (Color figure online)

The nested GCC constraint can be expressed with a decomposition of several standard GCCs: one for every deadline l_k. However, this GCCs decomposition hinders propagation as shown in [20]. An example in which the decomposition misses pruning opportunity is displayed in Fig. 5. In this example, there are already 4 variables set to value red in the range $[1, 16]$ constrained to contain at most 7 variables assigned to red. This imposes that the range before the first variable set to value red (range from 1 to 12) should contain at most 3 variables set to value red. This kind of unfeasible assignment would be detected by the flow-based GAC propagator of the nested_gcc from [20].

In practice, the strongest filtering algorithms are not always the winners on every problem. As explained in [15]: Maintaining a higher level of consistency takes more time; on the other hand, if more values can be removed from the domains of the variables, the search effort will be reduced and this will save time. Whether or not the time saved outweighs the time spent depends on the problem. In practice, many solvers (such as the very efficient OR-Tools [7]) use a default forward checking filtering (FWC) for the GCC. Our application problem contains large instances that will be solved with a Large Neighborhood Search (LNS). In an LNS setting, the strength of the filtering is also less important since the time spent at each node becomes the most critical to allow fast restarts and a good diversification. Our experience suggests that it is rarely the case that heavier propagation pays off when using LNS. Therefore we are interested to design an efficient forward checking propagation procedure for the nested_gcc.

In the following we design a FWC propagator achieving both a potentially stronger and faster pruning when compared to a naive decomposition of L FWC-GCCs. The improvement in pruning is obtained by a preprocessing step that strengthens the bounds of the cardinalities $lower^i_{l_k}$ and $upper^i_{l_k}$. The improvement in terms of running time is obtained by maintaining incremental counters avoiding the need to propagate every sub-GCC on each domain update. We present first the pre-computation step, then the FWC filtering procedure.

4.1   Bounds Pre-Computation

This step aims at tightening the bounds $lower^i_{l_k}$ and $upper^i_{l_k}$ specified by the user and minimizing the number of these to a minimal set. Two reasonings can be done:

1. between different ranges for the same value (e.g. the occurrences of red in range [1, 4] and range [1, 5]).
2. between the bounds for the different values specified at a same date t (e.g. the occurrences of red versus blue in range [1, 6]).

Per-value Deductions. The following forward and backward deductions can be made:

- Lower bounds: if there are at least *two* red in range [1, 4], then there are at least *two* red in range [1, 5] (forward), and at least *one* red in range [1, 3] (backward).
- Upper bounds: if there are at most *two* red in range [1, 4], then there are at most *three* red in range [1, 5] (forward), and at most *two* red in range [1, 3] (backward).

We can make those deductions based on the quantities lower_t^i and upper_t^i containing respectively the best-known lower and upper bounds on the occurrences of i for range [1, t]. This is done by traversing these values for each range once forward and once backward. The forward update of these values is defined as follows, t increasing from 1 to $n - 1$:

$$\text{lower}_t^i = \max \left\{ \begin{array}{l} \text{lower}_t^i \\ \text{lower}_{t-1}^i \end{array} \right. \qquad \text{upper}_t^i = \min \left\{ \begin{array}{l} \text{upper}_t^i \\ \text{upper}_{t-1}^i + 1 \end{array} \right.$$

Similarly, the backward update is defined as follows, i decreasing from n to 2:

$$\text{lower}_t^i = \max \left\{ \begin{array}{l} \text{lower}_t^i \\ \text{lower}_{t+1}^i - 1 \end{array} \right. \qquad \text{upper}_t^i = \min \left\{ \begin{array}{l} \text{upper}_t^i \\ \text{upper}_{t+1}^i \end{array} \right.$$

Inter-value Deductions. Intuitively, there are two types:

- For a given time t and for some paper type i, if the value lower_t^i is large, then the production of other types of paper before t is limited.
- For a given time t and for some paper type i, if the value upper_t^i is small, then the production of other types of paper before t must be compensated.

For example, for a period of length 5, if the sum of the deadlines for the other types is 3 ($\sum_{j \neq i} \text{lower}_5^j = 3$), then at most 2 units of red (type 1) can be produced, and similarly if the sum of the storage space ($\sum_{j \neq i} \text{upper}_5^j = 3$) for the other types is 3, then *at least* 2 units of red must be produced.

For every possible value i, and every possible index t, we define two quantities: lower_t^i and upper_t^i. These values are initially set to respectively, deadlines and stock constraints applying on range [1, t] for value i (or respectively 0 and $n = t$ if not defined). We aim at setting these values with the best-known respectively lower and upper bounds on the number of occurrences of i on range [1, t]. For every value i and every index t defining range [1, t], entries in arrays are defined as follows:

$$\text{lower}_t^i = \max \left\{ \begin{array}{l} \text{lower}_t^i \\ t - \sum_{j \neq i} \text{upper}_t^j \end{array} \right. \qquad \text{upper}_t^i = \min \left\{ \begin{array}{l} \text{upper}_t^i \\ t - \sum_{j \neq i} \text{lower}_t^j \end{array} \right.$$

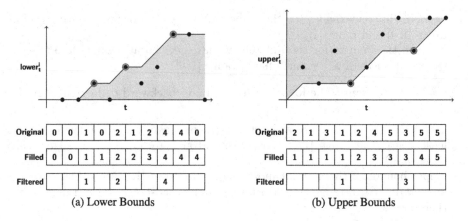

Original	0	0	1	0	2	1	2	4	4	0
Filled	0	0	1	1	2	2	3	4	4	4
Filtered			1		2			4		

Original	2	1	3	1	2	4	5	3	5	5
Filled	1	1	1	1	2	3	3	3	4	5
Filtered						1				3	

(a) Lower Bounds   (b) Upper Bounds

Fig. 6. Example of deduction of best-known lower and upper bounds for a value.

Example 1. An example of per-value pre-computation of lower bounds for a given value is shown in Fig. 6a. Initial lower bounds in the gray zone are updated since dominated by the other specified bounds. The arrays displayed in this example represent the quantities $lower_t^i$ at the different steps of the bound tightening. *Original* represents the original bounds specified by the user, *Filled* represents the bounds after application of the forward (left to right in the array) and backward (right to left in the array) updates described earlier.

After the tightening step of the bounds $lower_t^i$ and $upper_t^i$, the number of these can be minimized to only keep the useful bounds in a decomposition of the nested_gcc. On the example of Fig. 6a, the minimal set of useful bounds is indicated with a $\otimes$. Those are given in the *Filtered* array. A similar example to deduce the upper bounds for a given value is shown in Fig. 6b. The pre-computation is done only once, at the initialization of the constraint. As such, the equalities defining $lower_t^i$ and $upper_t^i$ are assignment statements (not constraints). It can be shown that the final minimal set of bounds obtained after (1) the per-value deductions, (2) inter-value deduction and (3) minimization of the set of bounds, is the unique smallest set of bounds that contains all the useful information initially specified in the quantities $lower_t^i$ and $upper_t^i$. Furthermore, the set of the times on which those final bounds apply is always a subset of the times at which a lower or upper bound was originally given.

4.2 Updating Locally

With the improved bounds we have pre-computed in the previous step, we could very well use a standard FWC-GCC constraint for every range that is involved in the bounds, and obtain an improved pruning. However, if there are p such ranges, it would result in $\mathcal{O}(p)$ amortized time complexity per domain update. We present here a propagator that performs updates in $\mathcal{O}(\log(p))$ amortized time and offers the same pruning. As a reference point, the pruning given by forward-checking GCCs is such that

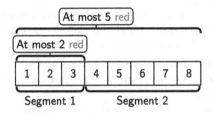

Fig. 7. Example of segment decomposition.

- when the number of variables whose domain still contains a given value decreases to the lower bound associated to it, these variables are assigned to the value.
- when the number of variables bound to a given value increases to the upper bound associated to it, this value is removed from all other variables.

The main challenge of this algorithm is to avoid checking those variable counts on every lower bound or upper bound when an update is received. In order to do that, for every value that we track and for both lower and upper bounds, we divide the variables into the segments that are formed by the bounds, and only count variables inside those segments. For example, if we have a maximum of 2 red in range $[1, 3]$ and a maximum of 5 red in range $[1, 8]$, we will separate the variables into the segments $[1, 3]$ (the first 3 variables) and $[4, 8]$ (the next 5 variables). We justify in the next paragraphs why local checks inside those segments are enough to detect and trigger the required pruning. This example is shown in Fig. 7.

Let us examine the differences between the bounds in our example: $5 - 2 = 3$. We will call this difference the *critical point* of the segment $[3, 7]$. If the number of variables bound to red in that segment reaches 3, then there will be at least 3 occurrences of red in that segment. As a consequence, if the pruning condition in range $[1, 3]$ is met, so that we have 2 variables bound to red in $[1, 3]$, then in total there will be at least 5 variables bound to red in the range $[1, 8]$, so we have to prune there as well. In other words, pruning in $[1, 3]$ can only happen if pruning in $[1, 8]$ also happens; and since in both cases pruning means removing the value red from all unbound variables, it becomes useless to track the upper bound on $[1, 3]$.

Conversely, if there are *less than* 3 variables bound to red in the segment $[4, 8]$, then pruning for the upper bound of range $[1, 8]$ will happen strictly after pruning happens in $[1, 3]$ (if ever). Indeed, pruning in $[1, 3]$ happens when 2 variables in that segment are bound to red, and at that point less than 5 variables would be bound to red in $[1, 8]$.

For the leftmost segment, since there is no bound on the left, we simply define the critical point as the bound on the right, in this case 2 for segment $[1, 3]$. In this segment, reaching the critical point by having 2 variables bound to red means reaching a pruning cases, so we have to remove the red value from the last variable. If the number of variables bound to red is strictly under the critical point, however, no pruning can be performed.

From these remarks we can notice that no pruning will happen in a segment until it reaches its critical point. All that is left is to precisely determine what to do when it is reached. Note that we have taken upper bounds as an example, but the critical point also makes sense for lower bounds: instead of counting the number of variables bound to the value, count the number of variables that have the value in their domain.

We can also observe a useful property of critical points: if we combine two consecutive segments, the distance to the critical point in the merged segment will be the sum of the distances to the critical points in the small segments. Indeed, when summing the critical points, the middle bound will cancel itself out; and the number of variables that are bound to a value or that have a value in their domain is clearly the sum of those numbers in the segments that are being merged.

4.3   Pruning Cases and Segment Merging

Let us now develop an updating strategy based on critical points. We split the variables into contiguous disjoint segments as described above. In the leftmost segment, pruning can happen only when its corresponding critical point has been reached. For other segments, if their respective critical point have not been reached, then no pruning can occur before some pruning happens on the left bound. When the critical point of a segment is reached, we can consider two different actions to perform, depending on whether the considered segment is the leftmost one or not.

First, if the segment is the leftmost segment, we have to trigger pruning in it. As none of the segment on its right has reached its critical point, no pruning should occur on those. Once the pruning has been applied to the leftmost segment, it is removed and its neighboring right segment, if it exists, is marked as the leftmost segment. To achieve fast pruning, we propose to maintain a list of unbound variables still containing a particular value in an array based reversible double linked list. This allows value removal in constant time (as there is one list per possible value). We refer to this list as the *unbound list*. When a critical point is reached, the pruning on a segment will only be applied on variables in the unbound list.

Second, if the segment is *not* the leftmost segment, then reaching the critical point makes the bound on the left of the segment completely redundant in terms of pruning with the bound on the right of the segment. Therefore, the bound on the left can be forgotten, and this segment can be merged with its left neighboring segment. Since distance to the critical point is additive, the larger segment will not have reached its critical point either. To keep the propagator efficient in terms of time complexity, we have to determine efficiently to which segment a variable belongs. We also have to determine an efficient way to merge segments. This problem can be solved easily using a union-find data structure [16].

Here is a description of the steps to perform when a variable x has been bound to a value v and it is inside an upper bound segment:

1. Remove the variable from the unbound list of v.
2. Find the segment containing the variable (*find* operation in our union-find data-structure).
3. Increase the counter of assigned variables bound to v in the segment.
4. If the critical point of the segment has been reached and the segment is the leftmost segment, remove v from all the variables in the segment. Then, if there exists a right neighbor segment, define it as the leftmost segment.
5. If the critical point of the segment has been reached and the segment is not the leftmost segment, merge the segment with its left neighbor segment (*union* operation in our union-find data-structure).

Similarly, the following steps are performed when a value v has been removed from a variable x and it is inside a lower bound segment:

1. Remove the variable from the unbound list of v.
2. Find the segment containing the variable (*find* operation in our union-find data-structure).
3. Decrease the counter of variables which domain contains v in the segment.
4. If the counter has reached the critical point of the segment and it is the leftmost segment, assign v to all the unbound variables in the segment. Then, if there exists a right neighbor segment, define it as the leftmost segment.
5. If the counter has reached the critical point of the segment and it is not the leftmost segment, merge the segment with its left neighbor segment (*union* operation in our union-find data-structure).

4.4 Complexity

The complexity analysis assumes one has access to the Δ change of the variables as for instance proposed in [1] for the OscaR solver also available in OR-Tools [7], or the advisors of Gecode [6].

Let us define u as the number of updates, that is, the sum of the number of value removals over the whole search. Note that when the constraint itself removes a value from a variable, it counts in u as well. We will also use n, the number of variables, and p, which as earlier is the number of distinct ranges involved in the bounds. Looking at the steps performed when a value has been removed/assigned, we can deduce the time complexity for a particular update. Note that even though step 4 can take $\mathcal{O}(n)$ for one particular update to be processed, the variables pruned also count as updates, so it remains amortized constant time per update.

When we combine all of this, we discover that the total complexity is the number of updates multiplied by the cost of a union-find operation. One would think that would give $\mathcal{O}(u \cdot \alpha(p))$ since there will be at most p segments in each union-find structure. However, as this is implemented in a CP framework, we are working with a reversible union-find structure. As such, a particular update could be repeated arbitrarily many times in different places in the search tree. This means we cannot use the amortized $\mathcal{O}(\alpha(p))$ complexity of union-find operations, but rather the $\mathcal{O}(\log(p))$ worst case. As a result, we obtain a time complexity in $\mathcal{O}(u \log(p))$ for the whole search, or an amortized complexity of $\mathcal{O}(\log(p))$ per update.

5   Results

We experiment the CP model on historical data from a tissue manufacturing site in Belgium. This historical data contains the amount and type of paper rolls produced from paper pulp over a couple of years. The historical electricity prices on the EU market over the same period are also available. Combining those two sources of data, we were able to produce instances as follows:

1. Randomly select two dates separated from a specified amount of days. This defines the time window tw representing the instance.
2. Collect over tw the historical type of paper roll produced every hour.
3. Collect over tw the historical European electricity prices every hour.
4. Collect over tw, for every paper type i the contiguous periods at which i is produced. Let $[t_1, t_2]$ be such an interval where product type i is produced continuously. A deadline is imposed to produce additionally at least $t_2 - t_1 + 1$ items for date $t_2 + \delta$.

The shifting of deadlines by δ gives some flexibility to the model for optimization. As we don't have the historical stock constraints, we only impose over the whole time window tw to produce the exact same type and numbers of rolls. We have generated 4 sets of 10 instances for planning of respectively 4, 6, 8 and 11 days (96, 144, 192 and 264 time periods).

In order to evaluate the efficiency of the new FWC procedure for nested_gcc, we compare several models including different propagation procedures. All these models are based on the one described in Sect. 3 and only differ by the propagation procedure for the nested_gcc constraints. We propose to compare three forward checking propagation procedures:

GCC-FWC. A decomposition of classic FWC-GCCs; one FWC-GCC for every range $[1, t]$ on which deadlines and stock constraints occur.

PreGCC-FWC. After a pre-computation of optimal bounds (as described in Sect. 4), a decomposition of classic FWC-GCCs; one FWC-GCC for every range $[1, t]$ on which optimal bounds occur.

NestedGCC-FWC. After a pre-computation of optimal bounds, the new FWC propagator described in Sect. 4.

These models and propagation procedures have all been implemented in the open-source solver OscaR [8]. The propagation procedures are compared with *performance profiles* as described in [17] to compare filtering algorithms using GCC-FWC as baseline. Our measures are obtained by replaying a search tree generated with the baseline approach. Performance profiles [2] are cumulative distribution functions of a performance metric τ. In this paper, τ is the ratio between the solution time (or number of backtracks) of a target approach (i.e. PreGCC-FWC or NestedGCC-FWC) and the one the baseline (i.e. GCC-FWC). For the resolution time metric, the function is defined as:

$$F_{\phi_i}(\tau) = \frac{1}{|\mathcal{M}|} \left| \left\{ M \in \mathcal{M} : \frac{t(replay(\mathrm{st}), M_i \cup \phi_i)}{t(replay(\mathrm{st}), M)} \leq \tau \right\} \right|$$

where $\mathcal{M}$ is the set of considered instances while $t(replay(\mathrm{st}), M \cup \phi_i)$ and $t(replay(\mathrm{st}), M)$ are the time (backtracks) required to replay the generated search tree respectively with our different models and the baseline. The function for the number of backtracks is similar. For this paper, the original search trees have been generated with the baseline model using a binary first-fail heuristic.

Figure 8a and b respectively provide the profiles for number of backtracks and resolution times for all 40 instances. In Fig. 8a, we can see that both approaches using the pre-computation step have a much smaller number of backtracks. Note that, as expected, once the new bounds have been computed, both PreGCC-FWC and NestedGCC-FWC offer the same pruning. We can also see that for about 35 % of the instances, these propagators were able to almost completely cut the search tree explored by GCC-FWC. Finally, we can observe that there are only a bit less than 15 % of the instances for which the propagators using pre-computed bounds are not able to achieve more pruning than GCC-FWC.

In Fig. 8b, we can see the profiles of resolution times for the different propagators. We can see that both PreGCC-FWC and NestedGCC-FWC are faster than GCC-FWC for about 90 % of the instances. The reason is the stronger filtering that is induced by the bounds-strengthening procedure. The 10 % of instances for which both these variants are slower than GCC-FWC are those on which they offer no additional pruning; and even in this case, they are at worst less than 1.5 time slower than GCC-FWC. We can see however that resolution times are similar for PreGCC-FWC and NestedGCC-FWC. After profiling the application, we have seen that the GCC constraints only take a small fraction of the resolution time on this problem (less than 2 %). Also the density of the number deadlines is not very large. This problem is thus not a good candidate to observe speedups with the more advanced FWC algorithm. We have tested artificial problems (not reported for space reason) with a larger number of deadlines. We observed a speedup between 2 to 3 times for the PreGCC-FWC.

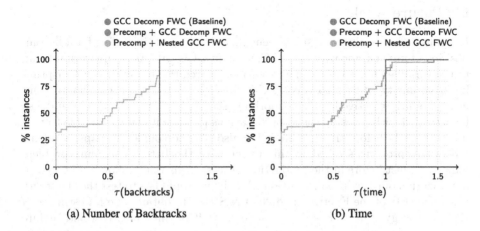

(a) Number of Backtracks   (b) Time

Fig. 8. Performance profiles of nested_gcc variants

5.1 Energy Consumption Minimization with LNS

This section aims at showing the potential improvement brought by our model over historical production plans. An LNS is used with our CP model from Sect. 3 over the historical data and we compare the reductions in terms of cost. The search strategy used is Conflict Ordering Search [3]. The LNS setting is the following: at each iteration, we select 80 % of possible values (e.g. paper types). Variables associated to these 80 % values are then relaxed. This is done to relax the production plan except some *blocks of production* over some specific paper types. The search is stopped if one of these two conditions is met:

1. 180 s have elapsed since the beginning of the restart.
2. 200 relaxations have been performed (with a maximum of 1000 backtracks).

Table 1 shows the ratio of objectives (initial/optimized value) obtained. We can see that the cost of transitions is on average significantly reduced. However, the variance over this objective ratio is high: the reduction of transition cost is really important on some instances but it decreases less on other instances. The ratio of the energy cost however has a small variance. On most of the instances, LNS is able to reduce energy costs by around 22.5 %. These results are promising but somewhat optimistic since it relies on a perfect knowledge of electricity future prices. Since forecasts can by definition be wrong, the gain could be reduced in practice.

Table 1. Ratio of historical and optimized objective values (historical/optimized).

	Global	Energy	Transition
Average	6.40	1.29	56.14
Variance	69.46	0.10	84, 211.22

6 Future work

It would be interesting to test the benefits of the bound tightening for a decomposition of nested_gcc with Bound Consistent GCC [10]. As future work we plan to use variable objective large neighborhood search [12] to obtain a better pruning from our two terms composing the objective or to compute a Pareto front using a multi-objective large neighborhood search [13]. The CP model could also be extended with stocking costs computations [5] since it is not desirable to produce too early before the deadlines. We also plan to couple the paper machine scheduling problem studied in this paper with the batch scheduling problem happening just before in the production process. This would allow an integrated optimization of the whole production. Finally we would like to test the electricity price forecasts of the Enertop module of N-SIDE[2] to obtain a better estimate of the real energy gain. It was not possible to do it in this work. It would require to feed the forecast module with external features (weather forecast, etc.) that we don't have for the historical data.

---

[2] http://energy.n-side.com/enertop-energy-flexibility-optimization/.

7 Conclusion

In this paper we described the problem of reducing energy costs in paper tissue production. To tackle this problem, we propose to reorganize a large part of the manufacturing process: the production of paper rolls from paper pulp. According to forecasts of electricity prices, paper rolls whose production require a larger amount of energy will be produced when prices are low. On the opposite, paper rolls whose production require a smaller amount of energy will be produced when prices are high. The problem is subject to many constraints; an important one is the order book that translates into hard production deadlines. To represent the problem, we propose a CP model including all the constraints. This model will be linked with other CP models corresponding to other steps of the production workflow. The deadline and stock constraints of the problem are expressed with nested_gccs. As the model will be solved with an LNS framework, it has to be scalable. We propose a new FWC propagation procedure for the nested_gcc. This new propagation procedure comports two main step. First, an optimal and minimal set of bounds is computed. This new set of bounds allow us to achieve additional pruning that wouldn't be achieved with initial bounds. Then, we propose a global FWC propagation procedure based on these bounds which has an amortized time complexity in $\mathcal{O}(\log(p))$ (where p is the number of ranges considered). The performances of our new propagation procedure was evaluated on instances generated from historical data. The preprocessing step tightening the cardinality bounds brought significant pruning for many instances.

References

1. de Saint-Marcq, V.l.C., Schaus, P., Solnon, C., Lecoutre, C.: Sparse-sets for domain implementation. In: Techniques for Implementing Constraint Programming Systems (TRICS) Workshop at CP 2013 (2013)
2. Dolan, E.D., Moré, J.J.: Benchmarking optimization software with performance profiles. Math. Program. **91**(2), 201–213 (2002)
3. Gay, S., Hartert, R., Lecoutre, C., Schaus, P.: Conflict ordering search for scheduling problems. In: Pesant, G. (ed.) CP 2015. LNCS, vol. 9255, pp. 140–148. Springer, Heidelberg (2015)
4. Hellsten, L., Pesant, G., van Beek, P.: A domain consistency algorithm for the stretch constraint. In: Wallace, M. (ed.) CP 2004. LNCS, vol. 3258, pp. 290–304. Springer, Heidelberg (2004)
5. Houndji, V.R., Schaus, P., Wolsey, L., Deville, Y.: The stockingcost constraint. In: O'Sullivan, B. (ed.) CP 2014. LNCS, vol. 8656, pp. 382–397. Springer, Heidelberg (2014)
6. Lagerkvist, M.Z., Schulte, C.: Advisors for incremental propagation. In: Bessière, C. (ed.) CP 2007. LNCS, vol. 4741, pp. 409–422. Springer, Heidelberg (2007)
7. OR-Tools Team, Laurent Perron. OR-TOOLS (2010). https://developers.google.com/optimization/
8. OscaR Team. OscaR: Scala in OR (2012). https://bitbucket.org/oscarlib/oscar
9. Pesant, G.: A regular language membership constraint for finite sequences of variables. In: Wallace, M. (ed.) CP 2004. LNCS, vol. 3258, pp. 482–495. Springer, Heidelberg (2004)

10. Quimper, C.-G., van Beek, P., López-Ortiz, A., Golynski, A., Sadjad, S.B.S.: An efficient bounds consistency algorithm for the global cardinality constraint. In: Rossi, F. (ed.) CP 2003. LNCS, vol. 2833, pp. 600–614. Springer, Heidelberg (2003)
11. Régin, J.-C.: Generalized arc consistency for global cardinality constraint. In: Proceedings of the Thirteenth National Conference on Artificial Intelligence, vol. 1, AAAI 1996, pp. 209–215. AAAI Press (1996)
12. Schaus, P., Variable objective large neighborhood search: a practical approach to solve over-constrained problems. In: 2013 IEEE 25th International Conference on Tools with Artificial Intelligence (ICTAI), pp. 971–978. IEEE (2013)
13. Schaus, P., Hartert, R.: Multi-objective large neighborhood search. In: Schulte, C. (ed.) CP 2013. LNCS, vol. 8124, pp. 611–627. Springer, Heidelberg (2013)
14. Simonis, H., Hadzic, T.: A resource cost aware cumulative. In: Larrosa, J., O'Sullivan, B. (eds.) CSCLP 2009. LNCS, vol. 6384, pp. 76–89. Springer, Heidelberg (2011)
15. Smith, B.M.: Modelling for constraint programming. In: Lecture Notes for the First International Summer School on Constraint Programming (2005)
16. Tarjan, R.E.: Efficiency of a good but not linear set union algorithm. J. ACM (JACM) **22**(2), 215–225 (1975)
17. Van Cauwelaert, S., Lombardi, M., Schaus, P.: Understanding the potential of propagators. In: Michel, L. (ed.) CPAIOR 2015. LNCS, vol. 9075, pp. 427–436. Springer, Heidelberg (2015)
18. Van Hentenryck, P., Carillon, J.-P., Generality versus specificity: an experience with ai and or techniques. In: AAAI, pp. 660–664 (1988)
19. Wtenhagen, R., Bilharz, M.: Green energy market development in germany: effective public policy and emerging customer demand. Energy Policy **34**(13), 1681–1696 (2006)
20. Zanarini, A., Pesant, G.: Generalizations of the global cardinality constraint for hierarchical resources. In: Van Hentenryck, P., Wolsey, L.A. (eds.) CPAIOR 2007. LNCS, vol. 4510, pp. 361–375. Springer, Heidelberg (2007)

Cyclic Routing of Unmanned Aerial Vehicles

Nir Drucker, Michal Penn, and Ofer Strichman$^{(\boxtimes)}$

Industrial Engineering and Management, Technion, Haifa, Israel
nirdru@gmail.com, {mpenn,ofers}@ie.technion.ac.il

Abstract. Various missions carried out by Unmanned Aerial Vehicles (UAVs) are concerned with permanent monitoring of a predefined set of ground targets under *relative deadline* constraints, which means that there is an upper bound on the time between two consecutive scans of that target. The targets have to be revisited 'indefinitely' while satisfying these constraints. Our goal is to minimize the number of UAVs required for satisfying the timing constraints. The solution to this problem is given in the form of cyclic (synchronized) routes that jointly satisfy the timing constraints. We develop lower- and upper-bounds on the number of required UAVs, show a reduction of the problem to a Boolean combination of 'difference constraints' (constraints of the form $x - y \geq c$ where $x, y \in \mathcal{R}$ and c is a constant), and present numerical results based on our experiments with several hundred randomly generated problems.

1 Introduction

Many defense- and civilian-related tasks targeted by Unmanned Aerial Vehicles (UAVs) are concerned with permanent monitoring of a predefined set of ground targets under *relative deadline* constraints, which means that there is an upper bound on the time between two consecutive scans of that target. The flight time between the targets and the time it takes to scan each target is given as part of the problem input. The targets have to be revisited 'indefinitely', then, while satisfying all these constraints. It is possible that more than one UAV is necessary in order to satisfy all the constraints, and our goal is to minimize this number. We term this problem Cyclic Routing of UAVs, or CR-UAV for short. The solution to this problem is given in the form of cyclic (synchronized) routes that together satisfy the timing constraints. This problem first appeared in our own technical report [19], and was recently shown to be Pspace-complete by Ho and Ouaknine [16], which implies that there is no polynomial bound on the solution route (such a bound would imply membership in NP). Ho shows in his thesis [15] a remarkable example, based on results in number theory, that only has an exponentially-long solution.

Relative deadlines may be related to the nature of the target and the speed in which the client needs to react to a particular scenario. One may imagine a long border patrolled by UAVs, where certain sensitive locations are associated with a relative deadline that is defined by the speed in which ground forces can react to an event detected by the UAV operator; or a situation in which a military

© Springer International Publishing Switzerland 2016
C.-G. Quimper (Ed.): CPAIOR 2016, LNCS 9676, pp. 125–141, 2016.
DOI: 10.1007/978-3-319-33954-2_10

monitors enemy gatherings, attempting to detect various changes when they occur. Civilian applications may include monitoring of facilities and monitoring of forests for fire. In each such application the relative deadline is calculated according to the relative value of shortening the time to react versus the cost of additional UAVs.[1]

The tasks discussed above are (seemingly endless) routines that can be solved with a cyclic plan. Only rarely it is necessary to deviate from such a plan. Loading preplanned flight routes are supported by modern UAV systems, but no one as far as we know used this capability for uploading optimal cyclic routes of fleets of UAVs, rather it is used for uploading ad-hoc flight plans. Automation of UAVs in various levels is an urgent need since the market, both the defense and civilian-related, is growing rapidly given the major progress in their capabilities and proven success in the last decade. As indicated in [2]: *"The field of air-vehicle autonomy is a recently emerging field, whose economics is largely driven by the military to develop battle-ready technology. Compared to the manufacturing of UAV flight hardware, the market for autonomy technology is fairly immature and undeveloped. Because of this, autonomy has been and may continue to be the bottleneck for future UAV developments, and the overall value and rate of expansion of the future UAV market could be largely driven by advances to be made in the field of autonomy"*. Later in the same article it is pointed out that one of the categories of automation is *"determining an optimal path for vehicle to go while meeting certain objectives and mission constraints, such as obstacles or fuel requirements"*. Somewhat related, concerning a review of the Pentagon for the 2011 budget it was noted in CNN that: *"The review also stresses learning better and more efficient ways to use the drones by improving operating effectiveness and using new technologies"* [8].

In the next section we formally define the CR-UAV problem. In Sects. 3 and 4 we prove respectively lower- and upper-bounds on the number of required UAVs. In Sect. 5 we propose a constraints model. We identify the set of constraints as belonging to the first-order theory of *difference constraints* [9,17], namely a Boolean combination of Boolean variables and constraints of the form $x - y \leq c$ where $x, y \in \mathbb{R}$ and c is a constant, and explain how they can be solved naturally with SMT (Satisfiability Modulo Theory) solvers [17], which we will describe in Sect. 6. Our empirical evaluation of this route is given in Sect. 6. We delay our discussion of related work to Sect. 7, because in order to be able to emphasize the differences between this and other problems that appear in the literature we first need to define it formally and discuss its complexity.

2   A Formal Definition of the CR-UAV Problem

Let V be the set of target areas.

---

[1] The problem was given to us by an industrial partner that develops software for the UAV industry. It has not yet materialized into a product.

2.1 Assumptions

We make several assumptions:

1. When the solution includes more than one UAV, each UAV flies in a different altitude. This allows us to ignore the issue of intersecting routes that may otherwise lead to collisions.
2. Scanning an area $v \in V$ can be done from any point in v.
3. For each pair of targets $v, v' \in V$, the flight time between v and v' is constant. Whereas in reality this is not precisely true because of wind etc., we expect the input figures to include a certain slack to accommodate for such fluctuations. Hence, we can assume that the flight time between areas is given to us as a matrix of constants.[2]
4. For each $v \in V$, the scanning time is large enough to allow any route within v, including turns. This simplifies the problem in two ways:
   - Since this assumption permits us to enter and leave the target area from any location, we can require the flight time figures to refer to the shortest routes between the source and target areas;
   - We can represent each target area v as a point. Hence v is a vertex.
5. The input data (e.g., the relative deadlines and the flight times) contains only integers or, equivalently, rationals. Clearly irrational flight times or relative deadlines are irrelevant in practice.

Since each target can be represented as a point, it is clear that we can view the CR-UAV problem as a graph problem. More specifically, it is a weighted, directed graph, with annotations at the vertices. The vertices are the targets of V, the weights on the arcs are the flight times and the annotations on the vertices are the relative deadlines. This view ignores the scanning time, but as we will show in Sect. 2.2, these can be integrated in the flight times.

2.2 Problem Inputs

In the rest of the article we refer to the elements of V not only as targets, but also as unique indices. Formally this duality can be avoided by defining a 1-to-1 function from a target area to an index, but we avoid it in order to keep the notation simple. We can now define the input to the CR-UAV problem:

1. **Scanning time**: An array ST of size $|V|$, such that for every $v \in V$, $ST[v]$ is the scanning time of v.
2. **Flight time**: A $|V| \times |V|$ matrix FT, such that for every pair $v, v' \in V$, $FT[v, v']$ is the Flying Time between v and v' (recall that by our assumption in Sect. 2.1, the flight time refers to the closest points in v, v').
3. **Relative deadline**: An array RD of size $|V|$, such that for every $v \in V$, $RD[v]$ is the maximum time allowed between consecutive scans of v.

---

[2] This matrix is typically symmetric, but we do not pose this as an assumption since our suggested solutions do not rely on this fact.

We assume that for each target v, $FT[v, v] = 1$. In the realm of our assumption that the input data is integral (see assumption #5), this does not impose any constraint on the solutions, but simplifies the modeling. The proof of this fact is given in Chap. 4 of [13].

Preprocessing of the Input. As a preprocessing step, we add the scanning-time to the flight time as follows. For each entry $FT[v, v']$ such that $v \neq v'$, we assign $FT[v, v'] + 0.5ST[v] + 0.5ST[v']$. Moving the 'cost' from the vertices to the edges simplifies the modeling later on and allows us to discard ST altogether. The following example demonstrates this transformation.

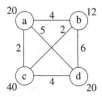

Fig. 1. As before numbers adjacent to vertices represent relative deadlines, and numbers on edges represent flight-times.

Example 1. Consider the following input, which is also depicted graphically in Fig. 1.

$$FT = \begin{pmatrix} 1 & 4 & 2 & 5 \\ 4 & 1 & 2 & 6 \\ 2 & 2 & 1 & 4 \\ 5 & 6 & 4 & 1 \end{pmatrix} \qquad ST = [2, 4, 6, 8] \qquad RD = [20, 12, 40, 20]$$

After the transformation, the FT matrix is: $FT = \begin{pmatrix} 1 & 7 & 6 & 10 \\ 7 & 1 & 7 & 12 \\ 6 & 7 & 1 & 11 \\ 10 & 12 & 11 & 1 \end{pmatrix}$.

For example, we added 3 to $FT[1, 2]$ because this is half of $(2 + 4)$, the accumulated scanning time of vertices a and b.   □

The time it takes to complete a cyclic route is equivalent before and after the transformation. For example, in Example 1 the cyclic route a,b,c takes (beginning from a) $4 + 4 + 2 + 6 + 2 + 2 = 20$ time units (note that this includes scanning time of all three target areas). Using the new matrix, the overall time is the same: $7 + 7 + 6 = 20$.

2.3   Objective

The objective is to find the minimal number n of UAVs and respective cyclic routes for each UAV, that satisfy the constraints.

2.4   Examples

Some example problems appear in Fig. 2. The numbers near the vertices are the relative deadlines, and the numbers near the edges are flight times. Since here the flight time in both directions is identical, the graphs are undirected. Assume that these problems have already been preprocessed as explained above, and hence the scanning time is ignored. Additional information about the solutions appear in the caption of the figure. Note that:

- in (i), there is no solution with one UAV following a *simple* cycle.
- in (ii), there is no solution with two UAVs starting each at a vertex.
- in (iii), there is no solution with two UAVs having non-intersecting routes.

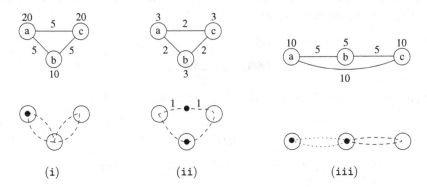

Fig. 2. Three examples of the CR-UAV problem (top), and possible solutions for them (bottom). The numbers above the vertices are the relative deadlines, and the numbers near the edges are flight times. Black circles denote UAV locations. In (i) the single UAV's route repeats a-b-c-b-... . In (ii) both UAVs take the same route, flying in the same direction (e.g., clockwise), where one of them starts in the middle of the distance between a and c. In (iii) the two UAVs have different routes (denoted by dotted and dashed lines, respectively) which intersect at b.

3   A Lower-Bound on the Number of UAVs

Let U denote the set of UAVs required for a solution. We now show a lower bound on $|U|$, the size of U.

We define the following notation. For a target $v \in V$, let

$$FT_{min}(v) = min_{\substack{\hat{v} \in V \\ \hat{v} \neq v}}\{FT[v, \hat{v}]\}. \tag{1}$$

In words, $FT_{min}(v)$ denotes the minimal weight on any outgoing edge of v. We use this notation to define:

Definition 1 (Isolated Vertex). A vertex $v \in V$ is *isolated* if $RD[v] \leq FT_{min}(v)$.

Intuitively, an isolated vertex is one that leaving it takes more time than its relative dead-line. Let $I \subseteq V$ denote the subset of isolated vertices.

We claim that:

Proposition 1. *A lower bound on $|U|$ is given by*

$$|I| + \left\lceil \sum_{v \in (V \setminus I)} \frac{FT_{min}(v)}{RD[v]} \right\rceil \leq |U|. \tag{2}$$

Proof. Let $T > 0$ be the time interval corresponding to a solution, i.e., the time it takes to complete one cycle.[3] Let $T_{sl}(v) \leq T$ be the total time spent at vertex v on self-loops. The figure below depicts such a time interval, where the boxes symbolize the time in which some UAV (not necessarily the same one) looped at v. The accumulated length of the boxes is $T_{sl}(v)$.

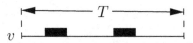

The number of UAV entries to v during T must be at least

$$\left\lceil \frac{T - T_{sl}(v)}{RD[v]} \right\rceil, \tag{3}$$

and hence the total flight time dedicated to v must be at least

$$\left\lceil \frac{T - T_{sl}(v)}{RD[v]} \right\rceil \cdot FT_{min}(v) + T_{sl}(v). \tag{4}$$

The overall flight time is given by aggregating (4) over V:

$$\sum_{v \in V} \left(\left\lceil \frac{T - T_{sl}(v)}{RD[v]} \right\rceil \cdot FT_{min}(v) + T_{sl}(v) \right). \tag{5}$$

This term must be lower than or equal to the total flight time of all UAVs during T, which is given by $T \cdot |U|$:

$$\sum_{v \in V} \left(\left\lceil \frac{T - T_{sl}(v)}{RD[v]} \right\rceil \cdot FT_{min}(v) + T_{sl}(v) \right) \leq T \cdot |U|. \tag{6}$$

We now separate the elements in the sum on the left according to whether $v \in I$:

$$\begin{aligned}
&\sum_{v \in I} \left(\left\lceil \frac{T - T_{sl}(v)}{RD[v]} \right\rceil \cdot FT_{min}(v) + T_{sl}(v) \right) + \\
&\sum_{v \in (V \setminus I)} \left(\left\lceil \frac{T - T_{sl}(v)}{RD[v]} \right\rceil \cdot FT_{min}(v) + T_{sl}(v) \right) \leq T \cdot |U|.
\end{aligned} \tag{7}$$

---

[3] Note that T is a cycle of the whole system, not just of one of the UAVs. In other words, the time it takes the system to return to the same state, where a state includes the location of the UAVs, the remaining time at the vertices until the relative deadlines expire, and finally the current target of each of the UAVs.

Let us focus on the first summation: since this expression is monotone in $T_{sl}(v)$ and $0 \leq T_{sl}(v) \leq T$ whereas the other variables are fixed, it is not hard to see that its value is in the range

$$\left[\sum_{v \in I} T, \quad \sum_{v \in I} \left\lceil \frac{T}{RD[v]} \right\rceil \cdot FT_{min}(v) \right]. \tag{8}$$

Hence the first sum in (7) can be lowered to $T \cdot |I|$, which gives us

$$T \cdot |I| + \sum_{v \in (V \setminus I)} \left(\left\lceil \frac{T - T_{sl}(v)}{RD[v]} \right\rceil \cdot FT_{min}(v) + T_{sl}(v) \right) \leq T \cdot |U|. \tag{9}$$

Furthermore, the second summation is larger than

$$\sum_{v \in (V \setminus I)} \frac{T - T_{sl}(v)}{RD[v]} \cdot FT_{min}(v) + T_{sl}(v), \tag{10}$$

(note that we removed the ceiling operator), which can be rewritten into

$$\sum_{v \in (V \setminus I)} \left(\frac{T \cdot FT_{min}(v)}{RD[v]} + T_{sl}(v) \cdot \left(1 - \frac{FT_{min}(v)}{RD[v]} \right) \right). \tag{11}$$

Note that by Definition 1, for every $v \in (V \setminus I)$ it holds that $\frac{FT_{min}(v)}{RD[v]} \leq 1$, which implies that the right operand is positive and consequently (11) is larger than

$$\sum_{v \in (V \setminus I)} \frac{T \cdot FT_{min}(v)}{RD[v]}. \tag{12}$$

Hence, based on (9) we have that

$$T \cdot |I| + \sum_{v \in (V \setminus I)} \frac{T \cdot FT_{min}(v)}{RD[v]} \leq T \cdot |U|. \tag{13}$$

Dividing by T and rounding up gives us the lower bound on $|U|$ as promised in the proposition:

$$|I| + \left\lceil \sum_{v \in (V \setminus I)} \frac{FT_{min}(v)}{RD[v]} \right\rceil \leq |U|. \tag{14}$$

$\square$

The Bound is Tight. Each of the three examples in Fig. 2 requires as many UAVs as specified by (14). As an example, in the right-most problem the center vertex is the only isolated vertex, and correspondingly the lower bound is given by $1 + \left\lceil \left(\frac{5}{10} + \frac{5}{10} \right) \right\rceil = 2$.

Covering Isolated Vertices. Definition 1 may tempt the reader to think that in an optimal solution a UAV should be dedicated to each isolated vertex. But the following example proves that this is not the case. The distances on the arcs approximately correspond to a metric. The center vertex (d), which has a relative deadline of 4, is isolated. Dedicating a UAV to it would also force us to dedicate a UAV for each of the other three vertices, hence requiring four UAVs all together. The suggested solution on the right, on the other hand, is based on three UAVS. Each of them cycles between a vertex on the perimeter and d, and they arrive to d at equal gaps of $\frac{10}{3}$ time-units.

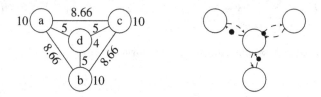

4   An Upper Bound on the Number of UAVs

A trivial upper bound on the number of UAVs is $|V|$. A less trivial upper-bound is given to us by considering the specific solution in which all UAVs follow the same route, evenly spaced. The shortest path going through all points is given to us by the traveling salesman problem (TSP). Let TSP, then, be a solution to this problem. Then an upper bound on the number of UAVs is

$$|U| \le \lceil \frac{TSP}{RD_{\min}} \rceil, \tag{15}$$

hence together we have a bound

$$|U| \le \min(|V|, \lceil \frac{TSP}{RD_{\min}} \rceil). \tag{16}$$

whereas an upper bound which in itself takes exponential time to solve seems not very useful, we note that (a) the application domain (CR-UAVs) has a relatively small number of targets to begin with, and (b) the famous result by Christofides [7] gives us an approximation of up to 1.5 from optimal in P-time, as long as the problem is defined over a metric, which is indeed true in our case.

Example 2. For the three examples in Fig. 2, the upper bounds are, left-to-right, 2,2 and 3. Note that for the middle graph (ii) the upper bound is also the lower bound.

5   A Constraints Model

Our modeling of the CR-UAV problem can be depicted with an array of size SN, where each entry is called a *slot*. Each such slot represents a visit to a vertex.

The value of SN represents the length of the route to be repeated indefinitely. Since we do not know this length in advance, solution strategies based on this model must search for a route starting with $SN = |V|$ and then increase it if a solution is not found. Since we do not have an upper-bound for SN, this method is *incomplete*, i.e., it is not guaranteed to terminate. Practically, in our experiments, we decide on some bound a-priori but if there is no solution up to that bound then we cannot know if it is because there is no solution or because the bound is not high enough.

We now show how the slots model can be used to solve the related satisfiability problem for a single UAV, i.e., a solution implies that a single UAV satisfies the input problem. In Sect. 5.2 we will extend it to multiple UAVs.

5.1 A Model for a Single UAV

The decision variables are:

- $O_{i,j}$: Boolean – for $i \in [1..SN]$, $j \in [1..V]$, $O_{i,j} = 1$ if and only if in slot i the UAV entered vertex j.
- S_i: Real – for $i \in [1..n]$ denotes the entry time to slot i.

The constraints are:

- Exactly one vertex is associated with each slot:

$$\forall i \in [1..SN], v \in V. \; O_{i,v} \implies \bigwedge_{\substack{\hat{v} \in V \\ \hat{v} \neq v}} \neg O_{i,\hat{v}}. \tag{17}$$

$$\forall i \in [1..SN]. \; \bigvee_{v \in V} O_{i,v}. \tag{18}$$

- Defining the accumulated time:

$$\forall i \in [1..SN], v_1 \in V, v_2 \in V. \, O_{i,v_1} \wedge O_{i+1,v_2} \implies S_{i+1} = S_i + FT[v_1, v_2]. \tag{19}$$

- Defining S_1:

$$\forall v_1 \in V, v_2 \in V. \, O_{SN,v_1} \wedge O_{1,v_2} \implies S_1 = FT[v_1, v_2]. \tag{20}$$

- Time between visits to the same vertex (see illustration in Fig. 3):

$$
\begin{aligned}
&\forall v \in V, i \in [1..SN]. \\
&\left(\bigvee_{l=1}^{i-1} O_{l,v} \wedge (S_i - S_l \leq RD[v]) \right) \vee && \textit{visited } v \textit{ in an earlier slot} \\
&\left(\bigvee_{l=i+1}^{SN} O_{l,v} \wedge (S_i + S_{SN} - S_l \leq RD[v]) \right) \vee && \textit{visited } v \textit{ in a later slot} \\
&\left(O_{i,v} \wedge \bigwedge_{l=1, l \neq i}^{SN} \neg O_{l,v} \wedge S_{SN} \leq RD[v] \right) && \textit{visited } v \textit{ only in slot } i.
\end{aligned}
\tag{21}
$$

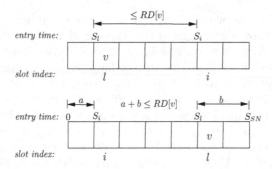

Fig. 3. Demonstrating the two cases handled by the two middle lines of (21). For each vertex $v \in V$ and each slot $i \in [1..SN]$, one of three cases has to hold: either there is a slot l, $l < i$, in which v is visited and the time from S_l to S_i is less or equal to $RD[v]$ (top diagram), there is a slot l, $l > i$, in which v is visited and the time from S_l to S_{SN} added to the time from 0 to S_i is less or equal to $RD[v]$ (bottom diagram), or v is visited only once in the cycle and $S_{SN} \leq RD[v]$ (not shown).

5.2 Multiple UAVs

A generalization of the solution given in Sect. 5.1 to multiple UAVs solves indirectly the primary objective as stated in Sect. 2.3, because one only needs to gradually increase the number of UAVs until a solution is found. Recall that there is always a solution with $|V|$ UAVs, which means that this process is guaranteed to terminate. However, since the solution for a given number of UAVs is incomplete, as explained in Sect. 5.1, then it is possible that our solution is not optimal since the search with a lower number of UAVs was stopped prematurely.

In order to generalize the model to multiple UAVs, we require that at each slot at least one UAV is reaching a new vertex, whereas other UAVs can be between vertices. For that we define a new variable $A_{u,i}$ that holds the time to destination i of UAV u. In contrast to the single UAV model, here a UAV u can have a route which contains only one vertex where $\forall i \in [1..SN] : A_{u,i} = 0$.

Additional variables for the multiple UAVs model:

- $\forall u \in U, i \in [1..SN], v \in V$. $O_{u,i,v}$ Boolean: $O_{u,i,v} = 1 \iff$ in slot i UAV u enters vertex v.
- $\forall u \in U, i \in [1..SN]$. $A_{u,i}$: Time left for UAV u to reach its new destination, when at slot i.

The constraints are:

- At least one UAV should enter a vertex in each slot:

$$\forall i \in [1..SN]. \bigvee_{\substack{u \in U \\ v \in V}} O_{u,i,v}. \tag{22}$$

– Each UAV can visit only one vertex at each slot:

$$\forall u \in U, i \in [1..SN], v \in V. \ O_{u,i,v} \implies \bigwedge_{\hat{v} \in \{V \backslash \{v\}\}} \neg O_{u,i,\hat{v}}. \tag{23}$$

– If a UAV u visits vertex v at time slot i then there is no other UAV $\hat{u}$ that reaches its vertex $\hat{v}$ before u visits v.

$$\forall u \in U, i \in [2..SN], v \in V, \hat{u} \in U, \hat{u} \neq u.$$
$$O_{u,i,v} \implies A_{u,i-1} \leq A_{\hat{u},i-1}. \tag{24}$$

– Same as above, for the first slot:

$$\forall u \in U, v \in V, \hat{u} \in U, \hat{u} \neq u. \ O_{u,1,v} \implies A_{u,SN} \leq A_{\hat{u},SN}. \tag{25}$$

– S_1 is non-negative (the values of other S_i variables will be larger owing to the constraints that follow):

$$S_1 \geq 0. \tag{26}$$

– S_i progression:

$$\forall u \in U, i \in [2..SN], v \in V.$$
$$O_{u,i,v} \implies S_i = S_{i-1} + A_{u,i-1}. \tag{27}$$

– Same, for the first slot:

$$\forall u \in U, v \in V.$$
$$O_{u,1,v} \implies S_1 = A_{u,SN}. \tag{28}$$

– If a UAV visits v_1 at slot i and v_2 at slot j and does not visit any other vertex in between, then the time to arrive at the destination should be set to the flying time between v_1 and v_2:

$$\forall u \in U, i \in [1..SN], v \in V, \hat{i} \in [i+1..SN], \hat{v} \in V.$$
$$(O_{u,i,v} \wedge O_{u,\hat{i},\hat{v}} \wedge (\neg \bigvee_{\substack{v_2 \in V \\ mid \in [i+1..\hat{i}-1]}} O_{u,mid,v_2})) \implies A_{u,i} = FT_{v,\hat{v}}. \tag{29}$$

– If a UAV visits v at slot i and v_2 at slot j and does not visit any other vertex after slot time j and before slot time i, then the arrival time should be set to the flying time between v_2 and v:

$$\forall u \in U, i \in [1..SN], v \in V, \hat{i} \in [i+1..SN], \hat{v} \in V.$$
$$(O_{u,i,v} \wedge O_{u,\hat{i},\hat{v}} \wedge (\neg \bigvee_{\substack{v' \in V \\ mid \in [1..i-1] \cup [\hat{i}+1..SN]}} O_{u,mid,v'})) \implies A_{u,\hat{i}} = FT_{\hat{v},v}. \tag{30}$$

– If a UAV visits v at slot i, then for each UAV $\hat{u}$, $A_{\hat{u},i}$ is equal to the difference between $A_{\hat{u},i-1}$ and $A_{u,i-1}$:

$$\forall u, \hat{u} \in U, \hat{u} \neq u, i \in [2..SN], v \in V.$$
$$O_{u,i,v} \implies A_{\hat{u},i} = A_{\hat{u},i-1} - A_{u,i-1}. \tag{31}$$

– Same, for the first slot:

$$\forall v \in V, u, \hat{u} \in U, \hat{u} \neq u.$$
$$O_{u,1,v} \implies A_{\hat{u},1} = A_{\hat{u},SN} - A_{u,SN}. \tag{32}$$

– Time between visits to the same vertex:

$$\forall v \in V, i \in [1..SN].$$

$(\bigvee_{l=1}^{i-1} (\bigvee_{u \in U} O_{u,l,v} \wedge S_i - S_l \leq RD[v])) \vee$	*visited v in an earlier slot*
$(\bigvee_{l=i+1}^{SN} (\bigvee_{u \in U} O_{u,l,v} \wedge S_i + S_{SN} - S_l \leq RD[v])) \vee$	*visited v in a later slot*
$(\bigvee_{u \in U} O_{u,i,v} \wedge \bigwedge_{\hat{u} \in U} \bigwedge_{l=1,l \neq i}^{SN} \neg O_{\hat{u},l,v} \wedge S_{SN} \leq RD[v])$	*visited v only in slot i.*

(33)

6   Experimental Results

To solve the mathematical model described in Sect. 5, we must bound the number of slots a-priori. For the experiments we chose the bound

$$\left\lceil \frac{RD_{\max} \cdot |U|}{FT_{min}} \right\rceil, \tag{34}$$

that is, the longest relative deadline divided by the shortest flight time, multiplied by the number of UAVs, and rounded up. Recall that there is no guarantee that this bound is sufficient, as explained in Sect. 5, which makes this method incomplete.

Since the formulation of the problem includes a Boolean structure beyond simple conjunctions, it is very natural to solve it with a Satisfiability Modulo Theories (SMT) solver. Satisfiability Modulo Theories (SMT) [17] is an extension of the classical propositional satisfiability problem to other decidable first-order theories, i.e., in addition to propositional variables the formula can contain predicates of some decidable theory T. For example, if T is linear arithmetic, then a formula such as $2x + 3y > 5 \vee \neg(3y - 5z \geq 6) \wedge (x - y < z)$ is a T formula. A standard framework to solve such formulas is called DPLL(T). It combines a propositional SAT solver (hence the name DPLL[4]), and a solver for a conjunction of T predicates, e.g., in the case of T being linear arithmetic that solver can be based on Simplex. This combination is far better than 'case splitting' (transforming the formula to disjunction normal form), because it enjoys SAT's capabilities to prune large parts of the search space by applying *learning* (adding constraints during the solution process, that block search paths that are known to not contain a solution) and other techniques that are known to be very effective in dealing with propositional formulas. There are several dozen SMT solvers and an annual competition between them called SMT-COMP. We experimented with two such solvers, YICES [14] and Z3 [12]. We only report on the results of Z3, however, because it completely dominates the results of YICES in terms of run-time.

---

[4] DPLL stands for the name of the authors in [10,11].

Over- and Under- Approximations. It is obvious that given a CR-UAV problem one can multiply all the relative deadlines and all the flight time by any constant fraction γ, and as long as the resulting figures are integers the new problem is isomorphic to the original one. We wanted to test, however, what happens if multiplying by γ results in fractions, and then we round the result in a way that guarantees either an over or under approximation (but not both). Keeping the approximation single-sided enables us to know when the answer can be trusted: with an overapproximating engine we only trust UNSAT answers, and with an underapproximating model we only trust SAT answers. To produce an overapproximating model we round up the relative deadlines, and round down the flight time. To produce an underapproximating model we do the opposite. The question is what is the price we pay in terms of correctness, and what is the benefit in run time. The results below include answers to these two questions.

The Input Problems. We generated random input problems, with varying topologies, flight times, relative deadlines and number of UAVs. More specifically the benchmarks were constructed according to the following parameters:

- Number of vertices $|V|$: 4..7.
- Number of UAVs $|U|$: 1..3.
- Flight-time FT: calculated as if we are on a metric[5], according to the following six topologies:
    1. *Line* – All vertices are ordered on a line with equal distance between them.
    2. *One Group* – One group of vertices, none of which is isolated.
    3. *Two Groups* – All vertices are ordered in two groups where the groups are far but the vertices within a group are near.
    4. *Three Groups* – All vertices are ordered in three groups where the groups are far but the vertices within a group are near.
    5. *Isolated location* – All vertices are grouped together except for one which is isolated (see Definition 1).
    6. *Cycle* – All vertices are ordered in a cycle.
- Relative deadline RD: we tested the following variants:
    1. $\forall v \in V : RD[v] = \sum_{v \in V} FT_{max}(v)$.
    2. $\forall v \in V : RD[v] = \sum_{v \in V} FT_{min}(v)$.
    3. $\forall v \in V : RD[v] = \sum_{v \in V} (FT_{max}(v) + FT_{min}(v))/2$.
    4. $\forall v \in V : RD[v] = FT_{max}(v)$.
    5. $\forall v \in V : RD[v] = FT_{min}(v)$.
    where $FT_{max}(v) = max_{\hat{v} \in V \atop \hat{v} \neq v}\{FT[v, \hat{v}]\}$. In none of the test cases we used the lower bound of Sect. 3 for early detection of the result.

---

[5] This implies that there can be non-integral and even irrational figures. We rounded in such cases the figures according to the overapproximation strategy explained above, with $\gamma = 1$.

This gave us 360 test-cases. From those we removed several trivial combinations (like 4 vertices with 3 groups), which left us with 300 benchmarks. We also generated over- and under-approximated versions of these problems as explained above, with $\gamma = 0.1$, to give us a total of 900 runs. All benchmarks are available from [1] for others to try.

Results. Our results show the following statistics:

- Over-approximation, with $\gamma = 0.1$: 0.3 % of SAT results are incorrect.
- Under-approximation, with $\gamma = 0.1$: 6.5 % of UNSAT results are incorrect.

(note that these statistics represent a property of the problem at hand with respect to a given γ, and not of the solving algorithm). Table 1 summarizes our results. The effect of approximation on run time is relatively small.

Table 1. Comparing the number of solved instances within a time limit of 10 min and the average run-time (instances that were not solved within 10 min were considered as solved in 10 min).

Method	γ	Z3	avg. run-time
Precise		578	26.8
Under	0.1	579	21.4
Over	0.1	580	24.9

7 Related Work

CR-UAV first appeared in our technical report [19] and thesis [13]. A problem closely related to CR-UAV for a *single* UAV is that of planning a cyclic *agent patrol* [4,5]. It tackles the problem of finding a route for a robot patrolling an enclosed area. The relative deadlines are related to the time it takes an adversary to break in, in specific vulnerable locations along the cyclic path. The goal defined there is to find whether there exists a cyclic route for the patrolling agent such that no break can go undetected. The solution given in the above reference is wrong, however, since it relies on a wrong theorem claiming that there exists a polynomial bound on the length of the (cyclic) path (which would imply NP-completeness)[6]. CR-UAV has slightly different constraints and a different goal. Whereas CR-UAV receives as input a full *FT* (flight-time) matrix, in the agent patrol problem some of these paths can be blocked (modeling a scenario in which the patrolling agent is restricted to a rail). Interestingly in [18] Fargeas et al. consider a very similar problem (in the context of UAVs!) but prove NP-completeness only for a bounded horizon.

---

[6] A counterexample to their theorem is given in the appendix of http://arxiv.org/abs/1411.2874.

Since CR-UAV is Pspace-complete [16], any problem in NP is of course irrelevant as a target for reduction. We leave for future research to check if approximations of such problems can be lifted to the CR-UAV problem. Let us nevertheless mention two seemingly related problems that are NP-complete, in order to emphasize the differences:

- *Deadline Traveling salesman Problem* (DTSP) and *TSP with time-window* (TSP-TW) [3] — Given a metric space on n nodes, with a start node r, deadline $D(v)$ for each vertex v, and a number $k \leq n$, find a path starting at r that visits k or more nodes by their deadlines. DTSP can be extended into the TSP-TW problem in which each node v also has a release time $R(v)$ and the goal is to visit k nodes within their time-windows $[R(v), D(v)]$. In both cases the fact that there is no requirement for repeated visits implies that the resulting tour is polynomially bounded, which puts them in NP.
- *Vehicle Routing Problem with Time Windows* (VRP-TW) [6] — there are n customers at n different points, to be served within a specified time window by several vehicles limited in capacity from one depot. The goal is to minimize the number of vehicles needed, such that each customer is reached within its time window while obeying the capacity constraints. As in the case of DTSP, it does not require repeated visits, and hence the length of the resulting tour is polynomially bounded, which puts this problem in NP.
- *Periodic Scheduling* — The periodic (cyclic) scheduling problem can be defined in various ways. In [20,21] events and activities are identically repeated at a constant rate. The periodic activities within a given common period can be considered as a "time window", reflecting the relative position of pairs of activities within the period. Each client i requests to be served for b_i consecutive time slots with no more than t_i time slots between them. The aim is to construct a schedule that minimizes the gap between the required periods and the actual scheduled ones. Suppose that we consider the decision problem of whether such a schedule is possible with a gap of 0. Still, there are several notable differences from the CR-UAV problem: first, in CR-UAV we are not restricted to one 'server' (UAV), whereas here there is only one and the goal is to schedule its service. Even if we consider the CR-UAV problem for a single UAV, there are still important differences: first, the flight-time in the CR-UAV problem adds a constraint on the visits (service time), which does not exist in periodic scheduling (this difference cannot be overcome by simply adding the flight time to the service time, because, recall, the flight time depends on the ordering of the targets in the route); second, and more importantly, the bound b_i implies, as in the previously mentioned problems, a polynomial bound on the result, which puts the problem in NP.

A more extensive literature review, including some problems related to UAVs, can be found, e.g., in [13,18].

8 Conclusions

We presented a problem of finding the minimal number of UAVs that are required in order to satisfy relative-deadline constraints. We showed a lower- and an upper-bound, a modeling of the problem for both a single and multiple UAVs, and presented our empirical evaluation.

References

1. The CR-UAV problem home-page. http://ie.technion.ac.il/~ofers/cruav/
2. The UAV web page. http://www.theuav.com/
3. Bansal, N., Blum, A., Chawla, S., Meyerson, A.: Approximation algorithms for deadline-tsp and vehicle routing with time-windows. In: Proceedings of the Thirty-Sixth Annual ACM Symposium on Theory of Computing, pp. 166–174. ACM (2004)
4. Basilico, N., Gatti, N., Amigoni, F.: Developing a deterministic patrolling strategy for security agents. In: Proceedings of the 2009 IEEE/WIC/ACM International Joint Conference on Web Intelligence and Intelligent Agent Technology, vol. 02, pp. 565–572. IEEE Computer Society (2009)
5. Basilico, N., Gatti, N., Amigoni, F.: Patrolling security games: definition and algorithms for solving large instances with single patroller and single intruder. Artif. Intell. **184–185**, 78–123 (2012)
6. Bräysy, O., Gendreau, M.: Vehicle routing problem with time windows, part i: route construction and local search algorithms. Transp. Sci. **39**(1), 104–118 (2005)
7. Christofides, N.: Worst-case analysis of a new heuristic for the travelling salesman problem. Technical report 388, CMU (1976)
8. CNN. Review shows dramatic shift in Pentagon's thinking, February 2010. http://edition.cnn.com/2010/POLITICS/02/01/us.pentagon.review/index.html
9. Cotton, S., Maler, O.: Fast and flexible difference constraint propagation for DPLL(T). In: Biere, A., Gomes, C.P. (eds.) SAT 2006. LNCS, vol. 4121, pp. 170–183. Springer, Heidelberg (2006)
10. Davis, M., Logemann, G., Loveland, D.: A machine program for theorem-proving. Commun. ACM **5**, 394–397 (1962)
11. Davis, M., Putnam, H.: A computing procedure for quantification theory. J. ACM **7**, 201–215 (1960)
12. de Moura, L., Bjørner, N.S.: Z3: an efficient SMT solver. In: Ramakrishnan, C.R., Rehof, J. (eds.) TACAS 2008. LNCS, vol. 4963, pp. 337–340. Springer, Heidelberg (2008)
13. Drucker, N.: Cyclic Routing of Unmanned Aerial Vehicles. Master's thesis, Technion, Industrial Engineering and Management (2014). Available from [1]
14. Dutertre, B., De Moura, L.: The Yices SMT solver. Technical report, SRI (2006)
15. Ho, H.-M.: Topics in Monitoring and Planning for Embedded Real-Time Sys-tems. Ph.D. thesis, CS, Oxford (2015). Sect. 6.2
16. Ho, H.-M., Ouaknine, J.: The cyclic-routing UAV problem is PSPACE-complete. In: Pitts, A. (ed.) FOSSACS 2015. LNCS, vol. 9034, pp. 328–342. Springer, Heidelberg (2015)
17. Kroening, D., Strichman, O.: Decision Procedures - An Algorithmic Point of View. Theoretical Computer Science. Springer, Heidelberg (2008)

18. Las Fargeas, J., Hyun, B., Kabamba, P., Girard, A.: Persistent visitation under revisit constraints. In: 2013 International Conference on Unmanned Aircraft Systems (ICUAS), pp. 952–957, May 2013

19. Drucker, M.P.N., Strichman, O.: Cyclic routing of unmanned air vehicles. Technical report, Technion, Industrial Engineering and Management (2014). IE/IS-2014-12. Also available from [1]

20. Patil, S., Garg, V.K.: Adaptive general perfectly periodic scheduling. Inf. Process. Lett. **98**(3), 107–114 (2006)

21. Serafini, P., Ukovich, W.: A mathematical model for periodic scheduling problems. SIAM J. Discrete Math. **2**, 550–581 (1989)

Parallelizing Constraint Programming with Learning

Thorsten Ehlers[1](✉) and Peter J. Stuckey[2,3]

[1] Department of Computer Science, Kiel University, 24098 Kiel, Germany
the@informatik.uni-kiel.de
[2] Department of Computing and Information Systems,
University of Melbourne, Melbourne 3010, Australia
pstuckey@unimelb.edu.au
[3] Victoria Laboratory, National ICT Australia, Melbourne, Australia

Abstract. Parallel Constraint Programming (CP) solvers typically split the search space in disjoint subspaces, and run solvers independently on these. This may induce significant overhead when solving optimization problems. Parallel Boolean Satisfiability (SAT) solvers typically run a portfolio of solvers, all solving the same problem but sharing some limited learnt clause information. In this paper we consider parallelizing a lazy clause generation (LCG) constraint programming solver, which is a constraint programming solver with learning. Since it is both a kind of CP solver and a kind of SAT solver it is not clear which approach to parallelization is likely to be most effective. We give examples of very different kinds of optimization problems we wish to parallelize and show that a hybrid approach to parallelization can provide a robust and high performing parallel LCG solver.

1 Introduction

Techniques for verification and optimization such as SAT, CP, SMT and MIP have greatly improved in the last decades, and are nowadays used in a wide range of applications. Besides algorithmic improvements, more and more powerful hardware has become available, giving an additional boost on sequential performance. But the time of this free lunch seems to be over, as clock rates and instructions per cycle are hardly improving anymore. In order to gain speedups from today's hardware, algorithms should be able to run in parallel. In this paper, we consider the parallelization of the LCG solver CHUFFED [5] for CP-based optimization problems. CHUFFED combines CP techniques such as search and strong propagation with techniques developed for SAT solving such as clause learning, restarts and activity based search.

T. Ehlers—Supported by a fellowship within the FITweltweit program of the German Academic Exchange Service (DAAD).

P.J. Stuckey—NICTA is funded by the Australian Government as represented by the Department of Broadband, Communications and the Digital Economy and the Australian Research Council through the ICT Centre of Excellence program.

© Springer International Publishing Switzerland 2016
C.-G. Quimper (Ed.): CPAIOR 2016, LNCS 9676, pp. 142–158, 2016.
DOI: 10.1007/978-3-319-33954-2_11

Whereas parallelization of CP solvers is usually based on some kind of search space splitting, parallelization of SAT solvers is usually based on some form of portfolio approach. Hence an interesting question arises for LCG solvers: should they use search space splitting or portfolio methods for parallelizing search? In this paper we investigate this question.

The contributions of this paper are

1. An analysis of the runtime-behavior of sequential solvers on optimization problems, showing extremely different characteristics of different problems.
2. An optimistic branching technique which allows for finding good solutions much earlier, which prevents superfluous work in search space splitting.
3. A comparison of search space splitting and work stealing, the common approach used in parallel CP, with a portfolio CP solver using techniques commonly used in parallel SAT.
4. A scalable, parallel LCG solver which allows for significant speedups on a wide range of benchmarks. Compared to the sequential solver, superlinear speedup is achieved in finding good solutions.

The structure of the paper is as follows. After discussing related work in Sect. 2 and presenting the architecture of our parallel solver (Sect. 3), we then examine the use of sequential optimization on two very different optimization problems in Sect. 4 and show the impact of basic approaches to parallelizing their solving. In Sects. 5 and 6, we present results on a suite of benchmarks for parallelizing using search space splitting implemented by work stealing, and an approach based on SAT-like portfolio solving. We then consider the effect of splitting the problem by objective value in Sect. 7. After combining these approaches to a stable and scalable solver in Sect. 8, we conclude in Sect. 9.

2   Related Work

The most common approach for solving CSP problems is to combine search with propagation [27]. The search is implemented as backtracking, and at each node of the search tree propagators are invoked to reduce variable domains with respect to the decisions made during branching. In case an inconsistent state is detected, i.e. some variable can take no possible value, the solver backtracks, and tries another variable assignment. Implementing fast and scalable parallel algorithms is noted as one of the large challenges in optimization [9].

2.1   Parallel CP

Parallel algorithms for CP typically split the search space, and run solver threads on disjoint subspaces [11]. This approach has been studied for several decades, and it is known that superlinear speedups are possible in some cases [14,22]. Most solvers use work stealing mechanisms to keep all solver threads busy [26], and significant speedups are reported for up to 512 threads [15]. Recent research tried to reduce the communication overhead in order to improve speedups for massively-parallel search. In [17], the authors suggest to split the search space by computing the discrepancy from a given search strategy. Thus, solver threads only need to know about their index to compute their chunk of the search space. Unfortunately, they report results for only a small number of experiments, and cannot prove optimality for them. Another approach is to split the search space before starting the search, and storing chunks of work in a master process [23,24].

If a good search strategy for a specific problem is known, this may be used to focus the parallel search on promising parts of the search space [6], and gain significant speedups.

It is known that some solvers are faster than others, depending on the problem instance. This fact was used in [2] to build a sequential portfolio, and later to create a parallel portfolio solver [1]. Significant speedups could be gained using a small number of threads, but it is not clear how scalable this approach is.

2.2 Parallel SAT

The satisfiability problem of propositional logic (SAT) can be seen as a special case of CP with binary variables, and constraints given in form of disjunctions, called clauses. It is typically solved using conflict driven clause learning (CDCL), an extension of the well-known DPLL algorithms, together with agile restarting strategies and activity based search [18]. These techniques allow for reusing information about parts of the search space which were proven infeasible, and restarting the search to emphasize important variables as well as recovering from bad decisions made close to the root of the search tree. Parallel algorithms for SAT either split the search space, or run different solver configurations in parallel [3]. The latter approach, typically referred to as portfolio approach, has proven very successful especially on structured instances. Recent research focusses on the exchange of learnt clauses between solver threads [7,12]. Unfortunately, the scalability of these solvers seems to be limited due to the sequential structure of resolution proofs [13].

2.3 Lazy Clause Generation

Lazy Clause Generation (LCG) [8] combines techniques from CP and SAT to solve Constraint Satisfaction Problems. If an inconsistency is detected during search, the reasons for this inconsistency are compiled into a clause, and added to the set of constraints. Thus, it is possible to reuse this knowledge in other parts of the search tree, which is extremely helpful if these clauses are good explanations for the failed search [28]. The LCG-based solver CHUFFED[1] [5] additionally supports the Variable State Independent Decaying Sum (VSIDS) branching heuristic that is commonly used in SAT solvers. This heuristic branches on variables first that have recently been involved in conflicts. CHUFFED can switch between activity based search and programmed search during runtime.

---

[1] https://github.com/geoffchu/chuffed.

3 Preliminaries

A Constraint Satisfaction Problem (CSP) problem ϕ is given by a triple (V, D, C), where $V = (v_1, \ldots, v_n) = vars(\phi)$ is a set of n variables on finite domains $D = (D_1, \ldots, D_n)$, and C is a set of predicates $c : D \mapsto \{\bot, \top\}$. We will assume integer variables, i.e. $v \in [lb(v), ub(v)] \cap \mathbb{Z}$ for all variables v. The set of feasible solutions is given by $S = \{x \in D \mid \forall c \in C.c(x)\}$. A Constraint Optimization Problem (COP) is a tuple (V, D, C, z) consisting of a CSP (V, D, C) and a variable z that takes the objective value. Throughout this paper we will only consider minimization problems, as maximization can be expressed by negating the expression that z is equated to. When adding further constraints, we will write $\phi_{|c} = (V, D, C \cup \{c\})$.

3.1 Parallel Solver Architecture

We have developed a parallel version of CHUFFED [5] which is used in all experiments. CHUFFED is a state-of-the-art lazy clause generation solver [20]. It comes with a Master-Slave-infrastructure [6], where communication is performed as message passing via MPI. When gaining parallelism by search space splitting, the master process sends conjunctions of literals, called jobs, to the slaves. We extended this scheme as follows to gain a more flexible solver.

- **Portfolio-Solving**, as in parallel SAT solving, can be achieved by sending empty jobs to each slave process, which allows them to search the whole search space. Diversification is gained by initializing the VSIDS-activities with random values.
- **Probing** on variable values: The master process can send jobs of the form $[x \circ c]$ to the slave processes, where $x \in vars(\phi)$, $c \in \mathbb{Z}$, and $\circ \in \{=, \neq, \leq, <, >, \geq\}$. This can, e.g., be used for guessing bounds on the objective value.
- **Learnt clauses** are sent to, and forwarded by the master process, if their length is sufficiently small. The threshold on the length can be adjusted dynamically to both maintain a sufficient communication between solvers, and prevent network congestion.
- **Adaptive** size of clause database: While CHUFFED has a fixed bound on the size of its clause database, we allow for a dynamic amount of received clauses. Whenever the learnt clause database is cleaned, we delete all received clauses with low activity.
- **Hybrid** approaches: It is possible to mix the modi operandi.

4 Optimization

As CP typically deals with decision problems, CP-based optimization is built around decision procedures. Algorithm 1 shows how, given a decision procedure DECIDE, an optimization algorithm can use this procedure to find an optimum solution. Running this algorithm will result in a sequence of solver calls, of which

---

Algorithm 1. Optimize CP

---

function OPTIMIZE(ϕ, z, lb, ub)
    $res \leftarrow (\bot, \infty)$
    while ($ub \geq lb$)
        $(sat, x) \leftarrow Decide(\phi_{|z \leq ub})$
        if (sat) $res \leftarrow (\top, z)$, $ub \leftarrow z - 1$
        else break
    return res

---

the last one returns UNSAT, proving that either no solution exists, or that the last solution found is optimal. In the remainder of this paper, we refer to the last call as the "proving optimality" part, and all other calls as the "search" part of the overall run.

In this section, we examine the behavior of these algorithms on two examples, both for the sequential and parallel case. The examples, cargo[2] and queens[3] were chosen from the MiniZinc Challenges 2013 and 2014, respectively. In all these experiments, the free search of CHUFFED was used: On every restart, CHUFFED flips between programmed search and VSIDS (restart_flip from [25]).

4.1 Parallel Optimization

We begin with the naïve approach, and simply reuse a parallel solver for optimization problems. Whenever one of the parallel solvers finds a solution, this is reported to the master. Furthermore, if the objective value of this solution is c, a unit clause containing the literal $[z < c]$ is sent to all solvers for stronger pruning. We parallelized solving using a portfolio approach, and using search space splitting with work stealing [6]. For the portfolio approach, the search is diversified by initializing variable activities with random values, based on different seeds for each solver thread, as is common in SAT portfolios.

Figure 1 shows the development of the objective value during the solver run on the cargo benchmark for $p \in \{2, 8, 64\}$ processes. Note that we re-use the master-slave architecture for this experiment, thus, one of these processes denotes the master. The speed-up we observe is very limited: 2.2 for 7 worker processes, and 4.2 for 63 workers. The reason for this is simple: All of the parallel processes find many solutions independently of each other, but they hardly benefit from new bounds found by other solvers. Similar results can be observed when gaining parallelism by splitting the search space, c.f. Fig. 2. Here, a lot of solutions are found in disjoint parts of the search space. Exchanging bounds on the objective among the worker threads leads to small improvements of the running time, but the overall speedup is disappointing (Fig. 3).

Our second running example, queens, shows a different behavior. Here, the running time is dominated by proving optimality, which is proving unsatisfiability.

---

[2] http://www.minizinc.org/challenge2013/probs/cargo/challenge04_1s_626.dzn.
[3] http://www.minizinc.org/challenge2014/probs/mqueens/n12.dzn.

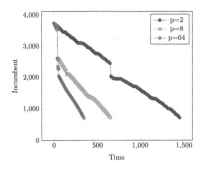

Fig. 1. cargo: results for portfolio parallel solving

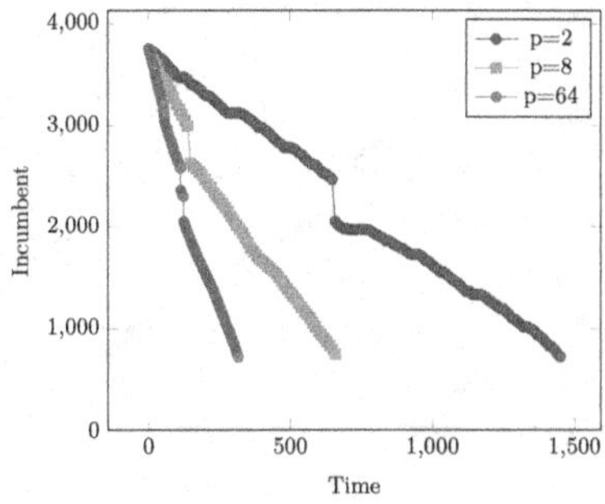

Fig. 2. cargo: results for search space splitting parallel solving

# CPUs	Workers	port	SSS
2	1	155s	155s
8	7	102s	22s
64	63	39s	11s

Fig. 3. queens: run times.

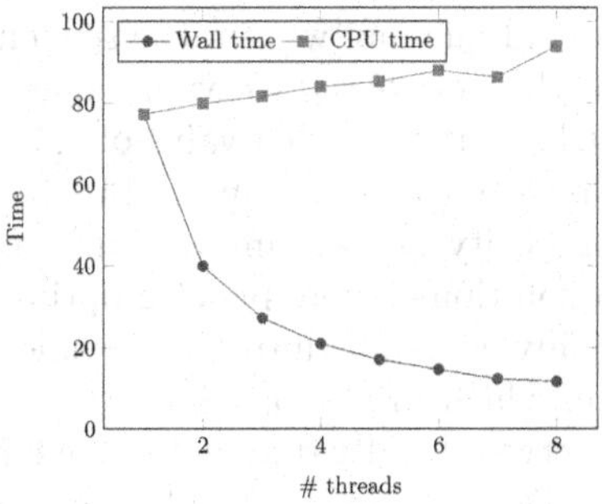

Fig. 4. queens: scaling behavior of Gecode

The portfolio solver shows some, but very limited speedup. Apparently, there is no small optimality proof here, and exchanging clauses among the solvers is of limited success due to the sequential structure of resolution proofs [13]. Search space splitting is much more promising here, with a speedup of 6.9 when using 7 workers threads instead of 1. Unfortunately, using more cores yields only limited additional speedup. Almost linear speedups can also be observed when running the parallel version of the CP solver Gecode on this benchmark, c.f. Fig. 4. Gecode also uses search space splitting to gain parallelism, and work stealing for load balancing.

We summarize these experiments with two main observations. First, search space splitting is superior to portfolio solving in terms of proving optimality. When run on many cores, it is crucial to split the work space in equally hard chunks to benefit from more parallel threads. Second, both approaches scale poorly when many suboptimal solutions can be found.

4.2 Sequential Optimization

To gain a better understanding of the results we found in the parallel setting, we discuss results obtained by running the sequential solver. For the two benchmarks

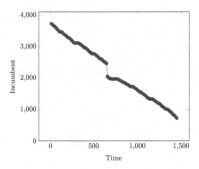
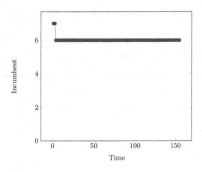

Fig. 5. `cargo`: slow convergence towards optimal result. **Fig. 6.** `queens`: slow proof of optimality.

`cargo` and `queens`, we run the sequential version of CHUFFED, and periodically record the best objective value found so far. The results for `cargo` can be seen in Fig. 5: starting with a value of $3,757$, the objective value is improved steadily, and drops to the optimum of 714 after $1,453$ s. In this case, the solver spends the vast majority of the running time improving the solution, and finds $2,654$ different solutions before proving optimality. Interestingly, the improvement speed is roughly constant during the whole run, and the final call to the decision procedure, which proves optimality, is not harder than previous calls. For `queens`, we observe a totally different behavior, c.f. Fig. 6. An optimum solution is found within 4 s, whereas the proof of optimality takes another 151 s.

This behavior is reflected by the difficulty of the respective decision problems. For both benchmarks, we added different bounds on the objective, and aborted the solver after finding the first solution. For `cargo` (Fig. 7), we ran this experiments for objective values in the interval $[700, 1000]$, i.e. for values close to the optimum objective value. The maximum running time observed was 30 s, and average running times of 3 s. A very interesting result of this experiment is that the running time close to the optimum solution is not higher than running times for higher bounds. So if we knew a good bound on the objective in advance, we might run the solver with a tightened bound, and find an optimal solution much faster. The `queens` benchmark shows a totally different behavior, c.f. Fig. 8. Here, the proof of optimality is hard, whereas both finding solutions and proving bounds tighter than the optimal value is extremely fast.

5 Search Space Splitting

As shown in Sect. 4, parallelisation by splitting the search space in disjoint parts allows for very good speedups, especially for proving optimality. Unfortunately, this approach comes with some drawbacks. After finding a solution, a sequential algorithm will continue its search, using a tighter bound on the objective value for further pruning. In parallel, worker threads may therefore search parts of the search space that would not be searched by a sequential algorithm, which may dramatically decrease the efficiency, as mentioned in [6].

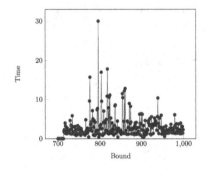

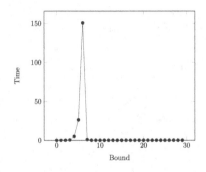

Fig. 7. `cargo`: time for decision problem, depending on bound on objective value.

Fig. 8. `queens`: time for decision problem, depending on bound on objective value.

In order to keep waiting times of worker threads low, it is common to store some jobs, i.e. chunks of the search space, at the master process. Whenever a worker finishes working on its part of the search space, a new chunk of work can be provided without waiting for another worker to provide work. In the worst case, this can lead to situations in which none of the workers searches the part of the search space containing the optimum solution. The decision how to split the search space is very important for gaining some benefit from this approach: Let $\phi = (V, D, C)$ denote an unsatisfiable CSP, and $C' \subset C$ a minimum unsatisfiable core, i.e. a set of constraints such that (V, D, C') is already unsatisfiable. Splitting on a variable that does not occur in C' will then be less likely to speed up the parallel solver. To overcome this problem, VSIDS activities can be used to choose variables for splitting the search space. As VSIDS focusses on variables that were involved in conflicts, this prevents branching on uninteresting variables. In our implementation, the master sends the empty job to one worker, which starts to work on this job. Whenever work has to be stolen, the master sends a request to one of the slaves, which creates new jobs according to its topmost branching decisions.

Example 1. *Assume a worker is working on a job given as $x_1 \wedge x_2$, and its topmost branching decisions are x_3 and x_4. If asked to provide two new jobs, it fixes its new job to $x_1 \wedge x_2 \wedge x_3 \wedge x_4$, and reports this to the master. In turn, the master creates new jobs $x_1 \wedge x_2 \wedge x_3 \wedge \bar{x}_4$ and $x_1 \wedge x_2 \wedge \bar{x}_3$.*

LCG solvers can reuse information about failed search to further prune the search space. Thus, the time required by a LCG solver to refute a part of the search space depends on previous search. If parallelism is gained by splitting the search space and running LCG solvers on disjoint parts of it, this may decrease the achievable speedup: whenever a worker receives a new chunk of the work space, it needs to learn clauses which are relevant to this new subspace, which might be the same as clauses for other chunks of the work space. Figure 9 shows the total number of conflicts occurring while solving `queens`. Here, the number of conflicts increases with additional processes. To reduce this burden, we exchange learnt clauses between solvers.

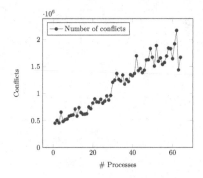

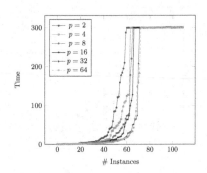

Fig. 9. `queens`: number of conflicts in search space splitting parallel solving without clause sharing

Fig. 10. `suite`: scaling of SSS with number of cores

Unless something different is stated for single experiments, this solver exchanges bounds on the objective, and short clauses. Short learnt clauses are exchanged between the solver processes. The bound on their size is adjusted dynamically such that approximately 10 % of the clauses in the database are received from other solvers, which gave good results in our tests. As in [12], we check the number of imported clauses regularly, and adjust the threshold on clause size to exchange if too many or too few clauses were received.

To evaluate parallelization approaches we created a set of 110 different benchmarks, `suite`, taken from MiniZinc challenges for 2013 and 2014.[4] We used a time limit of 5 min. We denote by SSS our search space splitting parallel solver. Figure 10 shows the scaling behavior for 2 to 64 processes. Significant speedups can be observed for up to 32 parallel processes. In Fig. 11, we compare the results for a parallel solver on 64 cores with the ones obtained using the sequential solver. The parallel solver clearly outperforms the sequential version. Furthermore, 12 more instances can be solved to optimality within 5 min.

In Fig. 12, we compare the running times with and without exchanging bounds on the objective between the solver processes. In some cases, e.g. if a good solutions can be found by all of the parallel solvers, there is only a small difference, whereas there is a huge difference for other instances, and 3 more benchmarks can be solved to optimality. In other words, it is crucial for the performance of a parallel LCG solver to find and communicate good bounds on the objective value as fast as possible.

Note that the benchmarks which timed out do not mean an equal result: As we are dealing with optimization problems, they often time out with different incumbent solutions. This issue will be further considered in Sect. 7.

---

[4] The exact set of instances is available at people.unimelb.edu.au/pstuckey/pchuffed.

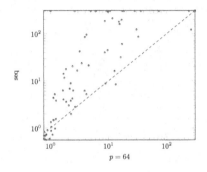

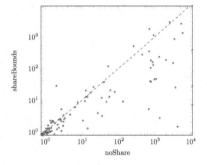

Fig. 11. `suite`: comparison between sequential and SSS with 64 cores

Fig. 12. `suite`: impact of bounds sharing in SSS with 64 cores

6 Portfolio

Portfolio solving is a common approach for parallel SAT solving. In this section, we investigate the behavior of a portfolio CP-solver with learning. As common in parallel SAT, the solvers are diversified by initializing their variable activities randomly. Additionally, we allow for some communication between the solver processes. As we are using a master-slave-architecture, a portfolio approach is simulated by sending the empty job (i.e. the empty conjunction) to each solver. For clause exchange, the same policy as in the SSS setting is used. We denote this solver as `port`. Figure 13 compares the running times of the sequential solver, and the portfolio solver on 64 cores, with a time limit of 5 min. Only little speedup can be observed for easy instances, whereas parallelism pays off for harder instances, and results in 10 more solved instances. The scaling behavior is shown in Fig. 14. Both significant speedups and an increased number of solved instances can be observed when using more CPU cores. On the other hand, 36 of the benchmarks time out, so either no optimum result was found, or the proof of optimality could not be completed.

In parallel SAT solving, it is a well-known fact that clause exchange is very helpful, especially for unsatisfiable instances. For parallel LCG, communication is also beneficial, but it appears harder to determine which, and how many clauses should be exchanged. Figure 15 compares the results of a portfolio solver on 64 cores with our adaptive clause exchange policy to a portfolio solver which only exchanges the incumbent objective value. Communicating learnt clauses yields a significant speedup and 6 more solved instances. Although this is a significant improvement, the power of clause exchange for parallel LCG appears limited. Further experiments showed that the exchange of clauses of size at most 2 speeds up the computation, whereas larger clauses do not always help, and may even significantly impede solving. As it appears difficult to determine the right choice of clauses to exchange, we used the conservative, adaptive approach. Recent work in SAT has emphasized the fact that many learnt clauses are not helpful

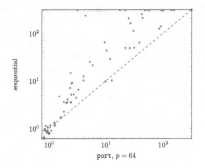

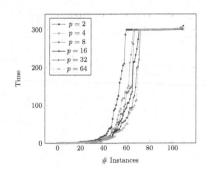

Fig. 13. `suite`: comparison between sequential and `port`

Fig. 14. `suite`: scaling of `port` with number of cores

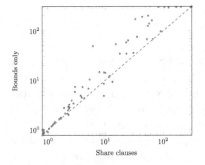

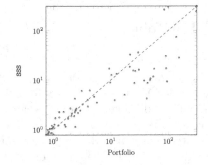

Fig. 15. `suite`: impact of clause sharing on `port` with 64 cores

Fig. 16. `suite`: comparison of SSS and `port`, 64 cores

for satisfiable formulas [19]. As the optimization process consists of solving a sequence of satisfiable problems, followed by one unsatisfiable one—the proof of optimality—this may be the reason for these results. Nevertheless, the portfolio solver is surprisingly strong. Using 64 cores, it can solve one instance that cannot be solved optimally by the search space splitting solver, c.f. Fig. 16.

7   Objective Probing

Both search space splitting and portfolio approaches yield good results. Nevertheless, they do not make use of the following observations from Sect. 4: in many cases finding a good solution is not harder than finding any solution. Finding good solutions early prunes the search, using the tighter objective bounds, and conversely, finding them late result in superfluous search, and may reduce the benefits of parallelism, c.f. Fig. 12. Hence, it appears promising to try to push a parallel solver towards finding good solutions quickly.

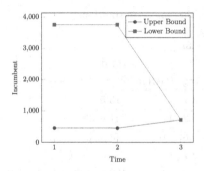 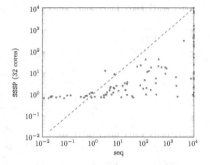

Fig. 17. `cargo`: objective probing portfolio solver

Fig. 18. `suite`: speedup in finding good solutions with objective probing SSSP

To do so, we guess bounds on the objective, and use some solver processes to probe whether there exists a solution satisfying the guessed bound, or not. A similar approach was already used for parallel Boolean optimization in [16], where the authors compute upper and lower bounds concurrently. When using n processes for objective probing, we use bounds

$$bound(i) = lb + \left\lfloor \frac{i(ub - lb)}{n} \right\rfloor, \tag{1}$$

where lb and ub denote lower and upper bounds on the objective value, and the ith process solves $\text{OPTIMIZE}(\phi, z, lb, bound(i))$.

Figure 17 shows the impact of this approach when solving the `cargo` benchmark. Compared to the naïve portfolio approach, c.f. Fig. 1, an impressive speedup of 483 is achieved, as the solver finds an optimum solution and proves its optimality within 3 s. Furthermore, probing objective values yields lower bounds on the objective. Thus, this approach allows for estimating the quality of solutions.

In the remainder of this section, we will discuss how to implement the objective probing, and show results. As we deal with optimization problems, we also consider the quality of solutions found. Therefore, we ran the sequential version of CHUFFED on each benchmark with a time limit of 3 h, and recorded the best solution found. Then, we tested how long it takes the parallel solver to find a better solution, or prove that no better solution exists.

7.1 Objective Probing in Search Space Splitting

For the search space splitting solver, we split the workers in three groups of equal size. Workers from the first group run on split parts of the search space as before. Workers from the second subset start by guessing an objective value according to Eq. 1. If this guess is refuted, i.e. a proof is found that no solution exists with

Table 1. `suite`: speedups when searching for good solutions.

#CPUs	SSS				SSSP			
	All		Hard		All		Hard	
	Avg	Median	Avg	Median	Avg	Median	Avg	Median
4	2.8	2	10.6	4.5	3.7	3.2	15.5	7.4
8	5	3.8	25.5	9.7	6.2	4	41.8	20.2
16	6.7	5.9	41.2	19.5	9.6	7.6	78.9	34.6
32	9.6	8	72.5	58.9	12.7	13.3	121.3	58.5
64	12.7	15	136.8	104	15.6	13.8	193.8	107

an objective value satisfying the bound, or it is implied, i.e. a better solution is found, they join the workers from the first group. The remaining threads behave like the ones from the second group, but re-guess bounds on the objective, until half of the given time limit is reached. This can be seen as a hyper-binary search on the objective, as the interval between lower and upper bound is split in several parts. We denote this solver as SSSP.

Figure 18 shows the impact of this technique on the search for good solutions. For very easy instances, which can be solved in less than one second, the parallel solver is slower than the sequential version. For harder instances, significant speedups can be observed. Table 1 shows the geometric average, and median speedups obtained on all benchmarks, and on hard ones. Here, a benchmark is considered hard if the sequential CHUFFED does not terminate within 300 s. The average speedup on all instances is sublinear, as many of them are too easy and do not allow for sufficient speedups by the parallel solver. Conversely, the speedup on hard instances is significant, and superlinear for every configuration. The configuration which uses objective probing, SSSP, reaches an average speedup of 193.8 on 64 cores. On 8 and 16 cores, it is even faster than SSS on 16 and 32 cores, respectively. On benchmarks of medium difficulty, the results are mixed. Figure 21 compares the results of the SSS and SSSP configuration, when using 64 cores. The SSSP configuration is significantly faster on some benchmarks, and slightly slower on some others. This is especially the case if probing fails in many cases, and thus only yields improved lower bounds on the objective instead of tighter pruning. Summarizing, combining a Search Space Splitting solver with objective probing gives an additional boost on the performance, especially for hard problems with a large value range for the objective value. Splitting the solvers in three groups of equal size works well on our benchmark suite, and it appears hard to find better choices that work well for all benchmarks, or adapt the group size dynamically.

7.2 Portfolio Solving and Objective Probing

In portfolio solving, guessing bounds on the objective value may be seen as an additional source of diversification for the solvers. As in the SSSP configuration,

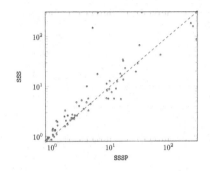

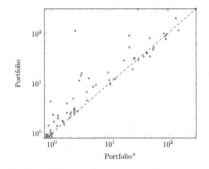

Fig. 19. suite: comparison between SSS and SSSP

Fig. 20. suite: comparison between port and portP

Table 2. suite: speedups when searching for good solutions.

#CPUs	port				portP			
	All		Hard		All		Hard	
	Avg	Median	Avg	Median	Avg	Median	Avg	Median
4	3.2	2	13.4	4.3	4.3	2.4	18.2	7.4
8	4.7	3.7	23.5	9	6.1	4.2	33.9	8.43
16	6.1	4.5	39.4	14.6	9.2	7	68.4	26.9
32	7.7	7.4	62.1	38	11.4	8.9	107.5	77.4
64	9.1	10.1	84.6	42	13.6	14.6	152	133.6

we split the solver processes in three groups. After objective probing is finished, the respective solvers continue running as in the normal portfolio configuration (Fig. 19).

On benchmarks of medium difficulty, this approach outperforms the common portfolio configurations, as can be seen in Fig. 20.

The reason for this behavior seems to be the following: The portfolio approach is fast in finding good solutions, but for proving optimality it does not scale as well as the search space splitting solver. Thus, the solving process is accelerated if better solutions are found early, but it is not slowed down too much if some workers spend computation time on proving lower bounds instead of participating in the proof of optimality.

Table 2 shows the speedups obtained when searching for good solutions. Again, superlinear average speedups can be observed for all configurations when considering only the hard benchmarks, reaching a maximum of 152 for the portfolio solver with objective probing, denoted Portfolio$^+$, and 64 cores. Here, the impact of objective probing is even larger than for the SSS solver. Interestingly, the difference is small on 8 cores, and grows larger when using more parallel workers, which may be a hint that the normal portfolio solver does not achieve sufficient diversification when using many cores. Furthermore, the median speedup on hard benchmarks is even higher than one obtained by the SSS solver.

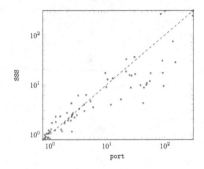

Fig. 21. suite: comparison between SSS and port, 64 cores

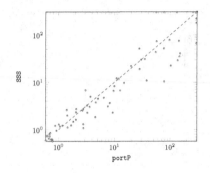

Fig. 22. suite: comparison between SSS and port, 8 cores

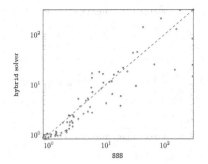

Fig. 23. suite: comparison between hybrid and SSS solver, 64 cores

#CPUs	all		hard	
	avg	median	avg	median
4	4.3	3.3	18.4	6.3
8	6.3	4.7	38.2	19.9
16	9.6	6.4	79.9	43
32	11.7	8.8	116	62.4
64	15.7	16	196	140

Fig. 24. suite: speedups for hybrid when searching good solutions

8 A Hybrid Solver

When comparing the results of the portfolio solver with those of the SSS solver, it becomes obvious that these approaches work differently well on different problems. As can be seen in the Figs. 21 and 22, SSS tends to perform better on average, both on 8 and 64 cores. Nevertheless, the SSS solver times out on some instances that can be solved by the portfolio solver, and vice versa. It appears therefore promising to combine both approaches to a (meta-)portfolio, which combines SSS, initial guesses on the objective value and a SAT-like portfolio solver. We therefore change the behavior of the SSSP-solver as follows. Workers from the second group, which finish working on the respective guessed objective values, continue working as portfolio solvers rather than joining the SSS solvers. Thus, they are capable of searching the whole search space instead of being fixed on one subspace, which maintains the strength of the highly agile VSIDS-based branching. We denote this as hybrid. Interestingly, this is especially advantageous when using just a few cores. Here, the number of solved instances is increased remarkably. The search for good solutions is improved significantly: Using 64 cores, the median of speedups increases from 13.8 to 16 on all instances, and from 107 to 140 on the hard ones (Fig. 23).

9   Conclusion

We presented results of different approaches to parallelize the LCG solver CHUFFED. A portfolio approach performs astonishingly well, especially when trying to find good solutions rather than proving optimality. Here, an approach based on search space splitting is more successful, although is does not scale as smoothly as classical CP solvers. To avoid redundant work and hence gain better speedups, it is important to communicate some information between the parallel solvers. The most important information is the best incumbent objective value, whereas the impact of exchanging longer clauses is limited (Fig. 24).

A hybrid solver which combines probing the objective value, portfolio solving and search space splitting yields significant speedups on a wide range of benchmarks. When trying to find better results than the sequential version of CHUFFED, the speedup obtained is significantly superlinear.

On the contrary, the speedups on unsatisfiable instances, e.g. when proving optimality, are sublinear, which matches results from parallel SAT solving.

Acknowledgements. The authors would like to thank Graeme Gange for the fruitful discussions, and Prof. Dirk Nowotka for providing the computational resources for the experiments conducted in this paper.

References

1. Amadini, R., Gabbrielli, M., Mauro, J.: A multicore tool for constraint solving. In: Yang, Q., Wooldridge, M. (eds.) Proceedings of the Twenty-Fourth International Joint Conference on Artificial Intelligence, IJCAI 2015, Buenos Aires, Argentina, 25–31 July 2015, pp. 232–238. AAAI Press (2015)
2. Amadini, R., Stuckey, P.J.: Sequential time splitting and bounds communication for a portfolio of optimization solvers. In: O'Sullivan [21], pp. 108–124
3. Bordeaux, L., Hamadi, Y., Samulowitz, H.: Experiments with massively parallel constraint solving. In: Boutilier [4], pp. 443–448
4. Boutilier, C. (ed.): IJCAI 2009, Proceedings of the 21st International Joint Conference on Artificial Intelligence, Pasadena, California, USA, 11–17 July 2009 (2009)
5. Chu, G.: Improving Combinatorial Optimization. Ph.D. thesis, University of Melbourne (2011)
6. Chu, G., Schulte, C., Stuckey, P.J.: Confidence-based work stealing in parallel constraint programming. In: Gent [10], pp. 226–241
7. Ehlers, T., Nowotka, D., Sieweck, P.: Communication in massively-parallel SAT solving. In: 26th IEEE International Conference on Tools with Artificial Intelligence, ICTAI 2014, Limassol, Cyprus, 10–12 November 2014, pp. 709–716. IEEE Computer Society (2014)
8. Feydy, T., Stuckey, P.J.: Lazy clause generation reengineered. In: Gent [10], pp. 352–366
9. de la Banda, M.G., Stuckey, P.J., Hentenryck, P.V., Wallace, M.: The future of optimization technology. Constraints 19(2), 126–138 (2014)
10. Gent, I.P. (ed.): CP 2009. LNCS, vol. 5732. Springer, Heidelberg (2009)

11. Gent, I.P., Jefferson, C., Miguel, I., Moore, N., Nightingale, P., Prosser, P., Unsworth, C.: A preliminary review of literature on parallel constraint solving. In: Proceedings PMCS 2011 Workshop on Parallel Methods for Constraint Solving (2011)

12. Hamadi, Y., Jabbour, S., Sais, L.: Control-based clause sharing in parallel SAT solving. In: Boutilier [4], pp. 499–504

13. Katsirelos, G., Sabharwal, A., Samulowitz, H., Simon, L.: Resolution and parallelizability: barriers to the efficient parallelization of SAT solvers. In: desJardins, M., Littman, M.L. (ed.) Proceedings of the Twenty-Seventh AAAI Conference on Artificial Intelligence, 14–18 July 2013, Bellevue, Washington, USA. AAAI Press (2013)

14. Lin, Y., Kumar, V.: Performance of and-parallel execution of logic programs on a shared-memory multiprocessor. In: FGCS, pp. 851–860 (1988)

15. Machado, R., Pedro, V., Abreu, S.: On the scalability of constraint programming on hierarchical multiprocessor systems. In: 42nd International Conference on Parallel Processing, ICPP 2013, Lyon, France, 1–4 October 2013, pp. 530–535. IEEE Computer Society (2013)

16. Martins, R., Manquinho, V., Lynce, I.: Parallel search for boolean optimization. In: RCRA International Workshop on Experimental Evaluation of Algorithms for Solving Problems with Combinatorial Explosion (2011)

17. Moisan, T., Quimper, C.-G., Gaudreault, J.: Parallel depth-bounded discrepancy search. In: Simonis, H. (ed.) CPAIOR 2014. LNCS, vol. 8451, pp. 377–393. Springer, Heidelberg (2014)

18. Moskewicz, M.W., Madigan, C.F., Zhao, Y., Zhang, L., Malik, S.: Chaff: engineering an efficient SAT solver. In: Proceedings of the 38th Design Automation Conference, DAC 2001, Las Vegas, NV, USA, 18–22 June 2001, pp. 530–535. ACM (2001)

19. Oh, C.: Between SAT and UNSAT: the fundamental difference in CDCL SAT. In: Heule, M., Weaver, S. (eds.) SAT 2015. LNCS, vol. 9340, pp. 307–323. Springer, Heidelberg (2015). doi:10.1007/978-3-319-24318-4_23

20. Ohrimenko, O., Stuckey, P., Codish, M.: Propagation via lazy clause generation. Constraints 14(3), 357–391 (2009)

21. O'Sullivan, B. (ed.): CP 2014. LNCS, vol. 8656. Springer, Heidelberg (2014)

22. Rao, V.N., Kumar, V.: Superlinear speedup in parallel state-space search. In: Nori, K.V., Kumar, S. (eds.) FSTTCS 1988. LNCS, vol. 338, pp. 161–174. Springer, Heidelberg (1988)

23. Régin, J.-C., Rezgui, M., Malapert, A.: Embarrassingly parallel search. In: Schulte, C. (ed.) CP 2013. LNCS, vol. 8124, pp. 596–610. Springer, Heidelberg (2013)

24. Régin, J., Rezgui, M., Malapert, A.: Improvement of the embarrassingly parallel search for data centers. In: O'Sullivan [21], pp. 622–635

25. Schrijvers, T., Tack, G., Wuille, P., Samulowitz, H., Stuckey, P.: Search combinators. Constraints 18(2), 269–305 (2013)

26. Schulte, C.: Parallel search made simple. Technical report TRA9/00, School of Computing, National University of Singapore, 55 Science Drive 2, Singapore 117599, September 2000, to appear

27. Schulte, C., Stuckey, P.J.: Efficient constraint propagation engines. Trans. Program. Lang. Syst. 31(1), 2:1–2:43 (2008)

28. Schutt, A., Feydy, T., Stuckey, P.J., Wallace, M.G.: Explaining the cumulative propagator. Constraints 16(3), 250–282 (2011)

Parallel Composition of Scheduling Solvers

Daniel Fontaine[1], Laurent Michel[1(✉)], and Pascal Van Hentenryck[2]

[1] University of Connecticut, Storrs, CT 06269-2155, USA
ldm@engr.uconn.edu
[2] University of Michigan, Ann Arbor, USA

Abstract. Recent work in model combinators, as well as projects like G12 and SIMPL, achieved significant progress in automating the generation of complex and hybrid solvers from high-level model specifications. This paper extends model combinators into the scheduling domain. This is of particular interest as, today, both Constraint Programming (CP) and Mixed-Integer Programming (MIP) perform well on scheduling problems providing different capabilities and trade-offs. The ability to construct hybrid scheduling solvers to leverage the strengths of both technologies as well as multiple problem encodings through high-level model combinators provides new opportunities. Complex parallel hybrids can be synthesized with minimal effort on the part of the user and provide substantial performance benefits over standalone solvers.

1 Introduction

Modern desktop and laptop systems are overwhelmingly parallel machines with 2–8 cores. At the same time, no combinatorial optimization approach dominates the field and it is often necessary to conduct extensive experiments to determine which techniques work best. Parallel tree search [15,17,21,26,28,31] has been under investigation for two decades and is possibly the sole effort to exploit small and large scale parallelism. Despite the advent of parallel hardware and the absence of dominating combinatorial techniques, surprisingly little has been done to produce *robust* parallel algorithmic combinatorial optimization techniques.

OBJECTIVE-CP [33], provides an architecture capable of filling this void. In [8], *runnables* and *model combinators* were introduced to greatly facilitate the creation of semantically meaningful composite solvers. *Runnables* represent optimization programs that combine a solver, a model and a search procedure.

Our purpose is to illustrate the benefits associated with this architecture for scheduling which has long been considered a strength of Constraint Programming (CP) solvers. Recent work on Failure-Directed Search [35] demonstrates that CP continues to provide state-of-the-art results. Work done in the last few years, e.g., [14] also shows that modern Mixed-Integer Programming (MIP) solvers using standard encodings are now competitive with, and sometimes superior to, commonly used CP scheduling solvers. Such results suggest that composite techniques leveraging both technologies are in order. To enable this effort, this paper extends *model combinators* [8] in the following ways:

© Springer International Publishing Switzerland 2016
C.-G. Quimper (Ed.): CPAIOR 2016, LNCS 9676, pp. 159–169, 2016.
DOI: 10.1007/978-3-319-33954-2_12

- *Runnables* are extended to handle modeling abstractions in the scheduling domain (concepts like tasks and resources).
- The ability to choose among several encodings for the same solver is supported (two MIP encodings are supported).
- Bound/solution communication is extended to handle specific limitations imposed by the Gurobi APIs.

2 Related Work

The approach in this paper is related to portfolio solvers such as CPHydra [19], Sunny-CP [2] and SATzilla [36]. CPHydra and Sunny-CP both use a k-nearest neighbor algorithm for scheduling a portfolio of algorithms while SATzilla uses runtime prediction models to dynamically choose an optimal algorithm. Proteus [11] introduced a system which makes use of a complex dynamic rule set to produce (potentially) multiple CP and SAT encodings and applies multiple solvers to each encoding with very good results. Hence, Proteus's use of multiple problem representations and different solving technologies makes it the most closely related system.

The approach in this paper has a number of very novel aspects. First, unlike all previous approaches, *our solvers are not competing with one and other, but instead, cooperating directly by exchanging bounds and solutions in real time.* This paper acknowledges that for many problems there may not be a single best algorithm and, therefore, provides a framework for solvers to cooperate in parallel, translating and sharing solutions between different encodings.

Furthermore, the portfolio approach often relies on complex machine learning techniques for choosing solvers and, in the case of Proteus, encodings. While portfolios can be effective, this paper argues in favor of high-level tools to transform, manipulate and compose solvers directly. This approach delivers a clearer picture of what solvers are running, why they are running, how they are performing and, ultimately, how they might benefit from different composition strategies.

Tools that facilitate the authoring of solver independent models were introduced over the last decade. The Comet Modeling Language (CML) strived to provide a full programming language in which models could be specified, manipulated and composed in sophisticated ways without the need for annotations. *G12* models written with *mini-Zinc* feature solver independent capabilities, model rewriting and even column generation and branch-and-price hybrids [24] via *Cadmium* [4]. *SIMPL* [37] is a high level modeling language based on the search-infer-relax philosophy. *Essence* [9] is designed for model specification and has recently been combined with *Conjure* [1] to automatically derive constraint models. Finally, work has already been done in providing a rich language of combinators within the context of search [27] and is revisited in [33].

```
1 id<ORModel> m = [ORFactory createModel];
2 // data setup ...
3 id<ORIntRange> J          = RANGE(m,0,nbJobs-1);
4 id<ORIntRange> M          = RANGE(m,0,nbMach-1);
5 id<ORIntMatrix> D         = [ORFactory intMatrix: m range: J : M];
6 id<ORIntMatrix> resource = [ORFactory intMatrix: m range: J : M];
7 // variables
8 id<ORTaskVarMatrix> task = [ORFactory tvMatrix:m range:J:M horizon:H duration:D];
9 id<ORIntVar> makespan     = [ORFactory intVar: m domain: RANGE(m,0,totalDur)];
10 id<ORTaskDisjunctiveArray> disjunctive = [ORFactory disjunctiveArray:m range: M];
11 // model
12 [m minimize: makespan];
13 for(ORInt i = J.low; i <= J.up; i++)
14    for(ORInt j = M.low; j < M.up; j++)
15       [m add: [[task at: i : j] precedes: [ task at: i : j+1]]];
16 for(ORInt i = J.low; i <= J.up; i++)
17    [m add: [[task at: i : Machines.up] isFinishedBy: makespan]];
18 for(ORInt i = J.low; i <= J.up; i++)
19    for(ORInt j = M.low; j <= M.up; j++)
20       [disjunctive[[resource at: i : j]] add: [task at: i : j]];
21 for(ORInt i=M.low; i <= M.up; i++)
22    [m add: disjunctive[i]];
```

Fig. 1. High-level technology-independent model in OBJECTIVE-CP.

3   Composition of Scheduling Solvers

3.1   Jobshop

An $n \times m$ jobshop problem is a set of n jobs $\{J_1, J_2, \ldots, J_n\}$ that must be processed on m machines $\{M_1, M_2, \ldots, M_m\}$. Each job J_i must be processed on every machine M_k requiring exclusive use of that machine for a processing duration $p_{i,k}$. The processing of job J_i on machine M_k is called the $task_{i,k}$. The sequence $(\sigma_1^i \ldots \sigma_m^i)$ represents a permutation of the machines indicating the processing order of tasks for job J_i. For example, if $(\sigma_1^i, \sigma_2^i, \sigma_3^i) = (2, 3, 1)$, this implies the tasks for job J_i are processed in the order $(task_{i,2}, task_{i,3}, task_{i,1})$. Note we use the same notation and MIP models as [14].

3.2   An Objective-CP Jobshop Model

With OBJECTIVE-CP, models are containers capturing constraints that must be satisfied as well as a relevant objective function. Figure 1 illustrates the creation of a high-level declarative model. Line 1 creates a model m while lines 3–4 create ranges for the jobs J and machines M. Lines 5–6 create matrices holding the processing time of the tasks as well the resources that any task requires. Line 8 creates a matrix $task$ holding all the tasks. Line 9 creates a variable representing the makespan of the instance and line 10 creates an array of disjunctive resources (as many as M). Lines 12–22 start by stating the objective function and creating the constraints. The loops state the job precedence constraints, the fact that the makespan follows the end of each job, enforces the duration of each task on its disjunctive resource and finally adds the disjunctive resources to the model.

Following [8], we emphasize that this model is purely descriptive, technology agnostic and captures a triplet $\langle X, C, O \rangle$ in which X is the set of variables, C

min *makespan*

$$\text{s.t.} \begin{cases} precedes(task_{i,j}, task_{i+1,j}) & \forall i \in M, \forall j \in J \\ finished\_by(task_{m,j}, makespan) & \forall j \in J \\ disjunctive(\{task_{\sigma_k^j,j} \mid j \in J, k \in 1 \ldots m, \sigma_k^j = r\}) & \forall r \in M \end{cases}$$

Fig. 2. Global constraint formulation

min *makespan*

$$\text{s.t.} \begin{cases} x_{i,j} \geq 0 & \forall j \in J, \forall i \in M \\ x_{\sigma_h^j,j} \geq x_{\sigma_{h-1}^j,j} + p_{\sigma_{h-1}^j,j} & \forall j \in J, h \in 2, \ldots, m \\ x_{i,j} \geq x_{i,k} + p_{i,k} - z_{i,j,k} * \mathcal{M} & \forall j, k \in J, k < j, i \in M \\ x_{i,k} \geq x_{i,j} + p_{i,j} - (1 - z_{i,j,k}) * \mathcal{M} & \forall j, k \in J, k < j, i \in M \\ makespan \geq x_{\sigma_m^j,j} + p_{\sigma_m^j,j} & \forall j \in J \\ z_{i,j,k} \in \{0,1\} & \forall i \in M, \forall j \in J, \forall k \in J \end{cases}$$

Fig. 3. Disjunctive formulation of Jobshop using big-M notation ($\mathcal{M}$).

is the set of constraints and O is an optional objective function. To exploit this model, it is necessary to *concretize* the model into a specific *program*.

Each technology imposes restrictions on what vocabulary can be used to describe models. For instance, a MIP requires linear inequalities over discrete and continuous variables only. OBJECTIVE-CP uses *model transformations* such as τ to rewrite models into refined forms that are equivalent but conform to the requirements of the technology. Namely, $M_1 = \tau(M_0)$ captures the rewriting of M_0 into an equivalent M_1. Once rewritten, models are mapped into a solver. OBJECTIVE-CP achieves this through a *concretization* function γ that delivers an executable program for a technology T, i.e., $P = \gamma_T(\tau(m))$. The reader is referred to [8] for the full details and the formalization. The same high-level model can be concretized several times into multiple solver instances. In particular, OBJECTIVE-CP supports the simultaneous concretization of one model into both a CP solver and a MIP solver, yielding two *independent* programs.

Scheduling Reformulations. The OBJECTIVE-CP *model reformulations* must be adapted to scheduling. The input is the model presented in Fig. 1. Three reformulation operators are provided:

- τ_{CP}: Transforms the high-level model into a suitable CP encoding (Fig. 2).
- $\tau_{MIP-Disjunctive}$: Uses a big-M encoding for the disjunctives (Fig. 3).
- τ_{MIP-TI}: Uses the time-indexed formulation.

The implementation of the reformulation operators uses rewriting rules for the global constraints similar to those found in [5]. It creates auxiliary variables and visits the global constraints to replace them with linear encoded equivalents. For instance, the big-M linear rewriting for disjunctive is:

```
1 linearize(disjunctive) ⇒
2   with: intvar z_{i,j} ∈ {0,1} ∀ t_i, t_j ∈ tasks(disjunctive), t_i ≠ t_j
3   in:   forall t_i, t_j ∈ tasks(disjunctive) ∧ t_i ≠ t_j :
4         post: start(t_i) + duration(t_i) ≤ start(t_j) + z_{i,j} * M
5         post: start(t_j) + duration(t_j) ≤ start(t_i) + (1 - z_{i,j}) * M
```

```
1 ORInt heuristic(id<CPProgram> cp,ORInt i) {
2   return [cp globalSlack: disjunctive[i]] + 1000*[cp localSlack:disjunctive[i]];
3 }
4 ...
5 id<ORRunnable> r0 = [ORFactory CPRunnable:m solve:^(id<CPProgram> program) {
6   [cp forall: M orderedBy:^ORInt(ORInt i) {return heuristic(cp,i);} do:^(ORInt i){
7      id<ORTaskVarArray> t = disjunctive[i].taskVars;
8      [cp sequence: disjunctive[i].successors
9              by: ^ORDouble(ORInt i) { return [cp est: t[i]]; }
10             then: ^ORDouble(ORInt i) { return [cp ect: t[i]];}];
11 }];
12 [cp label: makespan];
13 }];
```

Fig. 4. Basic Disjunctive scheduling search procedure.

Similar rules exist for other *global constraints* such as *precedes* and *finish_by* as well as different rules for the time-indexed formulation.

Custom Search. Models produced by the reformulation operators above are still purely descriptive and must be concretized into a solver and coupled with a search procedure (when necessary) to obtain *runnables*. The empirical results use two custom search procedures. Figure 4 depicts a custom procedure as well as how to create a Constraint Programming runnable for model m (from Fig. 1) with that specific procedure. The code uses the slack that exist on the disjunctive resources to select a machine and sequences the tasks of that machine first. The sequencing in lines 9–11 uses a lexicographic heuristic based on earliest start time and earliest completion time to rank the tasks.

The second search is Large Neighborhood Search [22]. A jobshop version is taken from [20]. Namely, it is an iterative process in which each iteration limits the number of failures to $3 * |J| * |M|$. When the limit is reached, LNS randomly selects two machines as well as a time window and fixes the precedence that exist between tasks *outside* the time window in the incumbent and re-optimizes.

3.3 Runnables

Runnables are combinatorial optimization programs augmented with new capabilities to communicate with each other. A runnable R can consume and produce *products* that represent artifacts such as upper bounds, lower bounds, or even entire solutions. These communication capabilities are asynchronous in nature and permit runnables to communicate across thread boundaries. Two runnables R_0 and R_1 running in two distinct threads T_0 and T_1 can cooperate transparently.

Given a runnable R_0 derived from a high-level model $M = \langle X, C, O \rangle$, a solution σ from R_0 is a mapping from X to $\mathbb{Z}$ associating to each variable in X a value that satisfies all the constraints in C. Observe how solutions are encoded in term of the original high-level model. It enables the transcoding of a solution to another runnable R_1 also derived from model M. Given a solution σ and a target runnable t, the fragment appearing below creates a new concrete solution σ' adapted to t's encoding. The loop on line 3 iterates over all the variables in σ and *decodes* with the call on line 4 the assignment to variable x

```
1 id<ORModel> m = ... // Def. of Jobshop Model
2 id<ORModel> LinearModel = [ORFactory linearize:m encoding:Disjunctive];
3 id<ORRunnable> r0 = [ORFactory CPRunnable: m solve: search ];
4 id<ORRunnable> r1 = [ORFactory MIPRunnable: LinearModel];
5 id<ORRunnable> parallel = [ORCombinator completeParallel: r0 with: r1];
6 [parallel run];
```

Fig. 5. Running a CP and MIP encoding of jobshop in parallel.

in term of its representation in t. If t relies on an encoding of x with domain $D(x) = \{0 \cdots n\}$ into $n + 1$ binary variables $x_0 \cdots x_n$, the decoder produces a collection of assignments to cover all the binary variables y_i of t (only one of which is assigned 1). Line 6 installs the solution σ' in t and returns it.

```
1 transcode(Runnable t,Solution σ) → σ'
2     σ' := ∅
3     forall x in vars(σ):
4         {⟨yᵢ ↦ vᵢ⟩} := decode(x,σ(x),t)
5         σ' := σ' ∪ {⟨yᵢ ↦ vᵢ⟩}
6     t.inject(σ')
7     return σ'
```

Gurobi cannot tighten its upper bound to a new incumbent bound f^* and instead must install and validate the entire solution. A Gurobi runnable must thus consume solutions from a callback invoked at each node of its search tree.

3.4 Combinators

Figure 5 illustrates the lines of code required to create a composite parallel solver. Line 2 creates the linear reformulation. Lines 3 and 4 create the CP and MIP runnable from the original formulation m and the selected linear reformulation $linearModel$[1]. Finally line 5 creates the parallel composite and lines 6 executes the resulting hybrid. Note how all the integration and communication aspects are fully automated. Indeed, the parallel combinator automatically takes care of the necessary plumbing to concurrently share the various products and transcode solutions as needed. Interested readers are referred to [8] for further details.

4 Case Studies

Basic Results. Experimental results are provided for various standard instances of the jobshop problem. Some instances remain very difficult to solve to optimality, even for modest sizes. Results are presented on various solvers described below:

MIP Gurobi 6.04 MIP with 2 threads and a disjunctive encoding.
CP OBJECTIVE-CP solver with 2 threads and a common *global slack* heuristic.
CP ‖ MIP A parallel composite with CP and MIP solvers.
LNS$_{CP}$ ‖ MIP A parallel composite with a CP-based LNS and a MIP solver.
LNS$_{CP}$ ‖ CP A parallel composite with a CP-based LNS and a plain CP solver.

---

[1] Line 3 refers to the search procedure defined earlier with a closure and named *search*.

Table 1. Experimental Results for CP and MIP solvers as well as three hybrids.

Instances	CP		MIP		$CP \parallel MIP$		$LNS_{CP} \parallel MIP$		$LNS_{CP} \parallel CP$	
	time	ub	time	ub	time	ub	time	ub	time	ub
Orb01(10 × 10)	145.38	1059*	600.0	1072	176.12	1059*	600.0	1071	**41.96**	**1059***
Orb02(10 × 10)	6.80	888*	19.06	888*	8.36	888*	18.97	888*	**6.33**	**888***
Orb03(10 × 10)	600.0	1015	600.0	1021	600.0	1015	**600.0**	**1005**	600.0	1015
Orb04(10 × 10)	8.17	1005*	63.07	1005*	16.33	1005*	53.33	1005*	**7.67**	**1005***
Orb05(10 × 10)	132.46	887*	74.20	887*	110.82	887*	70.92	887*	**70.35**	**887***
Orb06(10 × 10)	57.37	1010*	528.22	1010*	135.53	1010*	600.0	1010	**52.05**	**1010***
Orb07(10 × 10)	53.22	397*	43.64	397*	39.15	397*	18.65	397*	**11.23**	**397***
Orb08(10 × 10)	467.19	899*	99.86	899*	6.82	899*	84.41	899*	**4.57**	**899***
Orb09(10 × 10)	**5.31**	**934***	75.36	934*	9.41	934*	85.55	934*	5.31	934*
Orb10(10 × 10)	66.24	944*	51.20	944*	33.87	944*	28.34	944*	**5.31**	**944***
la31(30 × 10)	600.0	2801	600.0	2003	600.0	2109	30.82	1784*	**17.23**	**1784***
la36(15 × 15)	600.0	2059	600.0	1292	600.0	1297	600.0	1281	**136.96**	**1268***
la37(15 × 15)	600.0	1855	600.0	1454	600.0	1478	**13.62**	**1397***	13.97	1397*
la38(15 × 15)	600.0	1633	600.0	1230	600.0	1243	**600.0**	**1196**	600.0	1255
la21(15 × 10)	600.0	1129	600.0	1079	600.0	1097	600.0	1058	**600.0**	**1046**

The time-indexed formulation is omitted as it is not competitive (This is consistent with the Findings in [14]). All experiments are run on Mac OS X 10.10.5 with 4 GB or memory and an Intel Core 2 Duo 2.13 GHz. OBJECTIVE-CP's edge finder is based on [34]. Table 1 shows the best upper bound achieved and the running time in seconds for each solver within 10 min. The MIP solver uses, by default, 2 threads. For fairness, the CP solver uses a parallel tree search with 2 threads too. Composite solvers use 1 thread for the first runnable and 2 threads for the second. For instance, $CP \parallel MIP$ uses 1 thread for CP and 2 for the MIP. A single star (*) indicates the optimal bound was found and proved.

Table 1 confirms that MIP and CP are competitive as reported in [14] with MIP outperforming CP quality-wise (la31, la36, la37, la38, la21) and timewise (Orb05, Orb08) on 7 of the 15 benchmarks. More interestingly, the composite $CP \parallel MIP$ proved to be more robust than either CP or MIP

Table 2. Solvers with 1–4 threads.

Inst.	CP			MIP			CPS
threads	1	2	4	1	2	4	3
orb05	70.2	75.3	29.6	32.9	43.4	17.8	60.4
orb07	38.6	48.6	9.7	40.6	26.5	34.6	39.1
orb08	1.7	600	600	123.3	55.6	68.7	1.3
orb10	28.1	32.1	29.5	65.80	30.5	15.5	24.8
la10	0.2	0.2	0.2	600	600	600	0.3
la11	1.5	2.4	2.2	600	600	600	1.2

Table 3. Number of bounds sent.

	$LNS_{CP} \parallel MIP$		$CP \parallel MIP$		$LNS_{CP} \parallel CP$	
	→MIP	→LNS	→MIP	→CP	→CP	→LNS
orb01	159	39	37	42	98	41
orb02	93	30	43	11	6	38
orb03	144	36	57	21	5	57
orb04	87	34	131	24	56	80
orb05	70	25	68	34	51	47
orb06	161	44	55	18	22	43
orb07	55	19	62	29	49	28
orb08	202	59	197	35	88	118
orb09	70	19	78	19	47	40
orb10	118	36	88	36	69	50

alone often getting running times competitive with or better than (in 1 instance: orb07) the best standalone solver. The composite improves the quality of the solution over the standalone CP on all the la instances. The $LNS_{CP} \parallel MIP$ composite does even better managing to close la31, la37 and delivering the best incumbent on la38. Finally, the $LNS_{CP} \parallel CP$ composite is the best hybrid of this pack. It yields high quality bounds from the LNS search and restores completeness through its reliance on a complete parallel CP search. The running times are often the best and this composite now closes la36 and further improves la40.

Hybrid Communication. Table 3 reports the number of bounds and solutions exchanged between parallel solvers within the three hybrids on orb instances. It shows substantial inter-solver communication in all three hybrids. The MIP solver receives more bounds than it generates from both CP and LNS_{CP}. The $CP\text{-}LNS_{CP}$ hybrid shows a mix with roughly equal bound generation on some instances and one dominant solver (not always the same) on others.

Robust Runnables. What is, perhaps, unexpected in Table 1 is the behavior on benchmark like Orb08 where CP takes 467 s to prove optimality, MIP requires 99 s for the same result while the $CP \parallel MIP$ composite completes in a mere 7 s. The explanation lies in the parallel search. Conventional wisdom dictates that the number of threads ought to be equal to the number of cores. When CP (or MIP) is executing alone, it carries out a parallel tree search with 2 threads. When executing in the composite, the CP solver uses a sequential search while the MIP uses a parallel search (with 2 threads). The observed behavior is a simple lack of robustness of the parallel tree search. When the optimum is found, CP can prove optimality near instantly. Finding the optimum however, proves difficult. If a node on the path from the root to that optimum is shared with other threads, the discovery of the optimum may be postponed until that node is stolen, and a substantial delay may be incurred.

This phenomenon happens within MIP solvers too and is illustrated in Table 2 where the data was collected on a quad-core MacPro with a Xeon at 3.2 Ghz running OSX 10.11. For instance, the solving time for MIP on Orb10 improves as threads are added while it barely moves for the CP solver while Orb05 and Orb07 experience the opposite effect (adding threads hurt Gurobi). To explore this fairness question Table 2 reports on a few instances involving the CP and MIP solvers with 1, 2 and 4 threads as well as a new composite, dubbed CPS, which composes a sequential CP solver with a parallel tree search CP solver. The number of threads can have unsettling effects, sometimes improving or worsening the solving time. The ability to use the composite $CP \parallel CP(2)$ alleviates the problem. Indeed, sequential and parallel CP share their bounds.

5   Conclusion

This paper extended *model combinators* to scheduling and provided empirical evidence that *model combinators* are useful in this domain. The approach emphasizes end-user flexibility and fosters the development of composites with custom search strategies and non-trivial parallelization. The net result is a malleable platform in which one can express sophisticated algorithms going beyond portfolios. In addition, the composite solvers are more than the sum of their constituents and yield a synergistic integration with little to no user-visible complexity. Recent work suggest that MIP can compete with CP on certain classes of scheduling instances. The parallel solvers derived here demonstrate that one can routinely outperform standalone solvers at little to no costs to end-users.

References

1. Akgun, O., Miguel, I., Jefferson, C., Frisch, A., Hnich, B.: Extensible automated constraint modelling (2011)
2. Amadini, R., Gabbrielli, M., Mauro, J.: SUNNY-CP: a sequential CP portfolio solver. In: Proceedings of the 30th Annual ACM Symposium on Applied Computing, SAC 2015, pp. 1861–1867. ACM, New York (2015)
3. De Moura, L., Bjørner, N.: Satisfiability modulo theories: introduction and applications. Commun. ACM **54**(9), 69–77 (2011)
4. Duck, G.J., De Koninck, L., Stuckey, P.J.: Cadmium: an implementation of ACD term rewriting. In: Garcia de la Banda, M., Pontelli, E. (eds.) ICLP 2008. LNCS, vol. 5366, pp. 531–545. Springer, Heidelberg (2008)
5. Duck, G.J., Stuckey, P.J., Brand, S.: ACD term rewriting. In: Etalle, S., Truszczyński, M. (eds.) ICLP 2006. LNCS, vol. 4079, pp. 117–131. Springer, Heidelberg (2006)
6. Fazel-Zarandi, M.M., Beck, J.C.: Solving a location-allocation problem with logic-based benders' decomposition. In: Gent, I.P. (ed.) CP 2009. LNCS, vol. 5732, pp. 344–351. Springer, Heidelberg (2009)
7. Fontaine, D., Michel, L.: A high level language for solver independent model manipulation and generation of hybrid solvers. In: Beldiceanu, N., Jussien, N., Pinson, É. (eds.) CPAIOR 2012. LNCS, vol. 7298, pp. 180–194. Springer, Heidelberg (2012)
8. Fontaine, D., Michel, L., Van Hentenryck, P.: Model combinators for hybrid optimization. In: Schulte, C. (ed.) CP 2013. LNCS, vol. 8124, pp. 299–314. Springer, Heidelberg (2013)
9. Frisch, A., Harvey, W., Jefferson, C., Martínez-Hernández, B., Miguel, I.: Essence: a constraint language for specifying combinatorial problems. Constraints **13**, 268–306 (2008)
10. Hooker, J.N.: Logic-based benders decomposition. Math. Program. **96**, 33–60 (2003)
11. Hurley, B., Kotthoff, L., Malitsky, Y., O'Sullivan, B.: Proteus: a hierarchical portfolio of solvers and transformations. In: Simonis, H. (ed.) CPAIOR 2014. LNCS, vol. 8451, pp. 301–317. Springer, Heidelberg (2014)
12. Seldin, J.P., Hindley, J.R.: Lambda-Calculus and Combinators An Introduction, vol. 2. Cambridge University Press, Cambridge (2008)
13. Kadioglu, S., O'Mahony, E., Refalo, P., Sellmann, M.: Incorporating variance in impact-based search. In: Lee, J. (ed.) CP 2011. LNCS, vol. 6876, pp. 470–477. Springer, Heidelberg (2011)
14. Ku, W.-Y., Beck, J.C.: Revisiting off-the-shelf mixed integer programming and constraint programming models for job shop scheduling. Technical report, University of Toronto (2014). https://www.mie.utoronto.ca/research/technical-reports/reports/JSP.pdf

15. Michel, L., See, A., Van Hentenryck, P.: Transparent parallelization of constraint programming. INFORMS J. Comput. **21**(3), 363–382 (2009)
16. Michel, L., Van Hentenryck, P.: A decomposition-based implementation of search strategies. ACM Trans. Comput. Logic **5**(2), 351–383 (2004)
17. Moisan, T., Gaudreault, J., Quimper, C.-G.: Parallel discrepancy-based search. In: Schulte, C. (ed.) CP 2013. LNCS, vol. 8124, pp. 30–46. Springer, Heidelberg (2013)
18. Nasiri, M.M., Kianfar, F.: A guided tabu search/path relinking algorithm for the job shop problem. Int. J. Adv. Manuf. Technol. **58**(9–12), 1105–1113 (2012)
19. O'Mahony, E., Hebrard, E., Holland, A., Nugent, C., O'Sullivan, B.: Using case-based reasoning in an algorithm portfolio for constraint solving. In: 19th Irish Conference on AI (2008)
20. Pacino, D., Van Hentenryck, P.: Large neighborhood search and adaptive randomized decompositions for flexible jobshop scheduling. In: IJCAI, pp. 1997–2002 (2011)
21. Perron, L.: Search procedures and parallelism in constraint programming. In: Jaffar, J. (ed.) CP 1999. LNCS, vol. 1713, pp. 346–360. Springer, Heidelberg (1999)
22. Pisinger, D., Ropke, S.: Large Neighborhood Search. In: Gendreau, M., Potvin, J.-Y. (eds.) Handbook of Metaheuristics. International Series in Operations Research & Management Science, vol. 146, pp. 399–419. Springer, New York (2010)
23. Puchinger, J., Stuckey, P.J., Wallace, M., Brand, S.: From high-level model to branch-and-price solution in G12. In: Trick, M.A. (ed.) CPAIOR 2008. LNCS, vol. 5015, pp. 218–232. Springer, Heidelberg (2008)
24. Puchinger, J., Stuckey, P.J., Wallace, M.G., Brand, S.: Dantzig-wolfe decomposition and branch-and-price solving in G12. Constraints **16**(1), 77–99 (2011)
25. Refalo, P.: Linear formulation of constraint programming models and hybrid solvers. In: Dechter, R. (ed.) CP 2000. LNCS, vol. 1894, pp. 369–383. Springer, Heidelberg (2000)
26. Régin, J.-C., Rezgui, M., Malapert, A.: Embarrassingly parallel search. In: Schulte, C. (ed.) CP 2013. LNCS, vol. 8124, pp. 596–610. Springer, Heidelberg (2013)
27. Schrijvers, T., Tack, G., Wuille, P., Samulowitz, H., Stuckey, P.J.: Search combinators. In: Lee, J. (ed.) CP 2011. LNCS, vol. 6876, pp. 774–788. Springer, Heidelberg (2011)
28. Schulte, C.: Parallel search made simple. In: Proceedings of TRICS, a Post-Conference Workshop of CP 2000, Singapore, September 2000
29. Shaw, P.: Using constraint programming and local search methods to solve vehicle routing problems. In: Maher, M.J., Puget, J.-F. (eds.) CP 1998. LNCS, vol. 1520, pp. 417–431. Springer, Heidelberg (1998)
30. Stuckey, P.J., de la Banda, M.G., Maher, M.J., Marriott, K., Slaney, J.K., Somogyi, Z., Wallace, M., Walsh, T.: The G12 project: mapping solver independent models to efficient solutions. In: Gabbrielli, M., Gupta, G. (eds.) ICLP 2005. LNCS, vol. 3668, pp. 9–13. Springer, Heidelberg (2005)
31. Van Hentenryck, P.: Parallel constraint satisfaction in logic programming: preliminary results of CHIP within PEPSys. In: Sixth International Conference onLogic Programming, Lisbon, Portugal, June 1989
32. Van Hentenryck, P., Michel, L.: Synthesis of constraint-based local search algorithms from high-level models. In: Proceedings of the National Conference on Artificial Intelligence, 1(CONF 22), pp. 273–279 (2007)
33. Van Hentenryck, P., Michel, L.: The objective-CP optimization system. In: Schulte, C. (ed.) CP 2013. LNCS, vol. 8124, pp. 8–29. Springer, Heidelberg (2013)

34. Vilím, P., Barták, R., Čepek, O.: Extension of o(n log n) filtering algorithms for the unary resource constraint to optional activities. Constraints **10**(4), 403–425 (2005)
35. Vilím, P., Laborie, P., Shaw, P.: Failure-directed search for constraint-based scheduling. In: Michel, L. (ed.) CPAIOR 2015. LNCS, vol. 9075, pp. 437–453. Springer, Heidelberg (2015)
36. Lin, X., Hutter, F., Hoos, H.H., Leyton-Brown, K.: Satzilla: portfolio-based algorithm selection for sat. J. Artif. Int. Res. **32**(1), 565–606 (2008)
37. Yunes, T., Aron, I.D., Hooker, J.N.: An integrated solver for optimization problems. Oper. Res. **58**(2), 342–356 (2010)

Rail Capacity Modelling
with Constraint Programming

Daniel Harabor[1,2][✉] and Peter J. Stuckey[1,2]

[1] Victoria Research Lab, National ICT, West Melbourne, Australia
{daniel.harabor,peterj.stuckey}@nicta.com.au
[2] Department of Computing and Information Systems,
University of Melbourne, Melbourne, Australia

Abstract. We describe a constraint programming approach to establish the coal carrying capacity of a large (2,670 km) rail network in north-eastern Australia. Computing the capacity of such a network is necessary to inform infrastructure planning and investment decisions but creating a useful model of rail operations is challenging. Analytic approaches exist but they are not very accurate. Simulation methods are common but also complex and brittle. We present an alternative where rail capacity is computed using a constraint-based optimisation model. Developed entirely in MiniZinc, our model not only captures all dynamics of interest but is also easily extended to explore a wide range of possible operational and infrastructural changes. We give results from a number of such case studies and compare against an industry-standard analytic approach.

1 Introduction

Mining is one of the most important industries in Australia, and other parts of the world, and making mining supply chains efficient requires careful investment in the infrastructure that makes up the supply chain. The Bowen Basin in Central Queensland is home to 59 individual open-cut and underground mines. The large majority of all material is export coal with over 207 million tonnes having been produced in 2014. Once extracted, coal is railed from one of 37 different loadout points to one of 3 nearby coal ports. The set of all rail infrastructure serving the Bowen Basin is known as the Central Queensland Coal Network (CQCN).

Capacity planning in the context of the CQCN is an important and challenging topic. Investment decisions for infrastructure are typically highly expensive and have an effect over many years. In order to make the right decisions we need to model a range of competing alternatives and estimate in each case the maximum capacity (or throughput) of the rail network, typically measured in millions of tonnes of coal per annum (Mtpa). Key parameters that must be carefully considered include: the type of rolling stock, availability and performance of mines and ports, the number of lines in the network, the number and location of junctions and passing loops and operational constraints such as refuelling, crew changeover and temporal separation between trains. Figure 1 gives a small artificial example of an export coal supply chain. There are two typical approaches used to establish the capacity of rail in such a context:

© Springer International Publishing Switzerland 2016
C.-G. Quimper (Ed.): CPAIOR 2016, LNCS 9676, pp. 170–186, 2016.
DOI: 10.1007/978-3-319-33954-2_13

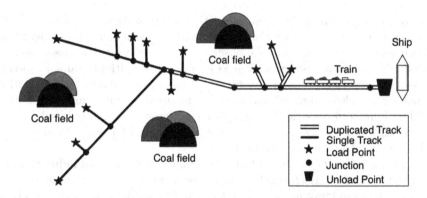

Fig. 1. The export coal supply chain. Raw material is extracted from large open-cut and underground mines. Once crushed and sorted, the coal is loaded onto trains and carried to unload points at a waterfront terminal. There the material is blended into various products and loaded onto ships for export. The rail component of such a supply chain comprises the load and unload equipment, rolling stock (locomotives and wagons), the physical rail network (lines and signals) and a set of operational parameters, in the form of rules, that govern how the infrastructure can be used in practice.

Analytic Models. This approach estimates the *theoretical capacity* of a rail system by creating simple mathematical models of operations that aim to saturate available infrastructure. A common approach is to consider capacity of a single line under e.g. fixed values for headway and travel time [3]; periodic traffic patterns [2] or; a set of fixed variables that represent mixed traffic and dwell times [12]. The primary advantage of these approaches is simplicity. The chief disadvantage is accuracy.

Simulation Models. Simulation methods can be used to model the physical infrastructure and the many operational requirements and constraints that arise in practice. An overview of such methods is given in [9]. In particular, tools such as OpenTrack [18] are intended to be very accurate but their primary strength is checking proposed schedules for feasibility; not deciding them in the first place. In cases where simulation models are extended to include a scheduling component, e.g. MultiRail [17], the typical approach is to add greedy algorithms to the simulation. The primary advantage of this approach is that many infrastructural and operational variables can be modeled together. The chief disadvantage is the time required to build the simulation and the quality of the decisions made within it.

In this paper we advocate a third approach, much less frequently used: building a CP-based *optimisation model* of the infrastructure system. While early examples of such works do exist (e.g. [13,16]), they are typically limited to small single-track networks with few junctions and trains. Alternatively one could consider a mixed integer programming (MIP) based optimisation model, and there are a number of such approaches e.g. [1,10]. These approaches are usually quite coarse grained, constraining capacities and using flow-based models, rather than actually building a scheduling model since time discretization is not feasible. This accords with experience in other minimum make span scheduling problems where CP is usually superior to MIP. We observe that recent years have seen massive increases in computing power as well as significant algorithmic gains in solving complex optimisation problems. Moreover, modelling and model transformation technology has also improved and the time required to create an optimisation model with modern constraint programming languages is much reduced [15]. To wit, we suggest that the time is ripe for switching to CP-based optimisation modelling for infrastructure planning.

To support this position, and at the request of a financial-industry partner with an interest in Queensland coal, we have created a scheduling-based constraint programming model of the CQCN. The model is written entirely in MiniZinc and offers many advantages: (i) the model describes all key infrastructural parameters of interest; (ii) the model considers decisions that actually reflect the best usage of the infrastructure; (iii) the model requires substantially less effort to produce than an equivalent simulation; (iv) the model makes it very easy to consider many "what if" situations. Indeed in many cases setting up such scenarios can be achieved by only changing input data.

We give a full description of the system and evaluate its performance in a range of freight-task scenarios. We also compare our model against a standard analytic approach to establishing rail capacity. Finally we apply the model to a number of "what if" infrastructural scenarios in order to demonstrate the flexibility of this approach and the benefits it can offer to industry planners.

2   The Central Queensland Coal Network

The Central Queensland Coal Network (CQCN) spans 2,670 km of rail track and is the primary means of transporting export coal volumes; from 37 regional load-out points in Queensland's Bowen Basin to the nearby ports of Gladstone, Hay Point and Abbott's Point. Owned and operated by Aurizon Pty Ltd, the CQCN can be naturally divided into four separate but centrally managed and connected rail systems. These are known as Blackwater, Goonyella, Moura and Newlands. Each system imposes different constraints on train operations and each is configured to feed coal volumes to a specific port. Table 1 gives an overview of the four rail systems in terms of some key parameters. This data is sourced from a range of publicly available system descriptions [3–8].

When attempting to establish the coal-carrying capacity of a network such as the CQCN industry planners first create an idealised model of rail operations. This model is used in two ways: (i) to compute a maximum throughput figure for the as-is network and; (ii) to explore a range of what-if scenarios where infrastructure is added or modified or in which different operational practices are employed. The main difficulty facing industry planners is the large number of variables that need to be modeled and accounted for. For example there are 49 separate load and unload points in the CQCN and more than 130 junctions where trains can be scheduled to operate. In addition there are various operational

Table 1. Key infrastructural parameters for the CQCN. Applicable units are Kt (kilo-tonnes) and Kt/h (kilo-tonnes per hour). NB: When reporting number of junctions, we count only intermediate locations (not endpoints) that appear on a mine-to-port path.

	Blackwater	Goonyella	Moura	Newlands
Track length	1108 km	978 km	261 km	320 km
Track type	Single + Duplic'd	Single + Duplic'd	Single	Single
# Junctions	60	41	17	18
Travel speed	80 km/h	80 km/h	80 km/h	80 km/h
Headway time	20 min	15 min	90 min	36 min
Shunt speed	10 km/h	10 km/h	10 km/h	10 km/h
Train payload (Max)	10.6 Kt	13.14 Kt	10.6 Kt	8.7 Kt
Wagon type	Hopper	Hopper	Hopper	Hopper
Wagon capacity	106 t	106 t	106 t	106 t
Wagon length	16.7 m	16.7 m	16.7 m	16.7 m
Load points	10	20	4	3
Load rate (Avg. max)[a]	4 Kt/hr	4 Kt/hr	4 Kt/hr	4 Kt/hr
Unload points	4 (Shared)	5	4 (Shared)	2
Unload rate (Avg. max)	5 Kt/h	5.5 Kt/hr	5 Kt/hr	5 Kt/hr

[a]We use as reference infrastructure equipment supplied by Techniplan to loadout points in the Goonyella system (at Carborough Downs and Isaac Plains)

requirements and constraints that can affect the efficacy of even idealised train services. These include: signalling, shunting, single track, crewing, refuelling, maintenance, and unexpected downtime.

3 Rail Capacity with Analytic Models

A common approach for analytically computing rail capacity is to combine a set of fixed operational parameters (train length, train payload, *headway* and *service time*[1]) together with simple models of relevant infrastructure. We create three such models to respectively characterise the maximum theoretical capacity of a single-track railway line, a mine loadout point and a port unload point:

---

[1] In industry terminology, headway refers to the minimum temporal separation between two trains traveling in the same direction on the same rail line. Meanwhile, service time is the time necessary to fully load or unload a train, including shunting.

$$A_{Line} = \frac{Total\ Time}{Headway\ Time} \times Train\ Payload \tag{1}$$

$$A_{Mine} = \frac{Total\ Time}{Load\ Time + Shunt\ Time} \times Train\ Payload \tag{2}$$

$$A_{Port} = \frac{Total\ Time}{Unload\ Time + Shunt\ Time} \times Train\ Payload \tag{3}$$

Parameters such as load, unload and shunt time are dependent on the exact characteristics of the train at hand and on the throughput capacity of load and unload points. Each of these can be varied to develop different scenarios. Where multiple parallel resources exist (e.g. duplicated rail lines or multiple loaders/unloaders) the models can likewise be extended appropriately. Every such analysis is obviously limited. For example the model A_{Line} assumes all trains are identical and always travel in the same direction. Meanwhile A_{Port} and A_{Mine} ignore the rail line altogether. Despite these drawbacks such methods are nevertheless attractive for their simplicity. Moreover, by computing analytic capacity from several different perspectives useful insights can often be attained. For example a very similar analytic approach to the one described here is currently used by Aurizon to "support pre-concept and concept studies" (in the CQCN) [3].

4   Rail Capacity with Optimisation Modelling

In order to establish rail capacity we will build a schedule of train trips to and from each mine. Since we are only creating a strategic model we will omit consideration of many operational matters (e.g. fleet-size and mix, crew pairing and rostering, variable travel times and any type of delay). We also do not model some existing dwell times; e.g. to facilitate refuelling and crew changeover, though these can be easily added. As such our results can be interpreted as assuming all trains are electric and autonomous.

Our model depends on two key parameters. The first of these, *loads per mine*, reflects the fact that we schedule the same number of round-trips from every mine site. It implicitly assumes that coal production is not a limiting factor any mine site.[2] The second parameter, *trains per mine*, reflects the fact that we assign a fixed number of dedicated trains to carry loads from each mine. This is not realistic (in practice the amount of rolling stock is usually limited) but appears quite reasonable for the purposes of infrastructural capacity estimation.

Next, rather than describe the entire rail network (which can be quite large), we simply model track segments between key junctions. These junctions are (i) load and unload points; (ii) rail yards where trains can be staged before/after servicing; (iii) junctions at the intersection of two or more branch lines; (iv) certain (hand chosen) passing loops which allow trains to share a single-track line. We also exploit the fact that in the CQCN (as in many rail networks) there is usually a single fixed path between each mine and the port. Every such path is computed a priori and made available as an input parameter to the model.

---

[2] With more data the model could be made more accurate in this regard.

Notice that the underlying problem we solve is just train scheduling. Our model supports a variety of constraints relevant to this context including minimum headway time, single-track constraints and optional waiting at selected junctions (including time allowances for stopping and starting).

4.1 MiniZinc

We now present a slightly simplified (for ease of exposition) version of our capacity planning model, written in MiniZinc [15]. The most important data are:

- a set of mines, MINE, where cargo originates.
- a set of junctions, JUNC, that split the rail network.
- the number of loads or round trips, LPM, to schedule from each mine.
- the number of trains available for each mine, TPM.
- a path, path, from each mine to the port, represented as list of at most maxleg junctions, using a dummy junction when we need less than maxleg.
- a set of locations, LOC ⊃ MINE of things of interest.
- a mapping from junctions to locations, junc_loc.
- an expected travel time from location $l1$ to location $l2$, travel_time$[l1, l2]$.

We represent the trips between mines and ports using the array TRIP. Full trips, designated FTRIP, are assigned even indexes while empty trips, ETRIP, have odd. We now introduce the key decision variables and discuss associated constraints.

Decision Variables: The key decisions are at the level of each mine and trip:

- mine_time, decides when a train leaves (full) or arrives (empty) at each mine.
- junction_time, decides when a train (full or empty) should arrive at each junction and at the port. Note that most trips will not arrive at all junctions.
- junction_wait, decides how long a train (full or empty) waits at a junction.

We measure time in minutes, though wait times are discretised to be divisible by 5. Time granularity could easily be changed in the model if required. We additionally employ an array of convenience variables, port_time, each of which is associated with a corresponding variable from the junction_time array. These redundant variables simply collect the times each train arrives at the port (full) and leaves the port (empty). Their definition makes use of a parameter, stop_allowance, which is the number of minutes required to bring the train to a full stop, minus the usual time it would take to travel the distance of the stop. There exists a corresponding term, start_allowance, that is defined similarly and encountered later in the model. The decision variable declarations are:

```
set of int: LEG = 1..maxleg;
set of int: XJUNC = JUNC union { dummy };
array[MINE,LEG] of XJUNC: path; % path of junctions from mine to port
set of int: TRIP = 0..2*LPM-1;
set of int: FTRIP = { 2*i | i in 0..2*LPM-1};    % full trips
set of int: ETRIP = TRIP diff FTRIP;             % empty trips
array[MINE,TRIP]      of var TIME: mine_time;    % time leaving/arriving mine
array[JUNC,MINE,TRIP] of var TIME: junction_time; % time arriving at junction
array[JUNC,MINE,TRIP] of var WAIT: junction_wait; % wait time at junction
array[MINE,TRIP]      of var TIME: port_time =   % time arriving/leaving port
  array2d(MINE,TRIP, [ junction_time[port,m,t] +
                       stop_allowance*(t in FTRIP) | m in MINE, t in TRIP ]);
```

Mine Loading Constraints: We require each full trip to be loaded and to depart in order. The first train can leave after loading and the remaining trains follow. After TPM departures trains can return but only in the same order.

```
forall(m in MINE, t in FTRIP)
  (if t = 0 then                                % first train
     mine_time[m,t] >= load_time[m] + start_allowance
   elseif t div 2 < TPM then                    % next few trains up to TPM
     mine_time[m,t] >= mine_time[m,t-2] + load_time[m] + start_allowance
                       + headway_time
   else
     mine_time[m,t] >= max(mine_time[m,t-2],mine_time[m,t-2*TPM+1])
                       + load_time[m] + start_allowance + headway_time
   endif);
```

Port Unloading Constraints: We require each empty trip to depart the port immediately after its full trip has unloaded, capturing the requirement that trains do not remain in the port after unloading. Note that our unload time includes a shunting component which is a function of the length of the train (this could also be modeled separately on a per-train basis).

```
forall(m in MINE, t in ETRIP)
  (port_time[m,t] = port_time[m,t-1] + unload_time + start_allowance);
```

Port Capacity Constraints: We ensure that no more trains are unloading at the port than there are dump stations, unload_capacity.

```
cumulative([port_time[m,t] | m in MINE, t in FTRIP],
           [unload_time | m in  MINE, t in FTRIP],
           [ 1 | m in MINE, t in FTRIP], unload_capacity);
```

Unused Junctions: We record a time for each trip at each junction, since there are not that many junctions, but of course almost no trips will visit all junctions. The unused junctions are set to have time and wait of 0.

```
array[MINE] of set of JUNC: junctions_for_mine =
  [ {path[m,l]|l in LEG where path[m,l] != dummy} | m in MINE];
array[JUNC] of set of MINE: mines_for_junction =
  [ {m | m in MINE where j in junctions_for_mine[m] }| j in JUNC ];
forall(m in MINE, t in TRIP, j in JUNC diff junctions_for_mine[m])
  (junction_time[j,m,t] = 0 /\ junction_wait[j,m,t] = 0);
```

Travel Time: Leg-to and Leg-from Mine: We model (separately) the travel time for full trips, from the mine to the first junction on its path to the port. In a similar way we also model travel time for empty trips, from the last junction in the path to the mine. Note how full trips constrain the times between junctions in the opposite order to empty trips.

```
forall(m in MINE, t in FTRIP)
  ( let { JUNC: j = path[m,1]; LOC: l = junc_loc[j]; } in
    junction_time[j,m,t] >= mine_time[m,t] + travel_time[m,1] );
forall(m in MINE, t in ETRIP)
  ( let { JUNC: j = path[m,1]; LOC: l = junc_loc[j]; } in
    mine_time[m,t] >= junction_time[j,m,t] + junction_wait[j,m,t]
                    + stop_allowance + travel_time[l,m] );
```

Travel Time: Inter-junction Legs: Travel time between adjacent junctions gives rise to a similar constraint.

```
forall(m in MINE, t in FTRIP)
  ( forall(s in 1..maxleg-1 where path[m,s+1] != dummy)
    ( junction_time[path[m,s+1],m,t] >= junction_time[path[m,s],m,t]
      + junction_wait[path[m,s],m,t]
      + travel_time[junc_loc[path[m,s]],junc_loc[path[m,s+1]]] ) );
forall(m in MINE, t in ETRIP)
  ( forall(s in 1..maxleg-1 where path[m,s+1] != dummy)
    ( junction_time[path[m,s],m,t] >= junction_time[path[m,s+1],m,t]
      + junction_wait[path[m,s+1],m,t]
      + travel_time[junc_loc[path[m,s+1]],junc_loc[path[m,s]]] ) );
```

Minimal Wait Times: A train needs to come to a complete stop to wait at a junction hence there is a minimal amount of time it is delayed by any wait.

```
forall(j in JUNC, m in MINE, t in FTRIP)
  ( junction_wait[j,m,t] = 0 \/
    junction_wait[j,m,t] >= stop_allowance + start_allowance );
```

Siding Capacity at Junctions: We constrain trains waiting at a junction j to be no more than the number of sidings at the junction, sidings[j].

```
forall(j in JUNC)
  ( cumulative([junction_time[j,m,t] | m in MINE, t in TRIP],
               [junction_wait[j,m,t] | m in MINE, t in TRIP],
               [ 1 | m in MINE, t in TRIP], sidings[j]) );
```

Headway Constraints at Junctions: Rather than using a disjunctive constraint to model that no two trains pass a junction in the same direction within headway time, since all the "durations" of these tasks are the same we simply use alldifferent. This is slightly stronger constraint than the disjunctive constraint but accurate enough for capacity planning.

```
forall(j in JUNC)
     (alldifferent([ junction_time[j,m,t] div headway_time
                   | m in mines_for_junction[j], t in FTRIP]) /\
      alldifferent([ junction_time[j,m,t] div headway_time
                   | m in mines_for_junction[j], t in ETRIP]));
```

4.2 Single Track Constraints

When there is only a single track between two locations we must ensure no two trains try to use the track while traveling in opposite directions. Though there are complex ways of modelling this using variable set up times we adopt a simpler approach where each train reserves the track for the entire time it is using it. By varying the granularity of the model (adding new junctions) we can limit the inaccuracy that derives from this overly restrictive constraint. This approach requires us to introduce the notion of *track segments* into the model.

A track segment `s in SEG` has: a start junction, `start_junc`, which may be `dummy` if the segment is a leaf; an end junction, `end_junc`; an (optional) set of mines that sit on that segment (usually in unmodelled mine-specific balloon loops), `mines_on_segment`; and a set of mines that use the segment on their path to and from the port, `mines_using_segment`.

Example 1. Consider the abstract rail network shown in Fig. 2 which includes junctions $j1$ and $j2$, mines $m1, \ldots, m5$ and unmodelled intersections $u1$, $u2$ and $u3$. The rail network consists of 2 segments: a leaf segment ending at $j2$ which includes the mines $m3$, $m4$ and $m5$, and a non-leaf segment from $j2$ to $j1$ which includes the mines $m1$ and $m2$. There are no (additional) mines that use the first segment on their path to the port, while the mines $m3$, $m4$ and $m5$ all use the second segment on their path to the port.   □

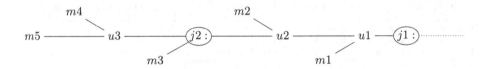

Fig. 2. Part of an (abstract) rail network.

Leaf Segments: Leaf segments connect mines to the rest of the network. We make sure that no train going to or from a mine in that segment overlap in time by using the travel time to/from the mine to the end junction of the segment.

```
array[SEG] of set of MINE: mines_on_segment;
forall(s in SEG where
start_junc[s] = dummy)
   ({ let { JUNC: j = end_junc[s]; LOC: l = junc_loc[j]; } in
      disjunctive([ if t in FTRIP then          % start time
                       mine_time[m,t]
                    else junction_time[j,m,t] + junction_wait[j,m,t] endif
                 | m in mines_on_segment[j], t in TRIP ],
                 [ if t in FTRIP then            % duration
                      travel_time[m,l]
                   else travel_time[l,s] endif
                 | m in mines_on_segment[s], t in TRIP ] ) );
```

Non-leaf Segments: Non-leaf segments are used to handle trains traveling between the start and end junctions of the segment. They also handle trains that travel from either of these junctions to a mine that sits on the segment. Notice that this constraint always uses the start-to-end travel time. There is an implicit assumption here that this duration is always less than the travel time to (or from) a mine that sits on the segment. For our data sets this is always the case, but the model would need adjustment if it were not the case.

```
array[SEG] of set of MINE: mines_using_segment;
forall(s in SEG where start_junc[s] != dummy)
  ( let { JUNC: sj = start_junc[s]; LOC: sl = junc_loc[sj];
          JUNC: ej = end_junc[s]; LOC: el = junc_loc[ej];
          set of MINE: M = mines_on_segment[s] union
                           mines_using_segment[s]; } in
    disjunctive([ if t in FTRIP then               % start time
                    junction_time[sj,m,t] + junction_wait[sj,m,t]
                  else junction_time[ej,m,t] + junction_wait[ej,m,t] endif
                | m in M, t in TRIP ],
                [ if t in FTRIP then               % duration
                    travel_time[sl,el]
                  else travel_time[el,sl] endif
                | m in M, t in TRIP ] ) );
```

4.3 Search Strategy

We use the Gecode [14] solver to tackle our models. The default autonomous search does not perform well so we employ the following simple hybrid which does: we use a dom/wdeg variable selection heuristic [11] but order the variables carefully so that tie-breaking in dom/wdeg chooses the variables in a sensible order. We have found the following simple ordering to be particularly effective: (i) decision variables that determine arrival and departure times from mine load-points appear first; (ii) decision variables that determine arrival and departure times from port unload points appear next; (iii) all other decision variables follow, in any order. Given decision variables that are ordered in a "good" way, we have found that Gecode can often identify near-optimal solutions very quickly.

5 Experiments

We use our optimisation model to explore a range of infrastructural scenarios, many of which are difficult to evaluate analytically. These scenarios are:

- Capacity of the current infrastructure.
- Capacity under the assumption of increased payloads per train.
- Capacity assuming the addition of new below-rail infrastructure[3]; e.g. additional signalling and duplicated rail lines.

---

[3] In industry terminology, *below-rail* refers to infrastructure controlled by the network owner, such as the physical track and signals. By comparison *above-rail* refers to infrastructure such as trains, wagons and other so-called rolling stock.

Table 2. Analytic evaluation of the theoretical capacity of each rail system in the CQCN. Each of the three models take as input operational parameters from Table 1.

Network	Theoretical capacity model			Additional parameters	
	A_{Mine}	A_{Port}	A_{Line}	Infrastructure availability	Line type
Blackwater	329.3	162.2	278.6	100 %	Single track
Moura	131.7		61.9	100 %	Single track
Goonyella	658.7	221.4	460.4	100 %	Single track
Newlands	98.7	81.0	127.02	100 %	Single track

Where possible we will compare our computational approach against the industry-standard analytic techniques discussed in Sect. 3. Recall that these simplified models are used to compute the *maximum theoretical capacity* of infrastructure. We will compare against these optimistic upper-bounds in order to evaluate the quality of solutions computed with our CP model. Capacity figures are always given in Mtpa: Millions of tonnes (of coal) per annum.

5.1   Infrastructural Capacity with Analytic Modelling

Recall that the analytic model from Sect. 3 focuses on different aspects of the network to the exclusion of all other factors. To mitigate this myopic bias we will compute analytic capacity from three points of view: ports, mines and the physical rail lines. Table 2 presents our results. We assume loading, unloading and travel all proceed without delay and that infrastructure is always available and always operates at maximum throughput. When modeling trains we use a range of established operational parameters including real-world headway times and industry maximums for train length and payload size in each rail system. The full set of all such parameters are given in Table 1 while results from this analysis are given in Table 2. We make several observations:

– The data suggests that water-front unload points (and not the rail network) is the most likely bottleneck in each rail system.
– The port bottleneck observation holds despite our (pessimistic) assumption of single-line track for every A_{Line} model. Note that while this assumption is true for Moura and Newlands there exist large portions of Blackwater and Goonyella that are duplicated. We continue to use the single-line assumption in these cases as the majority of mines are on spurs[4] that connect to the network via single-track branch lines.

5.2   Infrastructural Capacity with Optimisation

Next, we evaluate capacity in the CQCN using our scheduling-based optimisation model and the Gecode solver. As in the analytic case we employ the full range

---

[4] In industry terminology, a spur is a short branch usually leading to a private siding.

Table 3. CP-based rail capacity. We assume current CQCN operational parameters, as described in Table 1. Columns LPM and TPM respectively indicate the number of loads per mine (i.e. the size of the freight task) and the number of (dedicated) trains per mine. Figures denoted with * are provably optimal.

Network	Parameters		Network performance				
	LPM	TPM	Trains	Avg. Cycle time	Total wait	Port util	Capacity
B/Moura	15	2	13,464	17.4 h	0	87.8 %	142.7
Goonyella	15	2	15,130	19.8 h	0	89.2 %	198.9
Newlands	35	4	8,860	10.0 h	0	95.5 %	77.9*

of real-world parameters from Table 1 and assume that infrastructure is always available and operates at maximum throughput. The first solution is typically found in seconds and we allow the solver to run for up to a minute thereafter.

We evaluate the capacity of each rail system by measuring its steady-state performance and extrapolating out to a full year. To avoid warm-up and cool-down effects we ignore loading and unloading operations at the beginning and toward the end of the schedule. In particular we consider only port arrivals between the first and third quartiles of our planning horizon. Results are given in Table 3. We make several observations:

- In the case of the Newlands system we find that our optimisation approach is able to compute an exact figure for the maximum infrastructural capacity of rail. The figure (77.9 Mtpa) is within 5 % of the optimistic upper-bound established by the analytic model A_{Port}.
- In the case of Blackwater/Moura and Goonyella we compute approximate capacities which are within 10.8 % and 10.2 % of the upper-bound A_{Port}.
- In all three cases port utilisation is very close to or above 90 %. These figures suggest that the rail network is not the primary limiting factor for increased coal export volumes in the future. Rather, each system appears constrained by the infrastructural capacity of their respective ports.

For the experiments at hand the parameters LPM and TPM were hand-tuned on a per-model basis. If LPM is too small, the freight task can be finished quickly and before the system can reach a steady state. Alternatively, if LPM is too large the problem may grow to a size where our optimisation solver cannot compute a good solution in reasonable time. Similar observations are true for the parameter TPM. Given too few trains the port infrastructure can remain idle for long periods and its performance will not be indicative of potential capacity. On the other hand a TPM value that is too large can explode the search space, again making any solution difficult to find in a reasonable amount of time.

With LPM=15 and TPM=2 the size of the planning horizon is 7.3 days for Goonyella and 5.9 days fro Blackwater/Moura. We found these values sufficient to take reliable readings of network performance. In the case of the Newlands System the planning horizon with these parameters is too small to be useful

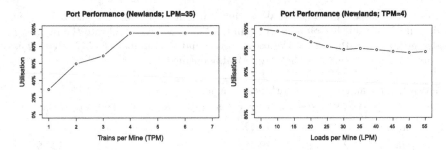

Fig. 3. Tuning LPM and TPM parameters for the Newlands System model. We isolate each parameter and vary its value. We measure the impact of each change by computing the percentage utilisation of port unloaders in each resultant scenario.

(<3 days). Figure 3 gives results from a range experiments in which we empirically identified appropriate values for Newlands. Notice that: (i) setting TPM > 4 does not make any difference to port utilisation but smaller values have a large impact; (ii) setting LPM < 30 is insufficient to reach the system's steady-state.

5.3   Case Study: Increased Payloads

One of the case studies asked for by our industry partner is to determine rail capacity under the assumption that all trains have fixed payloads. The proposed volumes are 10 Kt, 12 Kt and 14 Kt. Increased payload scenarios involve modeling trains which are longer or which comprise wagons that are more densely packed. Lacking data regarding alternative wagon configurations we opt to model longer trains. Note that both options may require additional below-rail infrastructure; either in the form of longer balloon loops (to support longer trains) or new load and unload equipment (configured to support densely packed trains).

To model trains with alternative payload configurations we simply modify a single value in the associated data file for each network and run the solver anew. No change to the optimisation model is needed. A similar data-driven change would also be sufficient to model the densely-packed scenario (in this

Table 4. Experiments using a range of alternative payload sizes. We measure capacity in three scenarios where all trains carry uniform payloads of 10, 12 and 14 Kt (kilo-tonnes) of coal. For context, we also give results from the current capacity scenario which considers fully-loaded trains of the maximal size currently permitted in each rail system (see Table 1). Figures in bold indicate best results (highest capacity) found.

	10 Kt Scenario		12 Kt Scenario		14 Kt Scenario		Current max scenario	
	Capacity	T. Len	Capacity	T. Len	Capacity	T. Len	Capacity	T. Len
B/Moura	139.6	1587 m	**145.6**	1904 m	138.4	2205 m	142.7	1670 m
Goonyella	197.5	1578 m	198.6	1904 m	197.8	2205 m	**198.9**	2071 m
Newlands	69.3	1578 m	61.9	1904 m	64.1	2205 m	**77.9**	1369 m

case we would need to modify wagon length and wagon capacity parameters in addition to payload size). All other parameters remain as in Sect. 5.2. Results from this experiment are given in Table 4. We observe that with few exceptions each increased/uniform payload scenario appears to make little difference to rail capacity beyond what can be achieved by running trains with the maximum currently permissible payload size. One exception is the Blackwater/Moura system where a small gain of 3 Mtpa can be achieved by running 12 Kt trains instead of the current maximum payload size of 10.6 Kt.

5.4   Case Study: Decreased Headway

Another possibility for increasing the capacity of a rail system is to decrease the cycle time (i.e. round-trip time) per train. Such scenarios could involve deploying additional infrastructure or technology to allow decreased headway (i.e. a smaller temporal separation) between trains or the introduction of new rolling stock that can travel at faster speeds. We model the decreased headway scenario here though new rolling stock is equally simple to analyse. In both cases we make changes only to parameter values. The optimisation model remains unchanged. Results are given in Table 5.

Table 5. Experiments using a range of fixed headway times. We evaluate their effectiveness in terms of capacity and port utilisation. For context, we compare these results against the capacity figures computed in Sect. 5.2 (row "Current"). Figures in bold indicate best results (highest capacity) found.

Headway (mins)	Blackwater/Moura		Goonyella		Newlands	
	Capacity	Port util	Capacity	Port util	Capacity	Port util
6	145.8	89.6 %	195.9	87.9 %	60.1	73.7 %
16	144.7	89.0 %	198.9	89.2 %	64.8	79.4 %
26	**147.6**	90.7 %	**202.6**	90.9 %	66.7	81.7 %
30	142.2	87.4 %	195.5	87.7 %	**80.5**	98.6 %
Current	142.7	87.8 %	198.9	89.2 %	77.9	95.5 %

In a range of experiments we observe that the total throughput of each rail system is largely invariant, even with reduced headway times. In the case of Blackwater/Moura system an increase of 3 % (vs. the Current Capacity scenario) appears achievable if we fix the headway time of all trains to 26 min. This value is larger than the 20 min currently used for junctions in the Blackwater system but much smaller than the 90 min used in Moura. A similar gain can be achieved in Newlands when headways are reduced to 30 min (cf. 36 currently). It is interesting to note that for the Goonyella system the best result is for 26 min (cf. 16 currently). We interpret this as suggestive that small amounts of extra waiting can help when there is a high degree of contention for rail resources.

5.5   Case Study: Track Duplication

For a final case study we consider the impact on rail capacity through the duplication of key sections of rail track. Introducing new line capacity into the system
reduces waiting and track contention and allows parallel travel in both directions
(i.e. simultaneously to and from the port). There are two aspects to such an
analysis: (i) we must identify which sections of track are most likely to yield the
greatest benefit; (ii) we must evaluate the effect of the proposed simulation. We
begin with an analysis of the Blackwater system.

Figure 4 shows the arrival frequency of trains at the most visited junctions
in the Blackwater system. A junction is a reasonable candidate for duplication if the arrival frequency of trains traveling in the same direction is close
to or less than the minimum headway time. We observe that while the busiest
single-line junctions (Dingo, Walton, Umolo and Bluff) have trains arriving every
27–28 min, the frequency in any single direction is almost twice that at 50 min.
As there is no contention we may thus infer that track duplication at these
points will not increase the infrastructural throughput of the system. We confirmed this hypothesis empirically. Similar results hold for each of the other rail
systems under consideration.

It is important to note that track duplication e.g. between Dingo and Bluff
may still make sense operationally. With only 30 min of idle time between
arrivals, and round-trip times of over 17 h (see Table 3), it is entirely possible
that unforeseen delays during loading, unloading or during travel on the network
could result in contention for track resources at these locations.

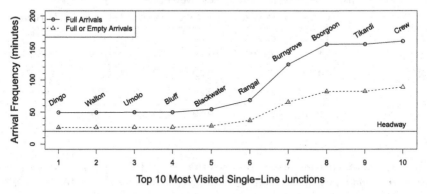

Fig. 4. Most visited single-line junctions in the Blackwater system. We give the average time difference between arrival times for full and empty trains at each junction.
Measurements are in minutes and reflect system performance during its steady state.

6   Conclusion

We evaluate the infrastructural capacity of four rail systems which together comprise the Central Queensland Coal Network. Similar capacity evaluation problems appear in a range of industrial settings but especially cases where bulk goods and freight containers must be railed between inland terminals and the waterfront. Effective models that capture the dynamics of a such a system are prized tools of industry planners.

We propose a new approach for rail capacity estimation using constraint programming with MiniZinc. Written in the form of a scheduling problem, our model is simple to develop, easy to extend and can be used to compute fast and accurate capacity estimates for large rail networks. Because it is data-driven the model makes it especially easy to evaluate a wide range of "what-if" scenarios of interest to industry planners. We give particular examples involving alternative train payloads, alternative headway times and track duplication scenarios.

There are many other scenarios of practical interest such as mixed train lengths and grade easing. We could extend our model to investigate these. We can also extend our model to capture further dynamics of the system like: scheduled downtime, different train speeds, refuelling operations and crew changeover. Most of these extensions appear quite straightforward to achieve.

We believe the principal lesson of this paper is that optimisation technology has matured to the point where we can quickly undertake detailed infrastructure modelling and analysis. Such capability is essential to inform long-term infrastructural investment decisions made by governments and large corporations.

Acknowledgements. We thank Eric Nettleton for useful discussions during the development of this work. NICTA is funded by the Australian Government through the Department of Communications and the Australian Research Council through the ICT Centre of Excellence Program.

References

1. Abdekhodaee, A., Dunstall, S., Ernst, A.T., Lam, L.: Integration of stockyard and rail network: a scheduling case study. In: Proceedings of the Fifth Asia Pacific Industrial Engineering and Management Systems Conference, Gold Coast, Australia (2004)
2. Abril, M., Salido, M.A., Barber, F., Ingolotti, L., Lova, A., Tormos, P.: A Heuristic technique for the capacity assessment of periodic trains. In: Proceedings of the 2005 Conference on Artificial Intelligence Research and Development, pp. 339–346. IOS Press, Amsterda (2005)
3. Aurizon Network Pty Ltd: 2014 network development plan (2014). http://www.aurizon.com.au/Downloads/AurizonNetworkDevelopmentPlan2014.pdf. Accessed 29 Sept 2015
4. Aurizon Network Pty Ltd: review of rail infrastructure and line diagrams for central queensland coal region, 30 December 2014. http://www.aurizon.com.au/network/central-queensland-coal-network/goonyella-system. Accessed 29 Sept 2015

5. Aurizon Network Pty Ltd: blackwater system information pack (issue 5.6) (2015). http://www.aurizon.com.au/network/central-queensland-coal-network/ blackwater-system. Accessed 29 Sept 2015

6. Aurizon Network Pty Ltd: goonyella system information pack (issue 6.4) (2015). http://www.aurizon.com.au/network/central-queensland-coal-network/ goonyella-system. Accessed 29 Sept 2015

7. Aurizon Network Pty Ltd: moura system information pack (issue 6.0) (2015). http://www.aurizon.com.au/network/central-queensland-coal-network/ moura-system. Accessed 29 Sept 2015

8. Aurizon Network Pty Ltd: newlands system information pack (issue 6.4) (2015). http://www.aurizon.com.au/network/central-queensland-coal-network/ newlands-system. Accessed 29 Sept 2015

9. Barber, F., Abril, M., Salido, M., Ingolotti, L., Tormos, P., Lova, A.: Survey of automated systems for railway management. Technical report DSIC-II/01/07, Department of Computer Systems and Computation, Technical University of Valencia (2007)

10. Boland, N.L., Savelsbergh, M.W.: Optimizing the hunter valley coal chain. In: Gurnani, H., Mehrotra, A., Ray, S. (eds.) Supply Chain Disruptions, pp. 275–302. Springer, London (2012)

11. Boussemart, F., Hemery, F., Lecoutre, C., Sais, L.: Boosting systematic search by weighting constraints. In: Proceedings of ECAI04, pp. 146–150 (2004)

12. Burdett, R., Kozan, E.: Techniques for absolute capacity determination in railways. Transp. Res. Part B: Methodol. **40**(8), 616–632 (2006)

13. Fukumori, K., Sano, H., Hasegawa, T., Sakai, T.: Fundamental algorithm for train scheduling based on artificial intelligence. Syst. Comput. Jpn **18**(3), 52–64 (1987)

14. Gecode: generic constraint development environment. www.gecode.org

15. Nethercote, N., Stuckey, P.J., Becket, R., Brand, S., Duck, G.J., Tack, G.: MiniZinc: towards a standard CP modelling language. In: Bessière, C. (ed.) CP 2007. LNCS, vol. 4741, pp. 529–543. Springer, Heidelberg (2007)

16. Oliveira, E., Smith, B.M.: A job-shop scheduling model for the single-track railway scheduling problem. Research Report Series-University of Leeds School of Computer Studies LU SCS RR (21) (2000)

17. OliverWyman: MultiRail planning suite (2012). http://oliverwyman.com. Accessed 24 Novem 2015

18. OpenTrack: simulation of rail networks (2015). http://www.opentrack.ch/. Accessed 24 Novem 2015

Scheduling Home Hospice Care
with Logic-Based Benders Decomposition

Aliza Heching[1] and John N. Hooker[2]([✉])

[1] Compassionate Care Hospice Group, New York, USA
aliza.heching@cchnet.net
[2] Carnegie Mellon University, Pittsburgh, USA
jh38@andrew.cmu.edu

Abstract. We propose an exact optimization method for home hospice care staffing and scheduling, using logic-based Benders decomposition (LBBD). The objective is to match hospice care aides with patients and schedule visits to patient homes, so as to maximize the number of patients serviced by available staff, while meeting requirements of the patient plan of care and scheduling constraints imposed by the patients and the staff. The Benders master problem assigns aides to patients and days of the week and is solved by mixed integer programming (MIP). The routing and scheduling subproblem decouples by aide and day of the week and is solved by constraint programming. We report preliminary computational results for problem instances obtained from a major hospice care provider. We find that LBBD is superior to state-of-the-art MIP and solves problems of realistic size, if the aim is to conduct staff planning on a rolling basis while maintaining continuity of the care arrangement for patients currently receiving service.

Keywords: Home health care problem · Routing and scheduling · Logic-based Benders decomposition · Home hospice care

1 Introduction

Home health care is one of the world's most rapidly growing industries, due primarily to cost advantages as well as aging populations. Home care allows patients to receive basic medical or hospice care in comfortable and familiar surroundings, rather than being transported or admitted to facilities that are expensive to operate. It also reduces the risk of acquiring drug-resistant infections that may spread in hospitals and nursing homes. The increasing availability of portable equipment and online consultation makes home care feasible for an ever wider variety of conditions.

The cost-effectiveness of home health care depends critically on the efficient dispatch of health care aides, whom we call *aides* for short. This poses the *home health care problem* (HHCP), which asks how home visits can be scheduled and staffed so as to make the best use of aides while meeting patient needs. Aides typically start their work shift at home or a central office, travel directly from

© Springer International Publishing Switzerland 2016
C.-G. Quimper (Ed.): CPAIOR 2016, LNCS 9676, pp. 187–197, 2016.
DOI: 10.1007/978-3-319-33954-2_14

one patient to the next, and return to home or office at the end of the shift. The shift may be subject to a number of legal or contractual restrictions, such as a maximum work time and the need for lunch/dinner breaks. Each medical or hospice service must be performed by an aide with the proper qualifications, and services may be restricted to specified days or time windows. It may be necessary for two or more aides to visit a patient at the same time, to carry out more complicated treatments.

We focus on hospice care, which has a few distinctive characteristics. Aides frequently provide personal and household services rather than medical treatment, or they may simply offer companionship. They tend to visit on a regular schedule over a period of several weeks, such as three times a week in the morning. It is often important for a given service to be provided by the same aide during every visit, so far as is possible. Staff planning is typically over a longer time horizon, perhaps several weeks.

Because of the regularity of visits and the need for staffing continuity, the primary challenge that arises in practice is to update the schedule and anticipate staffing needs as the patient population evolves. If patient turnover for the next few weeks can be forecast, then a schedule can be computed for the new population to determine what kind of work force will be required.

We therefore address the problem of recomputing the staff assignments and visitation schedule when a specified subset of the patients are replaced by new patients with known requirements. Due to the importance of continuity, we require that existing patients be served by the same aide on the same days as before, but allow for adjustments in the time of day. The models are easily modified to maintain the time of day as well, or to reschedule both the time of day and days of the week.

Due to the difficulty of the HHCP, nearly all existing solution methods are heuristic algorithms. Recent work can be found in [1–10]. The few exact methods include two branch-and-price methods [11,12] and a branch-and-bound method that relies on a traveling salesman algorithm [13].

We propose a very different exact method that uses logic-based Benders decomposition (LBBD) [14–18] and is well suited to scheduling on a rolling basis. An exact method offers the advantage that one can know with certainty whether a given work force can cover anticipated patient needs, and therefore when hiring additional staff is really necessary. We find that LBBD makes exact solution possible for applications of realistic size when the problem is to reschedule on a rolling basis, rather than schedule all the patients from scratch.

LBBD exploits a natural decomposition of the HHCP into an assignment component (allocation of patients to aides) and a routing and scheduling component (dispatching and routing of aides). It combines the complementary strengths of mixed integer programming (MIP) and constraint programming (CP), with MILP solving the assignment problem and CP solving the routing and scheduling problem.

LBBD is a generalization of classical Benders decomposition [19] in which the subproblem can be any combinatorial problem, not necessarily a linear programming problem. The Benders cuts are based on an inference dual of the subproblem, whose solution is regarded as a proof of optimality or infeasibility, rather than a linear programming dual. LBBD has reduced solution times by orders of magnitude relative to conventional methods in a variety of problems [14–18, 20–34]. In our solution of the HHCP, the Benders master problem assigns aides to patients and to days of the week on which these patients are serviced. The Benders subproblem is the routing and scheduling problem that results from the assignment obtained by solving the master problem. The subproblem decouples into routing and scheduling micro-problems that correspond to each aide and each day of the week. Infeasible micro-problems give rise to Benders cuts that are added to the master problem. The process repeats until all the micro-problems are feasible. Our primary methodological contribution is to identify a relaxation of the scheduling subproblem that, when included in the master problem, results in significantly faster solution.

The only previous application of LBBD to the HHCP of which we are aware is a heuristic method in an unpublished manuscript [4]. It solves the master problem with greedy heuristic and the subproblem with CP, while creating a schedule for only one day.

2 The Problem

The problem can be stated as follows. For each patient j there is a time window $[r_j, d_j]$ during which a visit to that patient must take place, as well as the visit duration p_j. It is assumed that each patient requires one type of visit. If a patient requires two or more types of visits, the patient is regarded as two or more distinct patients (with nonoverlapping time windows if the visits should not overlap). Aides must be qualified to serve assigned patients, but this requirement actually makes the problem easier to solve and is therefore not considered here.

Each aide i departs from home base b_i and returns to home base b_i' (which could be the same as b_i). The allowable shift hours of aide i are specified by a time window $[r_{b_i}, d_{b_i}]$ for departure from the origin base and a window $[r_{b_i'}, d_{b_i'}]$ for arrival at the destination base. Travel time between patient (or home base) j and patient j' is $t_{jj'}$.

We formulate the problem for a cyclic 7-day schedule with no visits on weekends. Each patient j requires v_j visits per week, with $v_j \in \{1, 2, 3, 5\}$. Twice-a-week visits must be separated by at least 2 days, and thrice-a-week visits by 1 day. The variables are designed to facilitate a decomposition scheme in which the scheduling subproblem is solved by constraint programming. Binary variable $\delta_j = 1$ when patient j is serviced, and binary variable $x_{ij} = 1$ when aide i is assigned to patient j. Binary variable $y_{ijk} = 1$ when aide i visits patient j on day k, so that $y_{ijk} \leq x_{ij}$ for all i, j, k. There are sequencing variables $\pi_{ik\nu}$ that represent the νth patient visited by aide i on day k. Variable s_{ijk} indicates the time at which aide i's visit to patient j starts on day k.

We maximize the number of patients that can be covered by a given work force. This not only determines whether the work force is adequate, but it tends to minimize idle time and driving time in an aide's schedule. The problem can be stated as follows:

$$\max \sum_j \delta_j \qquad\qquad (a)$$

$$\sum_i x_{ij} = \delta_j, \quad \sum_{i,k} y_{ijk} = v_j \delta_j, \ \text{all } j \qquad\qquad (b)$$

$$y_{ijk} \le x_{ij}, \ \text{all } i, j, k \qquad\qquad (c)$$

$$y_{ib_i k} = y_{ib_i' k} = 1, \ \text{all } i, k \qquad\qquad (d)$$

$$y_{ij,k+\tau} \le 1 - y_{ijk}, \ \tau = 1, 4 - v_j,$$
$$\text{all } i, j, k \text{ with } v_j \in \{2, 3\}, \ 1 \le k \le v_j + 1 \qquad (e) \quad (1)$$

$$\delta_j, x_{ij}, y_{ijk} \in \{0, 1\}, \ \text{all } i, j, k, \qquad\qquad (f)$$

$$n_{ik} = \sum_j y_{ijk}, \ \text{alldiff}\{\pi_{ik\nu} \mid \nu = 1, \ldots, n_{ik}\}, \ \text{all } i, k \qquad\qquad (g)$$

$$\pi_{ik\nu} \in \{j \mid y_{ijk} = 1\}, \ \text{all } i, k, \text{ and } \nu = 1, \ldots, n_{ik} \qquad\qquad (h)$$

$$\pi_{i1k} = b_i, \ \pi_{in_{ik}k} = b_i', \ \text{all } i, k \qquad\qquad (i)$$

$$r_j \le s_{ijk} \le d_j - p_j, \ \text{all } i, j, k \qquad\qquad (j)$$

$$s_{\pi_{ik\nu}} + p_{\pi_{ik\nu}} + t_{\pi_{k\nu}\pi_{k,\nu+1}} \le s_{\pi_{ik,\nu+1}}, \ \text{all } i, k, \text{ and } \nu = 1, \ldots, n_{ik} - 1 \quad (k)$$

Constraint (b) defines δ_j and ensures that every patient is visited by the same aide on the required number of days. Constraint (d) says that an aide's start and end home base must be visited every day. Constraint (e) controls the spacing of assigned days. Constraint (g) defines variable n_{ik} to be the number of patients assigned to aide i on day k and requires that the corresponding sequence variables take distinct values. Constraint (h) says that an aide's visits that are sequenced on a given day are in fact those assigned to the aide on that day. Constraint (i) ensures that the start and end home base are visited first and last, respectively. Constraint (j) enforces time windows. Constraint (k) ensures that a visit does not start before the aide can arrive from the previous visit.

When updating an existing schedule, we need only fix $y_{ijk} = 1$ when patient j remains in the population and is assigned to aide i on day k. To require that existing patients be served at the same time of day as before, their time windows can be set equal to the visit period. To allow existing patients to be served on different days of the week than before, we can fix the variables x_{ij} rather than y_{ijk}.

3   Benders Subproblem

The subproblem decouples into a separate micro-problem for each aide and each day. Each is a feasibility problem that checks whether there is a schedule that observes the time windows while taking account of the visit durations and travel times. If not, a Benders cut is generated as described below.

The subproblem formulation consists of the scheduling constraints in (1) after the daily assignment variables y_{ijk} are fixed to the values $\bar{y}_{ijk}$ they receive in the previous solution of the master problem. The micro-problem S_{ik} for each aide i and day k is

$$\text{alldiff}\{\pi_\nu \mid \nu = 1, \dots, \bar{n}_{ik}\}$$

$$\pi_1 = b_i, \ \pi_{\bar{n}_{ik}} = b_i'$$

$$r_j \le s_j \le d_j - p_j, \ \text{all } j \in P_{ik}$$

$$s_{\pi_\nu} + p_{\pi_\nu} + t_{\pi_\nu \pi_{\nu+1}} \le s_{\pi_{\nu+1}}, \ \nu = 1, \dots, \bar{n}_{ik} - 1$$

$$\pi_\nu \in P_{ik}, \ \nu = 1, \dots, \bar{n}_{ik}$$

where $P_{ik} = \{j \mid \bar{y}_{ijk} = 1\}$ and $\bar{n}_{ik} = |P_{ik}|$. If S_{ik} is infeasible, we generate a simple nogood cut $\sum_{j \in P_{ik}}(1 - y_{ijk}) \ge 1$ that prevents the same set of patients from being assigned to aide i on day k in subsequent assignments.

We can, in principle, generate stronger cuts by determining whether the same proof of infeasibility remains valid when smaller sets of patients are assigned to aide i on day k. Unfortunately, we do not have access to the mechanism by which CP solver proves infeasibility. We therefore tease out stronger cuts by re-solving S_{ik} for subsets of P_{ik}. S_{ik} can be rapidly re-solved because of its small size. We use the following simple heuristic, which has proved effective in several studies [18,20–22,34]. We initially set $\bar{P}_{ik} = P_{ik}$, and for each $j \in \bar{P}_{ik}$ we do the following: remove j from $\bar{P}_{ik}$, re-solve S_{ik}, and restore j to $\bar{P}_{ik}$ if the modified S_{ik} is feasible. This yields a Benders cut that results in significantly better performance:

$$\sum_{j \in \bar{P}_{ik}} (1 - y_{ijk}) \ge 1 \tag{2}$$

Whenever we derive a cut for a given aide i and day k of the week, we can generate a similar cut for every other day of the week. However, the resulting proliferation of cuts causes the solution of master problem to bog down. We found that an effective compromise is to sum the cuts for the remaining 4 weekdays. Thus for each cut (2), we also generate the cut

$$\sum_{k' \ne k} \sum_{j \in \bar{P}_{ik}} (1 - y_{ijk'}) \ge 4$$

4   Benders Master Problem

The basic master problem consists of constraints (a)–(f) of the original problem (1) and the Benders cuts generated in all previous iterations as described above. It also contains a relaxation of the subproblem, because computational experience in [22] and elsewhere indicates that including such a relaxation is crucial to obtaining good performance of LBBD.

We found the following *time window relaxation* to be effective. For each aide i, define a set $\{[r_{b_i}, \alpha_{i\ell}] \mid \ell \in L_i\}$ of *backward intervals* that begin with the start of the aide's shift, and a set $\{[\beta_{i\ell}, d_{b_i'}] \mid \ell \in L_i'\}$ of *forward* intervals that end with the termination of the shift. For each backward interval $\ell \in L_i$, let $J_{i\ell}$ be the set of visits whose time window $[r_j, d_j]$ is a subset of the interval, and define $J_{i\ell}'$ similarly for forward intervals. Let the *backward augmented duration* p_{ijk}' for a visit j, aide i and day k be the duration p_j plus the minimum transit time from the previous visit (which may be the origin base for the aide), and similarly for the *forward augmented duration* p_{ijk}''. That is,

$$p'_{ijk} = p_j + \min\left\{t_{b_ij}, \min_{j' \in Q_{ik}}\{t_{j'j}\}\right\}, \quad p''_{ijk} = p_j + \min\left\{\min_{j' \in Q_{ik}}\{t_{jj'}\}, t_{jb'_i}\right\}$$

where Q_{ik} is the set of visits that are already assigned aide i on day k, or that have not yet been assigned an aide. Thus the backward augmented duration is a lower bound on the time required to reach and carry out a visit, and similarly for the forward augmented duration.

We now observe that sum of the backward augmented durations of visits in $J_{i\ell}$ must be at most the width of backward interval ℓ, and similar for any forward interval:

$$\sum_{j \in J_{i\ell}} p'_{ijk} y_{ijk} \leq \alpha_{i\ell} - r_{b_i}, \ \ell \in L_i; \quad \sum_{j \in J'_{i\ell}} p''_{ijk} y_{ijk} \leq d_{b'_i} - \beta_{i\ell}, \ \ell \in L'_i \quad (3)$$

This because the visits and travel to each visit must fit between the beginning of the aide's shift and the end of the backward interval, and similarly for a forward interval. Inequalities (3), collected over all aides i, comprise a time window relaxation.

The backward and forward intervals should be chosen so that the visits that can take place within them have a large total duration relative to the width of the interval, as this results in tighter inequalities (3). In the test instances, the time windows of the visits span either most of the morning or most of the afternoon. It was therefore natural to use one backward interval ending at noon, and one forward interval beginning at noon, for each aide i. Thus $L_i = L'_i = \{1\}$ and $\alpha_{i1} = \beta_{i1} = $ noon for each i.

This is a weak relaxation when scheduling all patients from scratch, because the shortest travel time from the last (or next) visit is a weak bound on the actual travel time. However, it is more effective in the rolling problem, because the shortest travel time is computed only over patients who are already assigned aide i on day k or are unassigned.

5   Computational Results

We tested the LBBD algorithm on real-world data provided by a major hospice care firm. To obtain an initial schedule, we ran a greedy heuristic on an 80-patient population using 20 aides. Since the heuristic could schedule only 48 patients, we ran the LBBD algorithm on 60 of these patients, including 40 pre-scheduled by the greedy heuristic and 20 treated as new patients. LBBD scheduled all of the new patients using 18 aides. The resulting 60-patient schedule was used as a starting point for computational tests. It is better than a heuristic schedule but worse than an optimal one, as one might expect when scheduling on a rolling basis.

We compared the performance of LBBD and mixed integer programming (MIP) for different rates of patient turnover in the 60-patient population. One instance is generated for each number $m = 6, \ldots, 23$ of new patients, where the new patients are assumed to be the last m patients in the list of 60.

We designated 8 of the 18 aides as available to cover the new patients (along with their pre-assigned patients), because a minimum of 9 aides were required in nearly every instance. This allowed us to test computational performance near the phase transition for the problem.

We formulated an MIP model for the problem by modifying the well-known multicommodity flow model for the vehicle routing problem with time windows [35–37]. The model consists of (a)–(f) in (1) and the following:

$$w_{ijb'_ik} + \sum_{j' \neq j} w_{ijj'k} = w_{ib_ijd} + \sum_{j' \neq j} w_{ij'jk} = y_{ijk}, \ \text{all } i, j, k$$

$$w_{ib_ijk} + \sum_{j' \neq j} w_{ijj'k} = w_{ijb'_ik} + \sum_{j' \neq j} w_{ijj'k}, \ \text{all } i, j, k$$

$$s_{ij'k} \geq s_{ijk} + p_j + t_{jj'} - M_{jj'}(1 - w_{ijj'k}), \ \text{all } i, j, j', k$$

$$r_{b_i} \leq s_{ib_ik} \leq d_{b'_i}, \ r_j \leq s_{ijk} \leq d_j - p_j, \ \text{all } i, j, k$$

plus similar constraints in which j and/or j' is a home base. Here the binary variable $w_{ijj'k} \in \{0, 1\}$ represents flow and $M_{jj'} = \max\{0, \ d_j - p_j + t_{jj'} - r_{j'}\}$.

We implemented LBBD using the IBM ILOG CPLEX Optimization Studio version 12.6.2. The master problem was solved by CPLEX and the subproblem by the IBM ILOG CP Optimizer. The routing and scheduling micro-problems were formulated with a noOverlap constraint associated with sequencing and interval variables. We solved the MIP model using CPLEX. The CPLEX pre-solve routine removes variables in the MIP model and LBBD master problem that are fixed to 0 or 1 by preassignments. The solver was run in Windows 7 on a laptop with an Intel Core i7 processor and 7.75 GB RAM.

The results appear in Table 1. Since ILOG Studio does not report solution time for LBBD, the times shown are total elapsed clock times as indicated on the Studio console. They reflect overhead incurred in setting up the problem and retrieving the solution, which can be a significant fraction of total time for the smallest instances.

Both LBBD and MIP readily solve the smaller instances, but MIP suffers a combinatorial blowup when there are more than 14 or 15 new patients. MIP is disadvantaged by the fact that the number of variables grows quadratically with the number of new patients, while in LBBD it grows only linearly. LBBD therefore postpones the blowup significantly. Table 1 also shows that including a subproblem relaxation in the master problem is crucial to the performance of LBBD.

Patient records suggest that a 5–8 % turnover per week is typical in practice. LBBD therefore allows staff planning a month or so in advance for a patient population of 60. This is adequate for many real-world problem instances, particularly given that improvements in the LBBD model and subproblem relaxation are likely.

Table 1. Effect of patient turnover on computation times in a population of 60 patients and 18 aides, 8 of whom are available for new patients. The new patients replace an equal number of existing patients. Number of Benders iterations is shown, along with computation time (minutes : seconds). The last two columns show results for LBBD without a subproblem relaxation in the master problem.

New patients	Patients scheduled	LBBD		MIP	LBBD no relax	
		Iters.	Time	Time	Iters.	Time
6	60	2	**0:10**	0:39	17	1:17
7	60	3	**0:15**	0:39	18	1:23
8	60	7	**0:34**	0:49	22	1:49
9	59	7	**0:34**	0:41	20	1:38
10	59	6	**0:31**	0:43	20	1:41
11	59	6	**0:32**	0:41	31	2:52
12	59	9	**0:47**	0:45	30	2:54
13	59	24	2:15	**1:00**	51	6:53
14	59	29	**3:00**	20:27	63	9:18
15	59	37	**4:20**	11:40	72	11:57
16	59	39	**4:45**	142:08	87	16:26
17	59	39	**4:46**		129	36:39
18	59	38	**4:56**		126	30:00
19	59	75	**14:13**		138	48:01
20	58	75	**14:44**		141	63:49
21	58	87	**24:21**			
22	59	130	**48:00**			
23	59	159	**93:56**			

6   Conclusion

We find that logic-based Benders decomposition solves the home hospice care problem on a rolling basis more rapidly than state-of-the-art mixed integer programming, and it scales up to problems of realistic size. Unlike nearly all competing methods developed for this problem, it computes an optimal schedule and therefore allows planners to determine with certainty whether a given work force can meet projected patient requirements.

LBBD has the advantage that the routing and scheduling subproblems remain constant in size as the patient population grows, while the number of scheduling variables in MIP increases quadratically. The performance of LBBD also benefits from an effective time-window relaxation of the subproblem that we developed for inclusion in the master problem. LBBD is particularly well suited for scheduling on a rolling basis because continuity constraints strengthen this relaxation.

Due to the sensitivity of performance to the quality of the subproblem relaxation, future research will focus on identifying tighter relaxations, as well as incorporating constraints and objectives that more adequately reflect the complexity of the real-world problem.

References

1. Hertz, A., Lahrichi, N.: A patient assignment algorithm for home care service. J. Oper. Res. Soc. **60**, 481–495 (2009)
2. Trautsamwieser, A., Hirsch, P.: Optimization of daily scheduling for home health care services. J. Appl. Oper. Res. **3**, 124–136 (2011)
3. Nickel, S., Schröder, M., Steeg, J.: Mid-term and short-term planning support for home health care services. Eur. J. Oper. Res. **219**, 574–587 (2012)
4. Ciré, A., Hooker, J.N.: A heuristic logic-based Benders method for the home health care problem. Presented at Matheuristics 2012, Angra dos Reis, Brazil (2012)
5. Rendl, A., Prandtstetter, M., Hiermann, G., Puchinger, J., Raidl, G.: Hybrid heuristics for multimodal homecare scheduling. In: Beldiceanu, N., Jussien, N., Pinson, É. (eds.) CPAIOR 2012. LNCS, vol. 7298, pp. 339–355. Springer, Heidelberg (2012)
6. Allaoua, H., Borne, S., Létocart, L., Calvo, R.W.: A matheuristic approach for solving a home health care problem. Electron. Notes Discrete Math. **41**, 471–478 (2013)
7. Cappanera, P., Scutellà, M.G.: Joint assignment, scheduling and routing models to home care optimization: a pattern-based approach. Transp. Sci. **49**, 830–852 (2014)
8. Yalçındağ, S., Matta, A., Şahin, E., Shanthikumar, J.G.: A two-stage approach for solving assignment and routing problems in home health care services. In: Matta, A., Li, J., Sahin, E., Lanzarone, E., Fowler, J. (eds.) Proceedings of the International Conference on Health Care Systems Engineering. Proceedings in Mathematics and Statistics, vol. 61, pp. 47–59. Springer, New York (2014)
9. Mankowska, D.S., Meisel, F., Bierwirth, C.: The home health care routing and scheduling problem with interdependent services. Health Care Manage. Sci. **17**, 15–30 (2014)
10. Rest, K.D., Hirsch, P.: Daily scheduling of home health care services using time-dependent public transport. Flexible Services and Manufacturing Journal (published online 2015)
11. Dohn, A., Kolind, E., Clausen, J.: The manpower allocation problem with time windows and job-teaming constraints: a branch-and-price approach. Comput. Oper. Res. **36**, 1145–1157 (2009)
12. Rasmussen, M.S., Justesen, T., Dohn, A., Larsen, J.: The home care crew scheduling problem: preference-based visit clustering and temporal dependencies. Eur. J. Oper. Res. **219**, 598–610 (2012)
13. Chahed, S., Marcon, E., Sahin, E., Feillet, D., Dallery, Y.: Exploring new operational research opportunities within the home care context: the chemotherapy at home. Health Care Manage. Sci. **12**, 179–191 (2009)
14. Hooker, J.N.: Logic-based Benders decomposition. In: INFORMS National Meeting (1995)
15. Hooker, J.N., Yan, H.: Logic circuit verification by Benders decomposition. In: Saraswat, V., Hentenryck, P.V. (eds.) Principles and Practice of Constraint Programming: The Newport Papers, pp. 267–288. MIT Press, Cambridge (1995)

16. Hooker, J.N.: Logic-Based Methods for Optimization: Combining Optimization and Constraint Satisfaction. Wiley, New York (2000)
17. Hooker, J.N., Ottosson, G.: Logic-based Benders decomposition. Math. Program. **96**, 33–60 (2003)
18. Hooker, J.N.: Planning and scheduling by logic-based Benders decomposition. Oper. Res. **55**, 588–602 (2007)
19. Benders, J.F.: Partitioning procedures for solving mixed-variables programming problems. Numer. Math. **4**, 238–252 (1962)
20. Hooker, J.N.: A hybrid method for planning and scheduling. Constraints **10**, 385–401 (2005)
21. Hooker, J.N.: An integrated method for planning and scheduling to minimize tardiness. Constraints **11**, 139–157 (2006)
22. Ciré, A., Çoban, E., Hooker, J.N.: Logic-based Benders decomposition for planning and scheduling: a computational analysis. In: Barták, R., Salido, M.A. (eds.) COPLAS Proceedings, pp. 21–29 (2015)
23. Jain, V., Grossmann, I.E.: Algorithms for hybrid MILP/CP models for a class of optimization problems. INFORMS J. Comput. **13**, 258–276 (2001)
24. Harjunkoski, I., Grossmann, I.E.: Decomposition techniques for multistage scheduling problems using mixed-integer and constraint programming methods. Comput. Chem. Eng. **26**, 1533–1552 (2002)
25. Harjunkoski, I., Grossmann, I.E.: A decomposition approach for the scheduling of a steel plant production. Comput. Chem. Eng. **25**, 1647–1660 (2001)
26. Liu, W., Gu, Z., Xu, J., Wu, X., Ye, Y.: Satisfiability modulo graph theory for task mapping and scheduling on multiprocessor systems. IEEE Trans. Parallel Distrib. Syst. **22**, 1382–1389 (2011)
27. Lombardi, M., Milano, M., Ruggiero, M., Benini, L.: Stochastic allocation and scheduling for conditional task graphs in multi-processor systems-on-chip. J. Sched. **13**, 315–345 (2010)
28. Cambazard, H., Hladik, P.-E., Déplanche, A.-M., Jussien, N., Trinquet, Y.: Decomposition and learning for a hard real time task allocation problem. In: Wallace, M. (ed.) CP 2004. LNCS, vol. 3258, pp. 153–167. Springer, Heidelberg (2004)
29. Chu, Y., Xia, Q.: Generating Benders cuts for a class of integer programming problems. In: Régin, J.-C., Rueher, M. (eds.) CPAIOR 2004. LNCS, vol. 3011, pp. 127–141. Springer, Heidelberg (2004)
30. Maravelias, C.T., Grossmann, I.E.: Using MILP and CP for the scheduling of batch chemical processes. In: Régin, J.-C., Rueher, M. (eds.) CPAIOR 2004. LNCS, vol. 3011, pp. 1–20. Springer, Heidelberg (2004)
31. Maravelias, C.T., Grossmann, I.E.: A hybrid MILP/CP decomposition approach for the continuous time scheduling of multipurpose batch plants. Comput. Chem. Eng. **28**, 1921–1949 (2004)
32. Terekhov, D., Beck, J.C., Brown, K.N.: Solving a stochastic queueing design and control problem with constraint programming. In: Proceedings of the 22nd National Conference on Artificial Intelligence (AAAI 2007), vol. 1, pp. 261–266. AAAI Press (2007)
33. Benini, L., Bertozzi, D., Guerri, A., Milano, M.: Allocation and scheduling for MPSoCs via decomposition and no-good generation. In: van Beek, P. (ed.) CP 2005. LNCS, vol. 3709, pp. 107–121. Springer, Heidelberg (2005)
34. Ciré, A., Coban, E., Hooker, J.N.: Mixed integer programming vs. logic-based Benders decomposition for planning and scheduling. In: Gomes, C., Sellmann, M. (eds.) CPAIOR 2013. LNCS, vol. 7874, pp. 325–331. Springer, Heidelberg (2013)

35. Desrochers, M., Lenstra, J.K., Savelsbergh, M.W.P., Soumis, F.: Vehicle routing with time windows: optimization and approximation. In: Golden, B.L., Assad, A.A. (eds.) Vehicle Routing: Methods and Studies, pp. 65–84. North-Holland, Amsterdam (1988)
36. Desrochers, M., Laporte, G.: Improvements and extensions to the Miller-Tucker-Zemlin subtour elimination constraints. Oper. Res. Lett. **10**, 27–36 (1991)
37. Cordeau, J.F., Laporte, G., Savelsbergh, M., Vigo, D.: Vehicle routing. In: Barnhart, C., Laporte, G. (eds.) Handbook in Operations Research and Management Science, vol. 14, pp. 367–428. Elsevier, Amsterdam (2007)

A Global Constraint for Mining Sequential Patterns with GAP Constraint

Amina Kemmar[1(✉)], Samir Loudni[2], Yahia Lebbah[1], Patrice Boizumault[2], and Thierry Charnois[3]

[1] LITIO, University of Oran 1, EPSECG of Oran, Oran, Algeria
kemmar.amina@edu.univ-oran1.dz
[2] GREYC (CNRS UMR 6072), University of Caen, Caen, France
[3] LIPN (CNRS UMR 7030), University Paris 13, Paris, France

Abstract. Sequential pattern mining (SPM) under gap constraint is a challenging task. Many efficient specialized methods have been developed but they are all suffering from a lack of genericity. The Constraint Programming (CP) approaches are not so effective because of the size of their encodings. In [7], we have proposed the global constraint PREFIX-PROJECTION for SPM which remedies to this drawback. However, this global constraint cannot be directly extended to support gap constraint. In this paper, we propose the global constraint GAP-SEQ enabling to handle SPM with or without gap constraint. GAP-SEQ relies on the principle of right pattern extensions. Experiments show that our approach clearly outperforms both CP approaches and the state-of-the-art cSpade method on large datasets.

1 Introduction

Mining sequential patterns (SPM) is an important task in data mining. There are many useful applications, including discovering changes in customer behaviors, detecting intrusion from web logs and finding relevant genes from DNA sequences. In recent years many studies have focused on SPM with gap constraints [17,19]. Limited gaps allow a mining process to bear a certain degree of flexibility among correlated pattern elements in the original sequences. For example, [6] analyses purchase behaviors to reflect products usually bought by customers at regular time intervals according to time gaps. In computational biology, the gap constraint helps discover periodic patterns with significant biological and medical values [15].

Mining sequential patterns under gap constraint (GSPM) is a challenging task, since the *apriori property* does not hold for this problem: *a subsequence of a frequent sequence is not necessarily frequent*. Several specialized approaches have been proposed [6,10,19] but they have a lack of genericity to handle simultaneously various types of constraints. Recently, a few proposals [4,8,11,12] have investigated relationships between GSPM and constraint programming (CP) in order to provide a declarative approach, while exploiting efficient and generic solving methods. But, due to the size of the proposed encodings, these CP

© Springer International Publishing Switzerland 2016
C.-G. Quimper (Ed.): CPAIOR 2016, LNCS 9676, pp. 198–215, 2016.
DOI: 10.1007/978-3-319-33954-2_15

methods are not as efficient as specialized ones. More recently, we have proposed the global constraint PREFIX-PROJECTION for SPM which remedies this drawback [7]. However, this global constraint cannot be directly extended to support gap constraint.

In this paper, we introduce the global constraint GAP-SEQ enabling to handle SPM with or without gap constraint. GAP-SEQ relies on the principle of right pattern extension and its filtering exploits the prefix anti-monotonicity property of the gap constraint to provide an efficient pruning of the search space. GAP-SEQ enables to handle simultaneously different types of constraints and its encoding does not require any reified constraints nor any extra variables. Finally, experiments show that our approach clearly outperforms CP approaches as well as specialized methods for GSPM and achieves scalability while it is a major issue for CP approaches.

The paper is organized as follows. Section 2 introduces the prefix anti-monotonicity of the gap constraint as well as right pattern extensions that will enable an efficient filtering. Section 3 provides a critical review of specialized methods and CP approaches for sequential pattern mining under gap constraint. Section 4 presents the global constraint GAP-SEQ. Section 5 reports experiments we performed. Finally, we conclude and draw some perspectives.

2   Preliminaries

First, we provide the basic definitions for GSPM. Then, we show that the anti-monotonicity property of frequency of SPM does not hold for GSPM. Finally, we introduce right pattern extensions that will enable an efficient filtering for GSPM.

2.1   Definitions

Let $\mathcal{I}$ be a finite set of distinct *items*. The language of sequences corresponds to $\mathcal{L}_{\mathcal{I}} = \mathcal{I}^n$ where $n \in \mathbb{N}^+$.

Definition 1 (sequence, sequence database). *A sequence s over $\mathcal{L}_{\mathcal{I}}$ is an ordered list $\langle s_1 \ldots s_n \rangle$, where s_i, $1 \leq i \leq n$, is an item. n is called the length of the sequence s. A sequence database SDB is a set of tuples (sid, s), where sid is a sequence identifier and s a sequence denoted by $SDB[sid]$.*

We now define the subsequence relation $\preceq^{[M,N]}$ under $gap[M,N]$ constraint which restricts the allowed distance between items of subsequences in sequences.

Definition 2 (subsequence relation $\preceq^{[M,N]}$ under $gap[M,N]$). *$\alpha = \langle \alpha_1 \ldots \alpha_m \rangle$ is a subsequence of $s = \langle s_1 \ldots s_n \rangle$, under $gap[M,N]$, denoted by $(\alpha \preceq^{[M,N]} s)$, if $m \leq n$ and, for all $1 \leq i \leq m$, there exist integers $1 \leq j_1 \leq \ldots \leq j_m \leq n$, such that $\alpha_i = s_{j_i}$, and $\forall k \in \{1, ..., m-1\}, M \leq j_{k+1} - j_k - 1 \leq N$. In this context, the pair $(s, [j_1, j_m])$ denotes an **occurrence** of α in s, where j_1 and j_m represent the positions of the first and last items of α in s. We say that α is contained in s or s is a super-sequence of α under $gap[M,N]$. We also say that α is a $gap[M,N]$ **constrained pattern** in s.*

- Let $AllOcc(\alpha, s) = \{[j_1, j_m] \mid (s, [j_1, j_m])$ is an occurrence of α in $s\}$ be the set of all the occurrences of some sequence α under $gap[M, N]$ in s.
- Let $AllOcc(\alpha, SDB) = \{(sid, AllOcc(\alpha, SDB[sid])) \mid (sid, SDB[sid]) \in SDB\}$ be the set of all the occurrences of some sequence α under $gap[M, N]$ in SDB.
- Let $gap[M, \infty]$ and $gap[0, N]$ the **minimum and the maximum gap** constraints respectively. The relation $\preceq$ stands for $\preceq^{[0,\infty]}$ where the gap constraint is inactive.

For example, the sequence $\langle BABC \rangle$ is a super-sequence of $\langle AC \rangle$ under $gap[0, 2]$: $\langle AC \rangle \preceq^{[0,2]} \langle BABC \rangle$.

Definition 3 (prefix, postfix). Let $\beta = \langle \beta_1 \ldots \beta_n \rangle$ be a sequence. The sequence $\alpha = \langle \alpha_1 \ldots \alpha_m \rangle$ where $m \leq n$ is called the prefix of β iff $\forall i \in [1..m], \alpha_i = \beta_i$. The sequence $\gamma = \langle \beta_{m+1} \ldots \beta_n \rangle$ is called the postfix of s w.r.t. α. With the standard concatenation operator "concat", we have $\beta = concat(\alpha, \gamma)$.

The cover of a sequence α in SDB is the set of all tuples in SDB in which α is contained. The support of a sequence α in SDB is the cardinal of its cover.

Definition 4 (cover and support). Let α be a sequence. $cover_{SDB}^{[M,N]}(\alpha) = \{(sid, s) \in SDB \mid \alpha \preceq^{[M,N]} s\}$ and $sup_{SDB}^{[M,N]}(\alpha) = \#cover_{SDB}^{[M,N]}(\alpha)$.

Definition 5 ($gap[M, N]$ constrained sequential pattern mining (GS PM)). Given a sequence database SDB, a minimum support threshold $minsup$ and a gap constraint $gap[M, N]$. The problem of $gap[M, N]$ constrained sequential pattern mining is to find all subsequences α such that $sup_{SDB}^{[M,N]}(\alpha) \geq minsup$.

Table 1. A sequence database example SDB_1.

Sid	Sequence
1	$\langle ABCDB \rangle$
2	$\langle ACCBACB \rangle$
3	$\langle ADCBEEC \rangle$
4	$\langle AACC \rangle$

Example 1. Table 1 represents a sequence database of four sequences where the set of items is $\mathcal{I} = \{A, B, C, D, E\}$. Let the sequence $\alpha = \langle AC \rangle$. The occurrences under $gap[0, 1]$ of α in $SDB_1[2]$ is given by $AllOcc(\alpha, SDB_1[2]) = \{[1, 2], [1, 3], [5, 6]\}$. We have $cover_{SDB_1}^{[0,1]}(\alpha) = \{(1, s_1), (2, s_2), (3, s_3), (4, s_4)\}$. If we consider $minsup = 2$, α is a $gap[0, 1]$ constrained sequential pattern because $sup_{SDB_1}^{[0,1]}(\alpha) \geq 2$.

2.2 Prefix Anti-monotonicity of $gap[M, N]$

Most SPM algorithms rely on the *anti-monotonicity property of frequency* [1] to reduce the search space: all the subsequences of a frequent sequence are frequent as well (or, equivalently, if a subsequence is infrequent, then no super-sequence of it can be frequent). However, this property does not hold for the gap constraint, and more precisely for the maximum gap constraint. A simple illustration from our running example suffices to show that sequence $\langle AB \rangle$ is not a sequential pattern under $gap[0, 1]$ (for $minsup = 3$) whereas sequence $\langle ACB \rangle$ is a $gap[0, 1]$ constrained sequential pattern. As a consequence, one needs to use other techniques for pruning the search space. The following proposition shows how the *prefix anti-monotonicity property* introduced in [14] can be exploited to ensure the prefix anti-monotonicity of the gap constraint.

Definition 6 (prefix anti-monotone property [14]). *A constraint c is called prefix anti-monotone if for every sequence α satisfying c, every prefix of α also satisfies the constraint.*

Proposition 1. $gap[M, N]$ *is prefix anti-monotone.*

Proof. Let $\alpha = \langle \alpha_1 \ldots \alpha_m \rangle$ and $s = \langle s_1 \ldots s_n \rangle$ be two sequences s.t. $\alpha \preceq^{[M,N]} s$ and $m \leq n$. By definition, there exist integers $1 \leq j_1 \leq \ldots \leq j_m \leq n$, such that $\alpha_i = s_{j_i}$, and $\forall k \in \{1, ..., m-1\}, M \leq j_{k+1} - j_k - 1 \leq N$. As a consequence, the property also holds for every prefix of α. $\qquad\qquad\qquad\square$

Hence, if a sequence α does not satisfy $gap[M, N]$, then all sequences that have α as prefix will not satisfy this constraint. Section 4.2 shows how this property can be exploited to provide an efficient filtering.

2.3 Right Pattern Extensions

Right pattern extensions of some pattern p gives all the possible subsequences which can be appended at right of p to form a $gap[M, N]$ constrained pattern. According to Proposition 1, the set of all items locally frequent within the right pattern extensions of p in SDB can be used to extend p. In the following, we introduce an operator allowing to compute all the right pattern extensions of a pattern w.r.t. $gap[M, N]$.

Definition 7 (Right pattern extensions). *Given some sequence (sid, s) and a pattern p s.t. $p \preceq^{[M,N]} s$. The **right pattern extensions** of p in s, denoted by $Ext_R^{[M,N]}(p, s)$, is the collection of legal subsequences of s located at the right of p and satisfying $gap[M, N]$. To define $Ext_R^{[M,N]}(p, s)$, we need to define $BE^{[M,N]}(p, s)$ **basic right extensions**:*

$$BE^{[M,N]}(p, s) = \bigcup_{[j_1, j_m] \in AllOcc(p,s)} \{(j_m, \mathsf{SubSeq}(s, j_m + M + 1, min(j_m + N + 1, \#s)))\}$$

$$where \ \mathsf{SubSeq}(s, i_1, i_2) = \begin{cases} \langle s[i_1], ..., s[i_2] \rangle & if \ i_1 \leq i_2 \leq \#s \\ \langle \rangle & otherwise \end{cases}$$

Right pattern extensions $Ext_R^{[M,N]}(p,s)$ *is defined as follows:*

$$Ext_R^{[M,N]}(p,s) = \begin{cases} \{Sb \mid (j_m', Sb) \in BE^{[M,N]}(p,s) \land & if N \geq \#s \\ \quad j_m' = \min_{(j_m,Sb) \in BE^{[M,N]}(p,s)}\{j_m\}\} \\ \bigcup_{(j_m,Sb) \in BE^{[M,N]}(p,s)}\{Sb\} & otherwise \end{cases} \tag{1}$$

Formula (1) states exactly the set of all possible extensions of pattern p within s. In case where $(N \geq \#s)$, since that any extension from $BE^{[M,N]}(p,s)$ always reaches the end of the sequence s, thus all possible extensions can be aggregated within one unique extension going from the lowest starting position $j_m' = \min_{(j_m,Sb) \in BE^{[M,N]}(p,s)}\{j_m\}$. We point out that these cases $(N \geq \#s)$ cover the special case of no gap $gap[0,\infty]$.

The right pattern extensions of p in SDB is the collection of all its right pattern extensions in all sequences of SDB:

$$Ext_R^{[M,N]}(p,SDB) = \{(sid, Ext_R^{[M,N]}(p,s)) \mid (sid,s) \in SDB \land p \preceq^{[M,N]} s\} \tag{2}$$

Example 2. Let $p_1 = \langle AC \rangle$ be a pattern and the gap constraint be $gap[0,1]$. We have $AllOcc(p_1, SDB_1[2]) = \{[1,2], [1,3], [5,6]\}$. The right pattern extensions of p_1 in $SDB_1[2]$ is equal to $Ext_R^{[0,1]}(p_1, SDB_1[2]) = \{\langle CB \rangle, \langle BA \rangle, \langle B \rangle\}$. The right pattern extensions of p_1 in SDB_1 is given by $Ext_R^{[0,1]}(p_1, SDB_1) = \{(1, \{\langle DB \rangle\}), (2, \{\langle CB \rangle, \langle BA \rangle, \langle B \rangle\}), (3, \{\langle BE \rangle\}), (4, \{\langle C \rangle\})\}$.

Let the gap constraint be $gap[0,\infty]$. To compute $Ext_R^{[0,\infty]}(p_1, SDB_1[2])$, only the first occurrence of p_1 in $SDB_1[2]$ need to be considered (i.e. $[1,2]$) (cf. Definition 7). Thus, $Ext_R^{[0,\infty]}(p_1, SDB_1[2]) = \{\langle CBACB \rangle\}$. The right pattern extensions of p_1 in SDB_1 is equal to $Ext_R^{[0,\infty]}(p_1, SDB_1) = \{(1, \{\langle DB \rangle\}), (2, \{\langle CBACB \rangle\}), (3, \{\langle BEEC \rangle\}), (4, \{\langle C \rangle\})\}$.

We define $supext_{SDB}^{[M,N]}(\alpha, p)$ as the support of α within the right pattern extensions:

$$supext_{SDB}^{[M,N]}(\alpha, p) = \#\{(sid, s) \in SDB \mid \exists(sid, E) \in Ext_R^{[M,N]}(p,SDB), \\ \exists s' \in E, \langle \alpha \rangle \preceq s'\}. \tag{3}$$

Let $\mathcal{RF}_{SDB}^{[M,N]}(p)$ be the set of locally frequent items within the right extensions:

$$\mathcal{RF}_{SDB}^{[M,N]}(p) = \{v \in \mathcal{I} \mid \#\{sid \mid \exists(sid, E) \in Ext_R^{[M,N]}(p,SDB), \\ \exists \alpha \in E, \langle v \rangle \preceq \alpha\} \geq minsup\}. \tag{4}$$

Given a $gap[M,N]$ constrained pattern p in SDB, according to Proposition 1, items in $\mathcal{RF}_{SDB}^{[M,N]}(p)$ can be used to extend p. Proposition 2 establishes the support count of a sequence γ w.r.t. its right pattern extensions.

Proposition 2 (Support count). *For any sequence γ in SDB with prefix α and postfix β s.t. $\gamma = concat(\alpha, \beta)$, $sup_{SDB}^{[M,N]}(\gamma) = supext_{SDB}^{[M,N]}(\beta, \alpha)$.*

This proposition ensures that only the sequences in SDB grown from α need to be considered for the support count of a sequence γ. From Proposition 2, we can derive the following proposition to establish a condition to check when a pattern is a $gap[M,N]$ constrained sequential pattern.

Proposition 3. *Let SDB be a sequence database and a minimum support threshold minsup. A pattern p is a gap$[M, N]$ constrained sequential pattern in SDB if and only if the following condition holds: $\#Ext_R^{[M,N]}(p, SDB) \geq minsup$*

Example 3. Let *minsup* be 2 and the gap constraint be *gap*[0, 1]. From Example 2, we have $\#Ext_R^{[0,1]}(p_1, SDB_1) = 4 \geq minsup$. Thus, $p_1 = \langle AC \rangle$ is a *gap*[0, 1] constrained sequential pattern. The locally frequent items within the right pattern extensions $Ext_R^{[0,1]}(p_1, SDB_1)$ of p_1 are B and C with supports of 3 and 2 respectively. According to Proposition 2, p_1 can be extended to two *gap*[0, 1] constrained sequential patterns $\langle ACB \rangle$ and $\langle ACC \rangle$.

3 Related Works

Specialized Methods for GSPM. The SPM was first proposed in [1]. Since then, many efficient specialized approaches have been proposed [2,13,18]. There are also several methods focusing on gap constraints. Zaki [17] first proposed cSpade, a depth-first search based on a vertical database format, incorporating constraints on gap (min_gap and max_gap) and time windows (max_span). Other constraints on length, items and classes for classification datasets are also mentioned in the paper but they are not supported in the author's cSpade implementation. Ji et al. [6] and Li [9] studied the problem of mining frequent patterns with gap constraints. In [6], a minimal distinguishing subsequence that occurs frequently in the positive sequences and infrequently in the negative sequences is proposed, where the maximum gap constraint is defined. In [9], closed frequent patterns with gap constraints are mined. All these proposals, though efficient, lack genericity to handle simultaneously various types of constraints. Finally, Pei et al. [14] have proposed an algorithm based on prefix-growth which handles constraints that are prefix anti-monotone. These classes of constraints are stated a posteriori and are only used for -testing- solutions (without any pruning). For the particular case of the gap constraint, when a current prefix satisfies a constraint, no pruning is achieved and all possible "right-parts" have to be tested.

CP Methods for GSPM. There are few methods for SPM with gap constraints using CP. [11] have proposed to model a sequence using an automaton capturing all subsequences that can occur in it. The gap constraint is encoded by removing from the automaton all transitions that do not respect the gap constraint. [8] have proposed a CSP model for SPM with explicit wildcards[1]. The gap constraints is enforced using the regular global constraint. [12] have proposed two CP encodings for the SPM. The first one uses a global constraint to encode the subsequence relation (denoted global-p.f), while the second one (denoted decomposed-p.f) encodes explicitly this relation using additional variables and constraints in order to support constraints like gap. However, all these proposals usually lead to constraint networks of huge size. Space complexity is clearly

---

[1] A wildcard is a special symbol that matches any item of $\mathcal{I}$ including itself.

identified as the main bottleneck behind the competitiveness of these declarative approaches. In [7], we have proposed the global constraint PREFIX-PROJECTION for sequential pattern mining which remedies to this drawback. However, this constraint cannot be directly extended to handle gap constraints. This requires changing the way the subsequence relation is encoded.

The next section introduces the global constraint GAP-SEQ enabling to handle SPM with or without gap constraints. GAP-SEQ relies on the prefix anti-monotonicity of the gap constraint and on the right pattern extensions to provide an efficient filtering. This global constraint does not require any reified constraints nor any extra variables to encode the subsequence relation.

4   GAP-SEQ Global Constraint

This section is devoted to the GAP-SEQ global constraint. Section 4.1 defines the GAP-SEQ global constraint and presents the CSP modeling. Section 4.2 shows how the filtering can take advantage of the prefix anti-monotonicity property of the $gap[M, N]$ constraint (see Proposition 6) and of the right pattern extensions (see Proposition 5) to remove inconsistent values from the domain of a future variable. Section 4.3 details the filtering algorithm and Sect. 4.4 provides its temporal and spatial complexities.

4.1   CSP Modeling for GSPM

A *Constraint Satisfaction Problem* (CSP) consists of a set X of n variables, a domain $\mathcal{D}$ mapping each variable $X_i \in X$ to a finite set of values $D(X_i)$, and a set of constraints $\mathcal{C}$. An assignment σ is a mapping from variables in X to values in their domains. A constraint $c \in \mathcal{C}$ is a subset of the cartesian product of the domains of the variables that occur in c. The goal is to find an assignment such that all constraints are satisfied.

(a) Variables and Domains. Let P be the unknown pattern of size ℓ we are looking for. The symbol $\square$ ($\square \notin \mathcal{I}$) stands for an empty item and denotes the end of a sequence. We encode the unknown pattern P of maximum length ℓ with a sequence of ℓ variables $\langle P_1, P_2, \ldots, P_\ell \rangle$. Each variable P_j represents the item in the jth position of the sequence. The size ℓ of P is determined by the length of the longest sequence of SDB. The domains of variables are defined as follows: (i) $D(P_1) = \mathcal{I}$ to avoid the empty sequence, and (ii) $\forall i \in \{2 \ldots \ell\}, D(P_i) = \mathcal{I} \cup \{\square\}$. To allow patterns with less than ℓ items, we impose that $\forall i \in \{2..(\ell-1)\}, (P_i = \square) \rightarrow (P_{i+1} = \square)$.

(b) Definition of GAP-SEQ. The global constraint GAP-SEQ encodes both subsequence relation $\preceq^{[M,N]}$ under gap constraint $gap[M, N]$ and minimum frequency constraint directly on the data.

Definition 8 (GAP-SEQ global constraint). *Let $P = \langle P_1, P_2, \ldots, P_\ell \rangle$ be a pattern of size ℓ and $gap[M, N]$ be the gap constraint. $\langle d_1, \ldots, d_\ell \rangle \in D(P_1) \times \ldots \times D(P_\ell)$ is a solution of GAP-SEQ$(P, SDB, minsup, M, N)$ iff $sup_{SDB}^{[M,N]}(\langle d_1, \ldots, d_\ell \rangle) \geq minsup$.*

Proposition 4. GAP-SEQ$(P, SDB, minsup, M, N)$ *has a solution iff there exists an assignment* $\sigma = \langle d_1, ..., d_\ell \rangle$ *of variables of* P *s.t.* $\#Ext_R^{[M,N]}(\sigma, SDB) \geq minsup$.

Proof: This is a direct consequence of proposition 3.   □

(c) Other SPM constraints can be directly modeled as follows:
- *Minimum Size* constraint restricts the number of items of a pattern to be at least ℓ_{min}: $minSize(P, \ell_{min}) \equiv \bigwedge_{i=1}^{i=\ell_{min}}(P_i \neq \square)$
- *Maximum Size* constraint restricts the number of items of a pattern to be at most ℓ_{max}: $maxSize(P, \ell_{max}) \equiv \bigwedge_{i=\ell_{max}+1}^{i=\ell}(P_i = \square)$
- *Membership* constraint states that a subset of items V must belong (or not) to the extracted patterns. $item(P, V) \equiv \bigwedge_{t \in V} \mathtt{Among}(P, \{t\}, l, u)$ enforces that items of V should occur at least l times and at most u times in P. To forbid items of V to occur in P, l and u must be set to 0.

4.2   Principles of Filtering

(a) Maintaining a local consistency. SPM is a challenging task due to the exponential number of candidates that should be parsed to find the frequent patterns. For instance, we have $O(n^k)$ potential candidate patterns of length at most k in a sequence of length n. With gap constraints, the problem is even much harder since the complexity of checking for subsequences taking a gap constraint into account is higher than the complexity of the standard subsequence relation. Furthermore, the NP-hardness of mining maximal[2] frequent sequences was established in [16] by proving the #P-completeness of the problem of counting the number of maximal frequent sequences. Hence, ensuring *Domain Consistency* (DC) for GAP-SEQ i.e., finding, for every variable P_j, a value $d_j \in D(P_j)$, satisfying the constraint is NP-hard.

So, the filtering of GAP-SEQ constraint maintains a consistency lower than DC. This consistency is based on specific properties of the $gap[M, N]$ constraint and resembles forward-checking (regarding Proposition 5). GAP-SEQ is considered as a global constraint, since all variables share the same internal data structures that awake and drive the filtering. The prefix anti-monotonicity property of the $gap[M, N]$ constraint (see Proposition 6) and of the right pattern extensions (see Proposition 5) will enable to remove inconsistent values from the domain of a future variable.

(b) Detecting inconsistent values. Let $\mathcal{RF}_{SDB}^{[M,N]}(\sigma)$ be the set of locally frequent items within the right pattern extensions (see (4) in Sect. 2.3). The following proposition characterizes values, of a future (unassigned) variable P_{j+1}, that are consistent with the current assignment of variables $\langle P_1, \ldots, P_j \rangle$.

Proposition 5. *Let*[3] $\sigma = \langle d_1, \ldots, d_j \rangle$ *be a current assignment of variables* $\langle P_1, \ldots, P_j \rangle$, P_{j+1} *be a future variable. A value* $d \in D(P_{j+1})$ *occurs in a*

---
[2] A sequential pattern p is maximal if there is no sequential pattern q such that $p \preceq q$.
[3] We indifferently denote σ by $\langle d_1, \ldots, d_j \rangle$ or by $\langle \sigma(P_1), \ldots, \sigma(P_j) \rangle$.

solution for the global constraint GAP-SEQ(P, SDB, $minsup$, M, N) iff $d \in \mathcal{RF}_{SDB}^{[M,N]}(\sigma)$.

Proof: Assume that $\sigma = \langle d_1, \ldots, d_j \rangle$ is $gap[M, N]$ constrained sequential pattern in SDB. Suppose that value $d \in D(P_{j+1})$ appears in $\mathcal{RF}_{SDB}^{[M,N]}(\sigma)$. As the local support of d within the right extensions (see (3)) is equal to $supext_{SDB}^{[M,N]}(\langle d \rangle, \sigma)$, from Proposition 2 we have $sup_{SDB}^{[M,N]}(concat(\sigma, \langle d \rangle)) = supext_{SDB}^{[M,N]}(\langle d \rangle, \sigma)$. Hence, we can get a new assignment $\sigma \cup \langle d \rangle$ that satisfies the constraint. Therefore, $d \in D(P_{j+1})$ participates in a solution. □

From proposition 5 and according to the prefix anti-monotonicity property of the gap constraint, we can derive the following pruning rule:

Proposition 6. *Let $\sigma = \langle d_1, \ldots, d_j \rangle$ be a current assignment of variables $\langle P_1, \ldots, P_j \rangle$. All values $d \in D(P_{j+1})$ that are not in $\mathcal{RF}_{SDB}^{[M,N]}(\sigma)$ can be removed from the domain of variable P_{j+1}.*

Example 4. Consider the running example of Table 1, let $minsup$ be 2 and the gap constraint be $gap[1, 2]$. Let $P = \langle P_1, P_2, P_3, P_4 \rangle$ with $D(P_1) = \mathcal{I}$ and $D(P_2) = D(P_3) = D(P_4) = \mathcal{I} \cup \{\Box\}$. Suppose that $\sigma(P_1) = A$. We have $Ext_R^{[1,2]}(\langle A \rangle, SDB_1) = \{(1, \{\langle CD \rangle\}), (2, \{\langle CB \rangle, \langle B \rangle\}), (3, \{\langle CB \rangle\}), (4, \{\langle CC \rangle, \langle C \rangle\})\}$. As B and C are the only locally frequent items in $Ext_R^{[1,2]}(\langle A \rangle, SDB_1)$, GAP-SEQ will remove values A, D and E from $D(P_2)$.

4.3 Filtering Algorithm

Algorithm 1 describes the pseudo-code of GAP-SEQ filtering algorithm. It takes as input: the index j of the last assigned variable in P, the current partial assignment $\sigma = \langle \sigma(P_1), \ldots, \sigma(P_j) \rangle$, the minimum support threshold $minsup$, the minimum and the maximum gaps. The internal data-structure $\mathcal{ALLOCC}$ stores all the intermediate occurrences of patterns in SDB, where $\mathcal{ALLOCC}_j = AllOcc(\sigma, SDB)$, for $j \in \{1 \ldots \ell\}$. If $\sigma = \langle \rangle$, then $\mathcal{ALLOCC}_0 = \{(sid, [1, \#s]) \mid (sid, s) \in SDB\}$.

Algorithm 1 starts by computing the right pattern extensions Ext_R of σ in SDB by calling function GETRIGHTEXT (see Algorithm 2). Then, it checks whether the current assignment σ satisfies the constraint (line 2). If not, we stop growing σ and we return *False*. Otherwise, the algorithm checks if the last assigned variable P_j is instantiated to $\Box$ (line 4). If so, the end of the sequence is reached (since value $\Box$ can only appear at the end) and the sequence $\langle \sigma(P_1), \ldots, \sigma(P_j) \rangle$ is a $gap[M, N]$ constrained sequential pattern in SDB; hence, the algorithm sets the remaining $(\ell - j)$ unassigned variables to $\Box$ and returns *True* (5–6). If $(P_j \neq \Box)$, the set of locally frequent items, within the right pattern extensions Ext_R of σ in SDB, is computed by calling function GETFREQITEMS (line 7) and the domain of variable P_{j+1} is updated accordingly (lines 8–9).

Algorithm 2 gives the pseudo-code of the function GETRIGHTEXT. First, if σ is empty (i.e. $\#\sigma = 0$), all the sequences of SDB are considered as valid

---

Algorithm 1. FILTER-GAP-SEQ(SDB, σ, j, P, $minsup$, M, N)

Data: SDB: initial database; σ: current assignment $\langle\sigma(P_1),\ldots,\sigma(P_j)\rangle$; $minsup$: the minimum support threshold; $\mathcal{ALLOCC}$: internal data structure for storing occurrences of patterns in SDB; Ext_R: internal data structure for storing right pattern extensions of σ in SDB.

begin

1     $Ext_R \leftarrow$ GETRIGHTEXT(SDB, $\mathcal{ALLOCC}_{j-1}$, σ, M, N) ;

2   **if** ($\#Ext_R < minsup$) **then**

3   ⌊ **return** False ;

4   **if** ($j \geq 2 \wedge \sigma(P_j) = \square$) **then**

5   **for** $k \leftarrow j+1$ **to** ℓ **do**

6   ⌊ $P_k \leftarrow \square$;

    else

7        $\mathcal{RF} \leftarrow$ GETFREQITEMS(SDB, Ext_R, $minsup$) ;

8   **foreach** $a \in D(P_{j+1})$ $s.t.(a \neq \square \wedge a \notin \mathcal{RF})$ **do**

9   ⌊ $D(P_{j+1}) \leftarrow D(P_{j+1}) - \{a\}$;

10   **return** True ;

---

right pattern extensions; the whole SDB should be returned. Otherwise, the function GETALLOCC is called to compute the occurrences of σ in SDB (line 3). Then, the algorithm processes all the entries of $\mathcal{ALLOCC}_j$, one by one (line 5), and, for each pair $(sid, OccSet)$, scans the occurrences of σ in the sequence sid (line 7). For each occurrence $[j_1, j_m] \in OccSet$, the algorithm computes its right pattern extensions, i.e. the part of the sequence sid which is in the range $[j_m + M + 1, min(j_m + N + 1, \#s)]$ (line 8). If the new range is valid, it is added to the set Sb (line 10). After processing the whole entries in $OccSet$, the right pattern extensions of σ in the sequence sid are built and then added to the set Ext_R (line 11). The process ends when all entries of $\mathcal{ALLOCC}_j$ have been considered. The right pattern extensions of σ in SDB are then returned (line 12).

Function GETALLOCC computes incrementally $\mathcal{ALLOCC}_j$ from $\mathcal{ALLO}$-$\mathcal{CC}_{j-1}$. More precisely, lines (18–19) and (24–25) are considered when the first variable P_1 is instantiated (i.e. $\#\sigma = 1$), and consequently all of its initial occurrences should be found and stored in $\mathcal{ALLOCC}_1$ through the initialization step (lines 24–25). After that, $\mathcal{ALLOCC}_j(j > 1)$ is incrementally computed from $\mathcal{ALLOCC}_{j-1}$ through line (26).

Example 5. Consider the running example of Table 1, let the gap constraint be $gap[0, 4]$, and $\sigma = \langle ACB \rangle$. The occurrences of $\langle A \rangle$ in SDB_1 are stored in $\mathcal{ALLOCC}_1 = \{(1, \{[1, 1]\}), (2, \{[1, 1], [5, 5]\}), (3, \{[1, 1]\}), (4, \{[1, 1], [2, 2]\})\}$. From $\mathcal{ALLOCC}_1$, we get the occurrences of $\langle AC \rangle$: $\mathcal{ALLOCC}_2 = \{(1, \{[1, 3]\}), (2, \{[1, 2], [1, 3], \mathbf{[1, 6]}, \mathbf{[5, 6]}\}), (3, \{[1, 3]\}), (4, \{[1, 3], \mathbf{[2, 3]}, [1, 4], \mathbf{[2, 4]},\})\}$. But, as $BE^{[M,N]}(p, s)$ is only based on the final position j_m of each occurrence (see Definition 7), the occurrences with the same final position j_m (in bold in our example) are considered only once. Thus, $\mathcal{ALLOCC}_2 = \{(1, \{[1, 3]\}), (2, \{[1, 2], [1, 3], [1, 6]\}), (3, \{[1, 3]\}), (4, \{[1, 3], [1, 4]\})\}$.

Algorithm 2. GETRIGHTEXT(SDB, $\mathcal{ALLOCC}_{j-1}$, σ, M, N)

Data: SDB: initial database; $\mathcal{ALLOCC}_{j-1}$: occurrences of the partial assignment
$\langle \sigma(P_1), \ldots, \sigma(P_{j-1}) \rangle$ in SDB; σ: the current partial assignment $\langle \sigma(P_1), \ldots, \sigma(P_j) \rangle$;
$OccSet$: the positions of the first and last items of $\langle \sigma(P_1), \ldots, \sigma(P_{j-1}) \rangle$ in $SDB[sid]$;
Sb: the positions of the first and last items of the right pattern extensions of σ in
$SDB[sid]$.

begin
1   **if** ($\sigma = \langle \rangle$) **then**
2   **return** $\{(sid, (1, \#s)) | (sid, s) \in SDB\}$;
3    $\mathcal{ALLOCC}_j \leftarrow$ GETALLOCC(SDB, $\mathcal{ALLOCC}_{j-1}$, σ, M, N) ;
4    $Ext_R \leftarrow \emptyset$;
5   **foreach** *pair* $(sid, OccSet) \in \mathcal{ALLOCC}_j$ **do**
6      $s \leftarrow SDB[sid]$; $Sb \leftarrow \emptyset$;
7   **foreach** *pair* $[j_1, j_m] \in OccSet$ **do**
8        $j_1' \leftarrow j_m + M + 1$; $j_m' \leftarrow min(j_m + N + 1, \#s)$;
9   **if** ($j_1' \leq j_m'$) **then**
10          $Sb \leftarrow Sb \cup \{(j_1', j_m')\}$;
11      $Ext_R \leftarrow Ext_R \cup \{(sid, Sb)\}$;
12   **return** Ext_R ;

FUNCTION GETALLOCC (SDB, $\mathcal{ALLOCC}_{j-1}$, σ, M, N) ;
begin
13    $\mathcal{ALLOCC}_j \leftarrow \emptyset$; $inf \leftarrow 0$; $sup \leftarrow 0$;
14   **foreach** *pair* $(sid, OccSet) \in \mathcal{ALLOCC}_{j-1}$ **do**
15      $s \leftarrow SDB[sid]$; $newOccSet \leftarrow \emptyset$; $redundant \leftarrow false$; $i \leftarrow 1$;
16   **while** ($i \leq \#OccSet \wedge \neg redundant$) **do**
17        $[j_1, j_m] \leftarrow OccSet[i]$; $i \leftarrow i + 1$;
18   **if** ($\#\sigma = 1$) **then**
19          $inf \leftarrow 1$; $sup \leftarrow \#s$;
       else
20          $inf \leftarrow j_m + M + 1$; $sup \leftarrow min(j_m + N + 1, \#s)$;
21        $k \leftarrow inf$;
22   **while** ($(k \leq sup) \wedge (\neg redundant)$) **do**
23   **if** ($s[k] = \sigma(P_j)$) **then**
24   **if** ($\#\sigma = 1$) **then**
25              $newOccSet \leftarrow newOccSet \cup \{[k, k]\}$;
           else
26              $newOccSet \leftarrow newOccSet \cup \{[j_1, k]\}$;
27   **if** ($((sup = \#s) \wedge (\#\sigma > 1)) \vee (N \geq \#s)$) **then**
28              $redundant \leftarrow true$;
29          $k \leftarrow k + 1$;
30   **if** ($newOccSet \neq \emptyset$) **then**
31        $\mathcal{ALLOCC}_j \leftarrow \mathcal{ALLOCC}_j \cup (sid, newOccSet)$;
32   **return** $\mathcal{ALLOCC}_j$;

We avoid computing occurrences leading to redundant right pattern extensions thanks to the conditions $((sup = \#s) \wedge (\#\sigma > 1))$ in line (27). Moreover, when computing the right pattern extensions, instead of storing the part of subsequence $\langle s[j_1'], \ldots, s[j_m'] \rangle$, one can only store the positions of its first and last items (j_1', j_m') in the sequence sid. Finally, the filtering algorithm handles as efficiently the case *without gap constraints*. For each pair $(sid, OccSet)$, only the first occurrence $[j_1, j_m]$ in $OccSet$ is determined thanks to the condition $(N \geq \#s)$ in line (27).

Table 2. Dataset Characteristics.

Dataset	$\#SDB$	$\#\mathcal{I}$	Avg ($\#s$)	Max$_{s \in SDB}$ ($\#s$)	Type of data
Leviathan	5834	9025	33.81	100	book
PubMed	17527	19931	29	198	bio-medical text
FIFA	20450	2990	34.74	100	web click stream
BIBLE	36369	13905	21.64	100	bible
Kosarak	69999	21144	7.97	796	web click stream
Protein	103120	24	482	600	protein sequences

4.4 Temporal and Spatial Complexities of the Filtering Algorithm

Let $m=|SDB|$, $d=|\mathcal{I}|$, and ℓ be the length of the longest sequence in SDB. Computing $\mathcal{ALLOCC}_j$ from $\mathcal{ALLOCC}_{j-1}$ (see function GETALLOCC of Algorithm 2) can be achieved in $O(m \times \ell^2)$. The function GETRIGHTEXT (see Algorithm 2) processes all the occurrences of σ in each sequence of the SDB. The number of occurrences may exceed ℓ. However, as occurrences $(s, [j_1, j_m])$ with the same final position j_m are considered only once (see operator BE in Definition 7), there may exist at most ℓ of such occurrences in each sequence of the SDB in the worst case. So, the time complexity of function GETRIGHTEXT is $O(m \times \ell^2 + m \times \ell)$ i.e. $O(m \times \ell^2)$.

Proposition 7. *In the worst case, (i) filtering can be achieved in $O(m \times \ell^2 + d)$ and (ii) the space complexity is $O(m \times \ell^2)$.*

Proof: (i) The complexity of function GETRIGHTEXT is $O(m \times \ell^2)$. The total complexity of function GETFREQITEMS is $O(m \times \ell)$. Lines (8–9) can be achieved in $O(d)$. So, the whole complexity is $O(m \times \ell^2 + m \times \ell + d)$, i.e. $O(m \times \ell^2 + d)$. (ii) The space complexity of the filtering algorithm lies in the storage of the $\mathcal{ALLOCC}$ internal data structure. The occurrences $\mathcal{ALLOCC}_j$ of each assignment σ in SDB, with the length of σ varying from 1 to ℓ, have to be stored. Since it may exist at most ℓ occurrences of σ in each sequence sid, storing any $\mathcal{ALLOCC}_j$ costs in the worst case $O(m \times \ell)$. Since we can have ℓ prefixes, the worst space complexity of storing all the occurrences $\mathcal{ALLOCC}_j (j = 1..\ell)$, is $O(m \times \ell^2)$. □

5 Experiments

This section reports experiments on several real-life datasets [3,5] of large size having varied characteristics and representing different application domains (see Table 2). First, we compare our approach with CP methods and with the state-of-the-art specialized method cSpade in terms of scalability. Second, we show the flexibility of our approach for handling different types of constraints simultaneously.

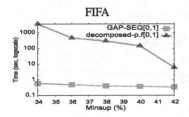 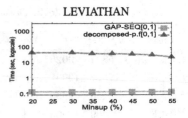

Fig. 1. Comparing `GAP-SEQ` with `decomposed-p.f` for GSPM: CPU times.

Table 3. `GAP-SEQ` vs. `decomposed-p.f` on FIFA dataset.

Dataset	Minsup (%)	#PATTERNS	CPU times (s)		#PROPAGATIONS		#NODES	
			GAP-SEQ	decomposed-p.f	GAP-SEQ	decomposed-p.f	GAP-SEQ	decomposed-p.f
FIFA	42	1	0.34	6.06	2	0	1	2
	40	5	0.37	144.95	10	778010	6	11
	38	10	0.4	298.68	20	2957965	11	21
	36	17	0.48	469.3	34	9029578	18	35
	34	35	0.59	–	70	–	36	–

Experimental Protocol. Our approach was carried out using the `gecode` solver[4]. All experiments were conducted on a processor Intel X5670 with 24 GB of memory. A time limit of 1 h has been set. If an approach is not able to complete the extraction within the time limit, it will be reported as (−). ℓ was set to the length of the longest sequence of SDB. We compare our approach (indicated by `GAP-SEQ`) with:

1. `decomposed-p.f`[5], the most efficient CP methods for GSPM,
2. `cSpade`[6], the state-of-the-art specialized method for GSPM,
3. the PREFIX-PROJECTION global constraint for SPM.

(a) GSPM: `GAP-SEQ` vs the most efficient CP method. We compare CPU times for `GAP-SEQ` and `decomposed-p.f`. In the experiments, we used the gap constraint $gap[0,1]$ and various values of $minsup$. Figure 1 shows the results for the two datasets FIFA and LEVIATHAN (results are similar for other datasets and not reported due to page limitation). `GAP-SEQ` clearly outperforms `decomposed-p.f` on the two datasets even for high values of $minsup$: `GAP-SEQ` is more than an order of magnitude faster than `decomposed-p.f`. For low values of $minsup$, `decomposed-p.f` fails to complete the extraction within the time limit.

Table 3 reports for the FIFA dataset and different values of $minsup$, the number of calls to the propagate function of `gecode` (col. 5) and the number of nodes of the search tree (col. 6). `GAP-SEQ` is very effective in terms of number of propagations. For `GAP-SEQ`, the number of propagations remains very small compared to `decomposed-p.f` (millions). This is due to the huge number of

---

[4] http://www.gecode.org.
[5] https://dtai.cs.kuleuven.be/CP4IM/cpsm/.
[6] http://www.cs.rpi.edu/~zaki/www-new/pmwiki.php/Software/.

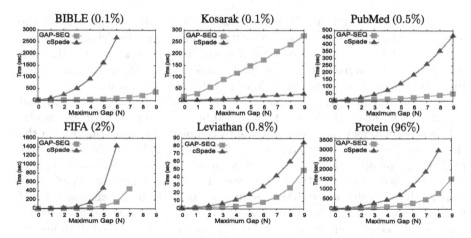

Fig. 2. Varying the value of parameter N in the gap constraint ($M = 0$): CPU times.

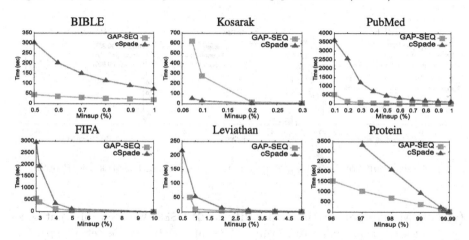

Fig. 3. Varying the value of *minsup* with the gap constraint $gap[0, 9]$: CPU times.

reified constraints used by decomposed-p.f to encode the subsequence relation. Regarding CPU times, GAP-SEQ requires less than 1s. to complete the extraction, while decomposed-p.f needs much more time to end the extraction (speed-up value up to 938).

(b) GSPM: GAP-SEQ vs the state-of-the-art specialized method. Second experiments compare GAP-SEQ with cSpade. We first fixed *minsup* to the smallest possible value w.r.t. the dataset used, and varied the maximum gap N from 0 to 9. The minimum gap M was set to 0. Figure 2 reports the CPU times of both methods. First, GAP-SEQ clearly dominates cSpade on all the datasets. The gains in terms of CPU times are greatly amplified as the value of N increases. On FIFA, the speed-up is 9.5 for N=6. On BIBLE, GAP-SEQ is able to complete the extraction for values of N up to 9 in 433s, while cSpade failed to complete the extraction for N greater than 6. The only exception is for the Kosarak dataset, where cSpade is efficient. For this dataset (which is the largest one both in terms of number of sequences and items), the size of the domains is important as compared to the other datasets. So, filtering takes much more time. This probably explains the behavior of GAP-SEQ on this dataset.

We also conducted experiments to evaluate how sensitive GAP-SEQ and cSpade are to *minsup*. We used the *gap*[0, 9] constraint, while *minsup* varied until the two methods were not able to complete the extraction within the time limit. Results are depicted in Fig. 3. Once again, GAP-SEQ obtains the best performance on all datasets (except for Kosarak). When the minimum support decreases, CPU times for GAP-SEQ increase reasonably while for cSpade they increase dramatically. On PubMed, with *minsup* set to 0.1 %, cSpade finished the extraction after 3, 500 s, while GAP-SEQ only used 500 s (speed-up value 7). These results clearly demonstrate that our approach is very effective as compared to cSpade on large datasets.

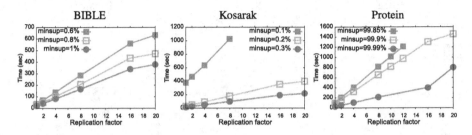

Fig. 4. Scalability of GAP-SEQ global constraint on BIBLE, Kosarak and Protein.

(c) GSPM: evaluating the scalability of GAP-SEQ. We used three datasets and replicated them from 1 to 20 times. The gap constraint was set to *gap*[0, 9], and *minsup* to three different values. Figure 4 reports the CPU times according to the replication factor (i.e. dataset sizes). CPU times increase (almost) linearly as the number of sequences. This indicates that GAP-SEQ achieves scalability while it is a major issue for CP approaches. The behavior of GAP-SEQ on Protein is quite different for low values of *minsup*. Indeed, for large sequences (such as in Protein), the size of $\mathcal{ALLOCC}$ may be very large and thus checking the gap constraint becomes costly (see Sect. 4.4).

(d) GSPM: handling various additional constraints. To illustrate the flexibility of our approach, we selected the PubMed dataset and stated additional constraints such as minimum frequency, minimum size, and other useful constraints expressing some linguistic knowledge as membership. The goal is to extract sequential patterns which convey linguistic regularities (e.g., gene - rare disease relationships) [3]. The size constraint allows to forbid patterns that are too small w.r.t. the number of items (number of words) to be relevant patterns; we set ℓ_{min} to 3. The membership constraint enables to filter out sequential patterns that do not contain some selected items. For example, we state that extracted patterns must contain at least the two items GENE and DISEASE. We used the $gap[0, 9]$ constraint, which is the best setting found in [3]. As no specialized method exists for this combination of constraints, we thus compare GAP-SEQ with and without additional constraints.

Table 4 reports, for each value of $minsup$, the number of patterns extracted and the associated CPU times, the number of propagations and the number of nodes in the search tree. Additional constraints obviously restrict the number of extracted patterns. As the problem is more constrained, the size of the developed search tree is smaller. Even if the number of propagations is higher, the resulting CPU times are smaller. To conclude, thanks to the GAP-SEQ global constraint and its encoding, additional constraints like size, membership and regular expressions constraints can be easily stated.

Table 4. GAP-SEQ under size and membership constraints on the PUBMED dataset.

$minsup$	#PATTERNS		CPU times (s)		#PROPAGATIONS		#NODES	
	gap	gap+size+item	gap	gap+size+item	gap	gap+size+item	gap	gap+size+item
1 %	14032	1805	19.34	16.83	28862	47042	17580	16584
0.5 %	48990	6659	43.46	34.6	100736	163205	61149	58625
0.4 %	72228	10132	55.66	43.47	148597	240337	90477	87206
0.3 %	119965	17383	79.88	59.28	246934	398626	151280	146601
0.2 %	259760	39140	143.91	100.09	534816	861599	329185	321304
0.1 %	963053	153411	539.57	379.04	1986464	3186519	1236340	1219193

(e) Evaluating the ability of GAP-SEQ to efficiently handle SPM. In order to simulate the absence of gap constraints, we used the ineffective $gap[0, \ell]$ constraint (recall that ℓ is the size of the longest sequence of SDB). We compared GAP-SEQ$[0, \ell]$ with PREFIX-PROJECTION and two configurations of cSpade for SPM: cSpade without gap constraint and cSpade with M and N set respectively to 0 and ℓ, denoted by cSpade$[0, \ell]$. Let us note that all the above methods will extract the same set of sequential patterns.

Figure 5 reports the CPU times for the four methods. First, cSpade obtains the best performance (except on Protein). These results confirm those observed in [7]. Second, GAP-SEQ$[0, \ell]$ and PREFIX-PROJECTION exhibit similar behavior, even if GAP-SEQ$[0, \ell]$ is slightly less faster. So, even if GAP-SEQ handles both cases (with and without gap), it remains very competitive for SPM. Third, GAP-SEQ$[0, \ell]$ clearly outperforms cSpade$[0, \ell]$ (except on Kosarak). This is probably due to the huge number of unnecessary joining operations performed by cSpade$[0, \ell]$.

To conclude, all the performed experiments demonstrate the ability of GAP-SEQ to efficiently handle SPM.

Finally, the gecode implementation of GAP-SEQ and the datasets used in our experiments are available online[7].

---

[7] https://sites.google.com/site/prefixprojection4cp/.

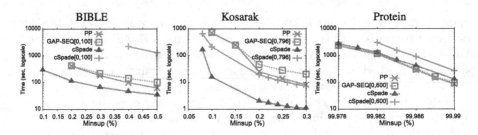

Fig. 5. Comparing GAP-SEQ with PREFIX-PROJECTION and cSpade for SPM on BIBLE, Kosarak and Protein.

6   Conclusion

In this paper, we have introduced the global constraint GAP-SEQ enabling to handle SPM with or without gap constraints. The filtering algorithm benefits from the principle of right pattern extensions and prefix anti-monotonicity property of the gap constraint. GAP-SEQ enables to handle several types of constraints simultaneously and does not require any reified constraints nor any extra variables to encode the subsequence relation. Experiments performed on several real-life datasets (i) show that our approach clearly outperforms existing CP approaches as well as specialized methods for GSPM on large datasets, and (ii) demonstrate the ability of GAP-SEQ to efficiently handle SPM.

This work opens several issues for future researches. We plan to handle constraints on set of sequential patterns such as closedness, relevant subgroup and skypattern constraints.

References

1. Agrawal, R., Srikant, R.: Mining sequential patterns. In: Yu, P.S., Chen, A.L.P. (eds.) ICDE, pp. 3–14. IEEE Computer Society (1995)
2. Ayres, J., Flannick, J., Gehrke, J., Yiu, T.: Sequential pattern mining using a bitmap representation. In: KDD 2002, pp. 429–435. ACM (2002)
3. Béchet, N., Cellier, P., Charnois, T., Crémilleux, B.: Sequential pattern mining to discover relations between genes and rare diseases. In: CBMS (2012)
4. Coquery, E., Jabbour, S., Saïs, L., Salhi, Y.: A SAT-based approach for discovering frequent, closed and maximal patterns in a sequence. In: ECAI, pp. 258–263 (2012)
5. Fournier-Viger, P., Gomariz, A., Gueniche, T., Soltani, A., Wu, C., Tseng, V.: SPMF: a Java open-source pattern mining library. J. Mach. Learn. Res. **15**, 3389–3393 (2014). http://jmlr.org/papers/v15/fournierviger14a.html
6. Ji, X., Bailey, J., Dong, G.: Mining minimal distinguishing subsequence patterns with gap constraints. In: ICDM 2005, pp. 194–201 (2005)
7. Kemmar, A., Loudni, S., Lebbah, Y., Boizumault, P., Charnois, T.: PREFIX-PROJECTION global constraint for sequential pattern mining. In: Pesant, G. (ed.) CP 2015. LNCS, vol. 9255, pp. 226–243. Springer, Heidelberg (2015)
8. Kemmar, A., Ugarte, W., Loudni, S., Charnois, T., Lebbah, Y., Boizumault, P., Crémilleux, B.: Mining relevant sequence patterns with CP-based framework. In: ICTAI, pp. 552–559 (2014)

9. Li, C., Wang, J.: Efficiently mining closed subsequences with gap constraints. In: Proceedings of the 2008 SIAM International Conference on Data Mining, pp. 313–322 (2008)
10. Li, C., Yang, Q., Wang, J., Li, M.: Efficient mining of gap-constrained subsequences and its various applications. Trans. Knowl. Discov. Data 6(1), 2:1–2:39 (2012)
11. Métivier, J.P., Loudni, S., Charnois, T.: A constraint programming approach for mining sequential patterns in a sequence database. In: ECML/PKDD Workshop on Languages for Data Mining and Machine Learning (2013)
12. Negrevergne, B., Guns, T.: Constraint-based sequence mining using constraint programming. In: Michel, L. (ed.) CPAIOR 2015. LNCS, vol. 9075, pp. 288–305. Springer, Heidelberg (2015)
13. Pei, J., Han, J., Mortazavi-Asl, B., Pinto, H., Chen, Q., Dayal, U., Hsu, M.: PrefixSpan: Mining sequential patterns by prefix-projected growth. In: ICDE, pp. 215–224. IEEE Computer Society (2001)
14. Pei, J., Han, J., Wang, W.: Mining sequential patterns with constraints in large databases. In: CIKM 2002, pp. 18–25. ACM (2002)
15. Wu, X., Zhu, X., He, Y., Arslan, A.N.: PMBC: pattern mining from biological sequences with wildcard constraints. Comput. Biol. Med. 43(5), 481–492 (2013)
16. Yang, G.: Computational aspects of mining maximal frequent patterns. Theoret. Comput. Sci. 362(1–3), 63–85 (2006)
17. Zaki, M.J.: Sequence mining in categorical domains: incorporating constraints. In: CIKM 2000, pp. 422–429 (2000)
18. Zaki, M.J.: SPADE: an efficient algorithm for mining frequent sequences. Mach. Learn. 42(1/2), 31–60 (2001)
19. Zhang, M., Kao, B., Cheung, D.W., Yip, K.Y.: Mining periodic patterns with gap requirement from sequences. TKDD 1(2) (2007)

A Reservoir Balancing Constraint with Applications to Bike-Sharing

Joris Kinable[(✉)]

Robotics Institute and Tepper School of Business,
Carnegie Mellon University, 5000 Forbes Ave., Pittsburgh, PA 15213, USA
jkinable@cs.cmu.edu

Abstract. A global CP constraint is presented which improves the propagation of reservoir constraints on cumulative resources in schedules with optional tasks. The global constraint is incorporated in a CP approach to solve a Single-Commodity Pickup and Delivery Problem: the Bicycle Rebalancing Problem with Time-Windows and heterogeneous fleet. This problem was recently introduced at the 2015 ACP Summer School on Constraint Programming competition. The resulting CP approach outperforms a Branch-and-Bound approach derived from two closely related problems. In addition, the CP approach presented in this paper resulted in a first place position in the competition.

1 Introduction

In scheduling problems, three types of resources can be distinguished: renewable resources (manpower, machinery), non-renewable resources (money, time, energy), and cumulative resources (containers, tanks, inventory). Cumulative resources are a special type of renewable resources which are produced and consumed by specific events in the schedule. Naturally, production events increase the availability of a resource, whereas consumption events decrease its availability. Resource constraints known as reservoir constraints restrict the minimum resp. maximum availability of a cumulative resource. A consumption event cannot be scheduled if there are not sufficient resources available. Likewise, a production event cannot be scheduled when the resource storage is at its maximum capacity.

Cumulative resources with reservoir constraints are used in a variety of scheduling and routing problems. In the Single-Commodity Pickup and Delivery Vehicle Routing Problem [10] a single commodity is produced at so-called supply nodes, and has to be distributed among a set of demand nodes. Practical applications arise for example in the distribution of left-over food from local restaurants [8], bike-sharing [4], and redistribution of self-service electric cars [6]. In the context of traditional scheduling problems, reservoir constraints are typically used to model storage restrictions or production limitations. Kolisch [11] presents a problem where spatial capacity constraints are used to model an assembly area with limited space. Similarly, Simonis and Cornelissens [17] present a case where reservoir constraints are extensively used in scheduling software for a herbicide

© Springer International Publishing Switzerland 2016
C.-G. Quimper (Ed.): CPAIOR 2016, LNCS 9676, pp. 216–228, 2016.
DOI: 10.1007/978-3-319-33954-2_16

production plant. The plant produces individual components of herbicides which are stored in intermediate buffer tanks with finite capacity. Different batch orders of pesticides require variable amounts of the individual components to be mixed. Consequently, multiple orders compete for the same shared resource.

In Constraint Programming (CP), dedicated global constraints exist to model cumulative resources. Verifying the feasibility of a (partial) schedule for a cumulative resource can be efficiently performed by calculating a resource-profile, which records, for each moment in time, the minimum and maximum utilization of the given resource [16]. The schedule is infeasible whenever, at any point in time, the consumption or production of the resource exceeds the capacity limitations imposed by the reservoir constraints. To calculate a resource profile for partial schedules, one has to compute a compulsory part for each task in the schedule. Accurately determining these compulsory parts is however difficult because tasks may be optional, or the time window during which the task has to be executed is proportionally large compared to the duration of the task. As a result, propagation of reservoir constraints may be very poor; determining whether a partial schedule is infeasible with respect to a reservoir constraint may only be possible late in the CP search when the start and end times of the majority of tasks have been fixed.

Simonis and Cornelissens [17] presented an approach to model reservoir constraints using well-known Cumulative constraints [2], but this approach is only applicable when there are no optional production and or consumption events. To mitigate this and some of the aforementioned issues, building on the seminal work by [12], this paper presents a new global constraint for cumulative resources. The constraint captures the intuitive notation that consumption events can only be scheduled when sufficient resources are produced, and vice versa. The constraint provides a tighter coupling between the *time* a resource event occurs, and the effective *change in the availability* of a resource.

In this work, we present the new global constraint in the context of a Bicycle Inventory Rebalancing problem which was launched as part of the 2015 ACP Summer School on Constraint Programming competition [1]. The CP model presented in this paper, strengthened with our new global constraint, resulted in a first place position in the aforementioned competition.

The remainder of this paper is structured as follows. First, Sect. 2 introduces the Bicycle Inventory Rebalancing Problem. A CP model, together with the new global constraint, is presented in Sect. 3. Finally, to assess the quality of our CP model, Sect. 4 presents an alternative solution approach for the Bicycle Inventory Rebalancing Problem based on a traditional Branch-and-Bound procedure. Computational results and discussion are provided in Sect. 5.

2   The Bicycle Rebalancing Problem

The city of Toronto (Canada) runs a bike-sharing system in which bicycles are made available for shared use to individuals. The bikes are kept at self-service terminals (stations) throughout the city. Individuals can rent a bike at a station and return it, after a certain amount of time, to the same or another station. Each station has a limited number of *docks* (places where bikes are positioned inside the station). Due to the fact that bikes are not necessarily returned to the station they originated from, certain stations may run out of bikes, or may have no empty docks left. Consequently, inventory rebalancing has to be performed periodically, thereby transporting bikes from stations with an excess of bikes to stations with a shortage of bikes. This problem, solved for a homogeneous fleet of vehicles, is known as the Bicycle Rebalancing Problem (BRP). The 2015 ACP Summer School on Constraint Programming competition launched a variation on this problem where each station has to be serviced within a given time window by a *heterogeneous* fleet of vehicles. We will denote this variant as BRP-TW.

The BRP-TW is formally defined as follows. Given is a directed, weighted graph $G(V, A)$, with vertex set $V = \{0, 1, \ldots, n\}$ and a set of arcs $A \subset V \times V$. Vertex 0 represents a depot where a fleet of trucks is parked; the remaining vertices represent bicycle stations. Each station $i \in V \setminus \{0\}$ has a positive or negative demand d_i for bikes: $d_i > 0$ represents an excess of bikes, $d_i < 0$ a shortage of bikes at station i. Bikes may be redistributed over the stations by a set of heterogeneous trucks K. Each truck has a capacity q_k and a usage cost per time unit c_k. Each station i has an associated time window $[a_i, b_i]$ during which the requested bikes have to be delivered or removed. A positive driving time t_{ij} is associated with each arc $(i, j) \in A$; the time required to service a station is negligible. Travel times t_{ij}, t_{ji}, are symmetrical, but do not necessary comply with the triangle inequality (e.g. $t_{ij} + t_{jk}$ may be smaller than t_{ik}). The objective is to calculate routes of minimum cost for each truck such that the demand of each station is met within its respective time window. The cost of a route is computed by the total number of time units the truck travels (excluding waiting time), multiplied by the usage cost of the truck c_k. A truck's route must start and end at the depot. To facilitate operations, a station may only be serviced once by a single truck during a given time interval. When the truck enters or leaves the depot, it may carry a positive number of bikes. To model that a busy station s may be visited multiple times a day, one can make a copy s of s and add it as a new station to the problem instance. Station s' has its own time window and demand. Travel times from/to station s are identical to those of station s. This simply results in an instance with one additional station, and can be solved without any modifications to the models proposed in the subsequent sections.

In what follows, let $\delta^+(i)$, $\delta^-(i)$ denote resp. the set of outgoing, incoming arcs into/from a node $i \in V$. Furthermore, it is assumed that there are no stations with a demand equal to zero. If such a station would exist, it can be removed in pre-processing phase. Similarly, we can remove arcs $(i, j) \in A$ if $a_i + t_{ij} > b_j$.

3   Constraint Programming Model

BRP-TW can be modeled as a CP problem through the use of interval variables [13,14]. An interval variable represents an interval during which an activity can be performed. More specifically, an interval variable α is a variable whose domain $dom(\alpha)$ is a subset of $\{\bot\} \cup \{[s, e)|s, e \in \mathbb{Z}, s \leq e\}$. An interval variable is fixed if its domain is reduced to a singleton, i.e. if α denotes a fixed interval variable:

- $\alpha = \bot$ if the interval is absent; the activity is not scheduled
- $\alpha = [s, e)$ if the interval is present.

An absent interval variable is ignored by any constraint or expression it is involved in. Such a constraint or expression would treat the absent interval variable as if it had never been specified to the constraint.

Each interval variable α has a start time $startOf(\alpha)$, an end time $endOf(\alpha)$, and a duration $dur(\alpha)$. Whenever an interval is present, it must hold that $endOf(\alpha) - startOf(\alpha) \geq dur(\alpha)$. As shorthand notation, an interval variable α is defined as a tuple: $\alpha = \{r, d, t, o\}$, specifying respectively the earliest start time of the interval, latest end time, minimum duration, and whether the interval is optional or obligatory. The constraints used in our model are summarized in Table 1.

To model the BRP-TW as a CP problem (Algorithm 1), three sets of interval variables are used. *Obligatory* interval variables v_0^k, $k \in K$, represent the start of the schedule for vehicle k. Next, *obligatory* intervals v_n^k for all $k \in K$ represent the end of the schedule for vehicle k. Here, $n = |V|$ represents a copy of the depot 0. As such, $t_{i,n} = t_{i,0}$ holds for all $i \in V$. Interval variables v_n^k are defined on the interval $[a_n, b_n] = [0, H]$, where H is some valid upper bound on the time horizon of the schedule. Finally *optional* intervals v_i^k for all stations $i \in V \setminus \{0\}$, $k \in K$ represent the servicing of station i by vehicle k.

In Algorithm 1, the constraint on line 7 ensures that each station is serviced exactly once by a single vehicle. The constraints on lines 9–11 sequence the visits to the stations for each vehicle, thereby ensuring that v_0^k and v_n^k are always resp. the first and last interval in the sequence. The objective function is defined on line 12. To simplify notation in the objective function, we use the shorthand t_{i,v_j^k} to denote $t_{i,j}$. Constraints 11–13 are the capacity constraints which ensure that the inventory of the vehicle is always between 0 (empty) and q^k. The constraint on line 15 is a new global constraint, described in the next subsection. Its purpose is to provide a tighter coupling between the resource and time constraints. Finally, the (redundant) constraints on line 17 strengthen the model by coupling the start and end time of each interval in the schedule for each vehicle.

3.1   Reservoir Balancing Constraint

The inventory of each vehicle in BRP-TW is known as a *reservoir resource*. Each reservoir has a minimum capacity of 0 and a maximum capacity of q^k, for all $k \in K$. Even though the constraints on lines 13–14 (Algorithm 1) are sufficient to manage the reservoir levels, they provide very little propagation because they do not consolidate at what *time*, how much of a particular resource is required to guarantee a feasible schedule. Building upon the work of Laborie [12], we present

Table 1. Description of CP constraints. All of these constraints, except the custom reservoirConstr, are available in IBM's CP Optimizer by default.

Constraint	Description
presenceOf(α)	Returns 1 if interval α is present, 0 otherwise
noOverlap($B, dist$)	Sequences the intervals in the set B. Ensures that the intervals in B do not overlap. Furthermore, the two-dimensional distance matrix $dist$ specifies for each pair of intervals a sequence dependent setup time. Absent intervals are ignored. Returns a sequence of the intervals in B
first(α, seq)	If interval α is present in sequence seq, it must be scheduled before any other interval in the sequence
last(α, seq)	If interval α is present in sequence seq, it must scheduled after any of the intervals in the sequence
pred(α, seq)	Returns the interval immediately preceding the interval α in the sequence seq, or $\perp$ is α is absent in seq
succ(α, seq)	Returns the interval immediately succeeding the interval α in the sequence seq, or $\perp$ is α is absent in seq
startOf(α)	Returns an expression representing the start time of interval α
endOf(α)	Returns an expression representing the end time of interval α
stepAtStart(α, h_{min}, h_{max})	Function in time t which returns a value between h_{min} and h_{max}, starting from time $t = $ **startOf**(α). The function returns 0 when α is absent, or before the start of α
reservoirConstr($\ldots$)	Custom global constraint, see Sect. 3.1 for details

a new reservoir constraint, the *Reservoir Balancing Constraint*, which connects the reservoir capacity constraints with the scheduling constraints.

Given are a set of resource events S which affect the capacity of a reservoir, and a precedence graph which provides a (partial) ordering of these events. The basic idea behind the reservoir constraint is to compute, for each event $x \in S$ in the precedence graph a lower and an upper bound on the reservoir level just before and just after x, and to compare these levels to the maximum and minimum capacities of the reservoir [12]. We extent upon the work by Laborie [12] by incorporating *optional* resource events (i.e. the presence status of the events is not necessarily fixed to present or absent) into the constraint, and by applying it to BRP-TW.

Following the notation used by Laborie [12], let P resp. C be the set of production, resp. consumption events, and let $S = P \cup C$. In case of BRP-TW,

---

Algorithm 1. CP model for BRP-TW.

Variable definitions:

1   $v_0^k = \{0, 0, 0, oblig.\}$ $\forall k \in K$

2   $v_n^k = \{0, H, 0, oblig.\}$ $\forall k \in K$

3   $v_i^k = \{a_i, b_i, 0, opt.\}$ $\forall i \in V \setminus \{0\}, k \in K$

4   $obj \in \{0, \infty\}$

Objective:

5   Min obj

Constraints:

6   **forall** $i \in V \setminus \{0\}$

7   $\left\lfloor \sum_{k \in K} \text{presenceOf}(v_i^k) = 1 \right.$

8   **forall** $k \in K$

9    $seq^k = \text{noOverlap}(\{\bigcup_{i \in V \cup \{n\}} v_i^k, t^k\}, t_{ij})$

10    $\text{first}(v_0^k, seq^k)$

11    $\text{last}(v_n^k, seq^k)$

12    $obj \mathrel{+}= c^k \sum_{i \in V} t_{i, \text{succ}(v_i^k, seq^k)}$

13    $cumulFunc^k = \text{stepAtStart}(v_0^k, 0, q^k) + \sum_{i \in V \setminus \{0\}} \text{stepAtStart}(v_i^k, d_i, d_i)$

14    $0 \leq cumulFunc^k \leq q^k$

15    $\text{reservoirConstr}(\{\bigcup_{i \in V \cup \{n\}} v_i^k\}, seq^k, cumulFunc, q^k)$

16   **forall** $i \in V \setminus \{0\} \cup \{n\}$

17    $\left\lfloor \text{startOf}(v_i^k) = \text{Max}\ \{a_i, \text{endOf}(\text{pred}(v_i^k, seq^k)) + t_{\text{pred}(v_i^k, seq^k), i}\} \right.$

$P = \{i \in V | d_i > 0\}$, $C = \{i \in V | d_i < 0\}$. Furthermore, let $B(x) \subset S$ be the events that have to be completed *strictly before* the *start of* event $x \in S$, $U(x) \subset S$ the set of events who's precedence relation with respect to x is undecided, i.e. an event $y \in U(x)$ can occur either before or after x. The relative change of the reservoir resource due to an event $x \in S$ is denoted by $q(x)$. $q_{min}(x)$, $q_{max}(x)$ are respectively the smallest and largest values in the domain of $q(x)$[1]. In case of BRP-TW, the relative change of the vehicle's inventory is modeled through the **stepAtStart**$(\alpha, h_{min}, h_{max})$ constraints (line 13, Algorithm 1), which implement a resource event x at the start of interval α, with $q_{min}(x) = h_{min}$, and $q_{max}(x) = h_{max}$. Finally, we can define the sets O, $\overline{O}$, $\widetilde{O}$, containing resp. the events which are present, absent, and the events who's presence status is undetermined. Obviously, the following relation holds: $O \cap \overline{O} = O \cap \widetilde{O} = \emptyset$. For the sake of generality, we assume that 0 and Q are fixed, finite bounds on the reservoir resource.

---

[1] Observe that when x is a consumption event, i.e. $q(x) < 0$, then by definition, $q_{max}(x)$ corresponds to the largest (least negative) value in the domain of $q(x)$.

For each event $x \in S$, we can now define an upper bound $L_{max}^<(x)$, and a lower bound $L_{min}^<(x)$ on the resource level just before x as follows:

$$L_{max}^<(x) = \sum_{y \in P \cap (B(x) \cup U(x)) \setminus \overline{O}} q_{max}(y) + \sum_{y \in C \cap B(x) \cap O} q_{max}(y) \tag{1}$$

$$L_{min}^<(x) = \sum_{y \in C \cap (B(x) \cup U(x)) \setminus \overline{O}} q_{min}(y) + \sum_{y \in P \cap B(x) \cap O} q_{min}(y). \tag{2}$$

Dead Ends and Presence Relations. Using the definitions from the previous subsection, a number of conditions can be specified under which the propagator of the Reservoir Balancing Constraint fails, or under which additional presence relations can be deduced. The propagator fails if $L_{max}^<(x) < 0$ or $L_{min}^<(x) > Q$, resulting in a backtrack. If $x \in O$, the propagator also fails if $L_{max}^<(x) + q_{max}(x) < 0$ or $L_{min}^<(x) + q_{min}(x) > Q$. Finally, if $x \in \tilde{O}$, we can post a constraint stating that x must be set to absent if $L_{max}^<(x) + q_{max}(x) < 0$ or $L_{min}^<(x) + q_{min}(x) > Q$, because the presence of event x would instantly result in a fail of the propagator.

Discovering New Precedence Relations. Let

$$\Pi_{min}^<(x) = - \sum_{y \in B(x) \cap ((P \setminus \overline{O}) \cup (C \cap O))} q_{max}(y)$$

be the smallest amount of resources that has to be produced *before* event x commences. Intuitively, if $\Pi_{min}^<(x)$ yields a positive value, then a number of production events must be scheduled before x, thereby producing at least $\Pi_{min}^<(x)$ resources. Let $P(x) = U(x) \cap P \setminus \overline{O}$ be the production events which can be scheduled either before or after event x. If there exists a $y \in P(x)$ such that:

$$\sum_{z \in P(x) \cap (B(y) \cup U(y))} q_{max}(z) < \Pi_{min}^<(x) \tag{3}$$

then a constraint stating that y must precede x can be posted. Intuitively, Condition (3) reads: take a production event $y \in P(x)$, which can be scheduled before or after x, and iterate over all remaining production events in $P(x)$ that can potentially precede y. If these events cannot produce at least $\Pi_{min}^<(x)$ resources, then naturally y must precede x. Observe that when $x \in O \cap C$, the right hand side of Condition (3) can be strengthened to: $\Pi_{min}^<(x) + q_{max}(x)$. Furthermore, when $x \in \tilde{O}$, an additional constraint can be posted, stating that presenceOf(x) implies the presenceOf(y).

Following a similar line of reasoning, define

$$\Pi_{max}^<(x) = \sum_{y \in B(x) \cap ((C \setminus \overline{O}) \cup (P \cap O))} q_{min}(y) - Q$$

as the least amount of resources that has to be consumed before x commences, and let $C(x) = U(x) \cap C \setminus \overline{O}$. If there exists a $y \in C(x)$ such that:

$$-\sum_{z \in C(x) \cap (B(y) \cup U(y))} q_{min}(z) < \Pi^{<}_{max}(x) \tag{4}$$

then a constraint can be posted, stating that y must precede x. When $x \in O \cap P$, the right hand side of Condition (4) can be strengthened to: $\Pi^{<}_{min}(x) + q_{min}(x)$. Furthermore, when $x \in \widetilde{O}$, an additional constraint can be posted, stating that presenceOf(x) implies the presenceOf(y).

4   Mixed Integer Programming Models

To compare the performance of our CP model in Sect. 5, we use a MIP model which essentially combines the models for a Pickup-and-Delivery problem with Time Windows [15] and a model for a related Bicycle Rebalancing Problem [4]. In contrast to this work, Dell'Amico et al. [4] do not consider time windows, and they assume a homogeneous fleet. Consequently, their problem is closer related to the Traveling Salesman Problem with Pickup and Deliveries as presented by Hernández-Pérez and Salazar-González [9].

For each arc $(i, j) \in A$, let binary variable x^k_{ij} denote whether vehicle $k \in K$ travels from station i to j. For each station $i \in V$, let variable C_i denote the time at which servicing station i completes. Finally, for each arc $(i, j) \in A$, variable f_{ij} counts the number of bikes in the vehicle that traverses arc (i, j). Obviously, $f_{ij} = 0$ if $x^k_{ij} = 0$ for all $k \in K$.

MILP :

$$\min \sum_{k \in K} c_k \sum_{(i,j) \in A} t_{ij} x^k_{ij} \tag{5}$$

$$\text{s.t.} \sum_{j \in \delta^+(0)} x^k_{0j} \leq 1 \qquad \forall k \in K \tag{6}$$

$$\sum_{k \in K} \sum_{j \in \delta^+(i)} x^k_{ij} = 1 \qquad \forall i \in V \setminus \{0\} \tag{7}$$

$$\sum_{i \in \delta^-(j)} x^k_{ij} = \sum_{i \in \delta^+(j)} x^k_{ji} \qquad \forall j \in V \setminus \{0\}, k \in K \tag{8}$$

$$C_j \geq C_i + t_{ij} - M_{ij}(1 - x^k_{ij}) \qquad \forall (i,j) \in A : j \neq 0, k \in K \tag{9}$$

$$\sum_{j \in \delta^+(i)} f_{ij} - \sum_{j \in \delta^-(i)} f_{ji} = d_i \qquad \forall i \in V \setminus \{0\} \tag{10}$$

$$\sum_{k \in K} \max\{0, d_i, -d_j\} x^k_{ij} \leq f_{ij} \qquad \forall (i,j) \in A : j \neq 0 \tag{11}$$

$$f_{ij} \leq \sum_{k \in K} \min\{q^k, q^k + d_i, q^k - d_j\} x^k_{ij} \qquad \forall (i,j) \in A : j \neq 0 \tag{12}$$

$$x^k_{ij} \in \{0, 1\} \qquad \forall (i,j) \in A, k \in K \tag{13}$$

$$a_i \leq C_i \leq b_i \qquad \forall i \in V \setminus \{0\} \tag{14}$$

M_{ij} is a constant, defined as: $M_{ij} = \max\{0, b_i + t_{ij} - a_j\}$. The objective function (5) minimizes the total distance traveled by each vehicle, weighted by the vehicle's cost. Constraints (6) ensures that a vehicle either stays at the depot (and hence is not used), or that it leaves the depot to service one or more stations. Constraints (7), (8) ensure that each station is serviced exactly once. Constraints (9), (14) ensure that each station is only serviced within its respected time window. In addition, Constraints (9) serve as subtour elimination constraints as they are a generalization of the Miller-Tucker-Zemlin subtour elimination constraints. Constraints (10)–(12) ensure that the desired amount of bikes are added or removed from each station, while simultaneously enforcing vehicle capacities. In particular, Constraints (11), (12) ensure that:

- the number of bikes in a vehicle $k \in K$ is always between 0 and q^k.
- if a vehicle collects bikes at node i, i.e. $d_i > 0$ then $f_{ij} \geq d_i$ after leaving node i, for some $j \in V$.
- if a vehicle delivers bikes at node j, i.e. $d_j < 0$ then the vehicle must have sufficient bikes in its inventory before reaching station i, i.e. $f_{ij} \geq -d_j$.
- and vice versa for the other direction.

As shown by Desrochers and Laporte [5], the bounds on the completion time variables C_i, $i \in V$ may be strengthened:

$$C_i \geq a_i + \sum_{k \in K} \sum_{j \in \delta^-(i)} \max\{0, a_j + t_{ji} - a_i\} x_{ji}^k \qquad \forall i \in V \setminus \{0\} \qquad (15)$$

$$C_i \leq b_i - \sum_{k \in K} \sum_{j \in \delta^+(i) \setminus \{0\}} \max\{0, b_i - b_j - t_{ij}\} x_{ij}^k \qquad \forall i \in V \setminus \{0\} \qquad (16)$$

Similarly, if for a given pair $i, j \in V$ both arcs (i, j) and (j, i) are contained in A then Constraint (9) may be replaced by a stronger equivalent:

$$C_j \geq C_i + t_{ij} - M_{ij}(1 - x_{ij}^k) + (M_{ij} - t_{ij} - \max\{t_{ji}, a_i - b_j\}) x_{ji}$$
$$\forall (i, j) \in A : j \neq 0, k \in K. \qquad (17)$$

4.1   Valid Inequalities

The family of clique inequalities described by Dell'Amico et al. [4] can be modified to our problem. Let $S(i, j, \bar{q}) = \{h \in \delta^+(j), h \neq i : |q_i + q_j + q_h| > \bar{q}\}$ for a given pair of nodes $i, j \in V$, $j \neq 0$, $(i, j) \in A$ and a capacity $\bar{q} \geq 0$. Similarly, let $T(i, j, \bar{q}) = \{h \in \delta^-(i), h \neq j : |q_i + q_j + q_h| > \bar{q}\}$ for a given pair of nodes $i, j \in V$, $i \neq 0$, $(i, j) \in A$ and a capacity $\bar{q}$. Finally, let $\overline{Q} = \{q^k, k \in K\}$ be the set of different vehicle capacities. Then the following inequalities are valid for BRP-TW:

$$\sum_{\substack{k \in K: \\ q^k \le \overline{q}}} \left(x_{ij}^k + \sum_{h \in S(i,j,\overline{q})} x_{jh}^k \right) \le 1 \tag{18}$$

$$\forall (i,j) \in A : j \ne 0, \overline{q} \in \overline{Q}, S(i,j,\overline{q}) \ne \emptyset$$

$$\sum_{\substack{k \in K: \\ q^k \le \overline{q}}} \left(\sum_{h \in T(i,j,\overline{q})} x_{hi}^k + x_{ij}^k \right) \le 1 \tag{19}$$

$$\forall (i,j) \in A : i \ne 0, \overline{q} \in \overline{Q}, T(i,j,\overline{q}) \ne \emptyset$$

The validity of inequalities (18) follows from the fact that (1) each station must be visited by exactly one vehicle and (2) a vehicle $k \in K : q^k \le \overline{q}$ does not have sufficient capacity to serve all three stations $i, j, h \in S(i,j,\overline{q})$. Similar for inequalities (19). Some of the inequalities (18), (19) are dominated by other inequalities in the same family, and can therefore be removed.

Separation of the clique inequalities (18), (19) is performed through complete enumeration. For every inequality in (18), (19) we evaluate the left hand side for a given solution $\overline{x}$. If the left hand side is strictly larger than one, the corresponding inequality is violated and is added to the problem.

Experiments were conducted with additional families of valid inequalities, such as the Fractional and Rounded capacity inequalities [9], but they did not have a positive impact on the performance of the MIP model. Hence they are omitted in this discussion.

5   Computational Results and Discussion

The CP and MIP models are implemented in resp. ILOG CPLEX and CP Optimizer (version 12.6.2), and executed with the default search parameters. The *Interval Sequence Inference Level* in the CP model has been set to extended.

A data set containing 13 problem instances was used during the 2015 ACP Summer School on Constraint Programming competition. According to the competition organizers, this data set contained both randomly generated instances, as well as instances based on Dumas' TSP-TW benchmark [7]. To provide more elaborate computational results in this paper, we added 11 random instances using the competition's random instance generator.

Table 2 summarizes the computational results. Out of a total of 24 instances, two instances could not be solved. For clarity, these two instances have been omitted from Table 2. Thus far it remains unknown whether these 2 instances (resp. 100 and 201 stations) are infeasible, or whether they are just particularly hard to solve. Table 2 shows for each instance the number of stations, and the number of vehicles. Furthermore, for the MIP approach, Table 2 provides the best upper (UB) and lower bound (LB) obtained after 30 min of computation time, the gap between these bounds and the actual computation time. Finally, for the CP approach, we show the upper bound, the gap between this upper bound and

Table 2. MIP vs CP. Instances 0–10 were provided during the ACP competition, the remaining instances have been generated to extend the computational results in this paper. Note that the competition instances (0–10) have been solved on a different (faster) system than instances (11–21).

Inst	$|V|$	$|K|$	MIP				CP		
			LB	UB	t(s)	Gap	UB	t(s)	Gap
0	5	1	137992	137992	0	0.00 %	137992	0	0.00 %
1	7	2	432270	432270	0	0.00 %	432270	1	0.00 %
2	28	6	424629	424629	193	0.00 %	518620	1800	18.12 %
3	16	4	770988	770988	1	0.00 %	770988	1800	0.00 %
4	14	4	317053	317053	1	0.00 %	317053	134	0.00 %
5	27	12	278734	453197	1802	38.50 %	443172	1800	37.10 %
6	42	10	353771	-	1803	-	895957	1800	60.51 %
7	81	10	488855	-	1800	-	2204450	1800	77.82 %
8	42	10	353771	-	1803	-	895957	1800	60.51 %
9	151	20	570806	-	1800	-	3348640	1800	82.95 %
10	50	6	107535	-	1804	-	255900	1800	57.98 %
11	55	10	336741	-	1807	-	981195	1800	65.68 %
12	49	10	407266	-	1803	-	655069	1800	37.83 %
13	41	10	279676	869799	1802	67.85 %	633738	1800	55.87 %
14	41	10	300006	737100	1803	59.30 %	573216	1800	47.66 %
15	51	10	477441	1410190	1803	66.14 %	965580	1800	50.55 %
16	49	10	471462	-	1803	-	1023970	1800	53.96 %
17	50	8	612040	-	1803	-	1142810	1800	46.44 %
18	37	8	501271	941084	1806	46.73 %	816506	1800	38.61 %
19	46	8	444028	-	1803	-	1024920	1800	56.68 %
20	51	8	497332	-	1803	-	1091130	1800	54.42 %
21	57	8	622815	-	1805	-	1414370	1800	55.97 %

the MIP lower bound, and the actual computation time. Whenever no feasible solution could be found within the allotted time, the optimality gap is assumed to be 100 %. As can be observed from Table 2, for the smaller instances, MIP outperforms CP: it solves these instances to optimality in a fraction of the computation time required by CP. However, for the larger instances, MIP is unable to find any feasible solutions. A similar trend was observed for alternative MIP based solution approaches used by other participants of the ACP competition.

Part of the decision problem is determining how many bikes each vehicle carries when it leaves the depot. This aspect makes the problem significantly harder to solve for a CP based approach. In particular, when we fix the number of bikes in the initial inventory for each vehicle, the CP method is able to obtain significantly better results in a shorter amount of time.

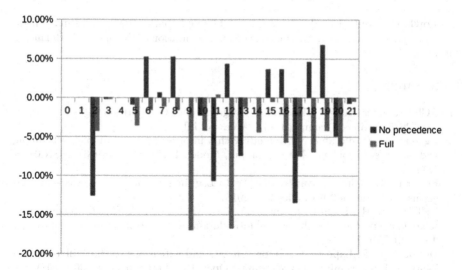

Fig. 1. Impact of reservoir constraint. Smaller percentage represents a bigger reduction of the optimality gap. (Color figure online)

Figure 1 analyses the impact of the Reservoir Balancing constraint presented in Sect. 3.1. We compare three situations: (1) the CP model without the reservoir constraint, (2) the CP model with a partially implemented reservoir constraint which does not generate precedence relations, (3) the CP model with the complete reservoir constraint. Figure 1 plots the increase/decrease of the optimality gap when we add resp. the partial reservoir constraint to the CP model (blue bars) or the complete reservoir constraint (orange bars). For example, for instance 5, we obtain optimality gaps of resp. 40.68 %, 39.77 %, 37.10 %, resulting in a decrease of the optimality gap (improvement) by resp. $39.77 - 40.68 = -0.91 \%$ (blue) and $37.10 - 40.68 = -3.58 \%$ (orange). As can be observed from Fig. 1, for all instances except instance 11, the Reservoir Balancing constraint improves the objective (orange bars). Only for instance 11 the optimality gap increases by 0.3 % when the Reservoir Constraint is added to the CP model. For the remaining instances, significant reductions in the optimality gap are observed, up to 17 % for instance 9. Whenever no precedence relations are deduced, the effectiveness of the Reservoir Balancing constraint is significantly reduced, and, due to its computational overhead, may have a negative impact on the solution quality. The latter may be observed from for example instances 6, 7, and 8.

In future work, the Reservoir Rebalancing Constraint may be extended by adding conditions on the earliest start and latest end times of an interval, based on the availability of a resource. In addition, specific search procedures for the BRP-TW may be developed, based on for example the station's demands. We conducted some preliminary tests with a search procedure which fixed the presence status of the v_i^k variables one by one, starting with stations with a large demand $|d_i|$, but this approach did not yield a notable improvement in the solution procedure. Finally, one may wish to incorporate constraints to improve the propagation of the sequence constraints (see Bergman et al. [3] for details).

Acknowledgement. I would like to thank Philippe Laborie (IBM) for his many helpful suggestions and comments regarding the implementation of the Reservoir Balancing constraint.

References

1. ACP Summer School Constraint Programming Competition (2015). http://acpss2015.uconn.edu/competition/. Accessed July 2015
2. Aggoun, A., Beldiceanu, N.: Extending chip in order to solve complex scheduling and placement problems. Math. Comput. Model. **17**(7), 57–73 (1993). ISSN 0895-7177
3. Bergman, D., Cire, A., van Hoeve, W.J.: Lagrangian bounds from decision diagrams. Constraints **20**(3), 346–361 (2015)
4. Dell'Amico, M., Hadjicostantinou, E., Iori, M., Novellani, S.: The bike sharing rebalancing problem: mathematical formulations and benchmark instances. Omega **45**, 7–19 (2014)
5. Desrochers, M., Laporte, G.: Improvements and extensions to the Miller-Tucker-Zemlin subtour elimination constraints. Oper. Res. Lett. **10**(1), 27–36 (1991)
6. Dror, M., Fortin, D., Roucairol, C.: Redistribution of self-service electric cars: a case of pickup and delivery. Research report RR-3543, INRIA, Projet PRAXITELE (1998). https://hal.inria.fr/inria-00073142
7. Dumas, Y., Desrosiers, J., Gelinas, E., Solomon, M.M.: An optimal algorithm for the traveling salesman problem with time windows. Oper. Res. **43**(2), 367–371 (1995)
8. Gunes, C., van Hoeve, W.-J., Tayur, S.: Vehicle routing for food rescue programs: a comparison of different approaches. In: Lodi, A., Milano, M., Toth, P. (eds.) CPAIOR 2010. LNCS, vol. 6140, pp. 176–180. Springer, Heidelberg (2010)
9. Hernández-Pérez, H., Salazar-González, J.J.: A branch-and-cut algorithm for a traveling salesman problem with pickup and delivery. Discrete Appl. Math. **145**(1), 126–139 (2004). ISSN 0166-218X
10. Hernández-Pérez, H., Salazar-González, J.J.: The one-commodity pickup-and-delivery traveling salesman problem: inequalities and algorithms. Networks **50**(4), 258–272 (2007). ISSN 1097-0037
11. Kolisch, R.: Integrated scheduling, assembly area-and part-assignment forlarge-scale, make-to-order assemblies. Int. J. Prod. Econ. **64**(13), 127–141 (2000)
12. Laborie, P.: Algorithms for propagating resource constraints in AI planning and scheduling: existing approaches and new results. Artif. Intell. **143**(2), 151–188 (2003)
13. Laborie, P., Rogerie, J.: Reasoning with conditional time-intervals. In: FLAIRS Conference, pp. 555–560 (2008)
14. Laborie, P., Rogerie, J., Shaw, P., Vilím, P.: Reasoning with conditional time-intervals. Part II: an algebraical model for resources. In: FLAIRS Conference (2009)
15. Ropke, S., Cordeau, J.-F., Laporte, G.: Models and branch-and-cut algorithms for pickup and delivery problems with time windows. Networks **49**(4), 258–272 (2007). ISSN 0028-3045
16. Schutt, A., Feydy, T., Stuckey, P.J.: Explaining time-table-edge-finding propagation for the cumulative resource constraint. In: Gomes, C., Sellmann, M. (eds.) CPAIOR 2013. LNCS, vol. 7874, pp. 234–250. Springer, Heidelberg (2013)
17. Simonis, H., Cornelissens, T.: Modelling producer/consumer constraints. In: Montanari, U., Rossi, F. (eds.) CP 1995. LNCS, vol. 976, pp. 449–462. Springer, Heidelberg (1995)

Optimization Models for a Real-World Snow Plow Routing Problem

Joris Kinable[1,2(✉)], Willem-Jan van Hoeve[2], and Stephen F. Smith[1]

[1] Robotics Institute, Carnegie Mellon University,
5000 Forbes Ave, Pittsburgh, PA 15213, USA
{jkinable,sfs}@cs.cmu.edu
[2] Tepper School of Business, Carnegie Mellon University,
5000 Forbes Ave, Pittsburgh, PA 15213, USA
vanhoeve@andrew.cmu.edu

Abstract. In cold weather cities, snowstorms can have a significant disruptive effect on both mobility and safety, and consequently the faster that streets can be cleared the better. Yet in most cities, plans for snowplowing are developed using simple allocation schemes that while easy to implement can also be quite inefficient. In this paper we consider the problem of optimizing the routes of a fleet of snow plowing vehicles, subject to street network topology, vehicle operating restrictions, and resource (salt, fuel) usage and replenishment constraints. We develop and analyze the performance of three different optimization models: a mixed-integer programming (MIP) model, a constraint programming (CP) model, and a constructive heuristic procedure that is amplified by an iterative improvement search. The models are evaluated on a set of snow plow routing problems of various sizes, constructed using Open Streets map data of Pittsburgh PA. Experimental results are presented that illustrate the differential strengths and weaknesses of each model, and suggest an alternative hybrid solution approach.

1 Introduction

Each year, many northern cities face significant expenditures pertaining winter road maintenance. Snow removal constitutes a significant part of these costs. For example, the city of Pittsburgh (USA) spent a staggering $4.3M on consumable resources (salt, deicing chemicals), $3.3M on personnel, and $800K on equipment (vehicles, plows, maintenance) during last year's winter season (2014/2015). In addition to these direct costs, a number of indirect costs can also be identified. Slippery roads deteriorate driving conditions thereby increasing the number of traffic accidents. Extensive utilization of snow plows, salt and chemicals damage the roads, corrode cars and metal bridges, and have an overall negative impact on the environment. Consequently, any ability to optimize winter road maintenance and deicing operations offers significant opportunities to realize substantial savings, to improve mobility and to reduce societal and environmental impact [4,14–16].

© Springer International Publishing Switzerland 2016
C.-G. Quimper (Ed.): CPAIOR 2016, LNCS 9676, pp. 229–245, 2016.
DOI: 10.1007/978-3-319-33954-2_17

In this work we study the real-world snow plow routing problem (SPRP) faced by the City of Pittsburgh PA where routes must be computed for a set of heterogeneous vehicles such that they collectively cover a geographical area, and comply with various resource constraints. Here, as in any snow plowing activity, each vehicle removes snow from the streets and simultaneously spreads a mixture of salt and chemicals for deicing purposes. Since each vehicle has only limited fuel and salt capacity, resources have to be periodically replenished. A number of resource depots are available throughout the city: these depots offer fuel, salt or both. The objective is to compute a schedule for each vehicle, which satifies resource constraints and minimizes the overall time it takes to clear all streets (i.e., the schedule makespan).

This work is part of a larger initiative to provide the city with an adaptive approach to snow plow route optimization and management. A route planning system is under development which will ultimately issue optimized turn-by-turn instructions to the vehicles in real-time during snow plowing operations, and dynamically revise these plans as unexpected events force changes. This paper lays the foundations for this project, by formally defining the problem and analyzing both exact (CP and MIP) and heuristic approaches for solving it. The heuristics presented are designed with scalability and adaptivity in mind, such that they can be adapted at a later stage of the project to modify schedules in response to dynamic events such as blocked roads, equipment problems and emergency requests.

The problem under consideration generalizes the well-known Chinese Postman Problem [8] and relates to other problems such as the Capacitated Arc Routing Problem [2,3] and Resource Constrained Project Scheduling. Although an extensive amount of research has been devoted to road maintenance and snow control, only a limited number of works has studied snow plow routing with resource constraints. For an excellent literature overview pertaining winter road maintenance problems in general, and related solution approaches, we refer to the survey series [9–12].

Salazar-Aguilar et al. [15] study a related routing problem where routes are computed in such a way that street segments with two or more lanes in the same direction are plowed simultaneously by different synchronized vehicles. This so-called 'tandem plowing' pushes snow from one lane to the next and eventually to the side of the road, thereby avoiding snow mounts building up between lanes. The problem in [15] is first defined through a MIP model. In addition, an efficient Adaptive Neighborhood Search approach is proposed. Although synchronized plowing has certain benefits, it is not being applied in Pittsburgh due to the added level of planning complexity that it implies. [15] primarily focuses on the plowing aspects; management of resources such as salt and fuel is not considered. The performance of their algorithms are evaluated on real-world data, including an instance from the city of Dieppe, New Brunswick, Canada. With a population of roughly 24,000 inhabitants, 462 intersections and 1,234 road segments, the city of Dieppe is less than one fifth the size of downtown Pittsburgh. Consequently, it is not obvious whether their approach can be scaled and adapted to our problem setting.

Perrier et al. [13] address another snow plow routing problem in urban areas. Each area is partitioned into a number of districts. Routes have to be determined for vehicles, parked at the district's depot, such that all road segments are serviced and all operational constraints are satisfied. Routes crossing these boundaries must be avoided from an administrative point of view. A similar situation arises currently in Pittsburgh where plows do not currently cross district boundaries. Although these artificial boundaries simplify the problem, they may also have an negative impact on the solution quality so these boundaries are not considered in this work. In addition to traditional routing constraints, Perrier et al. [13] consider road priorities, precedence relations between roads belonging to different priority classes, tandem plowing and limitations on the plows which can be used to service certain roads. The authors propose a multicommodity network flow structure to impose the connectivity of the route performed by each vehicle. Two heuristic approaches are presented: the first constructs routes in parallel by solving a multiple vehicle rural postman problem with side constraints, the second is a cluster-first route-second approach.

Gupta et al. [5] devise an iteration method to solve a snow plow routing problem on a network topography with a single depot. Per iteration, a trip, starting and ending at the depot and servicing a number of street segments is calculated. Every new iteration iteration, the street segments serviced in the previous iteration are removed from the network and a trip covering a (subset of) the remaining edges is calculated. The procedure repeats until all street segments have been serviced. The length of a single trip is limited by a maximum duration. Moreover, the total amount of salt required by the edges in a trip cannot exceed the truck's salt capacity. Although this problem bares strong similarities to our problem, the solution approach is not applicable because in our problem vehicles have to manage both salt and fuel resources, and not every depot offers both resources.

The remainder of this paper is structured as follows. First, Sect. 2 formally defines the problem and introduces notation. Next, Sects. 3 and 4 present a number of exact and heuristic models including a MIP model (Sect. 3.1), a CP model (Sect. 3.2), a constructive heuristic (Sect. 4.1) and a Late Acceptance improvement heuristic (Sect. 4.2). Finally, Sect. 5 compares the performance of these methods on real-world data, and draws some conclusions.

2   Problem Description

For a given network of streets and a fleet of snow plows, our SPRP consists of finding a route for each vehicle, such that the routes collectively cover the entire network. The objective is to minimize the duration of the longest route, i.e. to minimize the makespan of the schedule. The road network is modeled as a mixed multigraph. Vertices in the graph represent intersections in the road network, the arcs and edges represent resp. directed and undirected road segments. For instance, a road in between two intersections, consisting of 2 lanes in each direction, translates to 4 directed arcs in the graph. We will refer to these arcs as

Table 1. Parameters defining the snow plow optimization problem

Parameter	Description
V^R	Set of intersections
E^R	Set of two-way, single lane residential streets
A^R	Multi-set of directed lanes and one-way streets
K	Set of heterogeneous vehicles
$\mathbb{F}$	Fuel depots
$\mathbb{S}$	Salt depots
d_{ij}	Time or distance it takes to get from intersection i to intersection j, $i, j \in V^R$
f_{ij}^k	Fuel required to get from intersection i to intersection j, $i, j \in V^R$
s_{ij}^k	Salt required to get from intersection i to intersection j, $i, j \in V^R$
$0, n+1$	Resp. start and end depots of the trucks
$\overline{F}^k$	Maximum fuel capacity of vehicle $k \in K$
$\overline{S}^k$	Maximum salt capacity of vehicle $k \in K$
$\overline{C}$	Time horizon of the problem

unidirectional plow jobs. Unidirectional plow jobs are typically individual lanes of a multi-lane street, or one-way roads. In addition to unidirectional plow jobs, there also exist bidirectional plow jobs. Road segments in the latter category are small enough to be covered by a single pass of a snow plow, and the plow may come from either direction of the street. Typical examples of bidirectional plow jobs are streets in residential neighborhoods where cars are parked on each side of the road.

More formally, let $G^R(V^R, A^R \cup E^R)$ be a mixed multigraph where vertex set V^R represents the intersections, and E^R, A^R, the edges and arcs representing resp. the uni- and bidirectional street segments. For simplicity, it is assumed that graph G^R is strongly connected.

The roads are serviced by a heterogeneous fleet of snow plows K. Servicing a road segment $(i, j) \in A^R \cup E^R$ takes d_{ij} time. Vehicles may traverse road segments without servicing them. This is called deadheading. Due to the relatively low speed limits within the city, deadheading and servicing a road take equal amounts of time, independent of the road conditions. Each vehicle occasionally needs to refuel and resupply salt. A vehicle $k \in K$ has a fuel capacity $\overline{F}^k$ and salt capacity $\overline{S}^k$, $k \in K$. There are several depots throughout the city. Let $\mathbb{F}$ denote the set of fuel depots, $\mathbb{S}$ the set of salt depots. Some depots may supply both salt and fuel, hence $\mathbb{S} \cap \mathbb{F} \neq \emptyset$. The fuel (salt) consumption per street segment $(i, j) \in A^R \cup E^R$ using vehicle k is denoted by f_{ij}^k (s_{ij}^k). In addition to the fuel and salt depots, we define 0 and $n+1$ as the origin and destination depots where the vehicles are parked resp. before and after the trip. An overview of the various parameters is provided in Table 1.

3   Mathematical Models

In order to construct a MIP or CP model, we first define an auxiliary graph, using a set of unidirectional jobs $\vec{J}$ and bidirectional jobs $\tilde{J}$. For every $(u, v) \in A^R$ define a unidirectional plow job $j = (u, v) \in \vec{J}$, which takes $d_j = d_{uv}$ time to complete and requires resp. f_j^k fuel and s_j^k salt when serviced by vehicle $k \in K$. Similarly, for every $(u, v) \in E^R$ define a bidirectional plow job $j \in \tilde{J}$. Every bidirectional plow job $j \in \tilde{J}$ can be decoupled into two unidirectional plow jobs $\vec{j}, \overleftarrow{j}$, representing the different orientations of the job. Obviously, in order to service a bidirectional road, only $\vec{j}$ or $\overleftarrow{j}$ needs to be executed. Finally, define set J consisting of all jobs, i.e. $J = \vec{J} \cup \{\vec{j_i}, \overleftarrow{j_i} \mid i \in \tilde{J}\}$.

Let $i, j \in J$ be two different jobs, representing road segments $i = (u, v)$, $j = (s, t)$. Define d_{ij} as the time it takes to travel from intersection v to intersection s, *plus* the time required to complete job j. The travel time can be computed through a shortest path calculation in the routing graph G^R.

For each fuel depot $i \in \mathbb{F}$, a new ordered set of refuel jobs $F^i = 1, 2, \ldots$, is defined. Furthermore, let $F = \bigcup_{i \in \mathbb{F}} F^i$. A vehicle can refuel at a fuel depot $i \in \mathbb{F}$ by executing one of the fuel jobs $F^i = 1, 2, \ldots$ associated with depot i. Analogous for the salt depots $i \in \mathbb{S}$, we define sets $S^i = 1, 2, \ldots$, $S = \bigcup_{i \in \mathbb{S}} S^i$ representing salt resupply jobs.

We can now define our auxiliary graph, a directed, weighted multigraph $G(V_{0,n+1}, A)$ having vertex set $V_{0,n+1} = \{0\} \cup J \cup F \cup S \cup \{n+1\}$ and arc set A. For shorthand notation, denote $V = V_{0,n+1} \setminus \{0, n+1\}$, $V_0 = V_{0,n+1} \setminus \{n+1\}$, $V_{n+1} = V_{0,n+1} \setminus \{0\}$. Arc set A is defined as follows:

- there is an arc $(0, j)$ for all $j \in J \cup \{n+1\}$.
- there is an arc $(i, n+1)$ for all $i \in V_0$.
- there is an arc (i, j) for all $i, j \in J$, $i \neq j$.
- there are arcs $(i, j), (j, i)$ for all $i \in J, j \in F \cup S$.
- there is an arc (i, j) for all $i \in F \cup S, j \in J$.

Observe that any resource-feasible vehicle schedule for SPRP can be represented in the auxiliary graph through a simple path from vertex 0 to vertex $n + 1$.

3.1   MIP Model

A MIP model for SPRP can be constructed through the auxiliary graph. Let binary variables x_{ij}^k denote whether vehicle $k \in K$ travels from i to j, $(i, j) \in A$, *and* executes job j. Integer variables C^i record the time that job $i \in V_{0,n+1}$ is completed. In addition, C^{n+1} records the makespan of the schedule. Finally, integer variables F_i^k, S_i^k indicate resp. the fuel and salt supply levels of vehicle k after leaving node i. For notation purposes, let $\delta^+(i) = \{j \mid (i, j) \in A\}$ and $\delta^-(i) = \{j \mid (j, i) \in A\}$. Table 2 summarizes the various sets and parameters used in the MIP model.

The model, solvable using a traditional branch-bound-cut approach, is as follows:

$$P : \min \quad C^{n+1} \tag{1}$$

Table 2. Sets and parameters used in the MIP model

Param.	Description
V	$J \cup F \cup S$
V_0	$V \cup \{0\}$
V_{n+1}	$V \cup \{n+1\}$
$V_{0,n+1}$	$V \cup \{0, n+1\}$
d_{ij}^k	Setup time between job $i \in V_{0,n+1}$ and $j \in V_{0,n+1}$, $i \neq j$, plus the time required to perform job j, for vehicle k
f_{ij}^k	Fuel required to get from $i \in V_{0,n+1}$ to $j \in V_{0,n+1}$, $i \neq j$, plus the fuel required to perform job j, for vehicle k
s_{ij}^k	Salt required to get from $i \in V_{0,n+1}$ to $j \in V_{0,n+1}$, $i \neq j$, plus the salt required to perform job j, for vehicle k

$$\text{s.t.} \quad \sum_{j \in \delta^+(0)} x_{0j}^k = \sum_{i \in \delta^-(n+1)} x_{i,n+1}^k = 1 \qquad \forall k \in K \tag{2}$$

$$\sum_{j \in \delta^-(i)} x_{ji}^k = \sum_{j \in \delta^+(i)} x_{ij}^k \qquad \forall i \in V \tag{3}$$

$$\sum_{k \in K} \sum_{j \in \delta^+(i)} x_{ij}^k = 1 \qquad \forall i \in \overline{J} \tag{4}$$

$$\sum_{k \in K} \left(\sum_{j \in \delta^+(u)} x_{uj}^k + \sum_{j \in \delta^+(v)} x_{vj}^k \right) = 1 \qquad \forall i \in \overrightarrow{J}, u = \overleftarrow{j}_i, v = \overrightarrow{j}_i, \tag{5}$$

$$\sum_{k \in K} \sum_{j \in \delta^+(i)} x_{ij}^k \leq 1 \qquad \forall i \in F \tag{6}$$

$$\sum_{k \in K} \sum_{j \in \delta^+(u+1)} x_{u+1,j}^k \leq \sum_{k \in K} \sum_{j \in \delta^+(u)} x_{u,j}^k \qquad \forall i \in \mathbb{F}, u \in \{1, \dots, |F^i| - 1\} \tag{7}$$

$$\sum_{k \in K} \sum_{j \in \delta^+(i)} x_{ij}^k \leq 1 \qquad \forall i \in S \tag{8}$$

$$\sum_{k \in K} \sum_{j \in \delta^+(u+1)} x_{u+1,j}^k \leq \sum_{k \in K} \sum_{j \in \delta^+(u)} x_{u,j}^k \qquad \forall i \in \mathbb{S}, u \in \{1, \dots, |S^i| - 1\} \tag{9}$$

$$C^0 - M(1 - x_{0j}^k) \leq C^j - d_{0j}^k \qquad \forall (0, j) \in A, k \in K \tag{10}$$

$$C^i - M(1 - x_{ij}^k) \leq C^j - d_{ij}^k \qquad \forall (i, j) \in A, i \neq 0, k \in K \tag{11}$$

$$F_j^k \leq F_i^k - f_{ij}^k + \overline{F}^k(1 - x_{ij}^k) \qquad \forall i \in J \cup \{0\}, j \in J \cup \{n+1\}, k \in K \tag{12}$$

$$F_j^k \leq \overline{F}^k - f_{ij}^k x_{ij}^k \qquad \forall i \in F, j \in J \cup \{n+1\}, k \in K \tag{13}$$

$$S_j^k \leq S_i^k - s_{ij}^k + \overline{S}^k(1 - x_{ij}^k) \qquad \forall i \in J \cup \{0\}, j \in J \cup \{n+1\}, k \in K \tag{14}$$

$$S_j^k \leq \overline{S}^k - s_{ij}^k x_{ij}^k \qquad \forall i \in S, j \in J \cup \{n+1\}, k \in K \tag{15}$$

$$x_{ij}^k \in \{0, 1\} \qquad \forall (i, j) \in A, k \in K \tag{16}$$

$$0 \leq C^i \leq \overline{C} \qquad \forall i \in V \cup \{0, n+1\} \tag{17}$$

$$0 \leq F_i^k \leq \overline{F}^k \qquad \forall i \in V \cup \{0, n+1\} \tag{18}$$

$$0 \leq S_i^k \leq \overline{S}^k \qquad \forall i \in V \cup \{0, n+1\} \tag{19}$$

Constraints (2) define the starting and ending of the tour: every vehicle must start and end at the depot. Constraints (3) enforce flow preservation. Each unidirectional plow job must be performed exactly once ((4), (5)). Similarly, each bidirectional plow job must be executed, but only in one direction (5). Optional refueling/resupply jobs may be performed at most once (6), (8). Constraint (7) orders the refueling jobs: a refueling job $u \in F^i$ must be performed before $v \in F^i$, $v > u$, can be performed. This constraint reduces the amount of symmetry in the model. Constraint (9) is identical to Constraint (7) in the context of salt resupply jobs. Constraint (10), (11) relate the completion time variables to the nodes, while taking the setup times and job durations into consideration. Similarly Constraints (12), (13), (14) and (15) manage resp. the fuel and salt levels of the vehicles at each node. A vehicle leaves a refueling/resupply node with a full tank/salt supply.

3.2   CP Model

To model SPRP efficiently through CP, we will rely on interval variables [6,7]. An interval variable represents an interval during which an activity can be performed. For notation purposes, an interval variable will be denoted as a tuple $\alpha = \{r, d, t, [opt]\}$, where r denotes the earliest start time of the interval, d the latest finish time, t the minimum duration of the interval, and the optional parameter $[opt]$ indicates whether scheduling of the interval is optional. Optional intervals can be either present or absent in the final solution. An absent interval variable is ignored by any constraint or expression it is part of. The CP model presented in Algorithm 1 relies on three types of interval variables:

1. Job variables j_i for all $i \in V$ having duration d_i.
2. Assignment variables a_i^k for all $k \in K$, $i \in V_{0,n+1}$
3. Unidirectional plow job variables $\vec{j}_i, \overleftarrow{j}_i$ for all $i \in \bar{J}$ to distinguish the two possible orientations of bidirectional plow jobs.

A summary of the constraints used in Algorithm 1 is given in Table 3.

The objective of the model, minimize the makespan, is modeled through Constraints 5, 9. Constraint 6 states that every bidirectional plowing job has to be performed in only one direction and Constraint 7 ensures that every job is assigned to a single vehicle. Next, a number of constraints per vehicle are specified. Sequencing of the jobs on each vehicle is performed through Constraints 10–12. Resources are managed through a number of cumulative resource constraints (Constraints 13–16). Vehicles start with a full load of salt, performing a plow job i consumes s_i^k salt, and visiting a salt depot replenishes the salt resource (Constraints 13). For each truck, the salt level needs to remain between 0 and $\overline{S}^k$, the maximum salt capacity of the truck (Constraints 14). Similar constraints (15–16) are imposed for the fuel resource. In addition, Constraint 15 also takes the fuel consumption related to traveling in between jobs (deadheading) into account.

Finally, lines 17–20 specify a number of redundant constraints which are meant to improve the performance of the model. Constraints 17, 18 reduce the

Table 3. Description of CP constraints. All of these constraints are available in IBM ILOG CP Optimizer by default.

Constraint	Description
presenceOf(α)	Returns 1 if interval α is present, 0 otherwise
noOverlapSeq($B, dist$)	Sequences the intervals in the set B. Ensures that the intervals in B do not overlap. Furthermore, the two-dimensional distance matrix $dist$ specifies for each pair of intervals a sequence dependent setup time. Absent intervals are ignored. Returns a sequence of the intervals in B
first(α, seq)	If interval α is present in sequence seq, it must be scheduled before any other interval in the sequence
last(α, seq)	If interval α is present in sequence seq, it must scheduled after all other intervals in the sequence
succ(α, seq)	Returns the interval immediately succeeding the interval α in the sequence seq
pred(α, seq)	Returns the interval immediately preceding the interval α in the sequence seq
startOf(α)	Returns an expression representing the start time of interval α
endOf(α)	Returns an expression representing the end time of interval α
stepAtStart(α, h^-, h^+)	Function in time t which returns a value between h^- and h^+, starting from time $t = \textbf{startOf}(\alpha)$. The function returns 0 when t is absent, or before the start of α. When $h^- = h^+$, the shorthand **stepAtStart**(α, h) is used instead
alternative(α, B)	If interval α is present, then exactly one of the intervals in set B is present. The start and end of interval α coincides with the start and end of the selected interval from set B

amount of symmetry in the model by imposing an order on the refuel and resupply salt jobs. Constraint 20 links the start and end times of consecutive intervals.

A Note on Implementation. The CP model presented in Algorithm 1 is implemented in IBM ILOG CP Optimizer 12.6.2. To implement this model, a minor modification is required, as CP Optimizer has no direct way to implement the function **stepAtStart**($a_i^k, f_{i,\text{succ}[j_i, seq^k]}^k$) used in Constraint 15. To resolve this issue, a new variable $fuel_i^k$ is introduced into the model which records the fuel level of vehicle k after performing job i. Constraints 15–16 may now be replaced by the equivalent constraints from Algorithm 2.

4   Heuristic Models

4.1   Constructive Heuristic

The constructive heuristic uses a greedy approach to construct a feasible initial schedule. The heuristic works in two stages: stage one sequences all plow jobs while ignoring resource feasibility. Stage two makes the schedule feasible in terms of resources. The heuristic starts off with an empty schedule for every vehicle, that is, each vehicle has a schedule: $[0, n + 1]$. The heuristic iterates over all unscheduled *plow* jobs and schedules them one-by-one. To schedule a particular job, the heuristic evaluates for every vehicle all possible places to insert the job into its schedule. The impact of the job insertion onto the completion time of the vehicle's schedule is computed by factoring in the added travel time and job

---

Algorithm 1. CP model.

Variable definitions:

1   $j_i = \begin{cases} \{0, \infty, d_i\} & \text{if } i \in J \cup \bar{J} \\ \{0, \infty, d_i, opt\} & \text{if } i \in F \cup S \end{cases}$

2   $a_i^k = \begin{cases} \{0, 0, 0\} & \text{if } i = 0 \\ \{0, \infty, 0\} & \text{if } i = n+1 \\ \{0, \infty, d_i\} & \text{otherwise} \end{cases}$

3   $\vec{j}_i, \bar{\vec{j}}_i = \{0, \infty, d_i, opt\} \qquad \forall i \in \bar{J}$

4   $obj \in \{0, \infty\}$

Objective:

5   Min obj

6   alternative$(j_i, \{\bar{\vec{j}}_i, \vec{j}_i\}) \qquad \forall i \in \bar{J}$

7   alternative$(j_i, \bigcup_{k \in K} a_i^k) \qquad \forall i \in J \cup F \cup S$

8   **forall** $k \in K$

> **Objective Constraints:**
>
> 9 $obj \geq \text{endOf}(a_{n+1}^k)$
>
> **Sequencing Constraints:**
>
> 10 $seq^k = \text{noOverlapSeq}(\bigcup_{i \in J \cup F \cup S} a_i^k, [d_{ij} - d_j \mid (i,j) \in A])$
>
> 11   first(a_0^k, seq^k)
>
> 12   last(a_{n+1}^k, seq^k)
>
> **Salt Constraints:**
>
> 13   saltCumulFunc$^k = \text{stepAtStart}(a_0^k, \overline{S}^k) - \sum_{i \in J} \text{stepAtStart}(a_i^k, s_i^k)$
> $+ \sum_{i \in S} \text{stepAtStart}(a_i^k, 0, \overline{S}^k)$
>
> 14 $0 \leq \text{saltCumulFunc}^k \leq \overline{S}^k$
>
> **Fuel Constraints:**
>
> 15   fuelCumulFunc$^k = \text{stepAtStart}(a_0^k, \overline{F}^k) + \sum_{i \in F} \text{stepAtStart}(a_i^k, 0, \overline{F}^k)$
> $- \sum_{i \in J \cup F \cup S \cup \{0\}} \text{stepAtStart}(a_i^k, f_{i, \text{succ}[j_i, seq^k]}^k)$
>
> 16 $0 \leq \text{fuelCumulFunc}^k \leq \overline{F}^k$

Performance Constraints:

17   presenceOf$(j_v) \implies$ presenceOf$(j_u) \qquad \forall i \in \mathbb{F}, u \in \{1, \ldots, |F^i| - 1\}, v = u+1$

18   presenceOf$(j_v) \implies$ presenceOf$(j_u) \qquad \forall i \in \mathbb{S}, u \in \{1, \ldots, |S^i| - 1\}, v = u+1$

19   **forall** $k \in K$

20   startOf$(j_i) = \text{endOf}(\text{pred}(j_i, seq^k)) + t_{\text{pred}[j_i, seq^k], j_i}$
> $\forall i \in J \cup F \cup S \cup \{n+1\}$

Algorithm 2. CP model extension

1 **forall** $k \in K$

2 $\quad fuel_j^k \in \begin{cases} [\underline{F}^k, \overline{F}^k] & \text{if } j = a_0^k \\ [0, \overline{F}^k] & \text{if } j = a_i^k, i \in J \cup F \cup S \cup \{n+1\} \end{cases}$

3 $\quad fuel_{\text{succ}[j,seq^k]}^k =$
$\quad \begin{cases} fuel_j^k - f_{j,\text{succ}[j,seq^k]}^k - f_{\text{succ}[j,seq^k]}^k & \text{if } j = a_0^k \\ \overline{F}^k - f_{j,\text{succ}[j,seq^k]}^k - f_{\text{succ}[j,seq^k]}^k & \text{if } j = a_i^k, i \in J \cup F \cup S \cup \{0\} \end{cases}$

duration. In addition, a lower bound is calculated on the number of refuel and resupply trips the vehicle will have to make based on the amount of salt (fuel) the vehicle will need to complete its schedule. The number of refuel/resupply operations is then multiplied with the duration of a refuel/resupply job, thereby obtaining a lower bound on the time required to refuel and resupply. The actual driving time to a refuel or resupply depot is neglected in these calculations. Finally, recall that the bidirectional plow jobs can be performed from either direction. While evaluating a candidate position to insert the job, the heuristic chooses the best orientation of the plow job in respect to the jobs immediately preceding/succeeding the insert position.

After the plow jobs have been scheduled, phase two of the constructive heuristic will make the schedule resource feasible by inserting refuel and resupply jobs. For a given vehicle $k \in K$, the resupply salt jobs are inserted as follows. Let the plow jobs assigned to vehicle k in phase 1 be indexed from $0, \ldots, n$, and let j be the job for which $\sum_{i=0}^{j} s_i^k > \overline{S}_i^k$. That is, after $j - 1$ jobs, the vehicle runs out of salt and as such, cannot complete job j. In such cases, the heuristic schedules a resupply job between jobs $j - 1$ and j, thereby choosing the nearest resupply depot. This procedure is repeated until the schedule is feasible in terms of salt. Next, refuel jobs are inserted in a similar fashion. However, before inserting a new fuel job between jobs $j - 1$ and j, an extra check has to be performed to verify that after job $j - 1$ the vehicle has sufficient fuel to reach the nearest fuel depot. If not, we iterate backwards through the schedule, thereby searching for the nearest feasible position to insert a refuel job. A visual representation of the heuristic is given in Fig. 1.

4.2 Late Acceptance Improvement Heuristic

After executing the first phase of the constructive heuristic, a Late Acceptance (LA) heuristic [1] is used to improve the quality of the solution before phase 2 is initiated. To generate new solutions, the heuristic utilizes two simple search neighborhoods:

1. bestSwapMove: randomly choose a vehicle $k_1 \in K$, a job j_1 from the schedule of vehicle k_1 and a target vehicle k_2. For every possible plow job j_2 scheduled on vehicle k_2, and for every possible orientation of jobs j_1, j_2, evaluate the impact of swapping jobs j_1 and j_2.

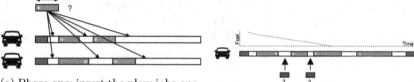

(a) Phase one: insert the plow jobs one-by-one, thereby selecting a insert position and potentially an orientation for the bidirectional jobs.

(b) Phase two: making the schedule resource feasible by inserting resource jobs

Fig. 1. Constructive heuristic

2. bestRemoveInsertMove: randomly choose a vehicle $k_1 \in K$, a job j_1 from the schedule of vehicle k_1 and a target vehicle k_2. For every possible insert position of the schedule of vehicle k_2 and for every possible orientation of job j_1, evaluate the impact of removing job j_1 from the schedule of k_1 and inserting it into k_2.

To move from one solution to a neighboring solution, we randomly select one of the two neighborhoods and evaluate the best candidate solution produced by this neighborhood. Following a standard LA approach, a move is accepted if its cost is better (or equal) to the cost of a solution L iterations ago, where L is a user-controlled parameter of the heuristic. The heuristic is terminated if (a) a maximum time limit is reached or (b) the incumbent solution has not been improved during 10000 consecutive iterations, where 10000 is determined empirically. Notice that when $L = 1$, the heuristic behaves as a greedy heuristic, only accepting improving moves. Selecting a larger value for L generally decreases the convergence rate of the heuristic, but reduces the chance of getting stuck in a local optimum.

5 Computation Experiments

5.1 Setup

Experiments are conducted on real world data, in collaboration with the city of Pittsburgh. Routing data is obtained through Open Street Maps (OSM). To extract data from a geographical area, including information about the roads, lanes, shapes, speed limits, traffic restrictions, etc., rectangular shaped snapshots are taken from an area on the map. In this experimental setup, we captured 21 different regions of Pittsburgh, varying from residential areas, downtown, rural areas, and business districts. Travel times between two neighboring intersections are computed by multiplying the length of the road with the maximum allowed driving speed.

Pittsburgh has 9 depots at different locations, 8 of which have salt, 5 of which have fuel. For experimental purposes, we use a small heterogeneous fleet of five vehicles to service each area. The smallest pickup-truck in our fleet has a capacity of 2 tons of salt and 26 gallons of fuel, whereas the largest plow has a capacity of 20 tons of salt and 75 gallons of fuel. Currently, the city utilizes about 1 ton of salt per mile, rendering salt the most constraining resource.

5.2 Results

Experiments have been conducted on 22 instances, which are summarized in Table 4. For each instance, the total number of plow jobs, percentage of bidirectional jobs, and total plowing distance (miles) is given. The MIP and CP models have been implemented using Cplex, resp. CP Optimizer 12.6.2. Experiments were run using default parameters and extended inference on the CP sequence variables.

Figure 2 compares the performance of CP and the LA Heuristic. Since each of these methods is warm-started with the solution obtained from the Constructive heuristic, we only show how much either of these approaches could improve the constructive solution. Runtimes for the CP approach were capped at resp. 10 min and 1 h. Similarly, the runtime of the LA Heuristic was capped at 10 min, or 10000 non-improving iterations, whichever came first. To measure the impact of the randomization in the LA Heuristic, 8 runs of the heuristic have been performed for each instance. The results of these runs are visualized by boxplots in Fig. 2.

The constructive heuristic produces an initial solution of reasonable quality in very little time, usually in the order of milliseconds for instances with less than 1000 jobs. For the smaller instances, up to 1000 jobs, the CP approach is capable of improving upon the constructive heuristic. For the larger instances, we noticed that the CP model ran out of memory and had to fall back on the much slower swap memory, thereby slowing down the CP approach tremendously. The largest instances, Residential Pittsburgh and inst18, could not be solved through CP on our machine due to insufficient memory. The LA approach produces good results in relatively little time. As can be observed from the largest instances, and most notably the Residential instance, the LA approach scales well. An additional advantage of this method is that the convergence rate can be adjusted, depending on the availability of computation time. Occasionally, as for instance inst12, the CP approach significantly outperforms the LA approach. The LA approach tracks for each vehicle how often it needs to resupply fuel and salt based on its resource consumption, and multiplies this with the average distance to a resupply depot to approximate the time spent on resupplying and refueling. This approximates becomes inaccurate when the travel time to a depot varies substantially, depending on the location of the vehicle. Calculating a more accurate approximation on the travel time to a depot, for instance by considering the position of the vehicle at the time it needs to resupply, would help mitigating this issue.

In addition to experiments with the CP model, a number of experiments were conducted with the MIP Model. For all but the smallest instance in our data set (Kaminst), the MIP model did not fit into our computer memory (16 GB + 30 GB swap). The latter is mainly attributed to the vast number of variables in each

Table 4. Instance data: number of jobs, percentage of jobs that are bidirectional, total distance to plow (mi).

Inst.	Jobs	Bidir.	Dist.	Inst.	Jobs	Bidir.	Dist.	Inst.	Jobs	Bidir.	Dist.
kaminst	45	38	3.4	inst5	631	74	55.2	inst13	529	59	47.3
downtown	724	38	38.1	inst6	632	68	52.9	inst14	498	64	37.2
mntWash	577	81	52.1	inst7	796	58	50.9	inst15	531	63	36.9
Residential	4073	64	315.5	inst8	500	31	42.8	inst16	498	92	47.5
inst1	233	61	27	inst9	481	38	38.4	inst17	499	75	42.6
inst2	346	93	38.6	inst10	574	64	41.5	inst18	1324	24	80.5
inst3	451	53	32.1	inst11	547	54	42.2				
inst4	287	87	22.9	inst12	339	91	30.7				

model, namely $|K||V|^2$ flow variables, and $2|K||V|$ resource variables. For the Kaminst instance, after a 1 h runtime, the MIP model (warm-started by the constructive heuristic) did not manage to improve upon its initial solution and had an optimality gap of 91.98 %. The large optimality gap is explained by the presence of the big-M constraints, where the ratio between M and the length of the jobs d_{ij}^k is very large.

Figure 3 shows more details for the 4 named instances in Table 4, and the spreading of the depots (blue squares). Each of these 4 instances represents a different geographical area in Pittsburgh, marked on the map in Fig. 5. From left to right: Mnt Washington, Downtown, Residential, Kaminst. The x-axis of the graphs in Fig. 3 shows the makespan of the schedule, converted to a HH::MM::SS format. At time 0, 0 % of the area has been serviced (y-axis), whereas, by the end of the schedule, 100 % of the area has been serviced. Some of the graphs, e.g. the Kamin instance, have a flat section at the beginning and end of the graph. This is where the vehicles travel from the nearest depot to the service area, and eventually back to the depot. The graphs have been generated using the same settings as before, unless mentioned otherwise.

Each graph shows the best CP solution, when one could be found, a solution from the constructive heuristic and LA improvement heuristic. For the LA heuristic, the graphs plot the average solution, as well as the diversity of solutions encountered. The MIP approach was unable to improve upon its warm-start solution, and is therefore not included in any of the graphs. As can be observed from the largest instance, the LA heuristic finds significant improvements over the constructive heuristic. Furthermore, when focusing on the robustness of the heuristic, the LA solutions show only a moderate variance in solution quality over multiple runs; the longer the heuristic runs, the smaller the variance.

Figure 4 presents a progress-over-time graph for the LA Heuristic for various list lengths L (see Sect. 4.2). Choosing L small results in an aggressive convergence, whereas higher values L allow a wider exploration of the search space at the cost of a slower convergence.

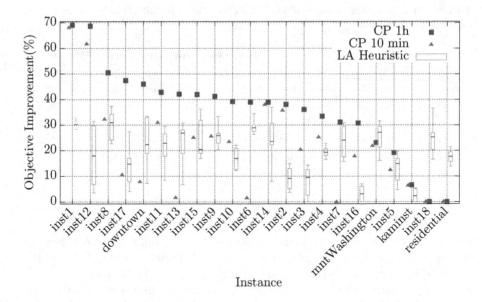

Fig. 2. Improvement over constructive heuristic: LA heuristic (8 iterations, 10 min runtime), CP (10 min resp. 1h runtime). Each of these methods is warm-started by the constructive solution.

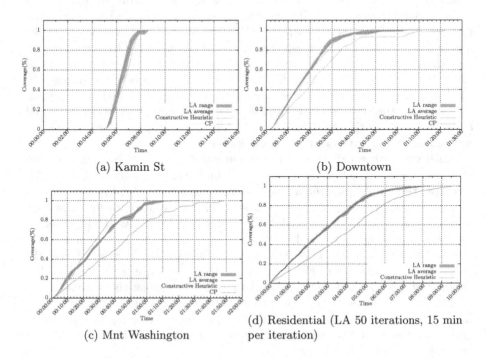

Fig. 3. Service over time

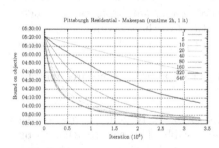

Fig. 4. Progress-over-time for various list lengths (Color figure online)

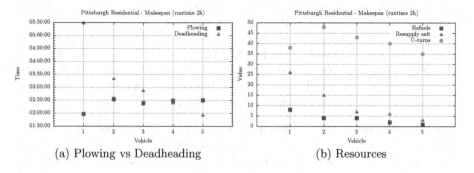

Fig. 5. Depots and names instances

(a) Plowing vs Deadheading

(b) Resources

Fig. 6. Solution details

Finally, Fig. 6 shows for the Residential instance the amount of plowing versus deadheading for every vehicle. The completion time of a vehicle schedule is obtained by summing these two values. As can be observed, the makespan of the schedule is dominated by the the completion time of the first vehicle. The capacity of this vehicle (1 ton salt) is significantly smaller than the capacity of the largest vehicle (20 ton). For such a large instance, the number of trips to a salt depot becomes significantly large, especially for smaller vehicles. Having a better approximation of the time required to travel to a depot would resolve this issue.

6 Conclusion

The constructive heuristic is capable of finding initial solutions of reasonable quality fast. The CP approach finds good solutions to instances up to a 1000 jobs, but does not scale well beyond that. The LA heuristic scales considerably better. A logical direction for further research would be to combine the LA heuristic and the CP approach in a Large Neighborhood Search. First, the LA heuristic is used to find a good global solution, after which the CP approach can be used to locally optimize small area's of the map in an iterative procedure.

Another research direction for this project involves online adaptations of the schedule. Unexpected events such as a blocked road, traffic congestion, emergency request etc., could necessitate modifications to the schedule. Again, the CP approach may be of use to 'repair' a small portion of the schedule, while leaving the remainder of the schedule intact.

Finally, from a model perspective, a number of additional features may be incorporated, including:

- Road priorities. The city assigns priorities to roads. In general, roads with high priorities should be serviced as fast as possible. This can be achieved by replacing the makespan objective by a weighted objective which minimizes the completion time per priority class.
- U-turns. Due to the size of the plows, having a large number of U-turns in a schedule is undesirable. As such, U-turns should be forbidden (hard-constrained) or penalized in the objective function.
- Road limitations. Some roads are too small or too steep to be plowed by the largest (and heaviest) vehicles. Similarly, in rural areas, the weight of large plows may exceed weight limitations on certain bridges. Consequently, a routing graph per vehicle category will be necessary. In addition, some plow jobs cannot be assigned to some of the heavier vehicles.

Road priorities are easily accounted for in the models presented, by assigning a priority class to each job and by using a weighted objective function which keeps track of the completion time of each priority class. Similarly, U-Turns can be penalized by increasing the setup time between a pair of jobs which would require a u-turn if one is performed immediately after the other. In case of a forbidden U-Turn, the setup-time will be significantly larger, representing the detour the truck has to make to get back, e.g. the time it takes to drive around the block.

References

1. Burke, E.K., Bykov, Y.: The late acceptance hill-climbing heuristic. Technical report, Department of Computing Science and Mathematics, University of Stirling (2012). http://www.cs.stir.ac.uk/research/publications/techreps/pdf/TR192.pdf
2. Eiselt, H.A., Gendreau, M., Laporte, G.: Arc routing problems, part I: the Chinese postman problem. Oper. Res. **43**(2), 231–242 (1995a). doi:10.1287/opre.43.2.231
3. Eiselt, H.A., Gendreau, M., Laporte, G.: Arc routing problems, part II: the rural postman problem. Oper. Res. **43**(3), 399–414 (1995b). doi:10.1287/opre.43.3.399
4. Environmental Protection Agency: Storm water management fact sheet, minimizing effects from highway deicing. Technical report EPA 832-F-99-016 (1999). http://water.epa.gov/scitech/wastetech/upload/2002_06_28_mtb_ice.pdf
5. Gupta, D., Tokar-Erdemir, E., Kuchera, D., Mannava, A.K., Xiong, W.: Optimal workforce planning and shift scheduling for snow and ice removal. Technical report Mn/DOT 2011-03, Center for Transportation Studies, University of Minnesota (2011). http://www.cts.umn.edu/Publications/ResearchReports

6. Laborie, P., Rogerie, J.: Reasoning with conditional time-intervals. In: FLAIRS Conference, pp. 555–560 (2008)
7. Laborie, P., Rogerie, J., Shaw, P., Vilím, P.: Reasoning with conditional time-intervals. Part II: an algebraical model for resources. In: FLAIRS Conference (2009)
8. Kwan, M.-K.: Graphic programming using odd or even points. Chinese Math 1, 273–277 (1962)
9. Perrier, N., Langevin, A., Campbell, J.F.: A survey of models and algorithms for winter road maintenance. Part I: system design for spreading and plowing. Comput. Oper. Res. 33, 209–238 (2006a)
10. Perrier, N., Langevin, A., Campbell, J.F.: A survey of models and algorithms for winter road maintenance. Part II: system design for snow disposal. Comput. Oper. Res. 33(1), 239–262 (2006b)
11. Perrier, N., Langevin, A., Campbell, J.F.: A survey of models and algorithms for winter road maintenance. Part III: vehicle routing and depot location for spreading. Comput. Oper. Res. 34(1), 211–257 (2007a)
12. Perrier, N., Langevin, A., Campbell, J.F.: A survey of models and algorithms for winter road maintenance. Part IV: vehicle routing and fleet sizing for plowing and snow disposal. Comput. Oper. Res. 34(1), 258–294 (2007b)
13. Perrier, N., Langevin, A., Amaya, C.-A.: Vehicle routing for urban snow plowing operations. Transp. Sci. 42(1), 44–56 (2008). doi:10.1287/trsc.1070.0195
14. Rubin, J., Garder, P.E., Morris, C.E., Nichols, K.L., Peckenham, J.M., McKee, P., Stern, A., Johnson, T.O.: Maine winter roads: salt, safety, environment and cost. Technical report, Margaret Chase Smith Policy Center, University of Maine (2010). http://umaine.edu/mcspolicycenter/files/2010/02/Winter-Road-Maint-Final.pdf
15. Salazar-Aguilar, M.A., Langevin, A., Laporte, G.: Synchronized arc routing for snow plowing operations. Comput. Oper. Res. 39(7), 1432–1440 (2012)
16. Usman, T., Fu, L., Miranda-Moreno, L.F.: Quantifying safety benefit of winter road maintenance: accident frequency modeling. Accid. Anal. Prev. 42(6), 1878–1887 (2010). doi:10.1016/j.aap.2010.05.008. ISSN 0001-4575

The TASKINTERSECTION Constraint

Gilles Madi Wamba[(✉)] and Nicolas Beldiceanu

TASC Team (CNRS/INRIA), Mines de Nantes, Nantes, France
{gilles.madi-wamba,nicolas.beldiceanu}@mines-nantes.fr

Abstract. Given a sequence of tasks $\mathcal{T}$ subject to precedence constraints between adjacent tasks, and given a set of fixed intervals $\mathcal{I}$, the TASKINTERSECTION$(\mathcal{T}, \mathcal{I}, o, inter)$ constraint restricts the overall intersection of the tasks of $\mathcal{T}$ with the fixed intervals of $\mathcal{I}$ to be greater than or equal ($o = $ '$\geq$') or less than or equal ($o = $ '$\leq$') to a given limit $inter$. We provide a bound($\mathbb{Z}$)-consistent cost filtering algorithm wrt the starts and the ends of the tasks for the TASKINTERSECTION constraint and evaluate the constraint on the video summarisation problem.

1 Introduction

More and more real world applications require taking into consideration a resource with a cost that is time dependent. A good example is electricity whose cost may change from one period to another [12]. Indeed, there exists an extension of the CUMULATIVE constraint [1,3] that takes into consideration such resource [13]. Nevertheless, that extension assumes fixed duration for the tasks and ignores precedence constraints between tasks. Recently, Kumar *et al.* [10] modelled this scheduling problem as a simple temporal problem extended with taboo regions. Taboo regions model periods of time where no job should be scheduled. The proposed algorithm evaluates the number of jobs scheduled in a taboo region rather than the intersection of the jobs with the taboo regions.

First this paper introduces the TASKINTERSECTION constraint, for concisely capturing scheduling problems with (1) varying 0–1 resource cost, with (2) variable duration tasks, and with (3) precedence constraints. Second it provides a dedicated cost filtering algorithm for the TASKINTERSECTION constraint. The provided filtering algorithm is bounds($\mathbb{Z}$) consistent wrt tasks start and end i.e. assuming fixed durations.

The TASKINTERSECTION constraint enforces the size of the intersection of a set of chained variable duration tasks with a fixed set of intervals to be greater than or equal ($o = $ '$\geq$') or less than or equal ($o = $ '$\leq$') than an integer variable. Tested on real instances of the video summarisation problem [4,6–8], it allows to improve by more than 20 % the solution.

In practice, assumptions are made on the start, on the duration and on the end of a task. For instance, (1) the start of a task is restricted by the availability of resources, (2) the duration of a task depends on resource properties, and (3) the end of a task is typically restricted by a deadline. The TASKINTERSECTION constraint is compatible with those assumptions and can be used to miminise or

© Springer International Publishing Switzerland 2016
C.-G. Quimper (Ed.): CPAIOR 2016, LNCS 9676, pp. 246–261, 2016.
DOI: 10.1007/978-3-319-33954-2_18

to maximise a resource usage. Given a fixed task start, its intersection with a set of fixed intervals increases with its duration. Indeed, since the task durations are variable, one can not just express the maximum intersection by taking a dual of the fixed intervals. In [14], the cost function to be minimized depends only on the end the taks.

Section 2 defines the TASKINTERSECTION constraint. Section 3 states and proves a necessary and sufficient condition for the feasibility of the constraint. Sections 4 and 5 are dedicated to the filtering algorithm and its implementation. Experimental results are presented in Sect. 6 and finally Sect. 7 concludes.

2 The TASKINTERSECTION Constraint

Definition 1 (Task). *A task t is described by its* start s_t, *its* duration d_t *and its* end e_t *variables.*

Definition 2 (Domain of a variable). *The domain of an integer variable var is denoted by $dom(var)$ and consists of one single interval $[\underline{var}, \overline{var}]$.*

Definition 3 (Feasible instantiation of a task). *A feasible instantiation of a task t is a triple (s_t, d_t, e_t) such that $d_t > 0$ and $s_t + d_t = e_t$.*

Definition 4 (Normalised task). *A task t is* normalised *iff:*

- $\exists d_t, d'_t \in \left[\underline{d_t}, \overline{d_t}\right], \; e_t, e'_t \in \left[\underline{e_t}, \overline{e_t}\right]$ *such that* $\underline{s_t} + d_t = e_t$ *and* $\overline{s_t} + d'_t = e'_t$,
- $\exists s_t, s'_t \in \left[\underline{s_t}, \overline{s_t}\right], \; e_t, e'_t \in \left[\underline{e_t}, \overline{e_t}\right]$ *such that* $s_t + \underline{d_t} = e_t$ *and* $s'_t + \overline{d_t} = e'_t$,
- $\exists s_t, s'_t \in \left[\underline{s_t}, \overline{s_t}\right], \; d_t, d'_t \in \left[\underline{e_t}, \overline{e_t}\right]$ *such that* $s_t + d_t = \underline{e_t}$ *and* $s'_t + d'_t = \overline{e_t}$.

From now on we assume all the tasks to be normalised.

Definition 5 (Normalised sequence of tasks). *A sequence $\mathcal{T} = (t_0, t_1, \ldots, t_{n-1})$ of tasks is* normalised *iff:*
(1) all tasks of $\mathcal{T}$ are normalised,   (2) $\forall t \in [0, n-2]: \underline{e_t} \leq \underline{s_{t+1}}, \; \overline{e_t} \leq \overline{s_{t+1}}$.

Definition 6 (Normalised sequence of intervals). *A sequence of intervals $\mathcal{I} = (r_0, r_1, \ldots, r_{m-1})$, where each interval r is described by two integer values ℓ_r, u_r, is* normalised *iff:*
(1) $\forall r \in [0, m-1]: \ell_r < u_r$,   (2) $\forall r \in [0, m-2]: u_r < \ell_{r+1}$.

Definition 7 (TASKINTERSECTION). *Given a normalised sequence $\mathcal{T}$ of n tasks, a normalised sequence $\mathcal{I}$ of m intervals, a comparison operator $o \in \{\leq, \geq\}$ and a variable inter, the TASKINTERSECTION$(\mathcal{T}, \mathcal{I}, o, inter)$ constraint holds iff*

$$\sum_{t=0}^{n-1} \left(\sum_{r=0}^{m-1} \max(\min(e_t, u_r) - \max(s_t, \ell_r), 0) \right) \; o \; inter.$$

In Definition 7, the value $\max(\min(e_t, u_r) - \max(s_t, \ell_r), 0)$ represents the intersection of task t with interval r ($[s_t, e_t] \cap [\ell_r, u_r]$).

Example 1. Consider the TASKINTERSECTION $\left(\begin{smallmatrix} (\langle s_0, d_0, e_0 \rangle, \langle s_1, d_1, e_1 \rangle, \langle s_2, d_2, e_2 \rangle), \\ ([5,9], [23,25], [30,40]), \quad \leq, \quad inter \end{smallmatrix} \right)$ constraint where:

$$\begin{cases} \text{dom}(s_0) = [2,8], \quad \text{dom}(d_0) = [3,15], \text{dom}(e_0) = [11,17], \\ \text{dom}(s_1) = [18,23], \text{dom}(d_1) = [5,6], \quad \text{dom}(e_1) = [23,29], \\ \text{dom}(s_2) = [31,40], \text{dom}(d_2) = [4,5], \quad \text{dom}(e_2) = [35,45], \end{cases}$$
and $\text{dom}(inter) = [0,5]$.

Figure 1 gives a solution for this TASKINTERSECTION constraint. Rectangles represent the three tasks instantiated in such a way that the total intersection is $9 - 5 = 4$ with the fixed intervals belongs to the domain of the intersection variable.

Proposition 1. *A reformulation of the* TASKINTERSECTION *is obtained by directly rewriting the relation* $\sum_{t=0}^{n-1} \left(\sum_{r=0}^{m-1} \max(\min(e_t, u_r) - \max(s_t, \ell_r), 0) \right)$ *o inter as a constraint (i.e. the sum of* $n \cdot m$ *terms where each term denotes the intersection between a given task and a given interval. We will use this reformulation in Sect. 6 for benching purposes.*

The filtering algorithm of this paper considers bounds($\mathbb{Z}$) consistency [5]. Assuming the domain of all variables of a constraint have no hole, bounds($\mathbb{Z}$) consistency ensures that the minimum and maximum value of a variable are part of a solution for that constraint. W.l.o.g. the comparison operator o is from now set to "$\leq$".

3   Checking Feasibility of the TASKINTERSECTION Constraint

Assuming no holes in the domains, this section provides a necessary and sufficient condition for the feasibility of the TASKINTERSECTION constraint. First, Proposition 2 gives a tight lower bound for the intersection of all the tasks with all the intervals. A tight lower bound is a lower bound archieved by constructing a fixed sequence of tasks $\mathcal{T}$ verifying Definition 5. Second, using this tight lower bound, Proposition 3 provides a necessary and sufficient condition to the TASKINTERSECTION constraint.

3.1   Tight Lower Bound for the Overall Intersection

To construct a tight lower bound, we proceed as follows:

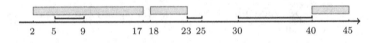

Fig. 1. A solution for the constraint of Example 1 with a total intersection of 4

Step 1 First, for each potential start s of the first task of $\mathcal{T}$ (i.e. task 0), we compute the minimum intersection between the intervals of $\mathcal{I}$ and all feasible instances of task 0 for which the start variable s_0 is fixed to value s.

Step 2 Second, assuming we have already computed the minimum intersection of tasks $0, 1, \ldots, t-1$ with the intervals of $\mathcal{I}$, we compute the minimum intersection of all tasks $0, 1, \ldots, t$ with $\mathcal{I}$ for the different possible starts of task t.

Step 3 Finally, the overall minimum intersection of tasks $0, 1, \ldots, n-1$ with the intervals of $\mathcal{I}$ is obtained after considering the last task.

We now detail those three steps. One key point to address is to directly take into account the variable task duration together with the constraint linking the start, the duration and the end of each task, in order to get a tight lower bound.

Step 1: Minimum Intersection of a Single Task. For a given potential start $s \in \text{dom}(s_t)$ of a task t they may exist more than one feasible instantiation of t with $s_t = s$. To any of these instantiations corresponds a value for the intersection of task t with $\mathcal{I}$. The function f_t gives the minimum of those intersections:

$$f_t(s) = \min_{d \in \text{dom}(d_t), e \in \text{dom}(e_t) | s+d=e} \left(\sum_{r=0}^{m-1} \max\left(\min(e, u_r) - \max(s, \ell_r), 0\right) \right)$$

Step 2: Minimum Intersection of Tasks 0,1, ..., t. For a task t (with $t \in [0, n-1]$) and a potential start $s \in \left[\underline{s_t}, \overline{s_t}\right]$, there may exist more than one feasible instantiation of the sequence of tasks $0, 1, \ldots, t$ with $s_t = s$. To any of these instantiations corresponds a value for the intersection of tasks $0, 1, \ldots, t$ with $\mathcal{I}$. The function g_t gives the minimum of those intersections. It is defined by Proposition 2.

Proposition 2. *For a task t and a start $s \in \left[\underline{s_t}, \overline{s_t}\right]$, the minimum intersection $g_t(s)$ of tasks $0, 1, \ldots, t$, where $s_t = s$, with the intervals of $\mathcal{I}$ is given by:*

$$g_t(s) = \begin{cases} f_0(s), & \text{If } t = 0, \\ f_t(s) + \min_{v_{t-1} \in [\underline{s_{t-1}}, \min(s-d_{t-1}, \overline{s_{t-1}})]} g_{t-1}(v_{t-1}) & \text{Otherwise.} \end{cases}$$

Example 2. Fig. 2b gives the curves of the functions g_t and Fig. 2a gives curves of the functions f_t, for each task t from Example 1.

Step 3: Overall Minimum Intersection. Function $g_t(s), s \in \left[\underline{s_t}, \overline{s_t}\right]$ computes the minimum intersection of all tasks $0, 1, \ldots, t$ with the fixed intervals $\mathcal{I}$, provided that $s_t = s$. Consequently, to find a lower bound for the overall minimum intersection of all n tasks, one needs to evaluate the minimum of g_{n-1} for all $s \in \left[\underline{s_{n-1}}, \overline{s_{n-1}}\right]$.

Example 3. From Fig. 2b of Example 2, we have $\min_{s \in [31,40]} g_2(s) = g_2(40) = 1$. Hence the lower bound for the overall minimum intersection of tasks 0,1 and 2 is 1.

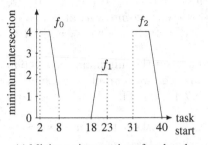
(a) Minimum intersection of each task

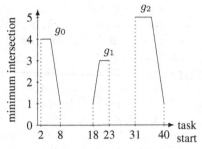
(b) $g_t, (0 \leq t \leq 2)$: minimum intersection of tasks $0, 1, \ldots, t$

Fig. 2. Minimum intersection functions f_t and $g_t (0 \leq t \leq 2)$ wrt all fixed intervals

3.2   Necessary and Sufficient Condition for Feasibility

Based on the g_t function introduced in Proposition 2 of Sect. 3.1 this section provides a necessary and sufficient condition for the TASKINTERSECTION constraint.

Proposition 3 (Necessary and sufficient condition). *Given a sequence T of n tasks and a sequence $\mathcal{I}$ of m intervals a necessary and sufficient condition for the* TASKINTERSECTION$(T, \mathcal{I}, \leq, inter)$ *constraint to hold is*

$$\min_{s \in [s_{n-1}, \overline{s}_{n-1}]} g_{n-1}(s) \leq \overline{inter}. \tag{1}$$

The necessary part follows from the definition of g_n. The proof of the sufficiency part consists of constructing a solution for the TASKINTERSECTION constraint. We first introduce Lemmas 1 and 2 regarding the characterisation of the suitable task duration required for minimising its intersection. Finally Lemma 3 shows how to construct a solution to the TASKINTERSECTION constraint with the task durations characterised by Lemma 2. This construction process of a solution will be illustrated in Example 4.

Notation 1. *Given a sequence $\mathcal{I}$ of intervals and a task t with start, duration and end respectively fixed to s, d and e, $f_t(s, d, e)$ denotes the intersection of task t with the intervals of $\mathcal{I}$.*

Lemma 1. *Let $\mathcal{I}$ be a set of m intervals and let t be a task. Given two feasible instances (s, d, e) and (s, d', e') of task t where $d \leq d'$, we have $f(s, d, e) \leq f(s, d', e')$.*

Proof.

$d \leq d' \Rightarrow e \leq e'$ (since $e = s + d$ and $e' = s + d'$)

$$\Rightarrow \sum_{r=0}^{m-1} \max(\min(e, u_r) - \max(s, \ell_r), 0) \leq \sum_{r=0}^{m-1} \max(\min(e', u_r) - \max(s, \ell_r), 0)$$

$$\Rightarrow f(s, d, e) \leq f(s, d', e').$$

□

Notation 2. *Given a task t, the minimum possible duration of those feasible instances of task t starting at time s is denoted d_s^{min}. To this minimum possible duration d_s^{min} corresponds a minimum possible end that we denote e_s^{min} ($s + d_s^{min}$).*

Lemma 2. *Given a task t, the minimum possible duration d_s^{min} of those feasible instances of task t starting at time s is $\max(\underline{d_t}, \underline{e_t} - s)$.*

Proof. Note that $s + \underline{d_t} \leq \overline{e_t}$, since we assume task t to be normalised.

1 If $s + \underline{d_t} \geq \underline{e_t}$ $\underline{d_t}$ is a feasible duration for the start s, and $\underline{d_t} = \max(\underline{d_t}, \underline{e_t} - s)$.
2 Otherwise if $s + \underline{d_t} < \underline{e_t}$, we must extend the minimum duration $\underline{d_t}$ from at least $\delta = \underline{e_t} - (s + \underline{d_t})$ to reach the earliest end $\underline{e_t}$ which leads to a minimum duration of $\underline{d_t} + \delta = \underline{e_t} - s$, which is equal to $\max(\underline{d_t}, \underline{e_t} - s)$.   □

The next Lemma shows how to construct a feasible solution for the TASKINTERSECTION constraint.

Lemma 3. *Let $\mathcal{T}$ be a sequence of n tasks and let $\mathcal{I}$ be a sequence of m intervals. Let $(\alpha_{n-1}, d_{\alpha_{n-1}}^{min}, e_{\alpha_{n-1}}^{min}), (\alpha_{n-2}, d_{\alpha_{n-2}}^{min}, e_{\alpha_{n-2}}^{min}), \ldots, (\alpha_0, d_{\alpha_0}^{min}, e_{\alpha_0}^{min})$ be an instantiation of tasks $n - 1, n - 2, \ldots, 0$ of $\mathcal{T}$, where α_t is the largest value such that:*

$$\begin{cases} f_t(\alpha_t) = \min_{s \in [\underline{s_t}, \overline{s_t}]} f_t(s), & \text{if } t = n - 1, \\ f_t(\alpha_t) = \min_{s \in [\underline{s_t}, \min(\alpha_{t+1} - \underline{d_t}, \overline{s_t})]} f_t(s) & \text{otherwise.} \end{cases}$$

If

$$\min_{\alpha_{n-1} \in [\underline{s_{n-1}}, \overline{s_{n-1}}]} g_{n-1}(\alpha_{n-1}) \leq \overline{inter} \tag{2}$$

then $(\alpha_{n-1}, d_{\alpha_{n-1}}^{min}, e_{\alpha_{n-1}}^{min}), (\alpha_{n-2}, d_{\alpha_{n-2}}^{min}, e_{\alpha_{n-2}}^{min}), \ldots, (\alpha_0, d_{\alpha_0}^{min}, e_{\alpha_0}^{min})$ is a solution for the TASKINTERSECTION$(\mathcal{T}, \mathcal{I}, \leq, inter)$ constraint.

Proof. We prove Lemma 3 in two steps.

First we show by induction on the task indices that $(\alpha_{n-1}, d_{\alpha_{n-1}}^{min}, e_{\alpha_{n-1}}^{min}), (\alpha_{n-2}, d_{\alpha_{n-2}}^{min}, e_{\alpha_{n-2}}^{min}), \ldots, (\alpha_0, d_{\alpha_0}^{min}, e_{\alpha_0}^{min})$ is a feasible instantiation of tasks $n - 1, n - 2, \ldots, 0$.

Second we show that if Eq. (2) is verified then the intersection of these fixed tasks with the intervals of $\mathcal{I}$ is less than or equal to $\overline{inter}$.

(1) • $[t = n - 1]$

   By hypothesis $\alpha_{n-1} \in [\underline{s_{n-1}}, \overline{s_{n-1}}]$. and by Lemma 2 $d_{\alpha_{n-1}}^{min} \in [\underline{d_{n-1}}, \overline{d_{n-1}}]$. By definition of $d_{\alpha_{n-1}}^{min}$, $e_{\alpha_{n-1}}^{min} \in [\underline{e_{n-1}}, \overline{e_{n-1}}]$. Hence $(\alpha_{n-1}, d_{\alpha_{n-1}}^{min}, e_{\alpha_{n-1}}^{min})$ is a feasible instance of task $n - 1$.
   • $[t < n - 1]$
   Assume that $(\alpha_t, d_{\alpha_t}^{min}, e_{\alpha_t}^{min}), (\alpha_{t+1}, d_{\alpha_{t+1}}^{min}, e_{\alpha_{t+1}}^{min}), \ldots, (\alpha_{n-1}, d_{\alpha_{n-1}}^{min}, e_{\alpha_{n-1}}^{min})$ is a feasible instantiation of tasks $t, t + 1, \ldots, n - 1$.
   To show that $(\alpha_{t-1}, d_{\alpha_{t-1}}^{min}, e_{\alpha_{t-1}}^{min}), (\alpha_t, d_{\alpha_t}^{min}, e_{\alpha_t}^{min}), \ldots, (\alpha_{n-1}, d_{\alpha_{n-1}}^{min}, e_{\alpha_{n-1}}^{min})$

is a feasible instantiation of tasks $t - 1, t, \ldots, n - 1$, we need to show that:

(a) the instantiation $(\alpha_{t-1}, d^{min}_{\alpha_{t-1}}, e^{min}_{\alpha_{t-1}})$ is feasible and

(b) $\alpha_{t-1} + d^{min}_{\alpha_{t-1}} \le \alpha_t$.

(a) Since $\alpha_{t-1} \in [\underline{s_{t-1}}, \min(\alpha_t - \underline{d_{t-1}}, \overline{s_{t-1}})] \subseteq [\underline{s_{t-1}}, \overline{s_{t-1}}]$ then

$d^{min}_{\alpha_{t-1}} \in [\underline{d_{t-1}}, \overline{d_{t-1}}]$ and $e^{min}_{\alpha_{t-1}} \in [\underline{e_{t-1}}, \overline{e_{t-1}}]$.

Hence $(\alpha_{t-1}, d^{min}_{\alpha_{t-1}}, e^{min}_{\alpha_{t-1}})$ is feasible.

(b) Since $\alpha_{t-1} \in [\underline{s_{t-1}}, \min(\alpha_t - \underline{d_{t-1}}, \overline{s_{t-1}})]$, then $\alpha_{t-1} \le \min(\alpha_t - \underline{d_{t-1}}, \overline{s_{t-1}})$. It follows that $\alpha_{t-1} + d^{min}_{\alpha_{t-1}} \le \alpha_t$.

(2) By construction,

$$\sum_{t=0}^{n-1} \left(f(\alpha_t, d^{min}_{\alpha_t}, e^{min}_{\alpha_t}) \right) = \min_{\alpha_{n-1} \in [\underline{s_{n-1}}, \overline{s_{n-1}}]} g_{n-1}(\alpha_{n-1})$$

Thus

$$\min_{\alpha_{n-1} \in [\underline{s_{n-1}}, \overline{s_{n-1}}]} g_{n-1}(\alpha_{n-1}) \le \overline{inter} \Rightarrow \sum_{t=0}^{n-1} \left(f(\alpha_t, d^{min}_{\alpha_t}, e^{min}_{\alpha_t}) \right) \le \overline{inter}$$

i.e. $(\alpha_{n-1}, d^{min}_{\alpha_{n-1}}, e^{min}_{\alpha_{n-1}}), (\alpha_{n-2}, d^{min}_{\alpha_{n-2}}, e^{min}_{\alpha_{n-2}}), \ldots, (\alpha_0, d^{min}_{\alpha_0}, e^{min}_{\alpha_0})$ is a feasible solution for the TASKINTERSECTION$(\mathcal{T}, \mathcal{I}, \le, inter)$. $\square$

Proposition 3 follows directly from Lemma 3.

Example 4. In the context of Example 3, this example illustrates how to construct a solution for the TASKINTERSECTION constraint. We have $\min_{s \in [31,40]} g_2(s) = g_2(40) = 1$, we find instantiations $(s_0, d_0, e_0), (s_1, d_1, e_1)$ and (s_2, d_2, e_2) such that $f(s_0, d_0, e_0) + f(s_1, d_1, e_1) + f(s_2, d_2, e_2) = g_2(40) = 1$.

$t = 2$: $\min_{s \in [31,40]} f_2(s) = f_2(40)$ i.e. $\alpha_2 = 40$. We thus compute $d^{min}_{\alpha_2}$ and $e^{min}_{\alpha_2}$ which are 4 and 44.

$t = 1$: From the curve of f_1 in Fig. 2a, $\min_{s \in [18,23]} f_1(s) = f_1(18) = 0$ i.e. $\alpha_1 = 18$. We thus compute $d^{min}_{\alpha_1}$ and $e^{min}_{\alpha_1}$ which are 5 and 23.

$t = 0$: From the curve of f_0 in Fig. 2a, $\min_{s \in [2,8]} f_0(s) = f_0(8) = 1$ i.e. $\alpha_0 = 8$. We thus compute $d^{min}_{\alpha_0}$ and $e^{min}_{\alpha_1}$ which are 3 and 11.

Hence $\{(8, 3, 11), (18, 5, 23), (40, 4, 44)\}$ is a solution, with $f(8, 3, 11) + f(18, 5, 23) + f(40, 4, 44) = \min_{s \in [\underline{s_3}, \overline{s_3}]} g_3(s) = g_3(40) = 1$.

4   Filtering the TASKINTERSECTION Constraint

This section shows how to filter the domains of the tasks start and end, in such a way that we get a feasible earliest (respectively latest) start and a feasible earliest (respectively latest) end time for each task wrt the maximum allowed intersection. Moreover, we adjust the minimum value of the intersection variable to a feasible value and also adjust the domain of the duration of the tasks by normalising them. All filtering will be derived from the necessary and sufficient condition given in Sect. 3, which is assumed to hold. We first describe the sets of values to filter out and characterise the corresponding filtering.

4.1   Characterising the Sets of Values to Filter Out

This section presents three propositions 4, 5 and 6 describing the sets of values to filter out from the domain of *inter* and from the domain of the start and end of each task.

Proposition 4. *The minimum feasible value of the intersection variable inter is* $\min_{s \in \left[\underline{s_{n-1}}, \overline{s_{n-1}} \right]} g_{n-1}(s)$.

Proof. This stems from Proposition 2.   □

To characterise whether a value s can be removed or not from the domain $\left[\underline{s_t}, \overline{s_t} \right]$ of the start s_t of a task t, (with $t \in [0, n-1]$), we first need to define the reverse of a TASKINTERSECTION constraint, and to introduce the minimum prefix/suffix intersection wrt the start of a task (Lemmas 4 and 5). Second, we use those minimum prefix/suffix intersections to evaluate the overall minimum intersection of all the tasks with the fixed intervals, given a potential task's start or end (Lemmas 6 and 7).

Definition 8 (Reverse of a TASKINTERSECTION constraint). *Let* $\mathcal{T} = (s_0, d_0, e_0), \ldots, (s_{n-1}, d_{n-1}, e_{n-1})$ *be a sequence of n tasks, and let* $\mathcal{I} = [\ell_0, u_0], \ldots, [\ell_{m-1}, u_{m-1}]$ *be a sequence of m fixed intervals. Given the* TASKINTERSECTION$(\mathcal{T}, \mathcal{I}, \leq, inter)$ *constraint, we define its* reverse *as the constraint* TASKINTERSECTION $(\mathcal{T}', \mathcal{I}', \leq, inter)$, *where to each task* (s_t, d_t, e_t) *of* $\mathcal{T}$ *(with* $t \in [0, n-1]$*) corresponds a task* $t' = (s'_t, d'_t, e'_t)$ *of* $\mathcal{T}'$ *defined by*

- $s_t' = \overline{e_{n-1}} - e_t$ • $d_t' = d_t$ • $e_t' = \overline{e_{n-1}} - s_t$

and to each interval $[\ell_r, u_r]$ *of* $\mathcal{I}$ *(with* $r \in [0, m-1]$*) corresponds an interval* $[\ell_r', u_r']$ *of* $\mathcal{I}'$ *defined by*

- $\ell'_r = \overline{e_{n-1}} - u_{m-1-r}$ • $u'_r = \overline{e_{n-1}} - \ell_{m-1-r}$

Lemma 4 (Minimum prefix/suffix intersection wrt the start of a task). *The minimum prefix (resp. minimum suffix) intersection wrt the start of a task t of* $\mathcal{T}$ *denoted* $P_t^{start}(s)$ *(resp.* $S_t^{start}(s)$*) is the minimum intersection of the tasks $0, 1, \ldots, t$ (resp. $t, t+1, \ldots, n-1$) with the m intervals of* $\mathcal{I}$, *provided task t starts at* $s \in \left[\underline{s_t}, \overline{s_t} \right]$. *We have* $P_t^{start}(s) = g_t(s)$ *and* $\underline{S_t^{start}(s)} = g_{t'}(s')$ *where* $s' = \overline{e_{n-1}} - e_s^{min}$.

Proof. The proof follows from the definition of function g (see Proposition 2).   □

Lemma 5. (Minimum prefix/suffix intersection wrt the end of a task).
The minimum prefix (resp. minimum suffix) intersection wrt the end of a task t of $\mathcal{T}$ denoted $\underline{P_t^{end}}(e)$ (resp. $\underline{S_t^{end}}(e)$) is the minimum intersection of the tasks $0, 1, \ldots, t$ (resp. $t, t+1, \ldots, n-1$) with the m intervals of $\mathcal{I}$, provided task t ends at $e \in \left[\underline{e_t}, \overline{e_t}\right]$. We have $\underline{P_t^{end}}(e) = \underline{P_{t'}^{start}}(e')$ and $\underline{S_t^{end}}(e) = \underline{S_{t'}^{start}}(e')$ where $e' = \overline{e_{n-1}} - e$.

Proof. The proof stems from definition of the reverse of a TASKINTERSECTION constraint. ☐

Lemma 6. *For a TASKINTERSECTION$(\mathcal{T}, \mathcal{I}, \leq, inter)$ constraint the minimum intersection of all the n tasks with the intervals of $\mathcal{I}$, provided that task t starts at $s \in \left[\underline{s_t}, \overline{s_t}\right]$, is equal to $\underline{P_t^{start}}(s) + \underline{S_t^{start}}(s) - f_t(s)$ and is denoted by $m_{s_t=s}^{start}$.*

Proof. Let $s \in \left[\underline{s_t}, \overline{s_t}\right]$ and assume that task t starts at s.

- The minimum intersection of tasks $0, 1, \ldots, t$ is given by $\underline{P_t^{start}}(s)$ and the minimum intersection of tasks $t, t+1, \ldots, n-1$ is given by $\underline{S_t^{start}}(s)$.

- Since the contribution $f_t(s)$ of task t occurs both in $\underline{P_t^{start}}(s)$ and in $\underline{S_t^{start}}(s)$, we subtract it once, thus:
$$m_{s_t=s}^{start} = \underline{P_t^{start}}(s) + \underline{S_t^{start}}(s) - f_t(s).$$
☐

Lemma 7. *For a TASKINTERSECTION$(\mathcal{T}, \mathcal{I}, \leq, inter)$ constraint the minimum intersection of all the n tasks with the intervals of $\mathcal{I}$, provided that task t ends at $e \in \left[\underline{e_t}, \overline{e_t}\right]$, is equal to $\underline{P_t^{end}}(e) + \underline{S_t^{end}}(e) - f_{t'}(e')$ where $e' = \overline{e_{n-1}} - e$ and is denoted by $m_{e_t=e}^{end}$.*

Proof. The proof is similar to the one of Lemma 6. ☐

Proposition 5. *For any task t (with $t \in [0, n-1]$) the minimum feasible value of s_t is α_{s_t} such that $\forall s \in [\underline{s_t}, \alpha_{s_t} - 1], m_{s_t=s}^{start} > \overline{inter}$ and $m_{s_t=\alpha_{s_t}}^{start} \leq \overline{inter}$. Similarly, the maximum feasible value of s_t is β_{s_t} such that $\forall s \in [\beta_{s_t} + 1, \overline{s_t}], m_{s_t=s}^{start} > \overline{inter}$ and $m_{s_t=\beta_{s_t}}^{start} \leq \overline{inter}$.*

Proof. The proof stems directly from Lemma 6. ☐

Proposition 6. *For any task t (with $t \in [0, n-1]$) the minimum value feasible value of e_t is α_{e_t} such that $\forall e \in [\underline{e_t}, \alpha_{e_t} - 1], m_{e_t=e}^{end} > \overline{inter}$ and $m_{e_t=\alpha_{e_t}}^{end} \leq \overline{inter}$. Similarly, the maximum feasible value of e_t is β_{e_t} such that $\forall e \in [\beta_{e_t} + 1, \overline{e_t}], m_{e_t=e}^{end} > \overline{inter}$ and $m_{e_t=\beta_{e_t}}^{end} \leq \overline{inter}$.*

Proof. The proof stems directly from Lemma 7. ☐

Example 5. This example shows how Propositions 4, 5 and 6 prune the *inter* variable as well as the start and end variables of each task of Example 1. From Fig. 2b, we have $\min_{s \in [\underline{s_2}, \overline{s_2}]} g_2(s) = g_2(40) = 1$. Proposition 4, filters out value 0 from the domain of *inter*, i.e. dom(*inter*) $= [1, 5]$. Applying Propositions 5 and 6 and normalising the tasks adjusts the minimum value of s_0 to 6, the maximum value of d_0 to 11, the minimum value of s_2 to 38 and the minimum value of e_2 to 42.

4.2   Characterising the Filtering

Proposition 7 of this section shows how bounds($\mathbb{Z}$) consistency is achieved on a
TASKINTERSECTION($\mathcal{T}, \mathcal{I}, \leq, inter$) constraint, with respect to the start of each
task when the task durations are fixed.

Proposition 7. *Assuming fixed durations, applying Propositions 4 and 5 on
the* TASKINTERSECTION($\mathcal{T}, \mathcal{I}, \leq, inter$) *constraint makes it bounds($\mathbb{Z}$) consistent wrt the start variables of the tasks.*

Proof. First, Proposition 4 ensures feasibility of the TASKINTERSECTION constraint. Second the quantities α_{s_t} and β_{s_t} of Proposition 5 resp. are the smallest
and the largest feasible values for the start of task t.   □

5   Implementation

The filtering algorithm of the TASKINTERSECTION constraint is decomposed in
3 parts:

- A first part evaluates the functions f_t introduced in Step 1 of Sect. 3.1.
- A second part computes the functions g_t introduced in Step 2 of Sect. 3.1.
- A third part uses function g_t in order to filter (1) the intersection variable of
  the TASKINTERSECTION constraint wrt Proposition 4, and (2) the start and
  the end of each task wrt Propositions 5 and 6.

Since the evaluation of functions f_t is the most involved part and for space
reason this section focusses on an efficient algorithm for computing the minimum
intersection f_t of a task. Note that the algorithms that implement Propositions 5
and 6 update the lower and upper limits of the start and end variables of each
task in one single step.

Computing f_t. By using two key ideas, this algorithm computes a piecewise
continuous curve that gives the value of $f_t(s)$, the minimum intersection of task
t wrt the fixed intervals provided task t starts at $s \in \left[\underline{s_t}, \overline{s_t}\right]$. The difficulty
for computing f_t is twofold: first we have to consider the feasibility constraint
$s_t + d_t = e_t$, second we want to avoid iterating over each value of $\mathrm{dom}(s_t)$ in
order to have a time complexity that only depends on the number of tasks and
on the number of intervals.

The first idea is that, if the position of the start of task t varies from one
unit from a start $s \in \left[\underline{s_t}, \overline{s_t}\right]$ to $s+1 \in \left[\underline{s_t}, \overline{s_t}\right]$, then the minimum intersection of
task t also varies from at most one unit, i.e. $|f_t(s) - f_t(s+1)| \leq 1$. This results
in a curve of slope varying between -1, 0 or 1. The algorithm creates a partition
$\mathcal{P} = (p_0, p_1, \ldots, p_k)$ of $\left[\underline{s_t}, \overline{s_t}\right]$, with $\underline{s_t} = p_0 < p_1 < \cdots < p_k = \overline{s_t} + 1$ and $k \geq 0$,
such that $f_t|_{[p_i, p_{i+1}[}$, the restriction of f_t to $[p_i, p_{i+1}[$, is either strictly increasing
(its slope is equal to 1), strictly decreasing (its slope is equal to -1) or constant
(its slope is equal to 0), and for any two consecutive subintervals $[p_i, p_{i+1}[$ and
$[p_{i+1}, p_{i+2}[$ the functions $f_t|_{[p_i, p_{i+1}[}$ and $f_t|_{[p_{i+1}, p_{i+2}[}$ do not have the same slope.

The second idea to find a subinterval's end p_{i+1} is as follows: there exists $\delta_i \in \mathbb{N}$ such that $f_t|_{[p_i,p_i+\delta_i[}(p_i + \delta_i) \neq f_t|_{[p_i+\delta_i,p_{i+2}[}(p_i + \delta_i)$. The value of p_{i+1} is thus given by $p_i + \delta_i$. To compute δ_i we first need to introduce three quantities $\delta_{i_s}, \delta_{i_d}$ and δ_{i_e} that we now define:

(1) When p_i belongs to an interval of $\mathcal{I}$, δ_{i_s} is the distance from p_i to the end of that interval (case (a) of Fig. 3), otherwise δ_{i_s} is the distance from p_i to the next interval's start when it exists ($case(b)$ of Fig. 3), or $+\infty$ when it doesn't (case (c) of Fig. 3).

$$\delta_{i_s} = \begin{cases} u_r - p_i & \text{if} \quad in(p_i) \wedge \ell_r \leq p_i < u_r, & (a) \\ \ell_r - p_i & \text{if} \quad \neg in(p_i) \wedge \text{interval } r \text{ is the first interval to the right of } p_i, & (b) \\ +\infty & \text{if} \quad \neg in(p_i) \wedge \nexists r \mid \ell_r > p_i. & (c) \end{cases}$$

where $in(p_i)$ is the function that returns true if there is a fixed interval r containing p_i, i.e. such that $\ell_r \leq p_i < u_r$, and false otherwise.

(2) δ_{i_d} is the difference between $d_{p_i}^{min}$ and $\underline{d_t}$: $\delta_{i_d} = d_{p_i}^{min} - \underline{d_t}$.

(3) When $e_{p_i}^{min}$ belongs to an interval of $\mathcal{I}$, δ_{i_e} is the distance from $e_{p_i}^{min}$ to the end of that interval (d), otherwise δ_{i_e} is the distance from $e_{p_i}^{min}$ to the next interval's start when it exists (e), or $+\infty$ if it does not exist (f).

$$\delta_{i_e} = \begin{cases} u_r - e_{p_i}^{min} & \text{if} \quad in(e_{p_i}^{min}) \wedge \ell_r \leq e_{p_i}^{min} < u_r, & (d) \\ \ell_r - e_{p_i}^{min} & \text{if} \quad \neg in(e_{p_i}^{min}) \wedge r \text{ is the first interval after } e_{p_i}^{min}, & (e) \\ +\infty & \text{if} \quad \neg in(e_{p_i}^{min}) \wedge \nexists r \mid \ell_r > e_{p_i}^{min}. & (f) \end{cases}$$

The value of δ_i is given by $\min(\delta_{i_s}, \delta_{i_e})$ or by $\min(\delta_{i_s}, \delta_{i_d})$ depending on whether the value of $d_{p_i}^{min}$ is equal to $\underline{d_t}$ or to $\underline{e_t} - p_i$:

$$\delta_i = \begin{cases} \min(\delta_{i_s}, \delta_{i_e}) & \text{if} \quad d_{p_i}^{min} = \underline{d_t}, \\ \min(\delta_{i_s}, \delta_{i_d}) & \text{otherwise (if } d_{p_i}^{min} = \underline{e_t} - p_i). \end{cases}$$

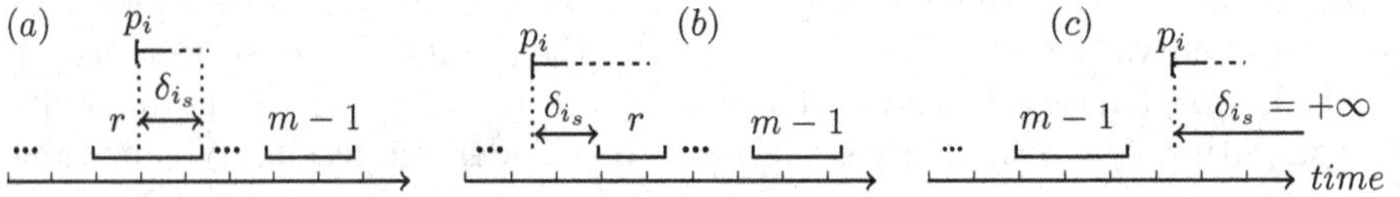

Fig. 3. Illustration of positions $(a), (b)$ and (c) of p_i and corresponding values to δ_{i_s}

After creating the partition $\mathcal{P}$ the algorithm computes $f_t|_{[p_i,p_{i+1}[}$, the restriction of f_t in $[p_i, p_{i+1}[$. To do so, the value of $f_t|_{[p_i,p_{i+1}[}(p_i) = f_t(p_i)$ is explicitly computed and used together with the slope of $f_t|_{[p_i,p_{i+1}[}$. Table 1 gives the slope values for $f_t|_{[p_i,p_{i+1}[}$ according to the positions of p_i, $e_{p_i}^{min}$ and $d_{p_i}^{min}$.

Once the partition is created and the slope of $f_t|_{[p_i,p_{i+1}[}$ is known for any subinterval $[p_i, p_{i+1}[$ of the partition, it is easy to obtain the constant part of the equation of the curve of f_t in $[p_i, p_{i+1}[$ knowing the value of f_t at p_i. The process is illustrated in Example 6.

Table 1. Different values taken by the slope of $f_t|_{[p_i,p_{i+1}[}$ relatively to the positions of p_i and $e_{p_i}^{min}$, given by values of $in(p_i)$ and $in(e_{p_i}^{min})$ as well as on the values of $d_{p_i}^{min}$

	$in(p_i) = \textbf{true}$		$in(p_i) = \textbf{false}$	
$in(e_{p_i}^{min}) = \textbf{true}$	-1 if $d_{p_i}^{min} = e_t - p_i$ 0 if $d_{p_i}^{min} = \underline{d_t}$		0 if $d_{p_i}^{min} = e_t - p_i$ 1 if $d_{p_i}^{min} = \underline{d_t}$	
$in(e_{p_i}^{min}) = \textbf{false}$	-1		0	

Example 6. We illustrate the algorithm sketched in Sect. 5 to compute functions f_0, f_1 and f_2, the minimum intersection of tasks 0, 1 and 2 of Example 1 with intervals $[5,9]$, $[23,25]$, $[30,40]$.

(1) For task 0, we first compute values for $p_0, d_{p_0}^{min}, e_{p_0}^{min}, in(p_0)$ and $in(e_{p_0}^{min})$.

$p_0 = \underline{s_0} = 2$,

$d_{p_0}^{min} = \max(\underline{e_0} - p_0, \underline{d_0}) = \max(11 - 2, 3) = 9$,

$e_{p_0}^{min} = p_0 + d_{p_0}^{min} = 2 + 9 = 11$,

$in(p_0) = \textbf{false}$ (interval i_0 is the first interval to the left of p_0),

$in(e_{p_0}^{min}) = \textbf{true}$ ($e_{p_0}^{min}$ is included in interval i_0),

Since $d_{p_0}^{min} = \underline{e_0} - p_0$, we have $\delta_0 = \min(\delta_{0_s}, \delta_{0_d})$,

$\delta_{0_s} = \ell_0 - p_0 = 5 - 2 = 3, \delta_{0_d} = d_{p_0}^{min} - \underline{d_0} = 9 - 3 = 6$, thus $\delta_0 = 3$,

$p_1 = p_0 + \delta_0 = 2 + 3 = 5$,

Since $in(p_0) = \textbf{false}$ and $in(e_{p_0}^{min}) = \textbf{true}$, then $slope_0 = 0$.

(2) We explicitly compute the value $f_0(p_0) = f_0(2) = 4$.
(3) The equation of $f_0|_{[p_0,p_1[}$ is given by $f_0|_{[p_0,p_1[}(s) = s \cdot slope_0 + (f(p_0) - p_0 \cdot slope_0)$. Hence $f_0|_{[2,5[}(s) = 4$.

We proceed similarly until $p_k = \overline{s_0} + 1$ and repeat the process for the remaining tasks. The results are presented in Table 2, which match the curves f_0, f_1 and f_2 of Fig. 2a.

Proposition 8. *Given n tasks and m intervals, the worst-case time complexity for computing all functions f_t (with $t \in [0, n-1]$) is $\mathcal{O}(nm)$.*

Proof. For a task t, the complexity of the algorithm is given by the number k of subintervals of the partition $\mathcal{P} = (p_0, p_1, \ldots, p_k)$ of $[\underline{s_t}, \overline{s_t}]$. For any $p_i, p_{i+1} \in \mathcal{P}$, $\exists \delta_i$ such that $p_{i+1} = p_i + \delta_i$. Given an interval r, and any $p_i \in \mathcal{P}$, δ_i is either

Table 2. Partition of $\left[\underline{s_t}, \overline{s_t}\right]$, slope and constant values for f_t restricted to each subintervals of the partitions, for each task t, $t \in [0, 2]$.

(a) Task 0: $s_0 = [2, 8]$

Subintervals	$[2, 5[$	$[5, 9[$
Slope	0	−1
Constant	4	9

(b) Task 1: $s_1 = [18, 23]$

Subintervals	$[18, 20[$	$[20, 24[$
Slope	1	0
Constant	18	2

(c) Task 2: $s_2 = [31, 40]$

Subintervals	$[31, 36[$	$[36, 41[$
Slope	0	−1
Constant	4	40

$\min(\delta_{i_s}, \delta_{i_e})$ or $\min(\delta_{i_s}, \delta_{i_d})$, i.e. $\delta_i \in \{u_r - p_i, \ell_r - p_i, u_r - e_{p_i}^{min}, \ell_r - e_{p_i}^{min}, \delta_{i_d}\}$. δ_{i_d} does not depend on the interval r and can take at most 2 values: 0 when $d_{p_i}^{min} = \underline{d_t}$ and $\underline{e_t} - p_i - \underline{d_t}$ when $d_{p_i}^{min} = \underline{e_t} - p_i$. Since there is a total of m intervals, the complexity for computing f_t is of order $4m + 2$. The overall complexity for all n tasks is thus $\mathcal{O}(nm)$.   □

6   Evaluation

We implement the algorithms of Sect. 5 in Choco [9]. Benchmarks were run on an Intel i7 (2.93 GHz) processor running under Mac OS X Yosemite. We conduct two types of benchmarks: a first type comparing the TASKINTERSECTION constraint wrt its reformulation (presented in Proposition 1) on random generated instances available at [11], a second type for testing the TASKINTERSECTION constraint in the context of the video summarisation problem [4,6,8] on real instances also used in [7].

6.1   Evaluation of the TASKINTERSECTION Constraint wrt its Reformulation

We generate random instances, of 50 tasks and 100 intervals each. For each randomly generated instance we use the necessary and sufficient condition stated in Sect. 3 to obtain the feasible lower bound for the total intersection, and fix the *inter* variable to that lower bound and try to find a solution. Then we relax more and more the maximum value of the *inter* variable by adding a percentage of that lower bound, creating 11 configurations: $\forall i \in [0, 10]$, configuration i corresponds to a relaxation of the *inter* variable by $100 - 10 \cdot i$ percent of the lower bound. For each configuration we generate 43 instances on which we run the tests with a time out of 10 min. First we perform a reliability test to evaluate how difficult it is for both approaches to find a solution for the 43 instances of each configuration before time out. Second we compute the average time needed for each configuration to find a solution, excluding the cases were no solution was found before the time out. On the one hand our algorithm always finds a solution for every instance of each configuration before the time out. On the other hand the reformulation finds increasingly fewer solutions as the *inter* variable is less relaxed, resulting in the decreasing curve depicted by Fig. 4a. From Figs. 4b and 4c depicting average times to find a solution, we observe that the average time needed by the reformulation increases significantly, as the *inter* gets relaxed by a smaller percentage of the optimal value. The curve representing the average time needed by our algorithm has an opposite slope. The explanation is that, the more we restrict the *inter* variable, the more values are removed from the start and end variables to satisfy the necessary and sufficient condition for the feasibility of the TASKINTERSECTION constraint.

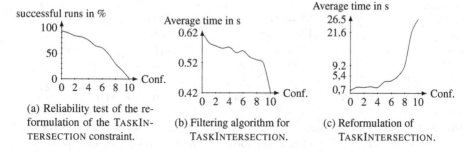

(a) Reliability test of the reformulation of the TASKIN-TERSECTION constraint.

(b) Filtering algorithm for TASKINTERSECTION.

(c) Reformulation of TASKINTERSECTION.

Fig. 4. Evaluation of the TASKINTERSECTION constraint wrt its reformulation for each configuration.

6.2 Evaluation on Real Instances of a Video Summarisation Problem

The video summarisation problem [4,6] consists of extracting video segments out of an input video under some constraints. Using Allen relations [2], Derrien *et al.* design in [7] the EXISTALLEN constraint as well as a propagator that they use in an application that generates video summary out of tennis match input video. A preprocessing step extracts several features from the input video (e.g. games, applause, speech) that are modelled as intervals. The problem is then to select video segments that will constitute the video summary, maximising the amount of applause in the summary under the following constraints:

(1a) a segment should not intersect a speech interval,
(1b) a segment should not intersect a game interval,
 (2) each selected segment should contain an applause interval,
 (3) the cardinality of the intersection between the segments and the dominant colour intervals should exceed one third of the summary.

We start from the model described in [7] where we rather use the TASKIN-TERSECTION constraint to enforce constraints (2) and (3). The summary should have a total duration between four and five minutes, and should be composed of ten video segments whose duration varies from 10 to 120 s. We run our model on the available 3 instances real dataset provided by Boukadida *et al.* [6]. To ensure we explore the same search tree as in Fig. 2 of [7] we consider a static search heuristic, selecting variables in lexicographic order and assigning them values in increasing order. The results of the comparative evaluation are reported in Fig. 5:

- For *Instance1* and *Instance2* the TASKINTERSECTION propagation algorithm finds the best solution right from the beginning and improves it by up to 25 % in both instances.
- For the last instance, the TASKINTERSECTION propagation algorithm firstly finds similar quality solutions and finally significantly improves them by up to 23 %.

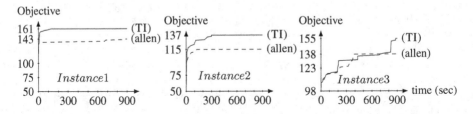

Fig. 5. Evaluation of the contribution of the model that include the TASKINTERSECTION constraint (TI) wrt the model described in [7] (allen) on the video summarisation problem; the plots report the evolution of the objective, i.e. the total duration of applauses, wrt the processing time within a time out of 900 s.

7   Conclusion

We introduce the TASKINTERSECTION for scheduling problems where variable duration tasks are subject to a chain of precedence constraints and where there is a resource with a 0–1 cost that varies over time. We provide a tight lower bound for the overall intersection and a filtering algorithm based on this bound. We evaluate this filtering algorithm on randomly generated instances and compare the results to a reformulation of the TASKINTERSECTION constraint. We also conduct a comparative evaluation on real instances of the video summarisation problem [7]. The positive outcome of these experiments actually shows that our filtering algorithm allows to find significantly better solutions in terms of cost, i.e. to increase by 20 % the total value of the applause.

Acknowledgement. This work has received a French state support granted to the CominLabs excellence laboratory and managed by the National Research Agency in the "Investing for the Future" program under reference Nb. ANR-10-LABX-07-01. We would like to thank the reviewers for their comments that help improve the manuscript.

References

1. Aggoun, A., Beldiceanu, N.: Extending CHIP in order to solve complex scheduling and placement problems. Math. Comput. Model. **17**(7), 57–73 (1993)
2. Allen, J.F.: Maintaining knowledge about temporal intervals. Commun. ACM **26**(11), 832–843 (1983)

3. Baptiste, P., Le Pape, C., Nuijten, W.: Constraint-Based Scheduling: Applying Constraint Programming to Scheduling Problems. International Series in Operations Research & Management Science, vol. 39. Springer Science & Business Media, Berlin (2012)

4. Berrani, S.A., Boukadida, H., Gros, P.: Constraint satisfaction programming for video summarization. In: 2013 IEEE International Symposium on Multimedia (ISM), pp. 195–202. IEEE (2013)

5. Bessiére, C.: Constraint propagation. Handbook of constraint programming, pp. 29–83 (2006). Chap. 3

6. Boukadida, H., Berrani, S.-A., Gros, P.: A novel modeling for video summarization using constraint satisfaction programming. In: Bebis, G., et al. (eds.) ISVC 2014, Part II. LNCS, vol. 8888, pp. 208–219. Springer, Heidelberg (2014)

7. Derrien, A., Fages, J.-G., Petit, T., Prudhomme, C.: A global constraint for a tractable class of temporal optimization problems. In: Pesant, G. (ed.) CP 2015. LNCS, vol. 9255, pp. 105–120. Springer, Heidelberg (2015)

8. Ekin, A., Mehrotra, R., et al.: Automatic soccer video analysis and summarization. IEEE Trans. Image Process. **12**(7), 796–807 (2003)

9. Fages, J.G., Prud'homme, C.: A free and open-source Java library for constraint programming (2015). http://choco.emn.fr/

10. Kumar, T.S., Cirillo, M., Koenig, S.: Simple temporal problems with taboo regions. In: AAAI. Citeseer (2013)

11. Madi Wamba, G.: Random generated instances of the taskintersection problem (2015). https://www.dropbox.com/sh/uwvn86rx7mxebty/AADyUAdnEWdOkm C8Xkcyjg3Ua?dl=0

12. Simonis, H., Hadzic, T.: A family of resource constraints for energy cost aware scheduling. In: Third International Workshop on Constraint Reasoning and Optimization for Computational Sustainability, St. Andrews, Scotland, UK, September 2010

13. Simonis, H., Hadzic, T.: A resource cost aware cumulative. In: Larrosa, J., O'Sullivan, B. (eds.) CSCLP 2009. LNCS, vol. 6384, pp. 76–89. Springer, Heidelberg (2011)

14. Sourd, F.: Optimal timing of a sequence of tasks with general completion costs. Eur. J. Oper. Res. **165**(1), 82–96 (2005)

A Stochastic Continuous Optimization Backend for MiniZinc with Applications to Geometrical Placement Problems

Thierry Martinez[1]([✉]), François Fages[1], and Abder Aggoun[2]

[1] Inria Paris-Rocquencourt, Team Lifeware, Rocquencourt, France
`thierry.martinez@inria.fr`
[2] KLS-Optim, Les Ulis, France

Abstract. MiniZinc is a solver-independent constraint modeling language which is increasingly used in the constraint programming community. It can be used to compare different solvers which are currently based on either Constraint Programming, Boolean satisfiability, Mixed Integer Linear Programming, and recently Local Search. In this paper we present a stochastic continuous optimization backend for MiniZinc models over real numbers. More specifically, we describe the translation of FlatZinc models into objective functions over the reals, and their use as fitness functions for the Covariance Matrix Adaptation Evolution Strategy (CMA-ES) solver. We illustrate this approach with the declarative modeling and solving of hard geometrical placement problems, motivated by packing applications in logistics involving mixed square-curved shapes and complex shapes defined by Bézier curves.

1 Introduction

MiniZinc [11] is a medium-level constraint modeling language which is becoming a standard in the Constraint Programming community. It is high-level enough to express most constraint problems easily, but low-level enough to be mapped onto existing solvers easily and consistently. This mapping is done through a flattening process which takes as input a MiniZinc instance and produces a FlatZinc instance. FlatZinc is a low-level solver input language designed to be easy to translate into the form required by a solver. It is chosen for that reason as target language for MiniZinc.

Currently, there exist FlatZinc backends for Mixed Integer Linear Programming (CPLEX, OR-tools[1], SCIP, . . .), Finite Domain Constraint Programming solvers (Choco[2], Eclipse[3], Gecode[4], JaCoP, Opturion-CPX[5], Oscar, SICStus

---

[1] https://code.google.com/p/or-tools/.
[2] https://github.com/chocoteam/choco-parsers.
[3] http://eclipseclp.org/doc/bips/lib_public/flatzinc/.
[4] http://www.gecode.org/flatzinc.html.
[5] http://www.opturion.com/cpx.

© Springer International Publishing Switzerland 2016
C.-G. Quimper (Ed.): CPAIOR 2016, LNCS 9676, pp. 262–278, 2016.
DOI: 10.1007/978-3-319-33954-2_19

prolog, ...), SAT solvers (MinisatID, ...) and recently Local Search (iZplus[6], Oscar-cbls [1]).

Most of FlatZinc implementations are thus dedicated to discrete domains. However, constraint optimization and decision problems over real numbers can be expressed in MiniZinc with high generality. Curently, such continuous constraint problems can be solved either using Linear Programming backends, with restrictions on the linearity of the constraints, or using interval arithmetic backends (e.g. G12ic, Eclipse fzn_ic).

In this paper, we study another kind of solver based on *stochastic continuous optimization* for solving FlatZinc instances over real numbers, using namely the *Covariance Matrix Adaptation Evolution Strategy* (CMA-ES) [6]. More specifically, we show how a FlatZinc instance over real numbers can be translated into a fitness function which can be directly used by CMA-ES to compute approximate solutions to the problem. The transformation we describe is quite general and applies virtually to any MiniZinc model over real numbers. The choice of CMA-ES among other evolutionary or particle swarm optimization algorithms is motivated by the absence of parameterization for this algorithm and by its performances on hard problems.

For discrete domains, there has been related work on the design of high-level constraint-based modeling languages for local search and genetic algorithms. The seminal work of Van Hentenryck and Michel on Comet [7,10] showed how a finite domain constraint model can be compiled into an objective function for local search metaheuristics, such as Tabu search, with default neighborhoods derived from the constraint model. In [1], Björdal et al. present a constraint-based local search backend for MiniZinc and show that it produces competitive results on the 2010 to 2014 MiniZinc challenges. In these systems, the local search solver is limited to finite domain constraints and use neighborhoods derived from the finite domains of the variables.

Here in the continuous domain, we illustrate our CMA-ES backend for FlatZinc with the solving of hard geometrical placement problems which, to the best of our knowledge, go beyond the state-of-the-art of declarative constraint modeling and solving. As a matter of fact, the only FlatZinc implementations listed on the MiniZinc web page that parse the FlatZinc instances presented in this paper are those based on exact methods using interval arithmetic (i.e. Eclipse fzn_ic and G12ic) but none of them can find solutions in reasonable computation time even for the examples presented here. In [9], we have already shown that the non-overlap constraint between squares, cubes, rectangles, boxes, triangles, polygons circles and spheres, can be associated with a *measure of overlap* between objects which can be used directly as a fitness function in CMA-ES for packing mixed shapes in a bin, with an interesting trade-off between generality and efficiency. The measure of overlap does not need to be the area of the intersection (and should not if one object can be included in another) but can be any measure equal to 0 in case of non-overlap, and capable of guiding the continuous optimization solver by measuring progress toward satisfaction [4]. On a benchmark of consecutive sizes circle packing

---

[6] http://www.minizinc.org/challenge2014/descriptionizplus.txt.

problems, we showed that CMA-ES finds solutions at 2 % of the best known costs obtained by running the three global optimization methods reported in Castillo et al. [3]. In [12], Salas and Chabert show that the overlap measures which were defined in an *ad hoc* manner in [9], can be computed by interval methods in IBEX[7] with a numerical algorithm that automatically measures the *penetration depth* of two objects of virtually any shape defined by conjunction and disjunction of non-linear inequalities. In this paper, we give general MiniZinc definitions for the penetration depths, or simpler overlap measures, between polygons, circles, and also *complex shapes defined by Bézier curves*, motivated by packing problems in the cosmetic and automotive industries. This illustrates the performance of MiniZinc-CMAES in terms of both declarative modeling and efficient (yet suboptimal) resolution of very hard geometrical packing problems with complex shapes and continuous rotations.

The rest of the paper is organized as follows. In the next section, we present the translation of a FlatZinc instance over real numbers in a fitness function over the reals, and the interface to the CMA-ES solver. In Sect. 4 we describe MiniZinc models of continuous packing problems involving continuous rotations, mixed square-curved shapes and complex shapes defined by Bézier curves. There we use some simple distance formulae for circles, Minkowski sums for the penetration depth between polygons [5] and De Casteljau's numerical algorithm for linearizing Bézier curves. In Sect. 5, we report on the performance results obtained through the compilation chain from MiniZinc, FlatZinc to CMA-ES, on complex shape packing problems. Finally, we conclude on the general perspective opened by this MiniZinc backend for continuous optimization and novel applications at the intersection of Optimization and Computer-Aided Design.

2   Compiling FlatZinc Instances over Real Numbers in Real-Valued Fitness Functions

In this section we describe our transformation of a FlatZinc instance containing arithmetic and trigonometric constraints over float variables in a fitness function which aggregates the costs of each constraint violation. This transformation is at the heart of the continuous optimization backend.

2.1   Arithmetic Expressions

The arithmetic expressions that constitute the constraint satisfaction problem need be rebuilt from the FlatZinc instance, since arithmetic sub-expressions and intermediary variables are introduced by the transformation from MiniZinc to FlatZinc. The constraints that result from this transformation are split in three groups:

1. inequality constraints, which are turned into *costs*,

---

[7] http://www.ibex-lib.org.

2. arithmetic and trigonometric constraints, which always appear to be directed and are turned into *functional expressions*,
3. and equality constraints, which can either be solved statically if there exists a topological sort of the constraint graph that makes the constraint directed, or turned into a cost otherwise.

Every variable X of the model is associated to an expression $[X]$ defined as $[X] = X$ if X is one of the search variables, or as the arithmetic or trigonometric expression deduced from the constraints on X. The notation is extended to float constants, $[f] = f$, and vectors of variables and/or float constants, i.e. for a vector $U = (X_1, \ldots, X_n)$, $[U]$ denotes the vector $([X_1], \ldots, [X_n])$.

Concerning the first group, the inequality constraints in FlatZinc are either strict or non-strict inequalities between linear expression of the form:

```
constraint float_lin_lt(U0,U1,K);
constraint float_lin_le(U0,U1,K);
```

where K is a constant, U_0 is a vector of constant coefficient and U_1 is a vector of variables. The semantics is respectively $U_0 \cdot U_1 < K$ and $U_0 \cdot U_1 \leq K$ where $\cdot$ denotes the scalar product. The cost we associate to such an inequality constraint c is

$$\text{cost}(c) = \max(0, [U_0] \cdot [U_1] - K)$$

where $[U_0]$ and $[U_1]$ are the expressions constructed from the arguments. For strict inequality, one could refine the cost to

$$\text{cost}(c) = \begin{cases} 1 + [U_0] \cdot [U_1] - K & \text{if } [U_0] \cdot [U_1] \geq K \\ 0 & \text{otherwise} \end{cases}$$

in order to ensure that the cost is null if and only if the constraint is satisfied. However, this is not necessary to guide the search for solutions since the constraints $a < b$ and $a \leq b$ are equivalent almost everywhere in a continuous setting.

Concerning the second group, every float variable X of the model is considered as a search variable, except if it appears in the result position of one of the following directed constraint:

```
constraint float_min(A, B, X);
constraint float_max(A, B, X);
constraint float_times(A, B, X);
constraint float_sqrt(A, X);
constraint float_cos(A, X);
constraint float_sin(A, X);
```

If X is in the result position of one of these constraints, then $[X]$ is defined as the expression that computes the associated value.

For the third group, FlatZinc linear equality constraints are of the form:

```
constraint float_lin_eq(U0,U1,K) :: defines_var(X) :: weight(w);
```

The FlatZinc compiler generates the annotation defines_var(X) which directs the constraint from its arguments to its result, in the case where the constraint results from a linear arithmetic expression. In that case, $[X]$ is defined as the expression that computes the linear combination.

For the equality constraints that result from reification, of the form:

```
constraint float_eq_reif(X,Y,B);
constraint float_lin_eq_reif(X,U,B);
```

we eliminate them statically if enough information is available. Otherwise, the model is currently rejected.

It is worth noting that this minimal handling of reification is needed to cope with the code generated by the MiniZinc compiler for partial functions like sqrt, where the formal argument is unified with the actual argument only if the function is defined for this argument. On the other hand, general reified equality constraints impose integrity constraints on the boolean variables. Turning an integrity constraint into a cost function causes rugged landscapes which may be difficult to explore, although CMA-ES is also a pretty good solver in this case. Since our benchmarks do not use reified constraints nor discrete variables, they are currently out of the scope of our MiniZinc backend.

2.2   Cost Aggregation

In our backend, the gathering of the cost functions can be tuned by using annotations. First, each constraint can be annotated with a weight which will affect the cost in the aggregation. By default, the weight of a constraint is 1.

```
annotation weight(float);
```

Second, every MiniZinc model contains one and only one *solve* instruction, which gives the objective. The *solve* item can be annotated with one of the following annotations which change the definition of the violation cost of the whole constraints. By default, weighted_sum is assumed.

```
annotation weighted_sum;
```

$$\text{violation\_cost} = \sum_c \text{weight}(c) \cdot \text{cost}(c)$$

```
annotation fuzzy;
```

$$\text{violation\_cost} = \max_c \text{weight}(c) \cdot \text{cost}(c)$$

```
annotation probabilistic;
```

$$\text{violation\_cost} = 1 - \prod_c \left(\frac{1}{1 + \text{cost}(c)}\right)^{\text{weight}(c)}$$

Third, if the FlatZinc instance is a constraint satisfaction problem, the fitness function is defined to be equal to the violation_cost.

```
solve satisfy;
```

If the FlatZinc instance requires to minimize an expression e, the fitness function is defined to be equal to $\alpha \cdot (1 + \text{violation\_cost}) + e$, where α is a coefficient large enough to dominate e. Then α can be set with the following annotation applied to the solve item.

```
annotation alpha(float);
```

By default, $\alpha = 10^{10}$.

3   Stochastic Continuous Optimization with CMA-ES

The Covariance Matrix Adaptation Evolution Strategy (CMA-ES[8]) [6] is one of the most powerful global optimization strategy for minimizing an objective function over the reals in a "black-box" scenario, i.e. without assuming any property about the objective function. This method is a multi-point method which uses a population of configurations (here valuations of the FlatZinc search variables, e.g. packings defined by the coordinates and orientations of the objects) to sample the search space, estimates the covariance matrix at each sampling, determines the next move in the most promising direction (e.g. translations and rotations of objects), and updates accordingly the *multi-variate normal distribution* for the next sampling (i.e. mean value and covariance of the variables).

CMA-ES behaves in effect like a second-order method where the landscape is estimated by sampling, according to some multi-variate normal distribution of the variables, which is itself updated during search in the most promising direction to adapt to the landscape, using an estimation of the second-order moment, the covariance matrix. When the objective function does not improve, CMA-ES can be restarted to find different local optima. We refer to [6] for more details on that stochastic optimization algorithm.

One advantage of CMA-ES is that it requires very little parameter tuning. All our benchmarks have been performed with the C implementation of CMA-ES using the same parameter set: a population size of 100, an initial standard deviation of 20 and a stopping criterion based on a difference less than 10^{-3} for the fitness function.

CMA-ES thus tries to minimize an arbitrary function $f : \mathbf{R}^n \to \mathbf{R}$, where n is the *dimension* of the search space (*i.e.* the number of FlatZinc search variables). The result is a vector $\boldsymbol{x} \in \mathbf{R}^n$ such that $f(\boldsymbol{x})$ is the smallest value encountered so far. The C implementation of CMA-ES expects that the function f has the following interface: `double f(double x[])`. The FlatZinc-to-CMA-ES back-end derives such a fitness function from the FlatZinc model according to the transformations described in Sect. 2.

---

[8] https://www.lri.fr/~hansen/cmaesintro.html.

4   MiniZinc Models of Geometrical Placement Problems

The problems addressed in this section are taken from the industry of cosmetics packaging. They consist of packing products with various shapes. In this application the objective is to pack a given quantity of a product in a minimum number of bins. The study of different forms in the industry of cosmetics packaging show that convex approximations of objects give poor results but that the objects can be modeled using a combination of Bézier curves. Existing Constraint Programming tools are limited and do not offer capabilities to solve such problems. We show here how to model and solve such problems in MiniZinc.

4.1   Overlap Measures Between Objects

The overlap measure should guide the optimization procedure towards a geometrical placement without overlap, i.e., ideally, the cost should decrease as long as the placement gets closer towards a placement without overlap. ϕ-functions [4] have been introduced for the same purpose of continuous optimization for geometrical placement problems, using decompositions in half-planes, triangles and circles. In this section, we describe three overlap measures for polygones and complex shapes delimited by Bézier curves using the intersection area, the penetration depth, or the sum of the pairwise distances between the intersection points of the borders.

Intersection Area. Object intersection area can be used as an overlap measure, which could seem quite natural. However, this area can be costly to compute and does not guide the optimization well. For example, when an object fully contains another one, the overlap measure remains in a plateau for every possible position where the contained object stays inside the container, giving no direction for getting outside the overlap zone (Fig. 1). In pratice, we do not use this measure.

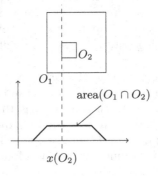

Fig. 1. Intersection area between two rectangles O_1 and O_2 in function of $x(O_2)$, with a plateau when O_2 is included in O_1.

Penetration Depth. The penetration depth is a common measure in computer-aided design [2]: the penetration depth between two objects is the smallest norm such that there exists a translation vector to apply to one of the two objects such to lead to a placement where the two objects do not intersect.

The penetration depth between two circles (C_1, r_1) and (C_2, r_2) is trivial to compute:

$$\text{pd}(C_1, r_1); (C_2, r_2)) = \max(0, r_1 + r_2 - C_1 C_2)$$

that is to say the difference between the sum of their radius and the distance between their centers (Fig. 2). This distance was already used for circles in [4,9].

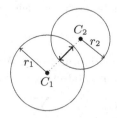

Fig. 2. Penetration depth between two circles (C_1, r_1) and (C_2, r_2)

More generally, the penetration depth between two objects O_1 and O_2 is equal to the distance between the origin $(0, 0)$ and the complementary of the Minkowski difference $O_1 \ominus O_2 = \{p_1 - p_2 \mid p_1 \in O_1, p_2 \in O_2\}$.

$$\text{pd}(O_1; O_2) = \min\{\|\boldsymbol{u}\| \mid \boldsymbol{u} \notin O_1 \ominus O_2\}$$

Indeed, this distance is by definition the smallest norm such that there exists a vector $\boldsymbol{u}$ such that $\boldsymbol{u} \notin O_1 \ominus O_2$, that is to say a vector such that for all $p_1 \in O_1, p_2 \in O_2$, $\boldsymbol{u} \neq p_1 - p_2$, thus we have $O_1 \cap (O_2 + \boldsymbol{u}) = \emptyset$.

The Minkowski difference of two polygons is a polygon, computable in a time quadratic to the number of edges, and the Minkowski difference of two convex polygons is a convex polygon, computable in a time linear to the number of edges [5] (Fig. 3).

Note that the penetration depth only consider translations, whereas the search space we consider for optimization may include rotation angles as additional "dimensions". [12] extends the Minkowski difference to consider object rotations as well. This extension is difficult to interpret geometrically and makes overlap measures depend on the choice of the origin for each object. We will restrict ourselves to Minkowski difference in the Euclidean space.

Sum of the Pairwise Distances Between the Intersection Points of the Borders. For the overlap measure between two objects O_1 and O_2 that have non polygonal shapes like those delimited by Bézier curves, or for heterogeneous shapes (for example, when mixing Bézier curves and polygons), we prefer to use another measure simpler to compute: the sum of the pairwise distances between the

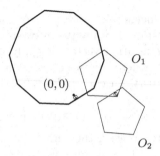

Fig. 3. Minkowski difference between two pentagons O_1 and O_2. The penetration depth between O_1 and O_2 is the distance between the origin and the border of the Minkowski difference.

intersection points of the borders ∂O_1 and ∂O_2. We suppose that $\partial(O_1) \cap \partial(O_2)$ is a finite set, applying infinitesimal offsets if necessary.

$$\mathrm{spd}(O_1; O_2) = \sum_{\substack{p_i, p_j \in \partial(O_1) \cap \partial(O_2) \\ i < j}} \|\overrightarrow{p_1 p_2}\|$$

There exist several methods to compute the intersections between two Bézier curves numerically [13] (Fig. 4). We use a dichotomic search by using de Casteljau's algorithm for splitting the curves. The dichotomic search can also be used to compute numerically the intersections between Bézier curves and circles.

Fig. 4. Intersection points of the borders of two objects delimited by Bézier curves.

The intersections between a Bézier curve and a segment can be computed algebraically. Indeed, by changing the frame, we can suppose without loss of generality that the segment lays on the abscissa axis. The Bézier curve (p_0, p_1, p_2) intersects the axis for every parameter t, $0 \leq t \leq 1$, such that $(1-t)((1-t)y_{p_0} + ty_{p_1}) + t((1-t)y_{p_1} + ty_{p_2}) = 0$: this is a second-order polynomial in t. For each solution t_0, it suffices to check that the abscissa $(1 - t_0)((1 - t_0)x_{p_0} + t_0 x_{p_1}) + t_0((1 - t_0)x_{p_1} + t_0 x_{p_2})$ belongs to the segment.

This measure does not fulfill the requirements of an ideal overlap measure: the measure is null when one object is included in the other and is not monotonic with respect to the penetration depth. However, it is locally monotonic in a neighborhood around overlap-free placements: in the context of a local search, by choosing an overlap-free initial placement (spreading the objects enough far ones from the others), this measure experimentally appears to be sufficient to preserve the overlap-freeness during the placement compaction process.

4.2   Continuous Packing Model

This section makes use of the overlap measures introduced above to express MiniZinc models for continuous packing of circles, arbitrary polygons with rotation and other complex shapes like rosettes delimited by Bézier curves. The penetration depth is used as measure of overlap between two circles and between two polygons, while the sum of distances between the intersection points is used for every other pair of shapes.

Circles. The following predicate expresses the constraint that the two circles $((x_1, y_1), r_1)$ and $((x_2, y_2), r_2)$ do not overlap.

```
predicate non_overlap_circles(
    var float: x1, var float: y1, var float: r1,
    var float: x2, var float: y2, var float: r2) =
    pow(x1 - x2, 2) + pow(y1 - y2, 2) > pow(r1 + r2, 2);
```

It is worth noticing that the inequality $(x_1 - x_2)^2 + (y_1 - y_2)^2 > (r_1 + r_2)^2$ is compiled into the cost function $(r_1 + r_2)^2 - (x_1 - x_2)^2 - (y_1 - y_2)^2$, which is monotonic with respect to the penetration depth $(r_1 + r_2) - \sqrt{(x_1 - x_2)^2 - (y_1 - y_2)^2}$, introduced in the previous section.

We consider a benchmark of circle placement problems [3] where there are n circles to place in a circular bin. The usual modelling [9] supposes that the circular bin is centered on the origin. The following function compute for each circle the minimum radius for the circular bin to contain the circle $(x, y), r)$.

```
function var float: bounding_circle_radius(
    var float: x, var float: y, var float: r) =
    sqrt(pow(x, 2) + pow(y, 2)) + r;
```

The search variables are the positions of the circle centers.

```
int: n;
array[1 .. n] of var float: x;
array[1 .. n] of var float: y;
```

The circle positions are constrained to be non-overlapping.

```
constraint forall(i in 1..n,j in i+1..n)(
    non_overlap_circles(x[i], y[i], radius(i), x[j], y[j], radius(j)));
```

The goal is to minimize the radius of the circular bin. Intermediary variables are introduced to store the minimal bounding radius for each circle to circumvent a limitation of the max function for arrays in MiniZinc that needs to know the bounds of the arguments (Fig. 5).

```
array[1 .. n] of var 0.0 .. 1000.0: bounding_radii;

constraint forall(i in 1 .. n)(
    bounding_radii[i] = bounding_circle_radius(x[i], y[i], radius(i)));

constraint bounding_radius = max(bounding_radii);

solve minimize bounding_radius;
```

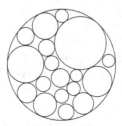

Fig. 5. Example of packing found by MiniZinc-CMAES for 18 circles of radii $i^{-1/2}$ for $1 \leq i \leq 18$ (circle packing benchmark of [3]).

It is worth noticing that applying min and max functions to overlap measures allows Boolean combinations of geometrical shapes to be expressed. For instance, the Fig. 6 shows a placement for 20 geometrical rosettes, where each rosette $\mathcal{R}((x, y), r)$ is defined as the union of six intersections between pairs of circles:

$$\mathcal{R}((x, y), r) = \bigcup_{i=1}^{6} \mathcal{C}((x + \cos(2 \cdot i \cdot \frac{\pi}{6}) \cdot r, y + \sin(2 \cdot i \cdot \frac{\pi}{6}) \cdot r), r)$$
$$\cap \, \mathcal{C}((x + \cos(2 \cdot (i+2) \cdot \frac{\pi}{6}) \cdot r, y + \sin(2 \cdot (i+2) \cdot \frac{\pi}{6}) \cdot r), r)$$

Objects with Rotations. The placement of each object in the subsequent examples is described by three search variables: the position on the x axis, the position on the y axis, and the rotation angle r.

```
set of int: position = 1 .. 3;
int: x = 1;
int: y = 2;
int: r = 3;
```

An object is described by a variable of type `array[position] of var float`, that is to say an array of three variables. The constants `x`, `y` and `r` are used as projectors: given an array `object`, components can be accessed as `object[x]`, `object[y]` and `object[r]`. (Note that MiniZinc has not yet support for records.)

Points are stored in an array of two coordinates. The points that describe the shapes of an object are expressed in a frame relative to the given object position and orientation. The function `image_of_point` defined below transforms the coordinates of a point to the global frame.

Fig. 6. Placement found by MiniZinc-CMAES for 20 geometrical rosettes, defined as unions of circle intersections.

```
set of int: coordinates = 1 .. 2;
function array[coordinates] of var float: image_of_point(
  array[position] of var float: object,
  array[coordinates] of float: point
) = [
  cos(object[r]) * point[x] - sin(object[r]) * point[y] + object[x],
  sin(object[r]) * point[x] + cos(object[r]) * point[y] + object[y]
];
```

Object positions are stored in a matrix.

```
set of int: objects = 1 .. n;
array [objects, position] of var float: object_positions;
```

The following function returns the position of an object given its index.

```
function array[position] of var float: object_position(int: object) =
  [object_positions[object, d] | d in position];
```

Polygons. We consider pentagons with the following vertex coordinates (relative to the object frame).

```
array[1..5, coordinates] of float: pentagon =
  [| 2.2024586, 58.90577
   | 18.54966,8.594238
   | 71.45033,8.594238
   | 87.79755,58.90576
   | 45.0,90.0 |];
```

These pentagons approximate the Bézier rosettes that we consider below: the vertices join the ends of the petals (Fig. 7). It is not exactly the convex hull since one petal goes outside the pentagon but it is close to (and the convex hull of the rosette is not polyhedric). However, this approximation is sufficient to observe the gain obtained in the placements by considering the precise Bézier rosettes instead of such approximations.

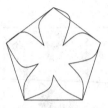

Fig. 7. The pentagon obtained by considering each petal's end as a vertex is not a correct approximation of a Bézier rosette.

We suppose that the following function computes the penetration depth between two (convex) polygons.

```
function var float: penetration_depth_between_polygons(
    array[int, coordinates] of var float: vertices0,
    array[int, coordinates] of var float: vertices1
);
```

The Minkowski difference between two polygons can be expressed with arithmetic constraints through reification: the constraint is cumbersome to write directly, but can be automatically generated, for example by a ClpZinc model [8]. Alternatively, the function can be implemented in the back-end as an auxiliary C function, which is the case of our current implementation (Fig. 8).

Fig. 8. Placement found by MiniZinc-CMAES for 29 pentagons

Bézier Curves. The rosettes that we consider are delimited by the 10 following quadratic Bézier curves (one curve by line, each curve is described by three control points).

```
set of int: curves = 1 .. 10;
set of int: quadratic_bezier_control_points = 1 .. 3;
array [
    1 .. card(curves) * card(quadratic_bezier_control_points),
    coordinates] of float: curve_points =
    [| 2.2024586, 58.90577 | 16.01051, 34.989525 | 31.150425, 40.5
     | 31.150425, 40.5 | 17.211597, 23.888353 | 18.54966, 8.594238
     | 18.54966, 8.594238 | 42.69821, 17.38359 | 45.0, 30.437695
     | 45.0, 30.437695 | 52.512943, 17.424889 | 71.45033, 8.594238
```

```
| 71.45033, 8.594238 | 73.003426, 26.34613 | 58.84958, 40.5
| 58.84958, 40.5 | 82.6242, 44.69211 | 87.79755, 58.90576
| 87.79755, 58.90576 | 70.514114, 71.00775 | 53.55951, 56.78115
| 53.55951, 56.78115 | 63.23457, 83.36317 | 45.0, 90.0
| 45.0, 90.0 | 29.181046, 76.72632 | 36.44049, 56.781155
| 36.44049, 56.781155 | 18.055468, 72.20802 | 2.2024586, 58.90577 |];
```

We suppose that the following function computes the sum of the distances between the intersections of two sets of curves. This function is implemented in the back-end as an auxiliary C function.

```
function var float: sum_of_distances_between_bezier_intersection_points(
    array[int, coordinates] of var float: curves0,
    array[int, coordinates] of var float: curves1
);
```

Fig. 9. Placement found by MiniZinc-CMAES for 10 Bézier rosettes

Mixing Bézier Curves and Rectangles. For computing the overlaps between Bézier curves and rectangles, we suppose that the following function computes the sum of the distances between the intersections of a Bézier curve and a rectangle. This function is implemented in the back-end (as an auxiliary C function), but the arithmetic could be expressed in MiniZinc as well (Fig. 9).

```
function var float:
    sum_of_distances_between_bezier_and_polygon_intersection_points(
    array[int, coordinates] of var float: curves,
    array[int, coordinates] of var float: vertices
);
```

The Fig. 10 shows a placement found for 16 Bézier rosettes and 16 rectangles. It is worth noticing that even if the optimization procedure has found a non-trivial placement, for instance for the rectangles and the rosettes in the bottom right of the figure, some visually obvious improvements of the placement of the left rosettes are not found in this run of CMA-ES.

Fig. 10. Placement found by MiniZinc-CMAES for 16 Bézier rosettes and 16 rectangles, in this run of CMA-ES which stays stick in a local minimum.

5   Evaluation Results of MiniZinc-CMAES

The following Tables 1 and 2 summarize the performance obtained with MiniZinc-CMAES, in terms of computation time, smallest area found, mean area and variance of the area among 50 runs of CMA-ES. For every example, results are averaged over 50 runs. All these results have been obtained with the default parameters of CMA-ES described in Sect. 3. It is worth noticing that smaller initial standard deviations tend to generate solutions with overlaps that the optimization fails to remove, and bigger standard deviations augment convergence times. Total time is the sum of the computation times for all the 50 restarts: for each problem, all the restarts have been computed in parallel on a cluster, one problem per core.

It is worth noticing that Eclipse and G12 with their interval constraint solvers can parse the MiniZinc models of the previous section that do not use predicates

Table 1. Computation time for placement of geometrical rosettes.

Roses	Total time (50 restarts)	Mean time	Smallest area found	Mean area	Variance (area)
10	17 min 32 s	21 s	25.266	27.303	1.893
11	31 min 13 s	37 s	27.369	29.692	1.585
12	41 min 18 s	49 s	28.761	32.511	3.675
13	1 h 1 min 39 s	1 min 13 s	32.936	34.987	2.648
14	1 h 17 min 37 s	1 min 33 s	33.816	37.714	2.926
15	1 h 43 min 19 s	2 min 3 s	37.233	41.085	3.57
16	2 h 6 min 57 s	2 min 32 s	39.729	43.86	5.891
17	2 h 26 min 23 s	2 min 55 s	41.883	46.113	5.947
18	3 h 11 min 20 s	3 min 49 s	43.582	49.123	13.828
19	3 h 47 min 24 s	4 min 32 s	46.74	52.594	10.769
20	5 h 8 min 42 s	6 min 10 s	49.006	54.89	8.579

Table 2. Computation time for placement of mixed shapes: Bézier's rosettes and rectangles.

Shapes	Total time (50 restarts)	Mean time	Best area	Mean area	Area variance
10 + 10	7 j 17 h 35 min 10 s	3 h 42 min 42 s	66 715.294	70 080.254	$3.784 \cdot 10^5$
11 + 11	12 j 5 h 34 min 30 s	5 h 52 min 18 s	72 846.432	78 495.343	$4.434 \cdot 10^5$
12 + 12	41 min 18 s	49 s	87 257.346	90 238.345	$5.499 \cdot 10^5$
13 + 13	13 j 23 h 13 min 39 s	6 h 42 min 16 s	85 492.98	$1.024 \cdot 10^5$	$6.985 \cdot 10^5$
14 + 14	17 j 9 h 27 min 23 s	8 h 20 min 56 s	$1.11 \cdot 10^5$	$1.398 \cdot 10^5$	$8.219 \cdot 10^5$
15 + 15	23 j 9 h 18 min 63 s	11 h 13 min 35 s	$1.078 \cdot 10^5$	$1.584 \cdot 10^5$	$1.281 \cdot 10^6$

defined as auxiliary C functions in the back-end. However the performances are very poor with results obtained only for 3 circles.

6   Conclusion

We have presented here a stochastic continuous optimization backend for MiniZinc models over real numbers. We have shown the benefits of this approach using the CMA-ES solver for continuous optimization on a series of geometrical placement problems motivated by industrial applications in logistics, involving mixed square-curve shapes, and also complex shapes defined by Bézier curves. Probably because of the novelty of these problems for complex shapes, we have not identified benchmarks for comparing the techniques presented here, but in [9] we showed that the solutions found with CMA-ES on circle packing were at just 2 % of the best solutions found with dedicated solvers.

The declarative modeling in MiniZinc combined to the solving using the transformation to CMA-ES described in this paper, does not come with any significant overhead and provides fully declarative solutions to very hard geometrical placement problems. The non-overlap constraint has a cost function based on the penetration depths between objects, using Minkowski sums for polygons, and a simpler measure of overlap for Bézier curves. A classical difficulty in the definition of the error function of a conjunction of constraints is the normalization of the error function for each constraint. This has been solved here by letting the modeller specify in MiniZinc the cost function if different from the default cost aggregation function (i.e. the sum of the costs).

The recourse to such a black-box optimization procedure for FlatZinc makes sense especially in presence of non-linear constraints, and in absence of integer variables, but the transformation we have given of a FlatZinc model in a non-negative real-valued cost function is quite general. We have focused on continuous placement problems, but our MiniZinc/CMA-ES can be applied in principle to any constraint model over real numbers. The examples taken here from industrial problems in logistics, including objects defined by Bézier curves, should contribute to open a new domain of application of constraint methods in computational geometry, at the intersection of optimization and computer-aided design.

Acknowledgements. This work has been funded by the ANR Blanc Simi2 Net-WMS-2 grant ANR-11-BS02-0005. We would like to thank all the partners of this project for fruitful discussions.

References

1. Björdal, G., Monette, J.-N., Flener, P., Pearson, J.: A constraint-based local search backend for minizinc. Constraints **20**(3), 325–345 (2015)
2. Cameron, S., Culley, R.: Determining the minimum translational distance between two convex polyhedra. In: Proceedings of the IEEE International Conference on Robotics and Automation, vol. 3, pp. 591–596, April 1986
3. Castillo, I., Kampas, F.J., Pintér, J.D.: Solving circle packing problems by global optimization: numerical results and industrial applications. Eur. J. Oper. Res. **191**(3), 786–802 (2008)
4. Chernov, N., Stoyan, Y., Romanova, T.: Mathematical model and efficient algorithms for object packing problems. Comput. Geom. **43**, 535–553 (2010)
5. Dobkin, D., Hershberger, J., Kirkpatrick, D., Suri, S.: Computing the intersection-depth of polyhedra. Algorithmica **9**, 518–533 (1993)
6. Hansen, N., Ostermeier, A.: Completely derandomized self-adaptation in evolution strategies. Evol. Comput. **9**(2), 159–195 (2001)
7. Hentenryck, P.V., Michel, L.: Synthesis of constraint-based local search algorithms from high-level models. In: Proceeding of the AAAI, pp. 273–278 (2007)
8. Martinez, T., Fages, F.: On translating minizinc constraint models into fitness function for evolutionary algorithms: application to continuous placement problems. In: Proceedings of the Sixth Workshop on Bin Packing and Placement Constraints, BPPC 2015, Associated to CP 2015, September 2015
9. Martinez, T., Vitorino, L., Fages, F., Aggoun, A.: On solving mixed shapes packing problems by continuous optimization with the cma evolution strategy. In: Proceedings of the First Computational Intelligence BRICS Congress, BRICS-CCI 2013, pp. 515–521. IEEE Press, September 2013
10. Michel, L., Van Hentenryck, P.: The comet programming language and system. In: van Beek, P. (ed.) CP 2005. LNCS, vol. 3709, p. 881. Springer, Heidelberg (2005)
11. Nethercote, N., Stuckey, P.J., Becket, R., Brand, S., Duck, G.J., Tack, G.R.: MiniZinc: towards a standard CP modelling language. In: Bessière, C. (ed.) CP 2007. LNCS, vol. 4741, pp. 529–543. Springer, Heidelberg (2007)
12. Salas, I., Chabert, G.: Packing curved objects. In: Proceedings of the 24th International Joint Conference on Artificial Intelligence, IJCAI 2015, Buenos Aires, Argentina (2015)
13. Sederberg, T.W.: Chapter 7, planar curve intersection. Technical report, Computer Aided Geometric Design Course Notes (2011)

Constructions and In-Place Operations
for MDDs Based Constraints

Guillaume Perez and Jean-Charles Régin[✉]

Université Nice-Sophia Antipolis,
CNRS, I3S UMR 7271, 06900 Sophia Antipolis, France
guillaume.perez06@gmail.com, jcregin@gmail.com

Abstract. This papers extends in three ways our previous work about efficient operations on Multi-valued Decision Diagrams (MDD) for building Constraint Programming models. First, we improve the existing methods for transforming a set of tuples, Global Cut Seeds or sequences of tuples into MDDs. Then, we present in-place algorithms for adding and deleting tuples from an MDD. Finally, we describe an incremental version of an algorithm which reduces an MDD. We show on a real-life application that in-place operations on MDDs combined with this incremental algorithm outperform classical operations. Furthermore, we give some experimental results showing that the creation algorithms we propose strongly improve upon existing ones.

1 Introduction

Table constraints are useful constraints for modeling and solving many real-world problems. They are explicitly defined by the set of elements of the Cartesian product of the variables, also called tuples, that are allowed. The complexity of arc consistency algorithm associated with table constraints mainly depends on the number of involved tuples. Thus, Cheng and Yap proposed to compress the tuple set of the constraint by using Multi-valued Decision Diagrams (MDD) leading to MDD-based constraints. They designed mddc, one of the first filtering algorithms establishing arc consistency for them [7,8]. Recently, we have presented MDD-4R, a new algorithm which improves mddc [17]. MDD-4R proceeds like GAC-4R (an efficient arc consistency algorithm for table constraints) and, unlike mddc, maintains the MDD during the search for a solution. MDD-4R outperforms table constraints when the compression is effective.

MDDs can also be directly used to express complex constraints that cannot be represented by Table constraints because the number of tuples would be exponential. We have introduced efficient algorithms for creating and reducing an MDD and some powerful algorithms for combining MDDs [18]. Thanks to these new algorithms, some experiments based on real-life applications have shown that the MDD approach becomes competitive with ad-hoc approaches like the filtering algorithms associated with the regular or the knapsack constraints. More precisely we have shown that modeling a complex problem by a succession

© Springer International Publishing Switzerland 2016
C.-G. Quimper (Ed.): CPAIOR 2016, LNCS 9676, pp. 279–293, 2016.
DOI: 10.1007/978-3-319-33954-2_20

of operations between MDDs may be a competitive approach with the design of a complex ad-hoc algorithm.

In this paper, we extends our work by showing that MDDs can also be used to efficiently implement partially compressed table constraints like the ones defined by Global Cut Seeds (GCS) or tuple sequences and by proposing some in-place algorithms for combining MDDs in order to avoid using intermediate MDDs, and by introducing an incremental reduction algorithm.

Table constraints can be specified either directly, by input from the user, or indirectly by synthesizing other constraints or subproblems [14,15]. They have been reinforced in order to deal either from tuple sets or from sequences of tuples [10,12,19]. This has two advantages: it improves their expressiveness and it reduces the number of tuples that are explicitly used and so decreases the practical complexity of the filtering algorithms because they mainly depends on that number.

GCSs and tuple sequences are partial compression of table constraints. This compression can be improved by transforming tables defined by GCS or tuple sequences into MDDs. In the first part of this paper we propose such transformations and we show that the obtained MDDs always uses less space than a set of GCS or tuple sequence for representing the same table. We will also present the first linear algorithm for building an MDD from a list of tuples.

Next, we consider in-place deletion and addition operations, that is operations that do not create a new MDD. Instead they directly modify the current MDD. In-place operations have three advantages: it avoids some memory consumption, it decreases the computation time, and it allows the design of more efficient reduction algorithms because they can be incremental. In this part, we show that the addition and the deletion of one tuple from an MDD can be efficiently done by using the method which consists of isolating the path of the MDD corresponding to the tuple in case of deletion and to the common prefix of the tuple in case of addition. These operations make addition/deletion operations easier on MDDs. Then, we generalize the algorithm for the addition/deletion of a set of tuples.

After each modification of an MDD the reduction operation must be applied and since the deletion or the addition of tuples may modify only a few nodes, we introduce IPREDUCE an incremental version of the reduction operation which allows us to reduce the complexity of the pair of operations formed by the modification and the reduction.

Before concluding, we present some experiments on a real life application showing some strong improvements brought by our algorithms notably compared to the ones previously proposed. We also empirically establish the advantages of the new creation algorithms we propose.

2   Background

MDD. Multi-valued decision diagram (MDD) is a data structure for representing discrete functions. It is a multiple-valued extension of BDDs [6]. An MDD, as used in CP [1,3,11,13,14], is a rooted directed acyclic graph (DAG) used to represent some multi-valued function $f : \{0...d-1\}^r \rightarrow \{true, false\}$, based on a given integer d. Given the r input variables, the DAG representation is designed to contain r layers of nodes, such that each variable is represented at a specific layer of the graph. Each node on a given layer has at most d outgoing arcs to nodes in the next layer of the graph (i.e. one per value). We will denote by $L[i]$ the nodes in layer i and by $\omega^+(x)$ the set of outgoing arcs of the node x. Each arc is labeled by its corresponding value. The final layer is represented by the true terminal node (the false terminal node is typically omitted). There is an equivalence between $f(v_1, ..., v_r) = true$ and the existence of a path from the root node to the true terminal node whose arcs are labeled $v_1, ..., v_r$. Nodes without any outgoing arc or without any incoming arc are removed.

MDD Constraint. In an MDD constraint, the MDD models the set of tuples satisfying the constraint, such that every path from the root to the true terminal node corresponds to an allowed tuple. Each variable of the MDD corresponds to a variable of the constraint. An arc associated with an MDD variable corresponds to a value of the corresponding variable of the constraint. Figure 1 gives the MDD representing the tuples {a,a}, {a,b}, {c,a}, {c,b} and {c,c}. For each tuple, there is a path from the root node (node 0) to the terminal node (node *tt*) which is labeled by the tuple values.

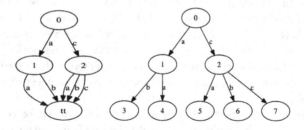

Fig. 1. An MDD (left graph) and a trie (right graph) representing the tuple set {{a,a},{a,b},{c,a},{c,b},{c,c}}

MDD Reduction. The reduction of an MDD is one of the most important operations. It consists of merging equivalent nodes, i.e. nodes having the same set of outgoing neighbors associated with the same labels. Usually, a reduction algorithm merges nodes until there is no more equivalent nodes. Most of the time, only reduced MDDs are considered mainly because they are smaller. Figure 5 exhibits an MDD having two equivalent nodes: *b* and *e*. These nodes will be merged by the reduction operation. Note that the reduction operation cannot increase the number of nodes or arcs. Recently, a new reduction algorithm with a linear space and time complexity has been proposed [18].

For convenience, we will denote by d the maximum number of values in the domain of a variable; and by (x, v, y) an arc from x to y labeled by v.

3 Transformations

In this section, we improve existing algorithms for building MDDs from tuple sets and we introduce new algorithms for building MDDs from compressed tuple sets.

3.1 From Trie to MDD

A trie is a data structure used by Gent et al. for compressing tuple sets [12]. Each path from the root to a leaf represents an allowed tuple. A trie representing a set of T tuples will have $|T|$ leaves. Each variable corresponds to a layer of the trie. A node has a maximum of d children, where d is the size of the domain of the corresponding variable of the node. An example of trie is given in Fig. 1. A trie can be transformed into an MDD by merging all the leaves into the terminal node tt and by applying the reduction operation [7].

3.2 From Table to MDD

A table is a data structure where each row represents a tuple and where each column corresponds to a value of a variable.

Cheng and Yap build an MDD from a table by defining a trie. Tuples are successively added to the trie. First, a common node is created: the root of the trie. Then paths starting from the root are created. The rooted subpaths common to several tuples are merged together in order to be represented only once. Afterwards, all the leaves are merged and the MDD is reduced. The drawback of this approach is the addition of a tuple, because we need to compute the common subpath of the tuple and the MDD. This operation can be performed in linear time only if we have d entries per node, so we increase the space complexity. Alternatively, we can keep a linear space complexity if we accept to increase the time complexity.

We propose a simple method with a linear time and space complexity: we lexicographically sort the table and we build the trie from the sorted table. Here is an example:

table						sorted table						trie				
a	a	c	a	a		a	a	b	a	b		a	a	b	a	b
a	b	a	b	b		a	a	b	a	c						c
a	a	b	a	c		a	a	c	a	a				c	a	a
a	a	b	a	b		a	b	a	a	b		b	a	a	b	
a	b	a	a	b		a	b	a	b	b				b	b	

This can be done efficiently because all tuples are consecutive and so there is no need to search for any position for a tuple: the last one is always the correct one. So we do not need the random access to children and this step can be achieved in linear time. In addition, the sort can be performed in linear time because a tuple can be viewed as numbers having r digits where a digit can take on up to d values. Thus we can sort a table containing t tuples in $O(r(t+d))$ by using a radix sort, which is linear in its size. Since, the merge of the leaves and the reduction can be performed in linear time, we obtain a linear time algorithm.

3.3   From GCS and Tuple Sequence to MDD

Compressed tuples improve the expressiveness of table constraints and reduce the complexity of the filtering algorithms. Therefore, it is interesting to represent them by MDDs in order to reinforce the compression.

A GCS (Global Cut Seed), is a compact representation of a tuple set [10]. A GCS is defined by a set of value sets: $\{\{v_{1,1}, v_{1,2}, ..., v_{1,k_1}\}, ..., \{v_{n,1}, v_{n,2}, ..., v_{n,k_n}\}\}$, where each value set corresponds to a variable. The Cartesian product of these sets defines the represented tuples. For instance, given $D = \{1,2,3,4\}$, the GCS $c = \{D, D, D, D\}$ represents the tuple set $\{\{1,1,1,1\}, \{1,1,1,2\}, ..., \{4,4,4,3\}, \{4,4,4,4\}\}$. One GCS may represent an exponential number of tuples. However all the tuples cannot be compressed by only one GCS. Two tuples can be represented by the same GCS if they have a Hamming distance equals to 1. For instance, the tuples $\{1,1,1\}$ and $\{1,1,2\}$ may be compressed into $\{1,1,\{1,2\}\}$. By contrast the tuples $\{1,1,1\}$ and $\{1,2,2\}$ have an Hamming distance equals to 2 and so cannot be represented by only one GCS. So, the compression of a table by a set of GCSs may require a huge number of GCSs. In order to remedy this problem, tuple sequences have been introduced [19]. They generalize GCSs.

A tuple sequence encapsulates a GCS and two tuples: t_{min} a minimum tuple, and t_{max} a maximum tuple. It bounds the lexicographic enumeration of the tuples of the GCS by these two tuples. For instance, let $D = \{1, 2, 3, 4\}$ then the tuple sequence $s = \{\{D, D, D, D\}, \{1, 2, 2, 2\}, \{3, 1, 3, 2\}\}$ represents the tuple set $\{\{1,2,2,2\}, \{1,2,2,3\}, ..., \{3,1,3,1\}, \{3,1,3,2\}\}$.

Since a tuple sequence is a generalization of a GCS, a method transforming a tuple sequence into an MDD could also be used for transforming a GCS into an MDD.

First, we propose an algorithm for representing one tuple sequence by an MDD. Then, we will show how we can deal with several tuple sequences. Let $s = (g, t_{min}, t_{max})$ be a tuple sequence. For transforming s into an MDD we introduce special nodes: wild card nodes. There is at most one wild card node per layer i which is denoted by $w[i]$. The wild card nodes are linked together. All the arcs outgoing from $w[i]$ are incoming arcs of node $w[i + 1]$ and all arcs outgoing $w[n - 1]$ are incoming arcs of tt.

The MDD representing s is built in three steps:

1. The paths corresponding to t_{min} and t_{max} are created.
2. Arcs from the nodes of the paths previously created to wild card nodes are created as follows. Consider the path created for t_{min}. For each layer i, let $val[i]$ be the value set of g for the layer i. For each value $a \in val[i]$ such that $a > t_{min}[i]$ we create an arc from the node n_i of the path representing t_{min} to the wild card node $w[i+1]$. We repeat this process for the path created for t_{max}. In addition, we add a particular treatment when a node is shared by the two initial paths: instead of considering all values of $val[i]$, we consider only the values in the interval $val[i] \cap]t_{min}[i], t_{max}[i][$.
3. From nodes $w[i]$ to node $w[i + 1]$ we add as many arcs as there are values in $val[i + 1]$.

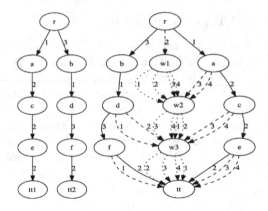

Fig. 2. Creation of an MDD from a tuple sequence

Figure 2 shows the resulting MDD. The left graph contains the two paths representing the minimum and maximum tuples. The right graph represents with dashed lines the added arcs to wild card nodes. For instance, for node a each value in $\{1,2,3,4\}$ greater than 2 labels an arc to node w_2. Arcs joining wild card nodes together and with tt are represented by dotted lines.

Let r be the number of involved variables. The number of nodes of the obtained MDD is bounded by $3(r-1)+2$. There are $2r$ arcs for the paths corresponding to t_{min} and t_{max}. There are at most $|val[i]|$ arcs from nodes of the t_{min} (resp. t_{max}) path to wild card nodes; There are $|val[i+1]|$ arcs from node $w[i]$ to node $w[i+1]$. Thus, there are at most $2\sum_{i=1}^{r}|val[i]|+2r$ arcs in the MDD. This is equivalent to the number of values of the tuple sequence.

Now, suppose that we have a set of tuple sequences. We can consider successively each tuple sequence and build for each sequence an MDD with the previous algorithm. Then, there are two possibilities. Either the tuple sequences are disjoint or not. The former case arises frequently (for instance when the tuple sequences represent a set of forbidden tuples). We just have to make the union of MDDs. This can be easily done because they are disjoint. The resulting MDD has a space complexity equivalent to the set of tuple sequences and we have:

Property 1. *A set of disjoint tuple sequences can be represented by an MDD having an equivalent space complexity.*

The latter case is more complex. A set of disjoint tuple sequences may be computed from a set of non disjoint tuple sequences and each disjoint tuple sequence can be represented by an MDD. Nevertheless, it may create an exponential number of tuple sequences [19] so an exponential number of MDDs.

4   Addition and Deletions of Tuples from an MDD

In this section, we define in-place algorithms for the addition/deletion of tuples from an MDD. Some work have been carried out for performing operations on BDDs. For instance, Bryant define some algorithms for applying different operators [5,6]. However, the described algorithms are not in-place (i.e. there is the creation of a resulting BDD) and it is not easy to generalize some algorithms designed for BDDs to MDDs mainly because some Booleans rules are no longer true when we have d values in the domain and because the complexity of some algorithms is multiplied by $O(d)$ when dealing with d values. Some generic algorithms have been proposed for applying operators on MDDs [2,18], but they are not in-place. An in-place algorithm has been given by Ciré and Hooker [9] but it only deal with partial assignments and has no incremental reduction.

4.1   Deletion of Tuples from an MDD

First, we give an algorithm for deleting one tuple from an MDD. Then, we generalize it for a set of tuples.

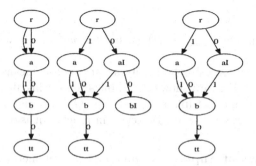

Fig. 3. Tuple {0,0,0} is removed from the left MDD. The isolation of the path corresponding to the tuple is performed (middle MDD) and then the reduction is applied (right MDD). Nodes aI and bI are created from nodes a and b during the path isolation.

The deletion of a tuple τ from an MDD is based on an operation named path isolation, which is a kind of local decompression. The idea is to build a specific path whose arcs are labeled by the values of τ. Furthermore, arcs equivalent to the ones of the isolated path are deleted from the MDD. After the isolation process, the MDD is reduced. Let $\tau[i]$ be the value for the variable $x[i]$. The isolation is performed in 3 steps:

Step 1. We identify $a_1 = (root, \tau[1], n_1)$ the arc of the first layer labeled by $\tau[1]$ the first value of the tuple. We create the node ne_1, the arc $(root, \tau[1], ne_1)$ and we delete the arc a_1. We set x_{mdd} (a node of the MDD) to n_1 and x_{path} (an isolated node) to ne_1.

Step 2. For each layer i from 2 to $r - 1$ we repeat the following operation. We identify $a_i = (x_{mdd}, \tau[i], n_{i+1})$ the outgoing arc from x_{mdd} labeled with $\tau[i]$. We create the node ne_{i+1} and the arc $(x_{path}, \tau[i], ne_{i+1})$. For each arc (x_{mdd}, w, y) such that $w \neq \tau[i]$ we create the arc (x_{path}, w, y). We set x_{mdd} to n_{i+1} and x_{path} to ne_{i+1}.

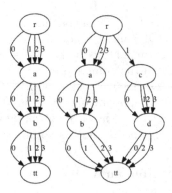

Fig. 4. The left MDD represents all the possible tuples for the values {0,1,2,3}. The right MDD represents the deletion of the GCS {1,{0,1,2,3},1} from the left MDD.

Step 3. For each arc (x_{mdd}, w, tt) such that $w \neq \tau[r]$ we create the arc (x_{path}, w, tt).

If at any moment we cannot identify an arc then it means that τ does not belong to the MDD. Figure 3 shows the application of this algorithm. The complexity of the deletion of a tuple is bounded by $O(rd)$ because for each isolated node we need to recreate its arcs. However, in practice it is often close to $O(d)$.

Deletion of a Set of Tuples. We propose a better method than repeating the previous algorithm for each tuple. We transform the set of tuples into an MDD and we subtract this new MDD from the initial one by following the same steps of the previous algorithm. We isolate nodes having a common path in both MDDs, then we remove the common arcs to the isolated nodes of the second last layer. At last, we call the incremental reduction algorithm.

Figure 4 shows the subtraction of the GCS {1,{0,1,2,3},1} from the MDD representing all the tuples possible for the values {0,1,2,3}. The GCS is isolated from the MDD. Then, the deletion of the arc labeled 1 of node d correspond to the deletion of only the tuples contained in the GCS. It is difficult to bound the complexity of the deletion of T tuples, because the MDD created from them may compress the information.

4.2   Addition of Tuples to an MDD

The addition of tuples into MDD follows the same principles as for the deletion. In this case, the isolated path contains arcs labeled by the values of the tuple that must be added. It is performed by applying the same steps as for the deletion.

First, we consider the addition of one tuple τ. The two first steps are very similar to the ones of the deletion. Excepted that at a point, there will be no more path in the MDD having the same subpath as τ. Otherwise, it would mean that τ is already in the MDD. Thus, at a certain moment we will not be able to identify any arc $(x_{mdd}, \tau[i], n_{i+1})$ as in step 2 in the deletion algorithm.

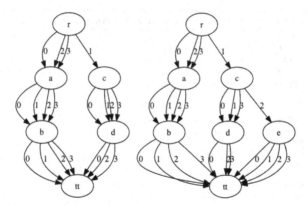

Fig. 5. The right MDD represents the addition of the tuple {1,2,1} to the left MDD, before the reduction.

When this case arises we can stop step 2 and directly create the path from the current isolated node to the terminal node. This path will be labeled by the values of τ for the remaining layers. Step 3 can be skipped. At last, we call the incremental reduction algorithm. The complexity of the addition of a tuple is in $O(rd)$ because for each isolated node we need to recreate its arcs.

Addition of a Set of Tuples. Let mdd_1 be the initial MDD. We transform the set of tuples into an MDD, named mdd_2. We add mdd_2 to mdd_1 by following the same steps as for the previous algorithm. We isolate nodes having a common path in both MDDs. When an arc belongs to mdd_2, we create an isolated node and we create an arc from the current isolated node to it. When an arc belongs only to mdd_1, we create an arc from the current isolated node to the node in mdd_1.

Figure 5 shows the effect of the addition of the tuple {1,2,1} in the MDD given in Fig. 4. We can see the usefulness of the path isolation for avoiding the addition of the tuples {1,{0,1,3},1}. The right MDD shows the impact of the reduction on the MDD: nodes e and b are merged because they have the same outgoing arcs. It is difficult to bound the complexity of the addition of T tuples, because the MDD created from them may compress the information.

Algorithm 1 is a possible implementation of the in-place deletion and addition operations.

4.3   Incremental Reduction

A reduction step is needed after the deletion/addition of tuples. Using a generic algorithm is costly because it will traverse all the nodes of the MDD and merge the equivalent ones. Since we consider that we add/delete tuples from an MDD which is reduced we can save some computations for the reduction applied after the operation. Only certain nodes have to be considered:

Algorithm 1. In-place deletion and addition algorithms

DELETION(L, mdd_1, mdd_2)

  for each $(root(mdd_1), v, y_1) \in \omega^+(root(mdd_1))$ **do**
    if $\exists (root(mdd_2), v, y_2) \in \omega^+(root(mdd_2))$ **then**
      ADDARCANDNODE($L, 1, root(mdd_1), v, y_1, y_2$)
      DELETEARC($root(mdd_1), v, y_1$)

  for each $i \in 1..r - 2$ **do**
    $L[i] \leftarrow \emptyset$
    for each $node\ x \in L[i - 1]$ **do**
      get x_1 and x_2 from $x = (x_1, x_2)$
      for each $(x_1, v, y_1) \in \omega^+(x_1)$ **do**
        if $\exists (x_2, v, y_2) \in \omega^+(x_2)$ **then**
          ADDARCANDNODE(L, i, x, v, y_1, y_2)

        else CREATEARC(L, i, x, v, y_1)

  for each $node\ x \in L[r - 1]$ **do**
    get x_1 and x_2 from $x = (x_1, x_2)$
    for each $(x_1, v, tt) \in \omega^+(x_1)$ **do**
      if $\nexists (x_2, v, y_2) \in \omega^+(x_2)$ **then**
        CREATEARC(L, r, x, v, tt)

  IPREDUCE(L)

ADDITION(L, mdd_1, mdd_2)

  for each $v \in \omega^+(root(mdd_1)) \cup \omega^+(root(mdd_2))$ **do**
    if $\exists (root(mdd_1), v, y_1) \in \omega^+(root(mdd_1))$ **then**
      if $\exists (root(mdd_2), v, y_2) \in \omega^+(root(mdd_2))$ **then**
        ADDARCANDNODE($L, 1, root(mdd_1), v, y_1, y_2$)
        DELETEARC($L, i, root(mdd_1), v, y_1$)

    else ADDARCANDNODE($L, 1, root(mdd_1), v, nil, y_2$)

  for each $i \in 1..r - 2$ **do**
    $L[i] \leftarrow \emptyset$ **for each** $node\ x \in L[i - 1]$ **do**
      get x_1 and x_2 from $x = (x_1, x_2)$
      // If x_1 is nil then $\omega^+(x_1)$ is empty
      for each $v \in \omega^+(x_1) \cup \omega^+(x_2)$ **do**
        if $\exists (x_1, v, y_1)\ \omega^+(x_1)$ **then**
          if $\exists (x_2, v, y_2) \in \omega^+(x_2)$ **then**
            ADDARCANDNODE(L, i, x, v, y_1, y_2)

          else CREATEARC(L, i, x, v, y_1)

        else ADDARCANDNODE(L, i, x, v, nil, y_2)

  for each $node\ x \in L[r - 1]$ **do**
    // If x_1 is nil then $\omega^+(x_1)$ is empty
    for each $v \in \omega^+(x_1) \cup \omega^+(x_2)$ **do**
      CREATEARC(L, i, x, v, tt)

  IPREDUCE(L)

ADDARCANDNODE(L, i, x, y_1, v, y_2)

  if $\nexists y \in L[i]\ s.t.\ y = (y_1, y_2)$ **then**
    y $\leftarrow$ CREATENODE(y_1, y_2)
    add y to $L[i]$
  CREATEARC(x, v, y)

Property 2. *After the application of an in-place operator, if two nodes are merged then one of these nodes must be an isolated node.*

Proof. Two nodes are merged if and only if they have the same set of outgoing neighbors associated with the same labels. Before the operation the MDD is reduced, so no two pairs of nodes can be merged. By induction from the terminal node to the root we prove the property: it is obvious for two nodes of the last layer. Then, from the definition of the path isolation, if a merge exists then it necessarily involved an isolated node because it was not possible to merge nodes of the MDD existing before the operation.   □

Thus, we can easily adapt pREDUCE algorithm [18] by rejecting a pack of nodes if it does not involve any isolated node. In addition, it is easy to identify isolated nodes because they belong to the list L of the in-place algorithms. The advantage of this approach is that the reduction step does not increase the complexity of the addition or deletion operations. This new algorithm is named IPREDUCE.

5   Experiments

The algorithms have been developed on top of or-tools 3158, a constraint programming solver developed by Google. The experiments have been executed on a MacBook Pro (Intel Core I7, 2.3 GHz, 8 GB memory).

Real Life Application. We consider the problem given in [16] which deals with Markov Sequence Generations on corpus having more than 10,000 words. The goal is to generate phrases having 24 words where all successions of 4 words come from the corpus and where there is no sequence of more than 8 words coming from the corpus. This problem can be modeled by using MDDs expressing sequences of words [18]. Values of variables are words of the corpus, so we have a huge number of values. From an initial MDD representing allowed sequences of 4 words, 20 intersections of MDDs are performed until obtaining mdd_r the final MDD. The main issue with this approach is the size of the MDDs because mdd_r has 1,208,219 nodes and 188,035,203 arcs. With the operators given in [18] were able to compute mdd_r in 425s. This requires to perform 20 intersections and 20 reductions of huge MDDs.

In this problem, twice a deletion followed by a reduction of the MDD are made. The results are given in the table below. Times are expressed in seconds. In the "Classic" columns, the algorithms given in [18] are used whereas the algorithm proposed in this paper are used in the "In-place" columns. These results clearly show the advantage of using the new algorithms. Using in-place algorithms instead of building intermediate MDDs reduces the memory consumption of the resolution of the whole problem from 52GB to 32GB.

Operations and Reduction. We propose to compare the performance of the classical and the in-place algorithms and the performance of the classical and the incremental reduction algorithms. We use random instances obtained from the

	Classic			In-place		
	deletion	reduction	total	deletion	reduction	total
First Operation	2	1.7	3.7	**1.3**	**0.9**	**2.2**
Second Operation	23.9	14.6	38.5	**1.5**	**6.3**	**7.8**

Table 1. Arity 12, domain size 10. Average deletion time (s) for random instances.

Instances	Classic		In-place	
	deletion	reduction	deletion	reduction
30*300K-300K	35,4	4.2	24.8	1.8
300K - 1K	5.3	0.7	1.2	0.6
90K-30K	2.1	0.2	1.6	0.2
300K-10	4.7	0.6	0.002	0.2

real life instances. The first number corresponds to the number of the tuples represented by the MDD whereas the second number is the number of tuples that are removed from the MDD. Table 1 gives the results we obtain. Our algorithms clearly improve the previous ones.

We also proposes a table summarizing the advantages of the different algorithms. We add results for the BDD and MDD packages proposed in [4,20] (See column Bryant). "P&R15" represents the results we previously obtained and "in-place" column corresponds to the new algorithms. Table 2 gives some resultats for MDD representing 10,000s tuples. Note that huge MDD are not tractable with some old methods.

From Tables to MDDs. We study the performance of the new creation algorithms. The times for sorting the elements are included into our results. First, we tested our algorithm on the instances of the XCSP competition. We give the results for the most representative ones. Sorted creation corresponds to our algorithm, unsorted creation is the classical creation.

These experiments show that it is always better to sort the table and use our creation algorithm.

Table 2. Arity 12, domain size 10. Average deletion time (ms) for random instances.

#tuples	#deleted	Bryant	P&R15	in-place
20000	1000	159	11.5	6
40000	2000	291	40	21
40000	20000	663	51	33
80000	40000	2643	174	114
40000	10	466	185	19

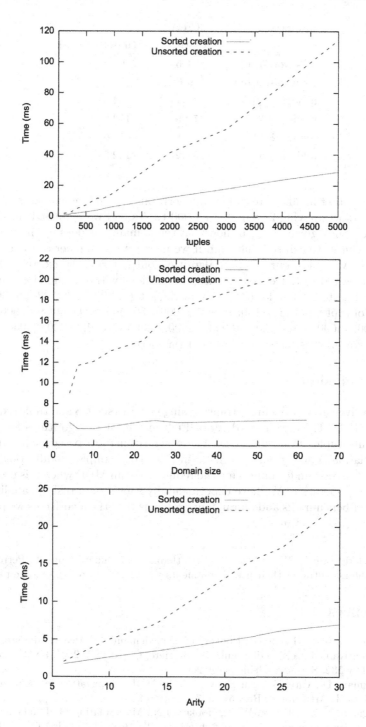

Fig. 6. Sorted vs unsorted creation

Instances	Creation	
	sorted (ms)	unsorted (ms)
crossword-m1c-ogd	**31.5**	66.2
crossword-m1c-uk-vg	**9.6**	23.1
nonogram-gp	**25.1**	34.5
rand-10-60-20-30	**70.9**	179.9
bdd-21-2713	**8.1**	11.6
bdd-21-133	**98.23**	122.3

On the other hand, we tested both algorithms on random instances. We have tested instances having 22 variables, 1,000 tuples and we increased the domain size. The results given in Fig. 6 show that the domain size does not influence the creation time. We can see that even if we increase the number of tuples or the number of variables, our creation algorithm outperforms the existing one. We have also tested instances for all the combinations with domain size in the set {2, 4, 8, 12, 20, 25, 30, 45, 60}, arity in the set {6, 10, 14, 18, 22, 25, 30} and number of tuples in the set{30, 100, 150, 200, 250, 300, 500, 700, 800, 900, 1000, 2000, 3000, 4000, 5000, 7500, 10000, 12500, 15000, 17500, 20000, 24000, 28000, 30000}. For all these cases, our method was better.

6 Conclusion

We have given an algorithm for transforming tuple sets, GCS and tuple sequences into an MDD. Then, we have described efficient in-place algorithms for adding or deleting tuples from an MDD. These algorithms are based on the idea of path isolation. Furthermore, we have introduced a simple modification of the pREDUCE algorithm for improving the reduction of an MDD when it is used after an in place operation. We have experimentally shown on a real life application, on a set of benchmarks and on random problems that the algorithms we propose outperform the existing ones.

Acknowledgments. We would like to thank very much Laurent Perron and Christophe Lecoutre for their useful comments which helped to improve the paper.

References

1. Andersen, H.R., Hadzic, T., Hooker, J.N., Tiedemann, P.: A constraint store based on multivalued decision diagrams. In: Bessière, C. (ed.) CP 2007. LNCS, vol. 4741, pp. 118–132. Springer, Heidelberg (2007)
2. Bergman, D., Cire, A., van Hoeve, W.-J.: MDD propagation for sequence constraints. J. Artif. Intell. Res. **50**, 697–722 (2014)
3. Bergman, D., van Hoeve, W.-J., Hooker, J.N.: Manipulating MDD relaxations for combinatorial optimization. In: Achterberg, T., Beck, J.C. (eds.) CPAIOR 2011. LNCS, vol. 6697, pp. 20–35. Springer, Heidelberg (2011)

4. Brace, K.S., Rudell, R.L., Bryant, R.E.: Efficient implementation of a BDD package. In: Proceedings of the 27th ACM/IEEE Design Automation Conference, pp. 40–45. ACM (1991)
5. Bryant, R.E.: Symbolic boolean manipulation with ordered binary decision diagrams. ACM Comput. Surv. **24**(3), 293–318 (1992)
6. Bryant, R.E.: Graph-based algorithms for boolean function manipulation. IEEE Trans. Comput. C **C35**(8), 677–691 (1986)
7. Cheng, K., Yap, R.: An MDD-based generalized arc consistency algorithm for positive and negative table constraints and some global constraints. Constraints **15**, 265–304 (2010)
8. Cheng, K.C.K., Yap, R.H.C.: Maintaining generalized arc consistency on ad hoc r-ary constraints. In: Stuckey, P.J. (ed.) CP 2008. LNCS, vol. 5202, pp. 509–523. Springer, Heidelberg (2008)
9. Ciré, A.A., Hooker, J.N.: The separation problem for binary decision diagrams. In: International Symposium on Artificial Intelligence and Mathematics, ISAIM 2014, Fort Lauderdale, FL, USA, 6–8 January 2014
10. Focacci, F., Milano, M.: Global cut framework for removing symmetries. In: Walsh, T. (ed.) CP 2001. LNCS, vol. 2239, pp. 77–92. Springer, Heidelberg (2001)
11. Gange, G., Stuckey, P., Szymanek, R.: MDD propagators with explanation. Constraints **16**, 407–429 (2011)
12. Gent, I., Jefferson, C., Miguel, I., Nightingale, P.: Data structures for generalised arc consistency for extensional constraints. In: Proceedings of AAAI 2007, pp. 191–197, Vancouver, Canada (2007)
13. Hadzic, T., Hooker, J.N., O'Sullivan, B., Tiedemann, P.: Approximate compilation of constraints into multivalued decision diagrams. In: Stuckey, P.J. (ed.) CP 2008. LNCS, vol. 5202, pp. 448–462. Springer, Heidelberg (2008)
14. Hoda, S., van Hoeve, W.-J., Hooker, J.N.: A systematic approach to MDD-based constraint programming. In: Cohen, D. (ed.) CP 2010. LNCS, vol. 6308, pp. 266–280. Springer, Heidelberg (2010)
15. Lhomme, O.: Practical reformulations with table constraints. In: ECAI, pp. 911–912 (2012)
16. Papadopoulos, A., Roy, P., Pachet, F.: Avoiding plagiarism in markov sequence generation. In: Proceeding of the Twenty-Eight AAAI Conference on Artificial Intelligence, pp. 2731–2737 (2014)
17. Perez, G., Régin, J.-C.: Improving GAC-4 for table and MDD constraints. In: O'Sullivan, B. (ed.) CP 2014. LNCS, vol. 8656, pp. 606–621. Springer, Heidelberg (2014)
18. Perez, G., Régin, J.-C.: Efficient operations on MDDs for building constraint programming models. In: International Joint Conference on Artificial Intelligence, IJCAI 2015, pp. 374–380, Argentina (2015)
19. Régin, J.-C.: Improving the expressiveness of table constraints. In: CP 2011, Proceedings Workshop ModRef 2011 (2011)
20. Srinivasan, A., Ham, T., Malik, S., Brayton, R.K.: Algorithms for discrete function manipulation. In: IEEE International Conference on Computer-Aided Design, ICCAD 1990. Digest of Technical Papers, pp. 92–95. IEEE (1990)

Balancing Nursing Workload
by Constraint Programming

Gilles Pesant[1,2(✉)]

[1] École Polytechnique de Montréal, Montreal, Canada
gilles.pesant@polymtl.ca
[2] CIRRELT, Université de Montréal, Montreal, Canada

Abstract. Nursing workload in hospitals has an impact on the quality of care and on job satisfaction. Understandably there has been much recent research on improving the staffing and nurse-patient assignment decisions in increasingly realistic settings. On a version of the nurse-patient assignment problem given a fixed staffing of neonatal intensive care units, constraint programming (CP) was shown to perform better than competing optimization methods. In this paper we take advantage of recent improvements to the CP approach to solve the integrated problem of staffing and nurse-patient assignment. We then consider a more difficult but also more realistic version of the problem in which patients are categorized into a small number of types and the workload associated with each type is nurse-dependent.

1 Introduction

Because of its impact on the quality of care, job satisfaction, and staff retention, nursing workload is a constant preoccupation in hospitals and it has received some recent attention in the scientific literature (e.g. [1,7,11]). Arguably the most important factor influencing nursing workload is patient acuity but others have been identified such as job interruption, patient turnover rate, and administrative paperwork [4]. If we define the workload of a nurse as the sum of the acuities of the patients he cares for, then we try to keep that value low but we also try to balance the individual workloads in order to avoid an overworked nurse and to show fairness between staff members.

Given a set of patients distributed in a number of units and an available nursing staff (previously determined as a result of *nurse rostering*), the *nurse staffing problem* consists of assigning an appropriate number of nurses to each unit. The *nurse-patient assignment problem* (NPA) then assigns patients to nurses. The number of patients per nurse may be as low as two or three in an intensive care unit [3] or around six in oncology and surgery units [10]. These two levels of assignments must be made so as to balance the resulting workloads. The typical time frame for the decision maker is one to two hours to perform staffing and 30 min to perform NPA [3].

This short paper focuses on solving the integrated nurse staffing and NPA problem in a neonatal intensive care setting (Sect. 2) and a version of the NPA with nurse-dependent patient acuities (Sect. 3).

© Springer International Publishing Switzerland 2016
C.-G. Quimper (Ed.): CPAIOR 2016, LNCS 9676, pp. 294–302, 2016.
DOI: 10.1007/978-3-319-33954-2_21

Table 1. CP models for the NPA (left) and staffing (right) problems

minimize σ s.t.

$\mathtt{spread}(\{w_j\}, \sum a_i/|N|, \sigma)$

$\mathtt{gcc}(\{n_i\}, \langle[p_{\min}, p_{\max}], \ldots, [p_{\min}, p_{\max}]\rangle)$

$\mathtt{binpacking}(\langle n_i \rangle, \langle a_i \rangle, \langle w_j \rangle)$

$w_j \geq w_{j+1}$ $j \in \{1, 2, \ldots, |N| - 1\}$

$n_i \in N$ $i \in P$

$w_j \in \mathbb{N}$ $j \in N$

$x_k \geq \lceil |P_k|/p_{\max} \rceil$ $k \in Z$

$x_k \leq \lceil |P_k|/p_{\max} \rceil + f$ $k \in Z$

$\sum_{k \in Z} x_k = |N|$

$\sum_{k \in Z} LB_{k, x_k} < ub$

$x_k \in \{1, \ldots, |N|\}$ $k \in Z$

2   Integrated Staffing and Nurse-Patient Assignment

The problem originally proposed by Mullinax and Lawley [3] asks for a balanced workload for nurses being assigned patients requiring various amounts of care (acuity) in a neonatal intensive care unit. Patients each belong to a zone, a nurse can only work in one zone, and there are upper limits both on the number of patients assigned to a nurse and on the corresponding workload.

Mullinax and Lawley solve that problem as a mixed integer linear program with a linear objective function minimizing the sum of differences between minimum and maximum workloads in each zone, which may lead to imbalance between zones. Schaus et al. [8,9] describe a constraint programming model minimizing the standard deviation of workloads globally using the spread constraint [6]. They significantly improve the quality of solutions and the computational efficiency, solving two-zone instances optimally. For larger instances they first compute a staffing decision heuristically by solving a continuous relaxation of that problem and then solve each zone separately, often finding provably optimal solutions. Ku et al. [2] applied mixed integer quadratic programming and constraint integer programming (CIP). The latter, coupled with a variable ordering heuristic prioritizing the staffing and workload variables, solves two-zone instances significantly faster than the previous CP approach. A stronger filtering algorithm (achieving domain consistency) was recently proposed by Pesant [5] for the spread constraint and evaluated empirically on the NPA (i.e. on individual zones). It was found to solve instances one to two orders of magnitude faster than the CIP approach.[1] Building on that performance we investigate solving the integrated staffing and nurse-patient assignment problem.

2.1   Nurse-Patient Assignment

Our CP model shown in Table 1 on the left is standard: given the set of nurses N, the set of patients P, the list of patient acuities $\langle a_i \rangle$, and the minimum and maximum number of patients per nurse $p_{\min}$ and $p_{\max}$ respectively, we use one variable n_i per patient i indicating which nurse it is assigned to and one variable w_j per nurse j indicating his workload. To the usual constraints we add static symmetry breaking among nurses, enforce domain consistency on the spread

---

[1] Personal communication from the authors of [2].

constraint, and use a simple static branching heuristic that first selects the w_j variables in lexicographic order and then the n_i variables in decreasing order of acuity (values are selected in lexicographic order) [5].

2.2 Staffing

In order to solve the integrated problem exactly in principle we need to consider every staffing configuration. Fortunately most configurations are of poor quality and we can eliminate them implicitly by computing a lower bound on the standard deviation from partial configurations. We first use the heuristic staffing from Schaus et al. [8] to solve the NPA in each zone to optimality in order to provide a good upper bound. We then express the staffing problem as a constraint satisfaction problem coupled with the computation of a lower bound. For a given zone $k \in Z$ with its total patient acuity A_k and number of nurses x_k, Schaus et al. describe a lower bound on its contribution to the standard deviation, which we adapt here:

$$\alpha(\lceil A_k/x_k \rceil - \mu)^2 + \beta(\lfloor A_k/x_k \rfloor - \mu)^2$$

where $\mu = \sum a_i/|N|$ is the mean workload, $\alpha = A_k + x_k(1 - \lceil A_k/x_k \rceil)$, and $\beta = x_k - \alpha$. Summing them over all zones, dividing the result by the total number of nurses, and then taking its square root provides a bound on the standard deviation. We pre-compute these lower bounds in each zone for every possible value of x_k and put them in a matrix LB. Let P_k represent the set of patients in zone k, ub an upper bound on the deviation (provided by the best solution so far), and $f = |N| - \sum_{i \in Z} \lceil |P_i|/p_{\max} \rceil$ the number of nurses that are free to be assigned to any zone. Table 1 (right) gives our CP model for the staffing problem: every solution of this model is a staffing from which we solve a CP model for the NPA in each zone.

2.3 Results

The benchmark instances used in the literature are inspired from a neonatal intensive care unit, with an upper limit of 3 newborns per nurse and of 105 for the total workload of a nurse. They were randomly generated by Schaus et al. [8] using a realistic statistical model proposed in [3]. They range from 2 to 20 zones and up to 102 nurses and 258 patients. All experiments were run on Dual core AMD 2.1 GHz processors with 8 GB of RAM, using IBM ILOG Solver 6.7 as the CP solver.

We solve all ten 2-zone instances in an average of 0.12 s and 319 fails compared to 2.07 s (on a similar processor) and 9254 fails for Schaus et al. [8] using their two-step approach in which they fix the staffing decision heuristically.[2] Our approach did not require to evaluate more staffing configurations: all others

---

[2] For these 2-zone instances they can show that their solutions are optimal for the integrated problem. Their initial model combining staffing and nurse-patient assignment took about two orders of magnitude more time.

Table 2. Results on the three larger instances

Zones	Nurses	Patients	Mean	SD	Schaus and Régin [9]		This paper		
					Fails	Time(s)	Fails	Time(s)	Staffings
6	31	78	84.58	4.20	12019	0.57	1387	0.37	1
15	71	198	81.95	5.33	38651	2.27	784	0.46	1
20	102	258	82.71	5.54	1176852	25.17	291286	27.04	5

were discarded during enumeration based on the lower bound calculation. The 2-zone instances used by Ku et al. [2] are not the same but were generated with the same parameters. Their best approach solves the instances to optimality in an average of 74.05 s on a faster processor.

For the 3-zone instances we require an average of 0.40 s and 1323 fails compared to 0.48 s and 16528 fails for Schaus and Régin [9] but for the latter the problem is decomposed by zone and optimized separately. Schaus and Régin could verify the optimality of all their solutions except in the case of Instance 7 — this is indeed the only instance for which we needed to explore a second staffing configuration and we can confirm that their solution is optimal. Ku et al. [2] do not report results beyond two zones.

Table 2 reports results on the three larger instances: we give the size of each instance, its mean workload, the optimal standard deviation on the workloads, the performance of the zone-decomposition approach of Schaus and Régin [9], and the performance of our approach for the integrated staffing and NPA problem including the number of staffing configurations we had to explore. We notice that our approach scales very well: indeed the increase in the number of zones does not increase the size of the problem within a zone and only the 20-zone instance took significantly more time because we needed to explore several staffing configurations (and solve more zones). Still, it is remarkable that the good quality of the lower bounds and of the initial configuration considered keep the number of eligible configurations so low. The optimality of the solution for the 20-zone instance was unknown until now: we confirm that it is. We thus close all current benchmark instances for this problem.

We created a new set of ten instances that are harder to solve to optimality in the sense that the best staffing may be different from the one obtained by solving the continuous relaxation. They were generated on 6 zones using the same parameters as Schaus et al. [8] except for the probability of success in the binomial distribution used to generate the acuity of patients which we increased from 0.23 to 0.33, yielding a wider span of acuities. Table 3 reports the performance of our approach on these instances. We see that for some of them a few staffing configurations had to be explored and, more importantly, that the optimal standard deviation is sometimes noticeably lower than the first one obtained with the heuristic staffing (shown in bold).

Table 3. Results on ten harder 6-zone instances

Nurses	Patients	Mean	First SD	Optimal SD	Fails	Time(s)	Staffings
34	80	94.88	**6.09**	**6.04**	3183	1.05	2
38	88	94.16	5.82	5.82	1133	0.40	1
40	89	92.38	**6.23**	**5.16**	13655	14.25	5
40	88	96.48	**5.84**	**5.79**	212304	26.38	4
37	88	93.00	4.30	4.30	1748	0.62	1
39	93	94.92	4.07	4.07	26740	2.71	1
36	83	93.94	5.57	5.57	3885	0.48	1
39	87	93.49	5.41	5.41	2875	0.81	1
37	83	92.70	**5.45**	**5.08**	2200	1.85	2
35	83	89.46	3.99	3.99	1295	0.56	1

3   Nurse-Dependent Patient Acuity

Patient classification systems (PCS) are commonly used in hospitals to estimate the amount of care needed by each patient. For example AcuityPlus® classifies patients into six types according to a weighted sum of 26 acuity indicators. Sir et al. [10] argue that, because of differences in experience, training, or preferences, nurses may not equally perceive the acuity associated with a given patient type. Through a survey of nurses in oncology and surgery units, they found that there could indeed be quite a bit of variation in perceived acuity between nurses and advocate that nurse-dependent patient acuity should be used when balancing nursing workloads. In this section we consider a variant of the NPA where patients are grouped according to their type and the acuity associated with each patient type is nurse-dependent.

Because the acuity of care provided for a given patient type is not perceived uniformly across nurses, we cannot know in advance what the total workload nor the mean workload will be. Hence there are really two dimensions to the quality of a solution: we wish to keep the total workload of the nursing staff low in order to offer better care and to admit new patients more seamlessly, but we also wish to balance the workload between nurses so as to be fair. In such a situation a useful decision support tool will provide the Pareto optimal front so that the decision maker has a small set of attractive solutions to work with.

Working with a variable mean is problematic for the spread constraint: its filtering algorithms either assume a fixed mean or sustain a significant increase in their time complexity. We propose to solve a succession of fixed mean-workload problems where we gradually increase that mean. Finding a good starting mean proved important for the efficiency of our approach: we initially minimize the mean workload by recasting our problem as a Generalized Assignment Problem.

3.1 Generalized Assignment Problem as Lower Bound on Total Workload

If we relax the balancing aspect of our problem and simply minimize the total workload, we can express it as a generalized assignment problem in which the patients are the tasks and the nurses are the agents, with a capacity to perform multiple tasks corresponding to the minimum and maximum number of patients per nurse. We solve it using the Hungarian algorithm by framing it as the following simpler assignment problem: we make as many copies of each nurse as the maximum number of patients he can care for but add a penalty to the assignment costs for the copies in excess of the minimum number of patients required (to ensure that the minimum is reached). The resulting assignment, with its total workload and its standard deviation from the mean workload, gives us an initial point from which to proceed.

3.2 Solving Fixed Mean-Workload Instances

Each COP we solve imposes a total workload fixed to one unit less than that of the previously found solution (except in the case of our first solution from the Hungarian algorithm, which we try to improve) and a standard deviation upper bounded by that of the previous solution. We branch using the default smallest-domain-first variable selection heuristic and lexicographic value selection heuristic. We stop this iterative process when either we reach a standard deviation of zero or an upper bound on the total workload.

3.3 Empirical Evaluation

The data used by Sir et al. [10] is proprietary but we generated instances using their reported findings. Specifically we consider five patient types (the sixth type was not sufficiently represented in their data) and use the mean acuity associated with each type for the oncology unit as reported in Fig. 6 of their paper to draw nurse-dependent acuities from a normal distribution with standard deviation equal to 2.5 (loosely extrapolated from Table 6). The number of patients of each type is generated using a Poisson distribution with an expected value chosen so that the average number of patients per nurse is close to six, which is consistent with oncology and surgery units. The typical size of such a unit is reported as about 30 patients and 5 nurses. We generated ten instances of that size and ten smaller ones with 3 nurses (and about 18 patients).

All experiments were run on Dual core AMD 2.1 GHz processors with 8 GB of RAM, using IBM ILOG Solver 6.7 as the CP solver. Each COP was given up to 5 min to run to completion. The plot on the left at Fig. 1 presents the individual Pareto optimal fronts for the ten smaller instances. Observe that each instance admits a solution with perfect balance (standard deviation equal to zero). The highest point of each front corresponds to the first solution provided by the Hungarian algorithm; note that sometimes we can find a solution of same mean but with much better balance (e.g. for instance A a standard deviation

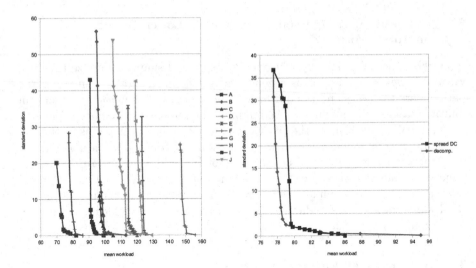

Fig. 1. Pareto solutions to the 3-nurse instances (left) and a comparison of solutions to one 5-nurse instance

Table 4. Solving the 3-nurse instances

	Instance	A	B	C	D	E	F	G	H	I	J
DC	time(s)	1.56	112.79	1.03	33.47	1.44	28.82	2.59	6.96	1.50	57.74
	fails	4007	43719	2327	71521	6894	54211	3165	36120	1699	24442
decomp.	time(s)	4.30	9.75	4.84	7.28	5.95	8.08	0.97	3.61	1.23	11.64
	fails	36305	93023	47939	118838	96734	85546	12922	81567	11200	95309

equal to 7.07 instead of 43.01). Table 4 compares the total computation time and number of fails to solve each 3-nurse instance between our model using the **spread** constraint and achieving domain consistency (DC) and a simpler model using linear and quadratic constraints instead (decomp.). Not surprisingly the stronger filtering of DC always exhibits fewer fails. However the latter is sometimes much slower on these instances. Upon closer inspection, there appears to be a strong correlation between the computation time and how long it takes to find solutions with a low standard deviation: instances B, J, D and F are the slowest to solve and also have several solution points in the top part of the plots at Fig. 1 (left); in contrast, instance A starts high but immediately finds solutions with a much lower standard deviation. This is not surprising because the time complexity of the domain filtering algorithm is influenced by the magnitude of the standard deviation. Regardless of this all these instances are solved well within the practical time frame of the hospital planner.

Moving on to the more challenging 5-nurse instances, Table 5 presents a similar comparison. The computation times jump up by a few orders of magnitude but DC is always faster here. (The difference in the number of fails, even more striking, is not shown in the table.) Note that the 5 min time limit is often

Table 5. Solving the 5-nurse instances

	Instance	A	B	C	D	E	F	G	H	I	J
DC	time(s)	3207	9638	20496	17078	6819	6444	4811	547	15505	5733
	last mean	86	141	108	130	126	113	102	104	115	91
	last SD	0	0	0	0	0	0	0	0	0	0
decomp.	time(s)	26181	21304	50298	31705	33027	48783	42723	15219	36663	54237
	last mean	95	142	127.6	138	141.8	138.4	125.6	108	128	116.6
	last SD	0	0	0.49	0	0.49	0.4	0.4	0	0.4	0.4

reached on these instances so the solutions found are not necessarily optimal for a given mean workload. Hence we do not provide a Pareto optimal front for them but show the typical behaviour of the two models on instance A in the right plot at Fig. 1: initially for higher bounds on the standard deviation the decomposition finds solutions more quickly but as the bound on the standard deviation gets tighter the trend reverses and DC performs better, which is consistent with the previous explanation of the variation in computation times on the 3-nurse instances. We also give in the table the mean and standard deviation of the last solution found by each: DC always finds a solution with perfect balance in the end but this is not always the case with the other model (it eventually terminates because of the upper bound on the total workload), and even when it does find a solution with perfect balance it is always strictly dominated by that of DC.

4   Conclusion

In this paper we considered the problem of balancing the workload of nurses. We closed the benchmark instances for the integrated staffing and nurse-patient assignment problem in the neonatal context and proposed a new set of instances that show better the advantage of our approach. The computation times are well within the usual time frame for this problem. We also considered a challenging variant of the nurse-patient assignment problem in which patient acuities are nurse-dependent. To be useful in practice for this problem, our approach should solve faster the instances considered, which are of realistic size. We could investigate better branching heuristics and it would also be interesting to evaluate the performance of a bound-consistent filtering algorithm for the `spread` constraint here, given what was observed with the domain-consistent filtering algorithm at larger standard deviations.

Acknowledgements. Financial support for this research was provided by Discovery Grant 218028/2012 from the Natural Sciences and Engineering Research Council of Canada.

References

1. Hertz, A., Lahrichi, N.: A patient assignment algorithm for home care services. JORS **60**(4), 481–495 (2009)

2. Ku, W.-Y., Pinheiro, T., Beck, J.C.: CIP and MIQP models for the load balancing nurse-to-patient assignment problem. In: O'Sullivan, B. (ed.) CP 2014. LNCS, vol. 8656, pp. 424–439. Springer, Heidelberg (2014)

3. Mullinax, C., Lawley, M.: Assigning patients to nurses in neonatal intensive. J. Oper. Res. Soc. **53**, 25–35 (2002)

4. Myny, D., Van Hecke, A., De Bacquer, D., Verhaeghe, S., Gobert, M., Defloor, T., Van Goubergen, D.: Determining a set of measurable and relevant factors affecting nursing workload in the acute care hospital setting: a cross-sectional study. Int. J. Nurs. Stud. **49**, 427–436 (2012)

5. Pesant, G.: Achieving domain consistency and counting solutions for dispersion constraints. INFORMS J. Comput. **27**(4), 690–703 (2015)

6. Pesant, G., Régin, J.-C.: SPREAD: a balancing constraint based on statistics. In: van Beek, P. (ed.) CP 2005. LNCS, vol. 3709, pp. 460–474. Springer, Heidelberg (2005)

7. Punnakitikashem, P., Rosenberber, J.M., Buckley-Behan, D.F.: A stochastic programming approach for integrated nurse staffing and assignment. IIE Trans. **45**(10), 1059–1076 (2013)

8. Schaus, P., Van Hentenryck, P., Régin, J.-C.: Scalable load balancing in nurse to patient assignment problems. In: van Hoeve, W.-J., Hooker, J.N. (eds.) CPAIOR 2009. LNCS, vol. 5547, pp. 248–262. Springer, Heidelberg (2009)

9. Schaus, P., Régin, J.-C.: Bound-consistent spread constraint. EURO J. Comput. Optim. **2**, 123–146 (2014)

10. Sir, M.Y., Dundar, B., Steege, L.M.B., Pasupathy, K.S.: Nurse-patient assignment models considering patient acuity metrics and nurses' perceived workload. J. Biomed. Inform. **55**, 237–248 (2015)

11. Sundaramoorthi, D., Chen, V.C.P., Rosenberger, J.M., Kim, S., Buckley-Behan, D.F.: A data-integrated simulation-based optimization for assigning nurses to patient admissions. Health Care Manag. Sci. **13**(3), 210–221 (2010)

Designing Spacecraft Command Loops
Using Two-Dimension Vehicle Routing

Eliott Roynette[1,2(⊠)], Bertrand Cabon[1], Cédric Pralet[2], and Vincent Vidal[2]

[1] Airbus DS Toulouse, Toulouse, France
{eliott.roynette,bertrand.cabon}@airbus.com
[2] ONERA – The French Aerospace Lab, 31055 Toulouse, France
{cedric.pralet,vincent.vidal}@onera.fr

Abstract. In the race to space, the reduction of the cost and weight of spacecrafts is one of the keys to progress further and further. In this perspective, a key component to consider in a spacecraft is the command system. The latter is a vital system which gives control over all the onboard devices, and which is materialized as a set of cables connecting devices with a set of controllers. To reduce the cost of development and the weight of this command system, there is a need to define optimization techniques for helping space engineers during the design phase. The objective of this paper is to present the problem tackled, which is a kind of Vehicle Routing Problem which we call a Two-dimension Vehicle Routing Problem, and to compare several solution techniques. One of these techniques is currently used for production.

1 Problem Description

In this paper, we tackle a problem from the space domain where the goal is to design the architecture responsible for controlling devices onboard spacecrafts. Spacecrafts typically contain hundreds to thousands of separate devices, which allow the mission assigned to the spacecrafts to be achieved.

A first possible way for being able to send commands to each of these devices during flight would be to define one command loop per device, linking the device with the central controller of the spacecraft. The main drawback of this approach would be a huge increase in weight, because physically each command loop is materialized as a cable. This is why space engineers developed command architectures where command loops are shared between devices. More precisely, the spacecrafts we consider use architectural components called *command matrices*, which are illustrated in Fig. 1(a). At an abstract level, a command matrix is a component made of rows and columns. One command loop is associated with each row and one command loop is associated with each column. Then, in the command architecture, each device d is placed in one row r and one column c of one matrix, and for sending a command signal to d, it suffices to send one signal on the command loop associated with r and one signal on the command loop associated with c. Doing so, the two emitted signals simultaneously reach device d positioned at the intersection between r and c. Several matrices are used

© Springer International Publishing Switzerland 2016
C.-G. Quimper (Ed.): CPAIOR 2016, LNCS 9676, pp. 303–318, 2016.
DOI: 10.1007/978-3-319-33954-2_22

instead of a single one mainly due to power and resistance constraints (placing too many devices on a single loop would generate too much resistance, reducing or destroying the functionality of the circuit). On real spacecrafts, command matrices typically range from size 2×2 to size 8×16 or even 16×16.

The main advantage of this command matrix architecture is that command loops, and therefore cables, are shared between devices. This reduces the weight and cost of the spacecraft. However, one difficulty is that for spacecraft reliability issues, some *segregation constraints* between devices must be satisfied, meaning that some devices are not allowed to be placed on the same row or on the same column of a command matrix. For instance, spacecrafts are often composed of sets of redundant devices, and the latter must not share a common command loop so that in case of failure of one loop, at least one device is still available. As a result, designing the placement of devices inside command matrices quickly becomes a combinatorial task.

Another feature of such an architecture is that each row (resp. each column) of a command matrix only defines the set of devices present on the command loop associated with that row (resp. with that column). On this point, one must also decide on the order of traversal of these devices by a physical cable, with the objective of minimizing the required cable length. For example, in Fig. 1(b), devices d_4, d_3, d_7, d_8 placed in row 2 of command matrix 1 are traversed in order $[d_3, d_7, d_8, d_4]$ in the physical command loop. Similarly, devices d_2, d_6, d_9 placed

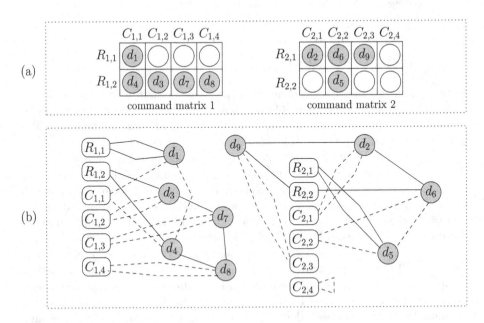

Fig. 1. Architecture for commanding devices: (a) allocation of positions in command matrices to a set of devices numbered from d_1 to d_9; (b) command loops physically used onboard the spacecraft; command loops associated with rows and columns are depicted using solid lines and dashed lines respectively

in row 1 of command matrix 2 are traversed in order $[d_9, d_2, d_6]$ by the cable associated with the command loop of this row.

The design problem considered in this paper consists in placing devices in command matrices and in deciding on the order in which these devices are traversed by cables, while satisfying segregation constraints and minimizing the total length (or weight) of cables used. In other words, the goal is to produce solutions such as the one depicted in Fig. 1. Also, in the problem we consider, the physical placement (x, y, z) of each device onboard the spacecraft is an input, as well as the length of a direct cable linking any two devices. In practice, this physical placement is produced based on other concerns such as thermal constraints or assembly constraints.

In the former approach used prior to this work, the problem was tackled using a two-step procedure, with (1) the dispatch of devices onto matrices, based on a genetic algorithm which tries to group devices which are near from each other in the spacecraft, and (2) the optimization of each order of traversal of devices in each command matrix, based on an home-made Traveling Salesman Problem (TSP) solver using Constraint Programming technology. This paper describes the new techniques developed to solve the problem based on a more global perspective. These new techniques have already been applied to one spacecraft which is currently active and to eight spacecrafts which are currently being built. From an industrial point of view, they led to a 15 % improvement in cable weight, and they allowed to reduce computation times to solve this problem from several days to a few hours.

The rest of the paper is organized as follows: the problem is first related to existing work (Sect. 2), then an integer linear programming formulation is proposed (Sect. 3), approximate search schemes are described (Sect. 4), and finally experimental results on realistic instances are provided (Sect. 5).

2   Problem Analysis

The command architecture design problem considered can be related with the design of electronic devices, which contain several electronic components which must be linked with a controller or with a power source. One key difference here is that the topology considered for spacecrafts is very different from that of electronic devices, as well as the architecture choice.

In another direction, the command architecture design problem can be related to *Vehicle Routing Problems* (VRPs [8]). In a VRP, the inputs are a set of customers placed at some positions and a set of vehicles placed at depots, and the goal is to find vehicle tours which start and end at depots, which visit all customers once, and which minimize the sum of the length of vehicle tours. In our design problem, devices can be seen as customers and each row (resp. each column) of a command matrix can be seen as a vehicle whose tour is the command loop of that row (resp. of that column). For each matrix, there is one depot per row and one depot per column. As these depots may be placed at different locations in the physical architecture of the spacecraft, the problem is actually a kind of *Multi-Depot Vehicle Routing Problem* (MDVRP). Moreover, as each loop associated with a row (resp. a column) cannot traverse more devices than the number of columns (resp. the number of rows) in a matrix, the problem can be related with the *Capacitated Vehicle Routing Problem* (CVRP), where vehicles have a limited capacity.

However, the design problem to solve has several differences with classical forms of VRPs. The first difference is that some devices (some customers) must be segregated in order to meet reliability constraints. Such kinds of segregation constraints were also considered in [4], where the goal was to compute vehicle routes for vehicles transporting hazardous materials with some constraints on materials which can be put together in a single vehicle. Another difference with standard VRPs is that each device (each customer) needs to be visited twice: once by a command loop associated with a row of a matrix, once by a command loop associated with a column of the same matrix. Moreover, due to the command matrix architecture, devices belonging to the same row cannot belong to the same column, therefore row allocation and column allocation are not independent. Because of this interaction between the VRP on the rows and the VRP on the columns, we call this problem the *Two-dimension Vehicle Routing Problem*. We are not aware of any previous work involving such a two-dimensional aspect, with or without additional segregation constraints and/or heterogeneous vehicle capacities and/or multiple depots. In the following, we propose a formulation for this new problem and we study different resolution techniques.

3   Integer Linear Programming Formulation

To formalize the command loop design problem, we consider the following input data:

- a set $\mathbf{D}$ of devices;
- a set $\mathbf{M}$ of command matrices; each command matrix m contains a set of rows $\mathbf{R_m}$ and a set of columns $\mathbf{C_m}$;
- the set $\mathbf{R}$ of command matrix rows, defined as $R = \cup_{m \in M} R_m$;
- the set $\mathbf{C}$ of command matrix columns, defined as $C = \cup_{m \in M} C_m$;
- the set $\mathbf{P}$ of positions available for placing devices, defined as $P = \cup_{m \in M} (R_m \times C_m)$; in other words, P contains all pairs (r, c) formed by a row and a column belonging to the same matrix; we assume that there are more positions than devices ($|P| \geq |D|$), otherwise the problem is directly inconsistent;
- a set of segregation constraints $\mathbf{Seg} \subseteq D \times D$; a pair (i, j) belongs to Seg when devices i and j must not belong to the same row or to the same column of a command matrix;
- the set of arcs $\mathbf{A^R} = (D \times D) \cup (R \times D) \cup (D \times R)$ containing all possible connections between successive devices on row command loops; for $r \in R$ and $i \in D$, pair (r, i) corresponds to the connection from the controller of the command loop of row r to device i, and pair (i, r) corresponds to the connection from device i to this controller;

- the set of arcs $\mathbf{A^C} = (D \times D) \cup (C \times D) \cup (D \times C)$ containing all possible connections between successive devices on column command loops; for $c \in C$ and $i \in D$, pair (c, i) corresponds to the connection from the controller of the command loop of column c to device i, and pair (i, c) corresponds to the connection from device i to this controller;
- a length function $\mathbf{L}$ such that for any arc $(i, j) \in A^R \cup A^C$, L_{ij} gives the length of a direct cable associated with arc (i, j); lengths specified by L are computed in a preprocessing step using a Floyd-Warshall algorithm on the graph which contains possible cable routes between devices.

Matrix Allocation. In order to model the problem, we first represent the allocation of devices to command matrices. To do this, we consider the following sets of variables:

- $\mathbf{pos_{ip}} \in \{0, 1\}$ ($i \in D, p \in P$), for representing whether device i is placed in matrix position p (value 1) or not (value 0);
- $\mathbf{row_{ir}} \in \{0, 1\}$ ($i \in D, r \in R$), for representing whether device i is placed in matrix row r (value 1) or not (value 0);
- $\mathbf{col_{ic}} \in \{0, 1\}$ ($i \in D, c \in C$), for representing whether device i is placed in matrix column c (value 1) or not (value 0).

Matrix allocation Constraints 1 to 6 are then imposed on these variables. Constraint 1 expresses that a device is allocated to exactly one position in command matrices. Constraint 2 expresses that each matrix position can be associated with at most one device. Constraints 3 and 4 define the row and the column associated with a device from the position at which this device is placed. Last, Constraints 5–6 impose that row and column choices must differ for devices which must be segregated. Note that the capacity constraints on the number of devices involved in a command loop is indirectly expressed by this set of constraints.

$$\forall i \in D, \sum_{p \in P} pos_{ip} = 1 \tag{1}$$

$$\forall p \in P, \sum_{i \in D} pos_{ip} \leq 1 \tag{2}$$

$$\forall i \in D, \forall r \in R, row_{ir} = \sum_{p \in P \,|\, p=(r,c)} pos_{ip} \tag{3}$$

$$\forall i \in D, \forall c \in C, col_{ic} = \sum_{p \in P \,|\, p=(r,c)} pos_{ip} \tag{4}$$

$$\forall r \in R, \forall (i, j) \in Seg, row_{ir} + row_{jr} \leq 1 \tag{5}$$

$$\forall c \in C, \forall (i, j) \in Seg, col_{ic} + col_{jc} \leq 1 \tag{6}$$

Cable Routing. In order to represent how devices are connected to each other, i.e. in order to represent vehicle routing constraints, we introduce two sets of variables:

- $x_{ij}^R \in \{0,1\}$ $((i,j) \in A^R)$, for representing whether arc (i,j) is traversed for connecting i and j on a row command loop;
- $x_{ij}^C \in \{0,1\}$ $((i,j) \in A^C)$, for representing whether arc (i,j) is traversed for connecting i and j on a column command loop.

Constraints 7 to 12 are imposed on these variables. These constraints are standard constraints used for representing vehicle routing problems. Concerning row connections, Constraints 7 and 8 are flow constraints expressing that for each device and each row controller, there is a unique incoming connection and a unique outgoing connection. Constraint 9 corresponds to sub-tour elimination, that is it forbids cable tours which do not contain any command loop controller. These sub-tour elimination constraints can be added incrementally when solving the problem, in order to avoid an exponential blow-up in the problem size. Constraints 10 to 12 impose similar constraints for columns.

$$\forall i \in D \cup R, \quad \sum_{j \,|\, (i,j) \in A^R} x_{ij}^R = 1 \tag{7}$$

$$\forall i \in D \cup R, \quad \sum_{j \,|\, (j,i) \in A^R} x_{ji}^R = 1 \tag{8}$$

$$\forall S \subseteq D \text{ s.t. } S \neq \emptyset \quad \sum_{(i,j) \in A^R \,|\, i,j \in S} x_{ij}^R \leq |S| - 1 \tag{9}$$

$$\forall i \in D \cup C, \quad \sum_{j \,|\, (i,j) \in A^C} x_{ij}^C = 1 \tag{10}$$

$$\forall i \in D \cup C, \quad \sum_{j \,|\, (j,i) \in A^C} x_{ji}^C = 1 \tag{11}$$

$$\forall S \subseteq D \text{ s.t. } S \neq \emptyset \quad \sum_{(i,j) \in A^C \,|\, i,j \in S} x_{ij}^C \leq |S| - 1 \tag{12}$$

Compatibility Between Matrix Allocation and Cable Routing. In order to represent the coupling between matrix allocation and cable routing, we introduce Constraints 13 to 20. Constraints 13 and 14 express that if device j is the successor of device i on the cable route associated with a row, then these two devices must be placed on the same matrix row. Constraint 15 imposes that if there is a connection from the controller of the command loop of row r to device i, then i must be placed on matrix row r. Similarly, Constraint 16 imposes that if device i is connected to the controller of the command loop of row r, then i must be placed on matrix row r. Constraints 17 to 20 impose similar specifications for columns. Note that some of these constraints are redundant.

$$\forall i,j \in D, \forall r \in R, \; row_{ir} + x_{ij}^R \leq row_{jr} + 1 \tag{13}$$

$$\forall i,j \in D, \forall r \in R, \; row_{jr} + x_{ij}^R \leq row_{ir} + 1 \tag{14}$$

$$\forall i \in D, \forall r \in R, \; x_{ri}^R \leq row_{ir} \tag{15}$$

$$\forall i \in D, \forall r \in R, \; x_{ir}^R \leq row_{ir} \tag{16}$$

$$\forall i, j \in D, \forall c \in C, \ col_{ic} + x_{ij}^C \leq col_{jc} + 1 \tag{17}$$

$$\forall i, j \in D, \forall c \in C, \ col_{jc} + x_{ij}^C \leq col_{ic} + 1 \tag{18}$$

$$\forall i \in D, \forall c \in C, \ x_{ci}^C \leq col_{ic} \tag{19}$$

$$\forall i \in D, \forall c \in C, \ x_{ic}^C \leq col_{ic} \tag{20}$$

Global Model. From all previous elements, the global model of the problem corresponds to the following integer linear program, in which the goal is to minimize the total length used for routing cables on row command loops and column command loops:

$$\text{minimize} \sum_{(i,j)\in A^R} x_{ij}^R L_{ij} + \sum_{(i,j)\in A^C} x_{ij}^C L_{ij} \tag{21}$$

subject to :

> Matrix allocation constraints (Constraints 1−6)
>
> Vehicle routing constraints (Constraints 7−12)
>
> Allocation/routing compatibility constraints (Constraints 13−20)
>
> $pos_{ip} \in \{0,1\}\, (i \in D, p \in P)$
>
> $row_{ir} \in \{0,1\}\, (i \in D, r \in R), col_{ic} \in \{0,1\}\, (i \in D, c \in C)$
>
> $x_{ij}^R \in \{0,1\}\, ((i,j) \in A^R), x_{ij}^C \in \{0,1\}\, ((i,j) \in A^C)$

Resolution. From the previous Integer Linear Program (ILP), it is possible to use standard solvers such as IBM ILOG CPLEX, by adding sub-tour elimination constraints step by step. As shown in the experiments (see Sect. 5), this approach does not scale well, even when considering only medium size instances. This is why we also defined approximate search schemes (see Sect. 4).

Constraint Programming Formulation. To model this problem, we also tried a pure Constraint Programming (CP) approach. For space limitation reasons, it is presented only at a global level. The CP model built contains:

- for each device $i \in D$, variables $row_i \in R$, $col_i \in C$, $pos_i \in P$, to respectively describe the row, the column, and the matrix position associated with i;
- for each element $i \in D \cup R$, one variable $rnext_i \in D \cup R$ representing the index of the element following i on its row (compared to the ILP model, introduction of integer variables instead of 0/1 variables);
- for each element $i \in D \cup C$, one variable $cnext_i \in D \cup C$ representing the index of the element following i on its column.

Over these variables, several basic constraints are imposed, including:

- *alldifferent* constraints over variables pos_i, to express that two devices cannot be located at the same matrix position;
- *element* constraints linking the row/column of a device with its position;
- constraints expressing that a device cannot follow itself on a row or column;

- row/column segregation constraints for device pairs $(i, j) \in Seg$;
- *element* constraints expressing that the row (resp. column) of a device i is the same as the row (resp. column) associated with $rnext_i$ (resp. $cnext_i$);
- *alldifferent* constraints over variables $rnext_i$ (resp. $cnext_i$).

To improve the power of constraint propagation, as in existing CP models of VRPs/TSPs, the model built also contains redundant variables $rprev_i$ (resp. $cprev_i$) to represent the element which precedes i on its row (resp. on its column), together with alldifferent constraints over these variables and constraints such as $rprev_{rnext_i} = i$ for every $i \in D \cup R$, and $cprev_{cnext_i} = i$ for every $i \in D \cup C$.

Last, the CP model built contains a set of symmetry breaking constraints. First, for every command matrix, it is possible to impose that any row (resp. any column) in this matrix always contains more devices than the next row (resp. the next column) in the matrix. When all matrices share the same dimension, it is also possible to enforce that the number of devices in matrix k cannot be less than the number of devices in matrice $k+1$. Such symmetry breaking constraints were also tested for the ILP model, where they were shown to degrade the results.

Many other CP models could be considered. With the CP model developed, we managed to find solutions, but their quality was not as good as the solutions obtained with ILP, which is why we focus on ILP in the rest of the paper.

4 Two Local Search Approaches

In this section, we present the two local search algorithms we developed and which permit to find good quality solutions within limited computing times. These algorithms start from a complete assignment of the decision variables of the problem and try to iteratively improve the current solution by applying actions towards promising regions of the search space. These actions are local moves that modify the current solution to neighbor solutions. We first present the neighborhood structure which is used, and then the two metaheuristics employed for driving local moves. These two metaheuristics are Simulated Annealing (SA [5]) and Iterated Local Search (ILS [6]).

4.1 Neighborhood Definition

The neighborhood we consider is obtained by combining two kinds of updates:

- *command matrix updates*, which update the allocation of devices to the rows and columns of the command matrices;
- *cable routing updates*, which update the way cables are routed to cover the set of devices belonging to a same row or to a same column.

More precisely, a local move in the neighborhood is performed as follows:

- select one device d placed at a position p, select one position $p' \neq p$ (possibly in a different matrix), and exchange the contents of p and p'; as a result, if position p' selected for reallocating d contains a device d', then after the move, position p contains device d' and position p' contains device d; if position p' contains no device, then after the move, position p contains no device and position p' contains device d;

- after the previous step, for all rows and columns whose associated set of devices is modified (at most 2 rows and 2 columns concerned), use a TSP routine for improving the routing of cables for these rows and columns.

For the second step, several strategies can be used for updating cable routes. A first possible strategy is to use a complete TSP solver for solving the TSP associated with each impacted row or column. However, this increases the duration required for performing a local move, and such a strategy is effective only for small size problems. A second possible strategy is to use a very fast mechanism which only looks for a good insertion position in cable routes for the device(s) whose rows and columns are updated (computing time linear in the maximum number of rows and columns). However, the cable routes obtained based only on such a greedy insertion rule may lead to a poor quality evaluation of the best cable routes, and therefore to a poor quality evaluation of candidate matrix updates.

To find a compromise between the speed of each move and the quality of the evaluation, we chose, after several experiments, an intermediate strategy which works well on the largest instances (16×16 command matrices). This strategy consists in running the standard 2-opt heuristic algorithm [3] on the TSP defined by each impacted row and column. Given a TSP tour $[d_1, d_2, \ldots, d_n]$ corresponding to a routing of cables between devices, the idea in 2-opt is to remove at each step two edges (d_i, d_{i+1}) and (d_j, d_{j+1}) from the tour (with $i < j$), and to reconnect the subtours created to consider the new valid tour $[d_1, \ldots, d_i, d_j, d_{j-1}, \ldots, d_{i+1}, d_{j+1}, \ldots, d_n]$ (head and tail of the initial tour kept, and part between d_{i+1} and d_j traversed in the other way around). The move is accepted only if the new tour obtained is shorter, which can be evaluated in constant time. In 2-opt, such moves are applied until no more 2-opt improvement is possible. The advantage of the 2-opt algorithm is that it is known to produce good quality routings within short computing times (complexity quadratic in the number of elements to be covered by the tour). Other approximate TSP resolution schemes could be considered such as 3-opt, k-opt [7] or or-opt [1]. The global idea is that faster TSP search permits to do more local moves, while a better quality TSP search permits to choose the most promising local moves.

4.2   Constraint Satisfaction and Criterion Evaluation

In practice, the local search algorithm is implemented by handling integer variables instead of 0/1 variables as in the ILP model. For instance, instead of manipulating variables $pos_{ip} \in \{0, 1\}$ for representing whether device i is placed at position p, we maintain integer variables pos_i representing the position of device i in matrices, as in the Constraint Programming model. Also, for realizing the local search algorithms, we do not consider symmetry breaking constraints because they reduce the accessibility between some neighbor solutions in the search space.

A first key property of the problem considered is that if device segregation constraints are discarded (Constraints 5–6), then it is easy to produce a first solution which satisfies all constraints of the problem. Indeed, it suffices to put device 1 in position 1, device 2 in position 2... and so on until all devices are put in matrices, and then it is possible to run any TSP algorithm on each non empty row and column obtained to get a routing of cables.

A second key property is that the local moves introduced in Sect. 4.1 preserve the satisfaction of all constraints, again except from segregation constraints whose satisfaction can be improved or deteriorated by each local move.

The satisfaction of segregation constraints being non-trivial, we relax the satisfaction of these constraints and manipulate a violation degree instead. To evaluate the overall quality of a solution, this violation degree is combined with the total cable length to get a global score. To do this, we maintain a score for each row r as:

$$scoreRow(r) = cableLengthRow(r) + nSegViolated(r) \times SEG\_COST$$

with $cableLengthRow(r)$ the total length of cables for row r, $nSegViolated(r)$ the number of segregation constraints violated on row r, and SEG_COST a constant factor set big enough for having the satisfaction of segregation constraints preferred to any improvement in cable length at the end of the local search. For each column c, a score $scoreCol(c)$ is defined similarly, and the total score associated with a solution sol is given by:

$$score(sol) = \sum_{r \in R} scoreRow(r) + \sum_{c \in C} scoreCol(c)$$

In other words, compared to the optimization criterion defined in Eq. 21, we add a constraint violated degree to the expression of the score. The score formulation provided also shows that the score can be decomposed by rows and columns, which allows incremental evaluations to be performed when changes occur only on a small part of the problem.

4.3 Simulated Annealing Metaheuristics

To obtain good sequences of local moves with the neighborhood structure defined, it is first possible to consider a standard Simulated Annealing algorithm [5]. The latter corresponds to Algorithm 1. It starts from an initial solution s_0 and at each step, it considers a random solution s' in the neighborhood of the current solution s. Solution s' is accepted as the new current solution if its score is better than the score of s. It is also accepted with some probability when s' deteriorates the score (see lines 6 to 8). Accepting deteriorating moves allows local minima to be escaped. The acceptance probability depends on the criterion deterioration Δ and on the temperature parameter $temp$ of the simulated annealing (use of the Metropolis rule $exp(-\Delta/temp)$). Initially, the temperature is high and many deteriorating moves can be accepted, whereas at the end of the search the temperature is low and almost only improving moves are accepted. In our case, the temperature is decreased step by step: as shown in lines 4–5, several iterations are performed with the same temperature, and the time spent with a particular temperature depends on the maximum time allowed and on the number of temperature steps required. The solution returned is the best solution found during all iterations.

As usual in local search, setting good values for parameters is not straightforward. In our settings, initial temperature *initTemp* is set using a fast automatic procedure which ensures that at the first temperature step, approximately 80 % of the local moves are accepted. To obtain such a setting, we start from a low temperature and progressively increase this temperature until we estimate that 80 % of the local moves are accepted. To obtain such an estimation, we perform 10000 local moves and we compute the number of these moves which would have been accepted by the simulated annealing acceptance rule. Following experimental evaluations, decreasing factor λ is arbitrarily set to 0.98 and the number of temperature steps *nTempSteps* is set to 1000. Last, the maximum time allocated to the search (*MaxTime*) is left free.

Algorithm 1. SimulatedAnnealing(*initTemp*,λ,*nTempSteps*,*MaxTime*)

Data: *initTemp*: initial temperature, λ: temperature reduction factor,
   nTempSteps: number of temperature reduction steps, *MaxTime*:
   maximum computing time allowed

1   $tempStep \leftarrow 1$;
2   $s \leftarrow firstSolution()$;
3 **while** $tempStep \leq nTempSteps$ **do**
4    $MaxTimeStep \leftarrow tempStep \cdot (MaxTime/nTempSteps)$;
5   **while** $currentTime() < MaxTimeStep$ **do**
6     $s' \leftarrow selectRandomNeighbor(s)$;
7     $\Delta \leftarrow score(s') - score(s)$;
8   **if** $(\Delta < 0) \vee (rand() < exp(-\Delta/temp))$ **then** $s \leftarrow s'$;
9    $temp \leftarrow \lambda \cdot temp$;
10    $tempStep \leftarrow tempStep + 1$;

4.4 Iterated Local Search Metaheuristics

A second approximate algorithm considered is the Iterated Local Search algorithm (ILS [6]), which has already been applied to Vehicle Routing Problems [4]. This algorithm iteratively tries to find local minima in the search space. More precisely, the ILS algorithm (Algorithm 2) iterates two phases:

- a *local search phase* during which the algorithm searches for a local optimum (lines 3 to 11);
- a *perturbation phase*, during which the local optimum found at the previous phase is perturbed by some random changes (line 12); usually, the perturbation strength must be sufficient for driving the search to another local optimum, and not too large for avoiding performing a kind of random restart.

The solution returned is the best solution found during all iterations.

Local Search Phase Setting. In our settings, the local search phase works on the neighborhood introduced in Sect. 4.1 and tries to select at each step the best neighbor in this neighborhood. As this neighborhood can be quite large (number of neighbors quadratic in the number of matrix positions), we use a neighbor selection strategy which tests only a subset of the candidate local moves. To do this, we first evaluate for each device d a contribution $contrib(d)$ to the global score. This contribution is given by:

$$contrib(d) = contribCableLength(d) + contribSegViolation(d)$$

with $contribCableLength(d)$ the sum of the length of incoming and outgoing cables for device d on its row and its column, and $contribSegViolation(d)$ the number of violated segregation constraints which involve device d times constant factor SEG_COST.

Then, we select a device d which has the highest contribution $contrib(d)$ to the score, and we analyze all neighbors in which d is moved to another position in command matrices (line 7). If the best neighbor found improves the global score, this best neighbor is kept as the new current solution and some device contributions are recomputed before going to the next iteration of the local search. If no best neighbor is found, then the algorithm considers another device d' not considered yet and which has the highest contribution $contrib(d')$. If all devices have already been considered, then a local minimum has been reached and the ILS procedure goes on with the perturbation phase.

Perturbation Phase Setting. For the perturbation phase (line 12), several strategies were tested, like doing an increasing number of random moves or doing many random moves. Experiments showed that for our problem, doing just one random move often suffices to go out from a local optimum. Also, in our problem, making several random moves at each perturbation phase slows down the local search phase because it then takes more time to converge to another local optimum.

Algorithm 2. IteratedLocalSearch($maxTime$)

Data: *MaxTime*: maximum computation time allowed
1 $s \leftarrow firstSolution()$;
2 **while** $currentTime() < MaxTime$ **do**
3 $\quad$ $Candidates \leftarrow D$;
4 $\quad$ **while** $(Candidates \neq \emptyset) \wedge (currentTime() < MaxTime)$ **do**
5 $\quad\quad$ $contribMax = \max_{e' \in Candidates} contribs(d')$;
6 $\quad\quad$ select d in $\{d' \in Candidates \mid contrib(d') = contribMax\}$;
7 $\quad\quad$ $s' \leftarrow selectBestNeighbor(s, d)$;
8 $\quad\quad$ **if** $score(s') < score(s)$ **then**
9 $\quad\quad\quad$ $s \leftarrow s'$;
10 $\quad\quad\quad$ $Candidates \leftarrow D$;
11 $\quad\quad$ **else** Remove d from $Candidates$;
12 $\quad$ $s \leftarrow selectRandomNeighbor(s)$;

---

5 Experiments

To compare the different resolution techniques proposed (ILP, SA, ILS), experiments were performed on real instances. We also developed a random instance generator which allowed us to test the different techniques on a wide set of realistic instances. This generator takes several parameters as an input: (1) the number of command matrices **nM** in the instance, (2) the number of rows **nRbM** and the number of columns **nCbM** in each command matrix; (3) the filling percentage of matrices **pctFill**; the number of devices in the instance generated is then automatically computed by $nD = \lfloor pctFill \cdot nM \cdot nRbM \cdot nCbM \rfloor$; distances between devices are generated by computing all pair shortest paths in a graph whose edge weights are given by a uniform random distribution (this way, device distances satisfy the triangle inequality, as in real instances); (4) the percentage **pctSeg** of devices which must be segregated in the instance generated; the number of segregation constraints in Seg is then given by $\lfloor pctSeg \cdot nD\,(nD-1)/2 \rfloor$; devices to be segregated are chosen randomly using a uniform distribution. Figure 2 gives an idea of how the problem becomes more and more constrained when parameters **pctFill** and **pctSeg** are increased. Such results are useful to help space engineers to define the dimensions of the command system.

Experimental Environment. The experimental results presented in the paper only concern randomly generated benchmarks. More precisely, for every fixed parameters settings, we generated at least 5 random instances. As shown in Figures 3, 4 and 5, we also made several experiments to evaluate the sensibility to the variation of some parameters (sensibility to the maximum CPU time allowed in Fig. 5a, and sensibility to the size of matrices in Fig. 5b). In a longer version, we could give more details concerning the influence of parameters settings, and also concerning the influence of algorithmic settings such as the initial temperature of the Simulated Annealing (more exploration of the search space

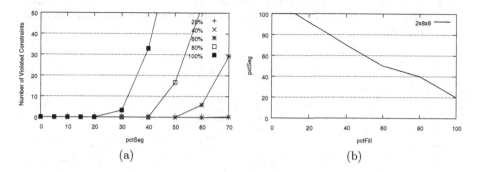

(a)   (b)

Fig. 2. (a) Number of violated segregation constraints in the best solution found in one hour, with parameters $nM = 2$ and $nRbM = nCbM = 8$ (the different lines correspond to distinct values of parameter *pctFill*); (b) consistency limit: no consistent solution found for instances corresponding to (pctFill, pctSeg) values located over the line

with a high temperature, faster production of first solutions with a low temperature). Experiments are performed on a processor Intel Xeon CPU W3530 2,80 GHz and 16 GiB of RAM.

Evaluation of the ILP Approach. To evaluate the ILP model presented in Sect. 3, we used IBM ILOG CPLEX 12.5. As shown in Fig. 3(a), based on the ILP model, finding optimal solutions and proving their optimality is only possible for very small instances. For example, when considering only 2 command matrices of size 4 × 4, there is already an exponential blow-up in computing times when the filling percentage increases. On this point, we believe that the two-dimension aspect of the VRP problems to solve and the segregation constraints make the search space much more combinatorial than in a standard VRP. However, the ILP approach allows us to have some optimum solutions which can be used to evaluate the efficiency of other algorithms on small instances. Such results are shown in Fig. 3(b), where we can see that the solutions obtained using SA and ILS are almost optimal for problems involving only 2 × 2 matrices which can

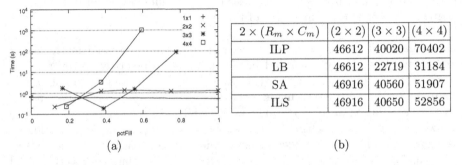

$2 \times (R_m \times C_m)$	(2×2)	(3×3)	(4×4)
ILP	46612	40020	70402
LB	46612	22719	31184
SA	46916	40560	51907
ILS	46916	40650	52856

(a)   (b)

Fig. 3. Some results for fixed parameters $nM = 2$ and $pctSeg = 10\%$: (a) computation time required to solve the ILP model proposed for small matrix sizes and for various filling percentages; (b) comparison between ILP (bestScore), ILP LowerBound (LB), SA, and ILS: best score obtained for $pctFill = 80\%$ with a maximum computing time $MaxTime = 300s$

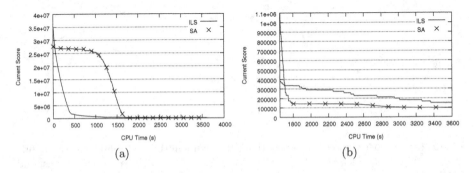

(a)   (b)

Fig. 4. (a) Evolution of the best known score during search, for parameter setting $nM = 8, nRbM = nCbM = 16, pctFill = 80\%, pctSeg = 10\%$; (b) zoom on (a)

be solved optimally by ILP. For matrices of size 4×4, Fig. 3(b) shows that the quality of the best solution found by ILP is rather poor compared to the approximate solutions produced by SA and ILS.

Evaluation of SA and ILS. Compared to ILP, the SA and ILS approaches are designed for tackling medium and large instances. Figure 4(a) gives a first comparison between SA and ILS on a large instance involving 8 matrices of size 16×16. This figure shows the evolution of the score associated to the best known solution in function of the computation time. It is possible to see that ILS finds an acceptable solution faster than SA (the steep decreasing of the score corresponds to the satisfaction of all segregation constraints). However, after some computing time, when the temperature of the SA algorithm decreases, SA manages to find an acceptable solution and this solution is better than the solution provided by ILS (see Fig. 4(b)). Such results can be explained by the fact that given the size of the problems considered, the convergence to a local optimum during the local search phase of ILS is rather slow, which entails that the ILS algorithm only visits a small number of local optima. On the opposite, the SA algorithm is looking first for the best area in the solution space without looking for the local optimum at each step, therefore it offers a greater diversity in the traversal of the search space.

Figure 5(a) shows that such conclusions are rather robust to the increase in the allowed computation time, since even for quite high values of the maximum computation time, the SA algorithm still produces better results. For low computation times, it is possible to see that ILS is still working on the satisfaction of the segregation constraints while SA is already improving the total length of cables. Additionally, Fig. 5(b) shows that the dominance of SA over ILS increases with the size of the problem, which again means that with the settings chosen,

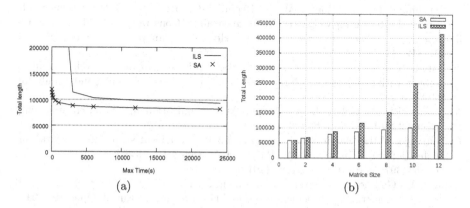

(a)   (b)

Fig. 5. Comparison between the best solutions produced by SA and ILS: (a) in function of the maximum time allocated to the solver, for fixed parameters $nM = 6, nRbM = nCbM = 16, pctFill = 80\%, pctSeg = 10\%$; (b) in function of the problem size, for fixed parameters $nRbM = nCbM = 16, pctFill = 80\%, pctSeg = 10\%$. In all case, the segregation constraints are satisfied.

SA performs a better exploration of the search space which seems to contain many local minima. Another possible explanation to these results is that in the settings chosen, SA adapts its behavior depending on the maximum computation time, thanks to the particular law chosen for decreasing the temperature. On the opposite, ILS does not exploit the maximum computation time information.

6   Conclusion

In this paper, we described the treatment of a design problem from the space domain which corresponds to a kind of Two-dimension Vehicle Routing Problem with segregation constraints. Several algorithms were proposed to solve this problem. Many other algorithmic settings could be considered, even for the Iterated Local Search scheme, but for the moment the best solution found is the Simulated Annealing approach. One of the perspective would be to use CP modeling elements dedicated to TSP problems [2]. A point is that the problem presented in this paper actually makes some simplifications compared to the real problem, which contains other aspects such as decisions on the width of cables used, constrained due to resistance issues. Additional work is required to optimize these other parts of the real problem. However, even with this approximation, it is still interesting to use the work presented in this paper. The Simulated Annealing (SA) is in production and it allowed to save one week in design time, 15 % in cable length, and approximately 10000 euros per satellite.

References

1. Babin, G., Deneault, S., Laporte, G.: Improvements to the or-opt heuristic for the symmetric traveling salesman problem. J. Oper. Res. Soc. **58**, 402–407 (2007)
2. Benchimol, P., van Hoeve, W.-J., Régin, J.-C., Rousseau, L.-M., Rueher, M.: Improved filtering for weighted circuit constraints. Constraints **17**, 205–233 (2012)
3. Croes, G.A.: A method for solving traveling salesman problems. Oper. Res. **6**, 791–812 (1958)
4. Hamdi, K., Labadie, N., Yalaoui, A.: An iterated local search for the vehicle routing problem with conflicts. In: 8th International Conference of Modeling and Simulation (MOSIM 2010) (2010)
5. Kirkpatrick, S., Gelatt Jr., C., Vecchi, M.: Optimization by simulated annealing. Science **220**(4598), 671–680 (1983)
6. Lourenço, H.R., Martin, O.C., Stützle, T.: Iterated local search: framework and applications. In: Gendreau, M., Potvin, J.-Y. (eds.) Handbook of Metaheuristics, pp. 363–397. Springer, New York (2010)
7. Rego, C., Gamboa, D., Glover, F., Osterman, C.: Traveling salesman problem heuristics: leading methods, implementations and latest advances. Eur. J. Oper. Res. **211**, 427–441 (2011)
8. Toth, P., Vigo, D.: Vehicle Routing: Problems, Methods, and Applications, 2nd edn. MOS-SIAM Series on Optimization (2014)

Constraint Programming Approach for Spatial Packaging Problem

Abdelilah Sakti$^{(\boxtimes)}$, Lawrence Zeidner, Tarik Hadzic, Brian St. Rock, and Giusi Quartarone

United Technology Research Center, East Hartford, USA
{SaktiA,ZeidneLE,HadzicT,strockb1,QuartaG}@utrc.utc.com

Abstract. The Spatial Packaging Problem (SPP) aims to solve a mixture of the 3D Packing Problem (3DPP) and the 3D Pipe-Routing Problem. The main feature that distinguishes the SPP from the traditional 3DPP is the interconnections that exist between its components. The SPP is more challenging because the shape and dimensions of the interconnections are unknown, and must be determined as part of the solution. In this paper, we propose a relaxation, a constraint programming model and a search heuristic to solve the SPP. We relax the SPP by using taxicab geometry and model it as a constraint satisfaction problem, then solve it by using a search heuristic based on interconnection volumes. The proposed approach has been evaluated on a challenging benchmark that reflects a range of aerospace and commercial applications varying in number of components and interconnections. The preliminary results show the effectiveness and efficiency of the proposed approach.

Keywords: Constraint programming · 3D Packing Problem · Spatial Packaging Problem · Pipe-routing problem · Taxicab geometry

1 Introduction

When developing a multi-component physical product, it is important to design its components and their interconnections in an optimal geometric manner, where the same functionality is provided in less volume, with shorter length of connections, and using simpler shapes. Achieving better geometric designs is important across different engineering industries. In building systems [2], the optimal design of heating, ventilation and air-conditioning products reduces their material and transportation costs, and simplifies their installation. In the electronics industry, the relative positions of devices and wires determine electrical interference or thermal overheating which causes component damage [9]. In the aerospace [6,22] and naval industries [20], the reduction of the system size, length of wires and pipes, and the duct volume significantly reduces the system weight, costs, and often enables performance improvements. Moreover, better geometric designs can significantly increase the manufacturability, assembly, and maintainability of systems. Yet, despite such an importance of optimized geometric

© Springer International Publishing Switzerland 2016
C.-G. Quimper (Ed.): CPAIOR 2016, LNCS 9676, pp. 319–328, 2016.
DOI: 10.1007/978-3-319-33954-2_23

layouts, the standard industrial-design processes are still manual. Skilled personnel – who understand the engineering applications, manufacturing, assembly, maintenance, and repair requirements – start from an existing product design and modify it to meet the new requirements. This process is time-consuming and results in inferior geometric layouts even with the use of computer-aided design tools which help design product packaging. Inefficiencies of these methods will only increase as product requirements grow more complex and required packaging volumes shrink.

The Spatial Packaging Problem (SPP) attempts simultaneously to *pack* components into a given volume and *route* interconnections between them. The *packing problem* (PP) [4,7,15,23] is well-known in the transportation sector [5], aiming to reduce the number of vehicles required. In the steel industry, optimal 3D cutting is important to minimize material cost. Similarly, 2D cutting appears in the textile, wood, glass, and paper industry where shapes of known dimensions are to be cut out of a strip with a given width, but variable length. The *routing problem* (RP) is commonly encountered in electronics [14], in vehicle routing [13,21], in artificial intelligence [11] and in computer networks [17]. The *pipe-routing problem* (PRP) [3,8,16,19,20] is a sub-class of the RP in which the locations are generally represented by components positioned in a 3D space that must be properly connected through pipes and wires used to transport material, energy, or information. The authors are not aware of any successful prior SPP solution approaches. To solve the SPP, we have previously tried both Mixed Integer Linear Programming (MIP) and Local Search (LS). We attempted to use a MIP-based approach - but unfortunately could not solve SPP instances with more than 3 interconnections. We concluded that this is partly due to the pipe-routing aspect of SPP and to the scalability of MIP techniques to 3D packing with a large number of items [4]. Recently, we contributed an instance of SPP as a challenge problem at the 110th European Study Group with Industry[1]. A follow-up effort is currently exploring LS as a solution technique for that instance. Based on these two experiences with SPP we believe that its complexity is driven primarily by the PRP (i.e., the number of interconnections) rather than the PP, i.e., the number of components. This led us to investigate Constraint Programming (CP) as an SPP solution method, due to its strength in routing.

In this paper, we propose to automate the generation of geometric designs for multi-component products, and demonstrate that CP is a viable solution approach. We formulate an industrially relevant version of the SPP in which only the components are known but interconnections have to be designed. The dimensions of components and interconnection diameters are known, but all other elements, such as 3D location and orientation of components, as well as the shape and the length of interconnections have to be determined by solving the SPP. We propose a relaxed version of the SPP using Taxicab Geometry [12] and a CP model of the SPP. We then show how to efficiently solve the feasibility version of a highly challenging benchmark through an effective search heuristic.

[1] http://www.macsi.ul.ie/esgi110/.

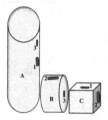

Fig. 1. Aircraft engine

Fig. 2. An SPP example with 3 components and 3 interconnections

2   The Spatial Packaging Problem

Figure 1 shows an aircraft-engine system in which hundreds of components must be linked using hundreds of interconnections. This is very challenging to design and impractical to optimize manually. The problem grows more difficult as the trend toward smaller-diameter engine cores drastically reduces the surface area, and thus the packaging volume (PV), within which these components and interconnections must fit. Automatically solving such problems would enable a designer to explore different solutions, compare them, and choose the one that best satisfies design requirements and multiple objectives.

The SPP determines 3D position and orientation of given *components* and designs *interconnections* between them so that each required pair of components is connected while all elements fit within a given (or smallest possible) packaging volume. Figure 2 illustrates a simple SPP example that involves three components (A, B, and C) which must be connected with three different interconnections (1, 2, and 3) so they can fit into the smallest possible packaging volume. The component A must be connected to component C via the connection 1 and to component B via the connection 3; also, the component B must be connected to component C via connection 2. Figure 3 shows a solution to the problem, determining the position of each component, the shape of each interconnection, and the packaging volume dimensions.

The *components* are represented as basic Euclidean-geometric forms (i.e., sphere, cylinder, or rectangular prism) since irregular shapes add mathematical-modeling complexity [10]. Their shapes and dimensions are provided as an input to the problem while their positions and orientations are determined as part of the solution. Some components need to be connected via *interconnections*, which are represented as a chain of cylinders with uniform diameter, each cylinder acting as a *connector*. For example, in Fig. 4 the shapes 3.1, 3.2, 3.3, and 3.8 are connectors that represent the interconnection 3. Interconnection diameter, start and end positions, at corresponding components, are given as part of the problem specification while their shape and position are determined as part of the solution. Interconnection length and the number of connectors should be *minimized*. The *packaging volume* (PV) is an enclosure of all components and interconnections. PV shape and dimensions can be either fully or partially specified as part

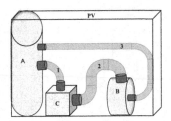

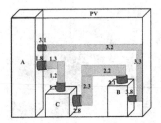

Fig. 3. An illustrative SPP solution in Euclidean geometry

Fig. 4. An illustrative SPP solution in the taxicab geometry

of the problem. PV dimensions can also be part of the optimization objectives: the total PV volume or some of its dimensions should to be minimized. Finally, no spatial interference between components, interconnections, or the packaging volume (PV) is allowed. The formulation above captures the key elements that if solved would be most important in industrial applications. In this paper, we focus on the feasibility version of the SPP in which the PV dimensions are fixed. Also, our approach assumes a fixed 3D orientation of input components, i.e., it does not attempt to rotate the components in search of feasible solutions.

Reduction to Taxicab Geometry. In this paper, we work on a specialization of the SPP through projection into the *taxicab geometry* [12] which transforms certain Euclidean-geometric forms into orthogonal forms (rectangular prisms). Figure 4 shows a taxicab solution corresponding to the Euclidean solution in Fig. 3. In this solution, all components, interconnections, and packaging volume consist of rectangular prisms. The interconnections become rectangular prisms with fixed width/height and variable length. The taxicab geometry defines Manhattan distance $d_T(P, Q) = |x_P - x_Q| + |y_P - y_Q|$ as a distance metric between the points (P and Q) which allows reduction of the possibly infinite number of pipe directions in Euclidean geometry to only 6 possible directions in 3D space. Only interconnection bends of 90° are allowed, and each connector is parallel to one of the 3 principal axes. This significantly reduces the computational complexity and simplifies modelling of SPP as a CP problem.

The taxicab relaxation is used to generate a diverse set of solutions as starting points for design. Each solution to the taxicab-relaxed SPP captures core aspects of a corresponding feasible Euclidean solution and can be transformed back to it for post processing, as an initial assignment for a local optimization. This is done by defining round cross-section pipes inside the rectangular interconnections, rounded bends at their corners, and rotated components inside the component boxes with internal connections.

3   Constraint Programming Model

We describe the input data, decision variables, and constraints that form the CP model of SPP. SPP can be seen as a special packing problem with strongly constrained items, some of which have variable dimensions.

Input Data. L_{pv}, W_{pv}, H_{pv} denote the maximum length, width, and height of the *packaging volume* (PV), respectively. C denotes the set of *components*, and for each $c \in C$, L_c, W_c, H_c denote the component length, width and height respectively. Op_c denotes the *origin point* of the component c, i.e., the corner with the smallest x, y and z coordinates. Cx denotes the set of all *interconnections* and MB denotes the maximum number of bends in any interconnection. For each interconnection $cx \in Cx$, d_{cx} denotes its diameter, S_{cx} its source component, and L_{scx}, W_{scx}, H_{scx} the length, width and height of the source point Sp_{cx} relative to the Op_c of the source component. Analogously, T_{cx} is a target component while L_{tcx}, W_{tcx}, H_{tcx} are the length, width, and height of the target point Tp_{cx} relative to the origin point of the target component.

Decision Variables and Constraints. For each component c in C, we define three assignment variables X_c, Y_c, and Z_c respectively that indicate the coordinates of the origin point Op_c of component c. For each interconnection $cx \in Cx$, variables $X_{scx}, Y_{scx}, Z_{scx}$ denote the coordinates of source point Sp_{cx} while $X_{tcx}, Y_{tcx}, Z_{tcx}$ denote the coordinates of the target point Tp_{cx}.

For each $cx \in Cx$ we define CT_{cx} to be the set of all possible *connectors* which form the interconnection. The number of connectors is limited by the maximum number of bends, MB. For each connector $ct \in CT_{cx}$, variable $E_{ct} \in \{0,1\}$ indicates if the connector *exists* as part of the interconnection cx. Each connector has an Op_{ct} which denotes a corner with the smallest x, y, and z coordinates. Variables X_{ct}, Y_{ct}, and Z_{ct} denote the coordinates of Op_{ct} while L_{ct}, W_{ct} and H_{ct} denote the dimensions of ct along the x, y, and z-axis. Geometric position of a connector is fully specified by an assignment to the above variables. We model the relationship of a connector to its neighbouring connectors by introducing a variable $dir_{ct} \in [1,6]$ which indicates the *relative direction* of connector ct with respect to its predecessor connector: 1 represents a direction that increases x-axis, 2: direction that decreases x-axis, 3 and 4: increase and decrease of y-axis, while 5 and 6 increase and decrease z-axis respectively.

Interconnection Constraints ensure that connectors form interconnections that connect appropriate start and end points at given components. $CT_{cx} = \{ct_1, \ldots, ct_{MB}\}$ denote the maximal set of connectors for an interconnection cx. Connectors in CT_{cx} have a fixed diameter d_{cx}. Depending on the direction of a connector ct_i, two of its dimensions must be fixed, e.g., if the direction increases or decreases x-axis coordinates, then its dimension along the y and z-axis are assigned to d_{cx}, i.e. $H_{ct_i} = W_{ct_i} = d_{cx}$. We further enforce that the minimal dimensions of each connector is d_{cx}. Furthermore, we enforce that the first and the last connector have all dimensions fixed to a diameter, i.e. they are *cubes* with the size equals to interconnection diameter, i.e. $L_{ct_1} = H_{ct_1} = W_{ct_1} = d_{cx}$, $L_{ct_{MB}} = H_{ct_{MB}} = W_{ct_{MB}} = d_{cx}$. The taxicab distance between the first connector ct_1 and the source-point Sp_{cx} must be equal to 0. The taxicab distance between the last connector ct_{MB} and the target-point Tp_{cx} must be equal to 0.

We model the variable number of connectors by allowing connectors preceding the last connector not to exist. Therefore, the connector ct_{MB-1} either does not exist, or its distance to the last connector ct_{MB} is 0. For each connector ct_i

other then the last connector ($i \neq$ MB), if connector does not exist ($E_{cti} = 0$) then its successor does not exist as well ($E_{cti+1} = 0$). If connector ct_i does not exist then all its coordinates and dimensions are equal to 0, i.e. $X_{ct} = Y_{ct} = Z_{ct} = L_{ct} = W_{ct} = H_{ct} = 0$. If connector ct_i exists, it is either connected to the last connector ct_{MB} or its successor ct_{i+1} exists. Each intermediate connector ct_i must start at the point where its predecessor stops. This is best understood as a restriction that the *ending cube* of a connector's predecessor with dimensions $d_{cx} \times d_{cx} \times d_{cx}$ is *sharing a surface* of dimensions $d_{cx} \times d_{cx}$ with the *starting cube* of the successor connector. Which surface is shared depends on the relative direction of the connector. For example, if ct_i is increasing across the z-axis, then the surface shared with ct_{i-1} is in a plane parallel with the x and y-axis. Finally, we enforce that if a connector ct_i exists, it can neither reverse nor maintain the direction of its predecessor since in the first case the connector would occupy the same area as its predecessor, while in the second case it could be replaced with a longer version of the predecessor connector.

In addition to the above constraints, we also enforce *Interference Constraints* which forbid overlap of rectangular prisms in 3D space. There is a 3D version of a general $diffn$ constraint [1] which is supported by several solvers and can also be implemented as $\frac{n \cdot (n-1)}{2}$ of pairwise interference constraints. This formulation is straightforward and we omit it due to space limitations.

4   Search Heuristic

A search heuristic significantly improved search-space exploration efficiency. Our search strategy selects branching variables and values based on the *volume* of components and interconnections. For a given component c, its *component volume* V_c denotes the volume in a standard geometric sense: $L_c \times W_c \times H_c$. For a given interconnection cx, its *interconnection volume* V_{cx} is defined as the sum of volumes of source and target component (S_{cx}, T_{cx}) as well as the volumes of all connectors directly attached to the source and the target components *regardless* of which interconnection they belong to. We refer to them as *attached-connectors*. An example of an interconnection volume is shown in Fig. 4: the volume of the interconnection 3, connecting the component A and B, is $V_3 = V_{3,A} + V_{3,B}$, where $V_{3,A}$ includes the volume of component A (V_A) as well as the volumes of connector $V_{3.1}$ and $V_{1.8}$. Note that connector 1.8 is not part of interconnection 3. Analogously, volume $V_{3,B}$ includes the volume of component B (V_B) as well as the volumes of connectors $V_{2.1}$ and $V_{3.8}$.

The heuristic proceeds as follows. It first sorts interconnections in a descending order based on their volumes. For each interconnection cx (in this order), the component with largest volume – that is attached to cx – is packed first. Using this order, the search heuristic packs each component with its attached-connectors; then, its interconnections within the existing set of already packed components are built: interconnections with a larger diameter are built first. To build an interconnection, the search heuristic starts by sequentially assigning the direction and the dimension of connectors from first to last.

5 Experimental Evaluation

In this paper, we propose a SPP benchmark that is the result of a long and careful process: we reviewed applications with product designers in a diverse set of United Technologies Corporation businesses, gathered a diverse set of relevant product architectures, and created a benchmark to represent their range of number of components and interconnections and their relative dimensions. The benchmark is typical of the SPP challenges encountered in commercial and defense applications within elevator, jet engine, helicopter, and aircraft subsystem businesses while also being a suitable non-proprietary challenge problem that can be shared broadly while complying with international trade regulations and intellectual property restrictions. A detailed description of the benchmark can be obtained directly from the authors.

The SPP benchmark describes ten components with volumes ranging from 4.000 to 36.000 units. These components are connected via fifteen interconnections with three different diameters: three interconnections with diameter of 5 units, nine interconnections with diameter of 10, and three interconnections with diameter of 20. Each component is involved in at least two, and at most four interconnections. We relaxed this benchmark using the Taxicab geometry and derived 212 SPP instances: we randomly generated fourteen sets of fifteen SPP instances with sizes ranging from 1 to 14, one size 0 instance, and one size 15 instance. Each instance, with size n, contains the ten components and n different interconnections that are randomly selected from the initial SPP benchmark.

5.1 Results

To better evaluate our heuristic, we compare it to a widely used heuristic for packing problems that chooses variables/values based on component volume (i.e., the highest-volume component is packed first) [4,15].

The SPP model and the search heuristic were implemented using the C# library Google OR-Tools solver [18]. We also implemented a search heuristic based on component volume. We performed our experiments on the 212 SPP instances using an Intel Core i7 (2.8 GHz, 8 GB Ram). Each run finishes with a success, when a feasible solution is found, or an out-of-time if it exceeds 1 min. The goal of generating a diverse set of taxicab solutions drives the 1-minute upper-bound for computing each solution.

Table 1 shows the results, grouped by number of interconnections, shown in the first column. The second column shows the number of instances per group. For each heuristic, we show the success rate (% Success) and the computation time (Avg. Time) in milliseconds, averaged over the instances that were successfully solved. Our heuristic, Heuristic Based Interconnection (HBI), outperforms a widely used packing-problem heuristic, Heuristic Based Component (HBC). HBI solved all the groups of instances with a high success rate, whereas HBC could not solve the benchmark with size 15 and its overall success rate is low, i.e., HBC solved only one instance with 14 interconnections and only 6 instances with 8 interconnections. For the solved instances, both heuristics found the solutions

Table 1. Percentage of success in terms of the number of interconnections

# inter.	# instances	H. based interconnection		H. based component	
		% success	Avg. time (ms)	% success	Avg. time (ms)
0 to 7	15	100	<715	83–100	<590
8	15	93	815	40	658
9	15	87	947	40	816
10	15	87	1.129	27	1.010
11	15	93	1.348	13	1.214
12	15	73	1.549	20	1.447
13	15	67	1.651	13	1.684
14	15	67	2.068	7	1.783
15	1	100	8.129	0	–

within a few seconds with a slight advantage for HBC. This can be explained by the complexity and the number of instances solved, i.e., HBC solved fewer instances than HBI and these instances might be less complex than the others that HBI solved.

The instances with fewer interconnections (<8) were solved in less than 715 ms with 100 % success rate. For sizes ranging from 8 to 14, the model and HBI were grater than 67 % successful and could not solve 1 to 4 instances. The third column shows that the number of unsolved instances increases in terms of the number of interconnections. This can be explained by the increasing complexity of the problem (i.e., more interconnections). For example, if each interconnection, on average, uses only 4 connectors, for 10 interconnections there would be 40 connectors to be determined and packed which increases the number of decision variables in the model.

6   Conclusion

The Spatial Packaging Problem (SPP) is a challenging problem that involves two NP-hard problems: the 3D Packing Problem (3DPP) and the 3D Pipe-Routing Problem. The main feature that distinguishes the SPP from the traditional 3DPP is the interconnections that exist between its components. This feature adds wholly new dimensions of complexity because the shape and dimensions of the interconnections are unknown, and must be determined as part of the solution. This paper presented the first constraint programming model and search heuristic to formulate and solve the SPP. The proposed approach relaxes the problem and models the SPP as a constraint-satisfaction problem, then solves it using a search heuristic based on interconnection volume. We presented experimental results for a challenging benchmark which indicate that constraint programming is a promising approach for the SPP.

Our future focus will be on: (1) developing a global interconnection constraint to significantly reduce the SPP search space, and (2) combining the proposed approach with local search to optimally solve the SPP.

References

1. Beldiceanu, N., Contejean, E.: Introducing global constraints in chip. Math. Comput. Model. **20**(12), 97–123 (1994)
2. Cagan, J., Clark, R., Dastidar, P., Szykman, S., Weisser, P.: HVAC CAD layout tools: a case study of university/industry collaboration. In: Proceedings of the ASME 1996 Design Engineering Technical Conferences and Computers in Engineering Conference (1996)
3. Christodoulou, S.E., Ellinas, G.: Pipe routing through ant colony optimization. J. Infrastruct. Syst. **16**(2), 149–159 (2010)
4. Crainic, T., Perboli, G., Tadei, R.: Recent advances in multi-dimensionalpacking problems (2012). http://www.intechopen.com/books/new-technologies-trends-innovations-and-research/recent-advances-in-multi-dimensional-packing-problems
5. Delgado, A., Jensen, R.M., Janstrup, K., Rose, T.H., Andersen, K.H.: A constraint programming model for fast optimal stowage of container vessel bays. Eur. J. Oper. Res. **220**(1), 251–261 (2012)
6. Drumheller, M.: Constraint-based design of optimal transport elements. In: Proceedings of the Seventh ACM Symposium on Solid Modeling and Applications, pp. 401–412. ACM (2002)
7. Dyckhoff, H.: A typology of cutting and packing problems. Eur. J. Oper. Res. **44**(2), 145–159 (1990)
8. Fan, X., Lin, Y., Ji, Z.: The ant colony optimization for ship pipe route design in 3D space. In: The Sixth World Congress on Intelligent Control and Automation, WCICA 2006, vol. 1, pp. 3103–3108 (2006)
9. Hamedani, P.K., Hessabi, S., Sarbazi-Azad, H., Jerger, N.E.: Exploration of temperature constraints for thermal aware mapping of 3D networks on chip. In: 2012 20th Euromicro International Conference on Parallel, Distributed and Network-Based Processing (PDP), pp. 499–506. IEEE (2012)
10. Jia, X., Williams, R.: A packing algorithm for particles of arbitrary shapes. Powder Technol. **120**(3), 175–186 (2001)
11. Karaboga, D., Basturk, B.: A powerful and efficient algorithm for numerical function optimization: artificial bee colony (ABC) algorithm. J. Glob. Optim. **39**(3), 459–471 (2007)
12. Krause, E.F.: Taxicab Geometry: An Adventure in Non-Euclidean Geometry. Courier Corporation, London (2012)
13. Laporte, G.: The vehicle routing problem: an overview of exact and approximate algorithms. Eur. J. Oper. Res. **59**(3), 345–358 (1992)
14. Lengauer, T.: Combinatorial Algorithms for Integrated Circuit Layout. Springer Science & Business Media, New York (2012)
15. Lodi, A., Martello, S., Monaci, M.: Two-dimensional packing problems: a survey. Eur. J. Oper. Res. **141**(2), 241–252 (2002)
16. Park, J.H., Storch, R.L.: Pipe-routing algorithm development: case study of a ship engine room design. Expert Syst. Appl. **23**(3), 299–309 (2002)

17. Park, V., Corson, M.: A highly adaptive distributed routing algorithm for mobile wireless networks. In: INFOCOM 1997: Proceedings of the IEEE Sixteenth Annual Joint Conference of the IEEE Computer and Communications Societies. Driving the Information Revolution, vol. 3, pp. 1405–1413, April 1997
18. Perron, L.: Operations research and constraint programming at Google. In: Lee, J. (ed.) CP 2011. LNCS, vol. 6876, p. 2. Springer, Heidelberg (2011)
19. Ren, T., Zhu, Z.L., Dimirovski, G.M., Gao, Z.H., Sun, X.H., Yu, H.: A new pipe routing method for aero-engines based on genetic algorithm. Proc. Inst. Mech. Eng. Part G J. Aerosp. Eng. **228**(3), 424–434 (2014)
20. Thantulage, G.I.: Ant colony optimization based simulation of 3D automatic hose/pipe routing. Ph.D. thesis, Brunel University School of Engineering and Design Ph.D. theses (2009)
21. Toth, P., Vigo, D.: Models, relaxations and exact approaches for the capacitated vehicle routing problem. Discrete Appl. Math. **123**(1), 487–512 (2002)
22. Van der Velden, C., Bil, C., Yu, X., Smith, A.: An intelligent system for automatic layout routing in aerospace design. Innovations Syst. Softw. Eng. **3**(2), 117–128 (2007)
23. Wäscher, G., Haußner, H., Schumann, H.: An improved typology of cutting and packing problems. Eur. J. Oper. Res. **183**(3), 1109–1130 (2007)

Detecting Semantic Groups in MIP Models

Domenico Salvagnin[1,2]([⊠])

[1] IBM Italy, Segrate, Italy
[2] DEI, University of Padova, Padua, Italy
salvagni@dei.unipd.it

Abstract. Current state-of-the-art MIP technology lacks a powerful modeling language based on global constraints, a tool which has long been standard in constraint programming. In general, even basic semantic information about variables and constraints is hidden from the underlying solver. For example, in a network design model with unsplittable flows, both routing and arc capacity variables could be binary, and the solver would not be able to distinguish between the two semantically different groups of variables by looking at type alone. If available, such semantic partitioning could be used by different parts of the solver, heuristics in primis, to improve overall performance. In the present paper we will describe several heuristic procedures, all based on the concept of partition refinement, to automatically recover semantic variable (and constraint) groups from a flat MIP model. Computational experiments on a heterogeneous testbed of models, whose original higher-level partition is known a priori, show that one of the proposed methods is quite effective.

1 Introduction

Mixed-integer-programming (MIP) is a powerful paradigm to solve many combinatorial optimization problems coming from both theory and applications. Despite the admittedly limited set of constructs that are allowed in the paradigm, namely linear inequalities and integer constrained variables, it turns out that surprisingly many optimization problems of practical interest can be exactly, or approximately, formulated as MIP models [27]. At the same time, MIP solvers improved so much in the last decades that MIP is considered nowadays a mature technology. The seemingly limited language of MIP was indeed (partly) instrumental to its success: although powerful enough to model many optimization problems, it was at the same time easy enough to allow for a development of a rich and general theory, and for the definition of a standard file format (namely MPS) since the very beginning. By modeling an optimization problem as a MIP and solving it with a MIP solver, one takes advantage from decades of past (and future) developments in solver technology.

On the other hand, the limited language of MIP is actually a double-edged sword: modeling real-world problems as MIPs is far from obvious—a thing that seasoned MIP modelers tend to forget—and, more importantly, part of the global

© Springer International Publishing Switzerland 2016
C.-G. Quimper (Ed.): CPAIOR 2016, LNCS 9676, pp. 329–341, 2016.
DOI: 10.1007/978-3-319-33954-2_24

structure, which could be exploited by problem-specific approaches, is lost when translated into a flat MIP model. In general, even basic semantic information about variables and constraints is hidden from the underlying solver.

Let us consider for example the prepack optimization problem [23]. This problem arises in inventory allocation applications, where the operational cost for packing the bins is comparable, or even higher, than the cost of the bins (and of the items) themselves. Assuming that automatic systems are available for packing, the required workforce is related to the number of different ways that are used to pack the bins to be sent to the customers. Pre-packing items into box configurations has benefits in terms of easier and cheaper handling; on the other hand, it can reduce the flexibility of the supply chain, leading to situations in which the set of items that are actually shipped does not match exactly the demands—such deviations are usually penalized in the objective function. Using the notation in [15], a mixed-integer nonlinear model for the prepack optimization problem reads:

$$\min \sum_{s \in S} \sum_{i \in I} (\alpha u_{is} + \beta o_{is}) \tag{1}$$

$$q_{bis} = x_{bs} y_{bi} \quad (b \in B; i \in I; s \in S) \tag{2}$$

$$\sum_{b \in B} q_{bis} - o_{is} + u_{is} = r_{is} \quad (i \in I; s \in S) \tag{3}$$

$$\sum_{i \in I} y_{bi} = \sum_{k \in K} k\, t_{bk} \quad (b \in B) \tag{4}$$

$$\sum_{k \in K} t_{bk} = 1 \quad (b \in B) \tag{5}$$

$$o_{is} \leq \delta_{is} \quad (i \in I; s \in S) \tag{6}$$

$$t_{bk} \in \{0, 1\} \quad (b \in B; k \in K) \tag{7}$$

$$x_{bs} \geq 0 \text{ integer} \quad (b \in B; s \in S) \tag{8}$$

$$y_{bi} \geq 0 \text{ integer} \quad (b \in B; i \in I) \tag{9}$$

where I is the set of types of products, S the set of stores, $K \subset Z_+$ the set of available bin capacities, and B is the set of box configurations. Parameters r_{is} are the actual demands, while δ_{is} are upper bounds on the amount overstocking. As for variables, integer variables y_{bi} encode products' packing into boxes, while integer variables x_{bs} encode the shipping of box configurations to stores. Understocking and overstocking are expressed by decision variables u_{is} and o_{is}. Then, we have additional binary variables t_{bk} to map box configurations to bin capacities and additional integer variables $q_{bis} = x_{bs}\, y_{bi}$ used to count the number of items of type i sent to store s through boxes loaded with configuration b. Finally, the nonlinear products that define variables q_{bis} are actually formulated in a MIP framework by adding artificial binary variables (say v and w) that basically provide the binary expansion of variables x and y and the corresponding products. We refer to [15] for more details on the model. Although far from complex, model (1)–(9) is a typical example of the ingenuity needed to model a real-world problem as a MIP.

From a high-level point of view, model (1)–(9) is made of several different sets of variables that are semantically distinct: for example x variables encode shipping decisions, while y encode packing decisions. On the same line, v, and w are artificial binary variables that are needed for the sole purpose of being able to encode the constraints of the model as linear inequalities, and are also semantically different. However, this semantic grouping is completely lost and hidden from the MIP solver, that basically sees only a bunch of integer and binary variables. In other words, for the purpose of solving, model (1)–(9) gets flattened as an arbitrary general MIP model like:

$$\min\{c^T z : Az \leq b, z_j \in \mathbb{Z} \; \forall j \in J \subseteq \{1, \ldots, n\}\} \tag{10}$$

Semantic partitioning is not the only piece of information which is lost in the flattening process: the overall specific structure of the model (or part of it) is usually lost too, as well as the mapping between variables and elements of the sets of indices—actually, the index sets used during modeling are not even part of the model that is submitted to the solver. As a matter of fact, modern MIP solvers have a rich arsenal of algorithms that basically try to *reverse-engineer* combinatorial substructures from a flat model like (10). Unfortunately, while these procedures are usually cheap and effective, they are still heuristic in nature and can be fooled by the many transformations that are applied to a given MIP formulation in the preprocessing phase.

In this paper we are interested in general-purpose heuristic procedures to recover, or approximate, the semantic partitioning of variables (and constraints) present in the original high-level model from a flat one. The paper is organized as follows: in Sect. 2 we will overview existing literature on the subject and provide motivations for our study. In Sect. 3 we will present several different partitioning algorithms and discuss their respective strengths and weaknesses. In Sect. 4 we will present some computational results on a heterogenous testbed of models, showing that some methods are indeed quite successful in this reconstruction. Conclusions are finally drawn in Sect. 5.

2   Related Work

Current state-of-the-art MIP technology lacks a powerful modeling language based on global constraints, a tool which has long been standard in constraint programming [32]. For this reason, it has become standard practice in MIP implementations to devise algorithms that basically try to *reverse-engineer* combinatorial substructures from a flat list of linear inequalities. In [3], a procedure for detecting network structures was presented; such structure, when present, is then used to improve cutting plane separation. In [34], a procedure for detecting permutation problems, i.e., problems that optimize and arbitrary objective function over the set of all possible permutations of a given ground set, was introduced, with the purpose of devising a specialized primal heuristic for this class of problems. Similarly, MIP solvers often have simple heuristics to detect whether the problem at hand admits a specialized solution algorithm—for example, a knapsack problem might be solved via dynamic programming—and switch to the latter according to some effort predictions.

In addition to algorithms that look for specific structures, like networks and permutations, modern MIP solvers also detect, usually during the preprocessing stage, general-purpose global structures that are used later in the process to improve the performance of the solver. Examples of global structures that are widely used include the clique table, the implication graph [36], and symmetries [28]. Those global structures are used to improve different parts of the solver, like domain propagation, cutting plane generation, and branching, see, e.g., [1]. Recently, in [18], the clique table and the implication graph have been used to define neighborhoods for a LNS primal heuristic.

Semantic partitioning naturally belongs to the class of general-purpose global structures, like the clique table or the symmetry group of the formulation. Such piece of information, if available, could be used in many different components of a MIP solver. In particular:

– *branching:* branching rules could be biased in order to prefer branching on variables of the same class. This could help when other branching scores are flat.
– *aggregation:* if a variable from a given class can be aggregated out, chances are that all variables in the same class can be aggregated out—this happens, for example, in many time-indexed formulations for scheduling problems [8]. This would eliminate part of the guess-work in the aggregation heuristics.
– *relaxation:* a partitioning of constraints based on semantics could open the way to automatic Lagrangian relaxations, where constraints from the same class are relaxed into the objective function.
– *primal heuristics:* semantic groups can be used to devise neighborhoods in LNS approaches, for biasing fix-and-dive heuristics and to implement general-purpose metaheuristics. An example is the alternate heuristics: given two subsets of variables, it consists in alternately solving the problem with the variables of one of the subsets fixed, and this is quite effective for some classes of problems like pooling [6] and prepack optimization [15].

An alternative approach to the one studied here consists in extending the solver to accept a higher-level model in the first place. In such an enriched environment, the user is allowed to model the problem using expressions that compactly encode complex substructures, and the solver, which is fully aware of those expressions, can take advantage of specialized methods. This is the de-facto standard in constraint programming, where *global constraints* are used exactly for this purpose. A further generalization of the concept is *metaconstraints* [12,20]: a metaconstraint is syntactically specified much as a global constraint is, but it is also amended with additional annotations that specify how it is to be relaxed, how it is supposed to do constraint propagation and to direct search (via branching) in case it is violated. Metaconstraints, pioneered in the modeling system SCIL [5], are partially supported in recent versions of high-level modeling systems, like AMPL [17], ECLiPSE [30], MiniZinc [37] and

SIMPL [40]: however they are fully exploited only if the underlying solver provides some native support for them, which is currently not the case for most MIP solvers. A notable exception is the open-source solver SCIP [2], whose *constraint handlers* are basically metaconstraints implemented at the C level.

While exploiting, as opposed to reverse-engineering, higher-level knowledge has clear benefits and should be the winning approach in the long run, we have to face the fact that currently most state-of-the-art MIP solvers accept only very limited extensions w.r.t. the regular MIP language, namely indicator constraints [25] and piece-wise linear functions [19]. In addition, a modeling language based on metaconstraints is not without its share of problems: as the solver automatically translates global constraints into a MIP model, it often creates auxiliary variables. Variables introduced by different metaconstraints might actually have the same meaning, but without some form of additional annotations, like the *semantic typing* proposed in [12], the solver is unable to recognize such relationships and produce a tight model, equivalent to what a human modeler might produce by hand. The issue of modeling with metaconstraints versus reverse-engineering them from a flat model is also discussed, among others, in [5,21,22,31].

Finally, there are connections between the subject of the current paper and symmetry detection [28]: intuitively, if two variables are symmetric, they also belong to the same semantic class, but the converse is not true. As such, semantic grouping generalizes the orbit partitioning that is obtained as a side product of symmetry detection, and it applies to a much wider range of problems, albeit with a completely different usage.

3   Detection Algorithms

Recovering the high-level partition of variables (and constraints) is inherently an ill-posed problem, as MIP solvers cannot truly have a notion of semantics. As such, we need to replace the notion of belonging to the same semantic class with something that is within the reach of the solver and can be inferred from the model alone. In the present paper, we propose an overall approach based on partition refinement, a basic tool in computer science. According to [4], partition refinement is defined as follows: given a set S, an initial partition π of S into pairwise disjoint blocks (also called *cells*) $\{B_1, \ldots, B_p\}$, and a function f on S^1, the task is to find the coarsest partition of S, say $\pi' = \{E_1, \ldots, E_q\}$, such that:

1. π' is consistent with π, that is each E_i is a subset of some B_j;
2. π' is compatible with f, which means that a and b in E_i imply $f(a)$ and $f(b)$ are in some E_j.

Partition refinement can be implemented in $O(n \log n)$ time, for arbitrary π and f, where $n = |S|$.

---

[1] In applications, such as graph automorphism and DFA minimization, function f is extended to dependent on a more complex domain than just S.

In our context, the partition we are interested in is obviously the one of variables and constraints in the model. The final outcome will depend on two choices, namely the initial partition π and function f. As for the initial partition, we can split variables according to their type and, optionally, according to whether they appear in the objective function or not. For constraints, we can start from their initial classification. Constraint classification [2,16] is a technique used to achieve faster constraint propagation: for example, variable bounds, or set covering constraints, can be propagated way more efficiently than an arbitrary linear constraint, and MIP solvers usually implement some form of classification in order to take advantage of that. More details about the constraint classes used in this paper are given in Sect. 4. As for function f, it must necessarily take into account how variables and constraints are structurally connected in model: a convenient tool is to encode the connections we are interested in a graph, and then define f accordingly—in this case function f usually encodes some form of *vertex invariant* in the graph. This is the approach taken, for example, in symmetry detection codes [13,14,29], that work by implicitly constructing an auxiliary graph and computing its automorphism group. Incidentally, partition refinement is a crucial building block of all graph automorphism packages. In those algorithms, function f is the so called *connection function*: given a vertex v and a set of vertices B, $f(v, B)$ gives the number of elements in B which are connected to v. In other words, each refinement step will pick a *target cell* B_i, an *inducing cell* B_j, and it will split the vertices in B_i according to their connection count w.r.t. B_j. Note that there is no actual choice of B: each time a cell is split, its pieces will act in turn as inducing cells, and the whole process is iterated until a fixed point is reached. An example of partition refinement according to the connection function is depicted in Fig. 1.

In the rest of this section, we will describe several partitioning schemes, that can all be cast into the refinement framework just described.

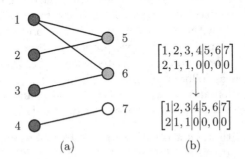

(a)   (b)

Fig. 1. Example of partition refinement according to connection function. The first row of each matrix on the right encodes the current partition, while the second gives the connection function w.r.t. cell [5, 6].

3.1 Simple Refinements

Some simple strategies rely entirely on carefully constructing an initial partition π, and take function f as the identity. In this case, it trivially holds that $\pi' = \pi$. The two strategies that we tried in this class are:

- type: partition variables by type alone, and constraints according to their initial classification.
- histogram: partition variables and constraints according to their so called *histogram*. For variables, this amounts to counting the number of constraints in which each variable appears, and partition based on this count—analogously for constraints.

It is quite obvious that method type will not be powerful enough in most cases, as it is often not possible to distinguish semantically different variables by type alone—model (1)–(9) is an example, as is network design with unsplittable flows. However, it is convenient to have the method as a baseline for benchmarking.

3.2 Iterative Refinements

The strategies in this class implement the full-blown partition refinement algoritm. They all start from the same initial partitioning described at the beginning of the section, but use different connection functions. The three strategies that we tried in this class are:

- fast. We construct a bipartite graph $G = (V, K, E)$, where V is the set of variables, K is the set of constraints classes, and there is edge (v_i, k_j) in the graph if and only if variable v_i appears in at least one constraint of class k_j. The connection function f is the regular connection function used in graph automorphism packages.
- recursive. We construct a bipartite graph $H = (V, C, E)$, where V is the set of variables, C is the set of constraints, and there is edge (v_i, c_j) in the graph if and only if variable v_i appears in constraint c_j. The connection function f is a modified version of the regular connection function, that ignores actual counts when deciding how to split a cell. In other words, a target cell B_i is split only distinguishing between its elements that connect to the inducing cell B_j from those that do not, without further refinement based on the actual counts. On the example in Fig. 1, refining cell $[1, 2, 3, 4]$ w.r.t. $[5, 6]$ would thus yield $[1, 2, 3 | 4]$ instead of $[1 | 2, 3 | 4]$.
- auto. We construct the same bipartite graph $H = (V, C, E)$ as in recursive, but use the regular connection function. Note that H is the very same bipartite graph that would be constructed to compute symmetries in a binary model [33], and auto then just performs the initial refinement step without doing the enumeration required to properly compute the set of generators.

It is worth noting that fast could have been equivalently defined as using the same connection function as recursive: given that the graph is bipartite, and one set of vertices, namely K, is already partitioned into blocks of size 1, no refinement can happen on K, and the connection count of a vertex $v \in V$ with a cell of K is always at most 1. As a result, fast can also be equivalently implemented with a specialized algorithm that just computes for each variable the subset of constraint classes it appears in and then just splitting V according to this piece of information, with a bucket-sort like procedure. Note also that fast does not produce a partitioning for constraints, although such a partition can be computed a-posteriori, with a second bucket sort in which the roles of constraints and variables are reversed.

The iterative refinement used in these methods is needed to be able to distinguish variables depending of the class of variables they connect to and not just basing on the kind of constraints they participate in: for example, in the prepack model, we have two sets of binary variables, namely w and v, that encode a binary expansion of two semantically different sets of variables (x and y respectively). Without further information, a method like fast would not be able to distinguish w from v, as those variables appear always in constraints of the same kind. At the same time, this behaviour can be an overkill and lead to an artificial split of variables. Consider, for example, a flow formulation on a layered graph[2]: the flow variables associated to the first (and/or last) layer are usually connected to some other variables in the model, while those associated with inner layers are not. Iterative refinement will not only distinguish outer layers from inner layers, but also recursively partition flow variables by layer, ending up with a semantic class for each and every layer.

A common characteristic of all methods is that they completely ignore the actual values of the coefficients in the model, but rather distinguish only between zero and non-zero values. This is in stark contrast with the symmetry detection case: actual values are needed to compute proper symmetries, but they are completely unsuited for semantic grouping, as variable semantics are independent of numerics. Similarly, actual connection counts are ignored for all methods but auto and histogram, although the situation is not as obvious as for values: in some cases connection counts could indeed help, but they would make the partitioning process too sensitive to trivial changes. For example, fixing a variable—and getting rid of it during presolve—would create an unwanted distinction between the variables that were connected to it and those that were not, while ideally it should be a neutral chance in most circumstances. Similarly, connection counts can be misleading when zero is a legitimate value for a parameter. For example, on instance markshare_5_0 from MIPLIB2010, which is basically a multidimensional subset sum problem, the constraint matrix is randomly generated and is *almost* fully dense: using exact counts would split the (very few) variables whose columns do not cover all rows from those who do.

Finally, all methods come in two variants: one in which the initial partition takes into account the objective function, by distinguishing whether a variable appears in it or not, and one in which no such distinction is made. It is not

---

[2] The same argument applies to the more common case of a general graph, but we will consider the layered case for simplicity.

obvious which one is more suited for the job: in some cases, like balanced sub-graph problems [24], the objective function is the only way to distinguish between vertex variables and edge variables, given the current constraint classification. On the other hand, as with the constraint matrix, zero might be a (rare but) legitimate value for some of the parameters of the model, say a cost, hence the resulting distinction would be artificial.

4 Computational Results

We implemented our code in C++, using IBM ILOG CPLEX 12.6.2 [25] as MPS reader. All tests have been performed on a PC with an Intel Core i5 CPU running at 2.66 GHz, with 8 GB of RAM. We collected a heterogeneous set of instances—most included in MIPLIB2010 [26]—whose high-level structure was either known or easily recoverable (by the author) from variables' and constraints' names, and used that as our testbed. As for constraint classification, we considered the following classes:

- *set covering:* inequality of the form $\sum_j x_j \geq 1$, involving binary variables only (possibly complemented).
- *set partitioning:* equality of the form $\sum_j x_j = 1$, involving binary variables only (possibly complemented).
- *set packing:* inequality of the form $\sum_j x_j \leq 1$, involving binary variables only (possibly complemented).
- *cardinality:* inequality of the form $\sum_j x_j \leq K$ or $\sum_j x_j \geq K$, involving binary variables only (possibly complemented).
- *cardinality equation:* equality of the form $\sum_j x_j = K$, involving binary variables only (possibly complemented).
- *variable bounds:* inequality of the form $ax \leq by$ or $ax \geq by$, with y binary.
- *mixed:* any inequality that does not fall in any of the classes above.
- *mixed equality:* any equality that does not fall in any of the classes above.

This classification is pretty basic, yet it can distinguish most of the constraint classes used in practice, and can be implemented (and executed) very efficiently with a single pass through the constraint matrix.

   We tested all the methods described in the previous section, both in their *objective* and *no-objective* variants: detailed results about the *objective* variant are available in Table 1. In the table, we provide basic statistics about each instance (namely, number of rows m and columns n), the known number of variable semantic groups (column g), and the number of groups identified by the proposed methods.

   Interpreting the results of the table is not straightforward, as a method may have incorrectly partitioned the set of variables even if it got the number of blocks right, so a deeper analysis of the outcome of the algorithms is needed. Here are some preliminary conclusions drawn from the synthetic numbers in the table plus a detailed analysis of the outcome of the algorithms on the individual instances. No method is always recovering the original semantic partitioning,

Table 1. Number of variable groups found by *objective* variant of partition refinement methods.

Instance	Size			Methods				
	n	m	g	Type	Fast	Recursive	Histogram	Auto
ash608gpia-3col [26]	3,651	24,748	2	2	2	2	18	1,217
csched007 [39]	1,758	351	4	3	4	1,758	4	1,758
dfn-gwin-UUM [41]	938	158	3	3	3	8	3	26
n3700 [38]	10,000	5,150	2	2	2	2	2	2
reblock166 [9]	1,660	17,024	1	1	1	1	28	1,660
Seymour [26]	1,372	4,944	1	1	1	1	126	1,155
Toll-like [24]	2,883	4,408	2	2	2	2	31	2,163
triptim1 [10]	30,005	15,076	16	4	13	30,055	55	30,055
uc-case3 [26]	37,749	52,003	7	3	6	12,898	190	14,410
Wachplan [26]	3,361	1,553	5	2	4	617	50	673
Prepack [15]	84,376	197,154	8	4	8	12	8	12
Multiactsched [35]	11,180	8,222	6	4	6	975	11	5590
Classification [11]	204	100	5	3	4	4	4	4
zib54-UUE-SAN [41]	240,240	81,134	2	2	2	2	2	150
Fac.location [7]	90,300	90,601	2	3	3	7	3	7

confirming that such a reverse-engineering is not a trivial task. In addition, it is pretty clear that all methods that rely on exact counts, namely histogram and auto, perform quite poorly, splitting the set of variables in way too many blocks. The phenomenon is quite clear, for example, on seymour, which is a pure set covering model, with all variables belonging to the same class. Surprisingly, also recursive, which does *not* keep exact counts, often splits the blocks too finely, in particular for scheduling models. The issue there (and on the multi-activity from [35]) is exactly the one described in the previous section concerning flow models.

Overall, fast qualifies as the best method: it is able to refine over the initial partitioning (type) when needed, and it never returns too fine a partition, achieving a reasonable approximation of the true number of blocks. The method is still not perfect though: for example, on the prepack model it cannot distinguish understocking from overstocking variables (and similarly for the multi-activity scheduling instance), and on the classification model it cannot distinguish the coefficients of the separating hyperplane from its right hand side–although whether the two are actually semantically different is debatable.

As for running times, all methods except recursive and auto always execute in a fraction of a second, while the other two can be relatively expensive, up to a few seconds. In any case, all methods required a negligible time w.r.t. the time that is needed to solve the instances, so that detection runtime is never an issue, and we are actually free to choose the method to apply based on success rather than on time.

Concerning the *no-objective* variant, the results are mixed: ignoring the objective function is not enough to fix the intrinsic weaknesses of recursive and histogram, and it only marginally affects fast, which is often able to infer the same partitioning with and without objective. On the one hand, ignoring the objective fixes the behaviour of fast on the facility location problems; on the other hand, it causes missing refinements on 5 other instances. Overall, taking the objective function into accounts seems slightly superior.

5 Conclusions

We described a family of procedures, all based on partition refinement, to heuristically recover the semantic grouping of variables from a MIP model. The problem is inherently ill-posed, given the lack of a truly semantic notion within MIP solvers, but partition refinement seems to capture decently well the concept of structurally equivalent variables, which is a proxy for semantic equivalence. Indeed, one of the proposed methods is quite successful at recovering the high-level variable partitioning of the model on an heterogeneous testbed of problems.

Still, no method is perfect, and for each example that supports a design decision, like whether to ignore the objective function or to iteratively refine the partition, there is a counterexample supporting the very opposite, confirming the fact that once semantic information is lost, it is quite difficult to recover in a robust way, if at all. As such, the experiments in the present paper are yet another piece of evidence that we should not dismiss so easily high-level knowledge about the optimization models that we want to solve.

References

1. Achterberg, T.: Constraint integer programming. Ph.D. thesis, Technische Universität Berlin (2007)
2. Achterberg, T.: SCIP: solving constraint integer programs. Math. Program. Comput. **1**(1), 1–41 (2009)
3. Achterberg, T., Raack, C.: The MCF-separator: detecting and exploiting multicommodity flow structures in MIPs. Math. Program. Comput. **2**(2), 125–165 (2010)
4. Aho, A., Hopcroft, J., Ullman, J.D.: The Design and Analysis of Computer Algorithms. Addison-Wesley, Reading (1974)
5. Althaus, E., Bockmayr, A., Elf, M., Jünger, M., Kasper, T., Mehlhorn, K.: SCIL - symbolic constraints in integer linear programming. In: Möhring, R.H., Raman, R. (eds.) ESA 2002. LNCS, vol. 2461, p. 75. Springer, Heidelberg (2002)
6. Audet, C., Brimberg, J., Hansen, P., Digabel, S.L., Mladenovic, N.: Pooling problem: alternate formulations and solution methods. Manag. Sci. **50**(6), 761–776 (2004)
7. Avella, P., Boccia, M.: A cutting plane algorithm for the capacitated facility location problem. Comput. Optim. Appl. **43**(1), 39–65 (2009)
8. Baker, K.R., Trietsch, D.: Principles of Sequencing and Scheduling. Wiley, New York (2009)

9. Bley, A., Boland, N., Fricke, C., Froyland, G.: A strengthened formulation and cutting planes for the open pit mine production scheduling problem. Comput. Oper. Res. **37**, 1641–1647 (2010)
10. Borndörfer, R., Liebchen, C.: When periodic timetables are suboptimal. In: Kalcsics, J., Nickel, S. (eds.) Operations Research Proceedings 2007, pp. 449–454. Springer, Heidelberg (2008)
11. Brooks, J.P.: Support vector machines with the ramp loss and the hard margin loss. Oper. Res. **59**(2), 467–479 (2011)
12. Cire, A., Hooker, J.N., Yunes, T.: Modeling with metaconstraints and semantic typing of variables. INFORMS J. Comput. (to appear)
13. Darga, P.T., Liffiton, M.H., Sakallah, K.A., Markov, I.L.: Exploiting structure in symmetry detection for CNF. In: Proceedings of the 41th Design Automation Conference, DAC 2004, San Diego, CA, USA, 7–11 June 2004, pp. 530–534 (2004)
14. Darga, P.T., Sakallah, K.A., Markov, I.L.: Faster symmetry discovery using sparsity of symmetries. In: Proceedings of the 45th Design Automation Conference, DAC 2008, Anaheim, CA, USA, 8–13 June 2008, pp. 149–154 (2008)
15. Fischetti, M., Monaci, M., Salvagnin, D.: Mixed-integer linear programming heuristics for the prepack optimization problem. Discrete Optimization (to appear)
16. Fischetti, M., Salvagnin, D.: Feasibility pump 2.0. Math. Program. Comput. **1**(2–3), 201–222 (2009)
17. Fourer, R., Gay, D.M., Kernighan, B.W.: AMPL: A Modeling Language for Mathematical Programming. Thomson, Stamford (2003)
18. Gamrath, G., Berthold, T., Heinz, S., Winkler, M.: Structure-based primal heuristics for mixed integer programming. In: Fujisawa, K., Shinano, Y., Waki, H. (eds.) Optimization in the Real World. Mathematics for Industry, vol. 13, pp. 37–53. Springer Japan, Tokyo (2016)
19. GUROBI: GUROBI 6.0 User's Manual (2015)
20. Hooker, J.N.: Integrated Methods for Optimization. Springer, Heidelberg (2006)
21. Hooker, J.N.: Logic-based modeling. In: Appa, G., Pitsoulis, M., Leonidas, S., Williams, H.P. (eds.) Handbook on Modelling for Discrete Optimization, pp. 61–102. Springer, US (2006)
22. Hooker, J.N.: Hybrid modeling. In: van Hentenryck, P., Milano, M. (eds.) Hybrid Optimization: the Ten Years of CPAIOR, pp. 11–62. Springer, Heidelberg (2011)
23. Hoskins, M., Masson, R., Gauthier Melançon, G., Mendoza, J.E., Meyer, C., Rousseau, L.-M.: The prepack optimization problem. In: Simonis, H. (ed.) CPAIOR 2014. LNCS, vol. 8451, pp. 136–143. Springer, Heidelberg (2014)
24. Hüffner, F., Betzler, N., Niedermeier, R.: Separator-based data reduction for signed graph balancing. J. Comb. Optim. **20**, 335–360 (2010)
25. IBM ILOG: CPLEX 12.6.2 User's Manual (2015)
26. Koch, T., Achterberg, T., Andersen, E., Bastert, O., Berthold, T., Bixby, R.E., Danna, E., Gamrath, G., Gleixner, A.M., Heinz, S., Lodi, A., Mittelmann, H., Ralphs, T., Salvagnin, D., Steffy, D.E., Wolter, K.: MIPLIB 2010 - mixed integer programming library version 5. Math. Program. Comput. **3**, 103–163 (2011)
27. Lodi, A.: Mixed integer programming computation. In: Jünger, M., Liebling, T.M., Naddef, D., Nemhauser, G.L., Pulleyblank, W.R., Reinelt, G., Rinaldi, G., Wolsey, L.A. (eds.) 50 Years of Integer Programming 1958–2008 - From the Early Years to the State-of-the-Art, pp. 619–645. Springer, Heidelberg (2010)
28. Margot, F.: Symmetry in integer linear programming. In: Jünger, M., Liebling, T.M., Naddef, D., Nemhauser, G.L., Pulleyblank, W.R., Reinelt, G., Rinaldi, G., Wolsey, L.A. (eds.) 50 Years of Integer Programming. Springer, Heidelberg (2009)

29. McKay, B.D.: Practical Graph Isomorphism (1981)
30. Milano, M.: Constraint and Integer Programming: Toward a Unified Methodology. Kluwer Academic Publishers, Norwell (2003)
31. Mitra, G., Lucas, C., Moody, S., Hadjiconstantinou, E.: Tools for reformulating logical forms into zero-one mixed integer programs. Eur. J. Oper. Res. **72**(2), 262–276 (1994)
32. Rossi, F., van Beek, P., Walsh, T. (eds.): Handbook of Constraint Programming. Foundations of Artificial Intelligence. Elsevier, Amsterdam (2006)
33. Salvagnin, D.: A dominance procedure for Integer programming. Master's thesis, University of Padova (2005)
34. Salvagnin, D.: Detecting and exploiting permutation structures in MIPs. In: Simonis, H. (ed.) CPAIOR 2014. LNCS, vol. 8451, pp. 29–44. Springer, Heidelberg (2014)
35. Salvagnin, D., Walsh, T.: A hybrid MIP/CP approach for multi-activity shift scheduling. In: Milano, M. (ed.) CP 2012. LNCS, vol. 7514, pp. 633–646. Springer, Heidelberg (2012)
36. Savelsbergh, M.W.P.: Preprocessing and probing for mixed integer programming problems. ORSA J. Comput. **6**, 445–454 (1994)
37. Stuckey, P.J., Tack, G.: MiniZinc with functions. In: Gomes, C., Sellmann, M. (eds.) CPAIOR 2013. LNCS, vol. 7874, pp. 268–283. Springer, Heidelberg (2013)
38. Sun, M., Aronson, J.E., McKeown, P.G., Drinka, D.A.: A tabu search heuristic procedure for the fixed charge transportation problem. Eur. J. Oper. Res. **106**, 441–456 (1998)
39. Yunes, T.: CuSPLIB 1.0: A library of single-machine cumulative scheduling problems (2009). http://moya.bus.miami.edu/tallys/cusplib/
40. Yunes, T., Aron, I.D., Hooker, J.N.: An integrated solver for optimization problems. Oper. Res. **58**(2), 342–356 (2010)
41. SNDlib (2006). http://sndlib.zib.de

Revisiting Two-Sided Stability Constraints

Mohamed Siala$^{(\boxtimes)}$ and Barry O'Sullivan

Insight Centre for Data Analytics, Department of Computer Science,
University College Cork, Cork, Ireland
{mohamed.siala,barry.osullivan}@insight-centre.org

Abstract. We show that previous filtering propositions on two-sided stability problems do not enforce arc consistency (AC), however they maintain bound($\mathcal{D}$) consistency (BC(D)). We propose an optimal algorithm achieving BC(D) with $O(L)$ time complexity where L is the length of the preference lists. We also show an adaptation of this filtering approach to achieve AC. Next, we report the first polynomial time algorithm for solving the hospital/resident problem with forced and forbidden pairs. Furthermore, we show that the particular case of this problem for stable marriage can be solved in $O(n^2)$ which improves the previously best complexity by a factor of n^2. Finally, we present a comprehensive set of experiments to evaluate the filtering propositions.

1 Introduction

Many real world problems involve matching preferences between two sets of agents while respecting some stability criteria. For instance, in *College Admissions* one needs to assign students to colleges while respecting the students' preferences over colleges, the colleges' preferences over students, as well as college quotas [3]. Gale and Shapley introduced the first polynomial time algorithm for solving this problem in their seminal paper [3]. Since then a number of algorithms have been proposed for solving variants of these problems. Such ad-hoc methods are unlikely to be reusable if there are minor changes to the problem.

Constraint programming (CP) is a rich framework for modelling and solving many combinatorial problems. Expressing problems involving preferences in CP is extremely beneficial for tackling variants that involve side constraints. We consider the notion of two-sided stability as a global constraint. We first make the observation that the previous CP propositions on two-sided stability problems (such as [4,7,9,10]) do not enforce Arc Consistency (AC), however they do maintain Bound($\mathcal{D}$) Consistency (BC(D)). We propose an incremental algorithm that achieves BC(D) with $O(L)$ time complexity where L is the length of the preference lists, thereby improving the previously best known complexity of $O(c \times L)$ (where c is the maximum quota). We also present, for the first time, an adaptation of the filtering to achieve AC on this global constraint with an additional cost of $n \times L$ (where n is the number of residents).

This research has been funded by Science Foundation Ireland (SFI) under Grant Number SFI/12/RC/2289.

© Springer International Publishing Switzerland 2016
C.-G. Quimper (Ed.): CPAIOR 2016, LNCS 9676, pp. 342–357, 2016.
DOI: 10.1007/978-3-319-33954-2_25

Based on the BC(D) propagator, we show that the hospital/resident problem with forced and forbidden pairs can be solved in polynomial time. Furthermore, we show that the particular case of this problem for stable marriage can be solved in $O(n^2)$ which improves the previously best complexity by a factor of n^2. Finally, we present a set of experiments to evaluate the filtering efficiency on randomly generated instances. The experimental results show compelling evidence that AC does further prune the search space as compared with BC(D), however, it considerably slows down the exploration of the search space.

The remainder of the paper is organized as follows. In Sect. 2 we give the definitions and the notation used throughout the paper. In Sect. 3 we show that the level of filtering of previous CP approaches is only BC(D). Next, we show an optimal implementation of a BC(D) algorithm running in $O(L)$ time in Sect. 4. We also show how to use the same algorithm to achieve AC. In Sect. 5 we discuss the complexity of the hospital/resident problem with forced and forbidden pairs. Finally, we present the experimental results in Sect. 6.

2 Definitions and Related Work

2.1 Constraint Programming

Let $\mathcal{X}$ be a set of integer variables. A *domain* for $\mathcal{X}$, denoted by $\mathcal{D}$, is a mapping from variables to finite sets of integers. For each variable x, we call $\mathcal{D}(x)$ the *domain of the variable* x. We use $\min(x)$ to denote the minimum value in $\mathcal{D}(x)$ and $\max(x)$ to denote the maximum value in $\mathcal{D}(x)$. Let $[x_1, \dots, x_k]$ be a sequence of integer variables. A *constraint* C defined over $[x_1, \dots, x_k]$ is a finite subset of $\mathbb{Z}^k$. The sequence $[x_1, \dots, x_k]$ is the *scope* of C (denoted by $\mathcal{X}(C)$) and k is called the *arity* of C. A *support* for C in a domain $\mathcal{D}$ is a k-tuple τ such that $\tau \in C$ and $\tau[i] \in \mathcal{D}(x_i)$ for all $i \in [1, \dots, k]$. The constraint C is *Arc-Consistent* (*AC*) in $\mathcal{D}$ iff $\forall i \in [1, \dots, k]$, $\forall j \in \mathcal{D}(x_i)$, there exists a support τ for C in $\mathcal{D}$ such that $\tau[i] = j$. C is *Bound (D) Consistent* (*BC(D)*) in $\mathcal{D}$ iff $\forall i \in [1, \dots, k]$, there exists two supports τ_1 and τ_2 for C in $\mathcal{D}$ such that $\tau_1[i] = \min(x_i)$ and $\tau_2[i] = \max(x_i)$ [1].

2.2 The Hospital/Resident Problem

Given a sequence $\mathcal{S}$ of distinct elements and $j \in \mathcal{S}$, we denote by $\mathcal{S}^{-1}[j]$ the index i such that $\mathcal{S}[i] = j$. We define a complete order $\underset{\mathcal{S}}{\prec\!\prec}$ on $\mathcal{S}$ as follows: $\forall k, l \in \mathcal{S}, \ k \underset{\mathcal{S}}{\prec\!\prec} l$ iff $\mathcal{S}^{-1}[k] < \mathcal{S}^{-1}[l]$. We will also use the notation $l \underset{\mathcal{S}}{\succ\!\succ} k$ when $k \underset{\mathcal{S}}{\prec\!\prec} l$. In the context of preferences, $k \underset{\mathcal{S}}{\prec\!\prec} l$ (respectively $k \underset{\mathcal{S}}{\succ\!\succ} l$) can be understood as k *better* (respectively *worse*) than l with respect to $\mathcal{S}$.

In the *Hospital/Resident* (*HR*) problem (called College Admissions in [3]), we are seeking the assignment of residents $r_1, \dots, r_{n_R}$ to hospitals $\mathcal{H}_1, \dots, h_{n_H}$. Each hospital h_j has a capacity c_j as the maximum number of assigned residents. Each resident r_i has a sequence of integers $\mathcal{R}_i$ ranking some hospitals in a strictly increasing order of preferences. That is, r_i prefers hospital h_k to hospital h_l iff $k \prec\!\!\prec_{\mathcal{R}_i} l$. Conversely, each hospital h_j is associated with a sequence of integers $\mathcal{H}_j$ ranking some residents in a strictly increasing order. We denote by $len^r{}_i$ the length of $\mathcal{R}_i$ and $len^h{}_j$ the length of $\mathcal{H}_i$.

Let $\mathcal{E} = \{(i, j) \mid i \in [1, n_R] \wedge j \in [1, n_H] \wedge i \in \mathcal{H}_j \wedge j \in \mathcal{R}_i\}$ the set of acceptable pairs. A *matching* $\mathcal{M}$ is a subset of $\mathcal{E}$ where $|\{j \mid (i, j) \in \mathcal{M}\}| \leq 1 \ \forall i \in [1, n_R]$ and $|\{i \mid (i, j) \in \mathcal{M}\}| \leq c_j, \ \forall j \in [1, n_H]$. A resident r_i is said to be *unassigned* in $\mathcal{M}$ iff $|\{j \mid (i, j) \in \mathcal{M}\}| = 0$. Similarly, a hospital h_j is *under-subscribed* in $\mathcal{M}$ iff $|\{i \mid (i, j) \in \mathcal{M}\}| < c_j$. A pair $(i, j) \in \mathcal{E} \setminus \mathcal{M}$ is said to be *blocking* $\mathcal{M}$ iff the following two conditions are true:

1. r_i is unassigned in $\mathcal{M}$ or $\exists k \in [1, n_H]$ such that $(i, k) \in \mathcal{M}$ and $j \prec\!\!\prec_{\mathcal{R}_i} k$
2. h_j is under-subscribed in $\mathcal{M}$ or $\exists l \in [1, n_R]$ such that $(l, j) \in \mathcal{M}$ and $i \prec\!\!\prec_{\mathcal{H}_j} l$

A matching $\mathcal{M}$ is said to be *stable* iff there is no blocking pair for $\mathcal{M}$.

The Hospital/Resident (HR) problem is to find a stable matching for a given instance. The stable marriage problem (SM) is a particular case of HR where $c_j = 1$ for all $j \in [1, n_H]$. We assume, without loss of generality in the remainder of the paper, that a resident r has a hospital h in its preference list iff h has r in its preference list. In this case, the length of the preference lists $L = \sum_{i=1}^{n_R} len^r{}_i = \sum_{j=1}^{n_H} len^r{}_j$.

Gale and Shapley proposed an $O(L)$ algorithm for solving the HR problem [3]. The algorithm, known as the *resident-oriented Gale/Shapley algorithm (RGS)* returns the unique matching where each resident is assigned to the best possible hospital that it can be assigned to in any stable matching. A similar algorithm for hospitals (i.e. *hospital-oriented Gale/Shapley algorithm (HGS)*) exists and has the same complexity $O(L)$. RGS and HGS operate by removing residents/hospitals from preference lists. The intersection of the reduced lists (returned by RGS and HGS) is called the GS-lists. The GS-lists are important since every stable matching is included in it [5].

Theorem 1. *From [5]*

1. *The number of assigned residents per hospital is the same in all stable matchings.*
2. *If a resident r_i is unassigned in one stable matching then it is unassigned in all stable matchings.*
3. *If a hospital h_j is under-subscribed in one stable matching then it is assigned exactly the same residents in all stable matching.*

We will use the following notation:

- $HUnder = \{j \mid h_j$ is under-subscribed in all stable matchings$\}$.
- $HUnder_j = \{i \mid r_i$ is assigned to h_j in all stable matchings$\}$ where $j \in HUnder$
- $HFull = [1, n_H] \setminus HUnder$.
- $RUnassigned = \{i \mid r_i$ is unassigned in all stable matchings$\}$.
- $RFree = \{i \mid r_i \notin RUnassigned$ and $\{j \mid i \in HUnder_j\} = \emptyset\}$.

2.3   Related Work in Constraint Programming

We first describe one of the CP models for the HR problem proposed in [7]. Each resident r_i is associated with an integer variable x_i where $\mathcal{D}(x_i) = [1, .., len^r{}_i] \cup \{n_H + 1\}$. Each hospital h_j is associated with $c_j + 1$ integer variables $y_{j,k}$ ($k \in [0..c_j]$) where $\mathcal{D}(y_{j,0}) = 0$ and $\mathcal{D}(y_{j,k}) = [k, len^h{}_j] \cup \{n + k\}$. Assigning x_i to $n_H + 1$ is understood as the resident r_i being unassigned. Assigning x_i to a value $a \in [1, len^r{}_i]$ is semantically equivalent to assigning r_i to its ath favourite hospital. Similarly, assigning $y_{j,k}$ to a value $b \in [1, len^h{}_j]$ means that the bth favourite resident to h_j is assigned to the kth position of h_j. If $y_{j,k}$ is assigned to $\{n + k\}$ then the kth position for hospital h_j is not assigned to any resident. These variables are subject to the following constraints (in these constraints $p_{i,j}$ denotes the rank of hospital h_j in $\mathcal{R}_i$ and $q_{i,j}$ denotes the rank of the resident r_i in $\mathcal{H}_j$):

$$y_{j,k} < y_{j,k+1} \ (\forall j \in [1, n_H], \forall k \in [1, c_j - 1]) \tag{1}$$

$$y_{j,k} \geq q_{i,j} \implies x_i \leq p_{i,j} \ (\forall j \in [1, n_H], \forall k \in [1, c_j], \forall i \in \mathcal{H}_j) \tag{2}$$

$$x_i \neq p_{i,j} \implies y_{j,k} \neq q_{i,j} \ (\forall i \in [1, n_R], \forall j \in \mathcal{R}_i, \forall k \in [1, c_j]) \tag{3}$$

$$(x_i \geq p_{i,j} \wedge y_{j,k-1} < q_{i,j}) \implies y_{j,k} \leq q_{i,j} \ (\forall i \in [1, n_R], \forall j \in \mathcal{R}_i, \forall k \in [1, c_j]) \tag{4}$$

$$y_{j,c_j} < q_{i,j} \implies x_i \neq p_{i,j} \ (\forall j \in [1, n_H], \forall i \in \mathcal{H}_j) \tag{5}$$

We refer to this encoding as Γ. Enforcing AC on Γ yields to a domain that is equivalent to the GS-lists [7]. This property is important, however, it does not necessarily rule out all inconsistent values. The authors of [7] showed an efficient implementation of this encoding using one constraint. Their filtering runs in $O(c \times L)$ (where $c = \max\{c_j | j \in [1, n_H]\}$) and does not further prune the domains. In fact, in terms of the level of propagation, all previous work in the literature including SM [4,7,9,10] focus on showing how their encodings maintains the GS-lists and never investigate the question of completing the filtering.

Note also that the notion of GS-lists is not well-defined during search. That is, for instance, when few residents are assigned/unassigned to some specific hospitals at a given node of the search tree.

3   Characterizing the Level of Consistency

We show in this section that the previous CP models are not complete and enforce only BC(D).

Example 1 (Counter-Example). Consider the case where $n_R = n_H = 4, c_1 = c_2 = c_3 = c_4 = 1, \mathcal{R}_1 = [3, 2, 1], \mathcal{R}_2 = [4, 1, 3, 2], \mathcal{R}_3 = [2, 4, 3], \mathcal{R}_4 = [1, 3, 4], \mathcal{H}_1 = [1, 2, 4], \mathcal{H}_2 = [2, 1, 3], \mathcal{H}_3 = [3, 2, 4, 1], \mathcal{H}_4 = [4, 3, 2]$. The domain is initialised as follows: $\mathcal{D}(x_1) = \mathcal{D}(x_3) = \mathcal{D}(x_4) = \{1, 2, 3, 5\}, \mathcal{D}(x_2) = \{1, 2, 3, 4, 5\}, \mathcal{D}(y_{1,0}) = \mathcal{D}(y_{2,0}) = \mathcal{D}(y_{3,0}) = \mathcal{D}(y_{4,0}) = 0, \mathcal{D}(y_{1,1}) = \mathcal{D}(y_{2,1}) = \mathcal{D}(y_{4,1}) = \{1, 2, 3, 5\}$, and $\mathcal{D}(y_{3,1}) = \{1, 2, 3, 4, 5\}$. Constraint 2 with $y_{1,1} >= 1$ enforces $x_1 \leq 3$, hence removing the value 5 from $\mathcal{D}(x_1)$. A similar propagation is performed on Constraint 2 with $y_{2,1} \geq 1, y_{3,1} \geq 1$, and $y_{4,1} \geq 1$ and the value 5 is removed from

$D(x_2), D(x_3)$, and $D(x_4)$. Now consider Constraint 4 with $x_1 \geq 1 \wedge y_{3,0} < 4$. This enforces $y_{3,1} \leq 4$. The same constraint is triggered with $x_2 \geq 1 \wedge y_{4,0} < 3$, $x_3 \geq 1 \wedge y_{2,0} < 3$, and $x_4 \geq 1 \wedge y_{1,0} < 3$. Therefore the value 5 is removed from $D(y_{1,1}), D(y_{2,1}), D(y_{3,1})$, and $D(y_{4,1})$. No more propagation is needed. However, assigning x_2 to 3 does not belong to any solution.   □

Example 1 shows that AC on Γ is not sufficient to provide complete filtering. Note that since all capacities $c_j = 1$ this example confirms the property even for the particular case of stable matching.

In the rest of the paper we use 2-SIDEDSTABILITY$(\mathcal{X}, \mathcal{A}, \mathcal{B}, \mathcal{C})$ to denote the global constraint modelling 2-sided stability. More precisely, for a given HR problem:

- $\mathcal{X}$ is the set of variables $x_1, \ldots, x_{n_R}$ defined the same way in Γ,
- $\mathcal{A} = \{\mathcal{R}_1, \ldots, \mathcal{R}_{n_R}\}$
- $\mathcal{B} = \{\mathcal{H}_1, \ldots, \mathcal{H}_{n_H}\}$
- $\mathcal{C} = \{c_1, \ldots, c_{n_H}\}$

We show that AC on Γ enforces BC(D) on any domain $\mathcal{D}$.

Lemma 1. *If Γ is AC then*

1. $\forall i \in RUnassigned, \ D(x_i) = \{n_H + 1\}$.
2. $\forall j \in HUnder, \ \forall i \in HUnder_j, \ D(x_i) = \{k\}$ *such that* $k = \mathcal{R}_i^{-1}[j]$.
3. $\forall j \in HUnder, \ \forall k \in [1, |HUnder_j|], \ D(y_{j,k}) = \{a_k\}$ *such that* a_k *is the kth favourite resident to h_j whose index is in $HUnder_j$.*
4. $\forall j \in HUnder, \ \forall k > |HUnder_j|, \ D(y_{j,k}) = \{n + k\}$.
5. $\forall j \in HFull, \ \forall i \in [1, n_R]$, *if* $\exists k \in D(x_i)$, *such that* $j = \mathcal{R}_i[k]$, *then* $i \in RFree$.
6. $\forall i \in RFree, \ \forall j \in [1, n_H]$ *if* $\exists k \in D(x_i)$, *such that* $j = \mathcal{R}_i[k]$, *then* $j \in HFull$.
7. $|RFree| = \sum_{j \in HFull} c_j$.

Proof. The lemma is a direct consequence of Theorem 1 and the fact that AC on the initial domain is a superset of any domain returned by AC in the search tree. Recall that AC on the initial domain is equivalent to the GS-lists.   □

Lemma 2. *If Γ is AC then for all $i \in [1, n_R], \ 1 \leq k < \max(x_i)$, and $h = \mathcal{R}_i[k]$, we have $h \in HFull$.*

Proof. Consider the domain $\mathcal{D}^*$ obtained after enforcing AC on the initial domain. We know that this domain corresponds to the GS-lists. Therefore, assigning all variables to their maximum in $\mathcal{D}^*$ is a support (i.e. the hospital-oriented stable matching). Hence for all $1 \leq k < \max(\mathcal{D}^*(x_i))$, and $h = \mathcal{R}_i[k]$, $h \in HFull$ (from the definition of stability). Therefore, if Γ is AC on any arbitrary domain, then we know that $\max(x_i) \leq m$, and consequently for all $1 \leq k < \max(x_i)$, and $h = \mathcal{R}_i[k]$, we have $h \in HFull$.   □

Lemma 3. *If Γ is AC then $\forall j \in HFull, \ |\{i \mid \mathcal{R}_i[\min(x_i)] == j\}| = c_j$.*

Proof. Let $\Phi_j = \{i \mid \mathcal{R}_i[\min(x_i)] == j\}$. We first show that $|\Phi_j| \leq c_j$. Suppose by contradiction that $\exists j \in HFull, |\Phi_j| > c_j$. For all $k \in [1, |\Phi_j|]$, we define r_{a_k} to be the kth favourite resident to h_j whose index is in Φ_j (i.e. $a_k \in \Phi_j$).

We show by induction that $\forall k \in [1, c_j], \max(y_{j,k}) \leq \mathcal{H}_j^{-1}[x_{a_k}]$. For $k = 1$, since Constraint 4 of Γ is AC, and $y_{j,0} = 0 < \mathcal{H}_j^{-1}[x_{a_1}]$ then $\max(y_{j,1}) \leq \mathcal{H}_j^{-1}[x_{a_1}]$. Suppose that the property holds for $k \in [1, c_j - 1]$. We have $\max(y_{j,k}) \leq \mathcal{H}_j^{-1}[x_{a_k}]$. Therefore, Constraint 4 of Γ enforces $\max(y_{j,k+1}) \leq \mathcal{H}_j^{-1}[x_{a_{k+1}}]$ since $\mathcal{H}_j^{-1}[x_{a_k}] < \mathcal{H}_j^{-1}[x_{a_{k+1}}]$.

Consider now $k \in [c_j + 1, |\Phi_j|]$. We have $\max(y_{j,c_j}) < \mathcal{H}_j^{-1}[x_{a_k}]$ (since $\mathcal{H}_j^{-1}[x_{a_{c_j}}] < \mathcal{H}_j^{-1}[x_{a_k}]$). Therefore Constraint 5 of Γ removes $\mathcal{R}_{a_k}^{-1}[j]$ from $\mathcal{D}(x_{a_k})$ which contradicts the fact that $\mathcal{R}_{a_k}[\min(x_{a_k})] = j$. Hence $|\Phi_j| \leq c_j$.

Using Properties 5, 6, and 7 of Lemma 1 we obtain

$$\sum_{j \in HFull} |\Phi_j| = \sum_{j \in HFull} c_j.$$

Therefore, $|\Phi_j| = c_j$. $\square$

Lemma 4. *If Γ is AC then $\forall j \in HFull, |\{i \mid \mathcal{R}_i[\max(x_i)] == j\}| = c_j$.*

Proof. We show first that $\forall j \in HFull, \forall k \in [1, c_j]$, if $\mathcal{H}_j[\min(y_{j,k})] = i$, then $\mathcal{R}_i[\max(x_i)] = j$. Suppose by contradiction that there exists j, k, i such that $\mathcal{H}_j[\min(y_{j,k})] = i$ and $\mathcal{R}_i[\max(x_i)] \neq j$. Then Constraint 2 enforces $\max(x_i) \leq \mathcal{R}_i^{-1}[j]$. Therefore we have $\max(x_i) < \mathcal{R}_i^{-1}[j]$. Thus, Constraint 3 removes $\mathcal{H}_j^{-1}[i]$ from $\mathcal{D}(y_{j,k})$ which contradicts the hypothesis. Hence we have $\forall j \in HFull, \forall k \in [1, c_j]$, if $\mathcal{H}_j[\min(y_{j,k})] = i$, then $\mathcal{R}_i[\max(x_i)] = j$.

Observe that propagating Constraint 1 enforces $\min(y_{j,1})$, $\min(y_{j,2})$, $\ldots$, $\min(y_{j,c_j})$ to have different values. Therefore $\forall j \in HFull, |\{\min(y_{j,k}) \mid k \in [1, c_j]\}| = c_j$. Thus $|\{i \mid \mathcal{R}_i[\max(x_i)] == j\}| \geq c_j$.

Since $\mathcal{R}_i[\max(x_i)] == j$ is true only when $i \in RFree$, and for all $i \in RFree$, $\mathcal{R}_i[\max(x_i)] \in HFull$ then $\sum_{j \in HFull} |\{i \mid \mathcal{R}_i[\max(x_i)] == j\}| = |RFree| = \sum_{j \in HFull} c_j$. Therefore, $\forall j \in HFull, |\{i \mid \mathcal{R}_i[\max(x_i)] == j\}| = c_j$. $\square$

We now introduce a sufficient and necessary condition for stability in Theorem 2.

Theorem 2. 2-SidedStability$(\mathcal{X}, \mathcal{A}, \mathcal{B}, \mathcal{C})$ *is satisfiable iff*

$$\forall\, 1 \leq j \leq n_H, \sum_{i=1}^{n_R} (\mathcal{R}_i[x_i] == j) \leq c_j \,\wedge$$

$\forall\, 1 \leq i \leq n_R, \forall\, 1 \leq j \leq n_H + 1, x_i = j \implies \forall k \in [1, j[, if\ h = \mathcal{R}_i[k], then$ $\sum_{m=1}^{n_R} (\mathcal{R}_m[x_m] == h) = c_h \wedge \forall l \underset{\mathcal{H}_h}{\succ} i, \mathcal{R}_l[x_l] \neq h$.

Proof. ($\Rightarrow$) Let $\mathcal{M}$ be a stable matching and suppose that the variables are assigned accordingly. Clearly, by construction, $\forall\, 1 \leq j \leq n_H, \sum_{i=1}^{n_R} (\mathcal{R}_i[x_i] == j) \leq c_j$. Let x_i be assigned to j, $1 \leq k < j$, and $h = \mathcal{R}_i[k]$. We show that $\sum_{m=1}^{n_R} (\mathcal{R}_m[x_m] == h) = c_h \wedge \forall l \underset{\mathcal{H}_h}{\succ} i, \mathcal{R}_l[x_l] \neq h$.

If we suppose by contradiction that $\sum_{m=1}^{n_R}(\mathcal{R}_m[x_m] == h) \neq c_h \vee \exists l \underset{\mathcal{H}_h}{\succ\succ} i, \mathcal{R}_l[x_l] = h$, then, $\mathcal{H}_h$ is under subscribed or $\exists (l, h) \in \mathcal{M}$ and $i \underset{\mathcal{H}_h}{\prec\prec} l$.

Therefore, (i, h) is blocking $\mathcal{M}$. Hence the contradiction.

($\Leftarrow$) Consider an assignment of the the variables $x_1, .., x_n$ satisfying the property. We show that the corresponding matching $\mathcal{M}$ is stable. If $\mathcal{M}$ is not stable then there exists a blocking pair (a, b). There are two cases to consider:

- r_a is unassigned in $\mathcal{M}$: In this case $x_a = n_H + 1$, hence $\forall h \in [1, n_H], \sum_{l=1}^{l=n_R}(\mathcal{R}_l[x_l] == h) = c_h \wedge \forall l \underset{\mathcal{H}_h}{\succ\succ} a, \mathcal{R}_l[x_l] \neq h$. Therefore h_b cannot be under-subscribed and there does not exist $l \in [1, n_R]$ such that $(l, b) \in \mathcal{M}$ and $a \underset{\mathcal{H}_b}{\prec\prec} l$.

- $\exists k \in [1, n_H]$ such that $(a, k) \in \mathcal{M}$ and $b \underset{\mathcal{R}_a}{\prec\prec} k$: In this case $x_a = e$ where $e = \mathcal{R}_a^{-1}[k]$, hence for all $w \underset{\mathcal{H}_a}{\prec\prec} k, \sum_{m=1}^{n_R}(\mathcal{R}_m[x_m] == w) = c_w \wedge \forall l \underset{\mathcal{H}_w}{\succ\succ} a, \mathcal{R}_l[x_l] \neq w$. Since $b \underset{\mathcal{R}_a}{\prec\prec} k$, then h_b cannot be under-subscribed and there does not exist $l \in [1, n_R]$ such that $(l, b) \in \mathcal{M}$ and $a \underset{\mathcal{H}_b}{\prec\prec} l$.

Therefore $\mathcal{M}$ is stable.   □

Lemma 5. *If Γ is AC then assigning all variables to their minimum value is solution.*

Proof. We use Theorem 2 to prove that assigning all variables to their minimum satisfies the constraint. We already know that $\forall j \in [1, n_H], |\{i \mid \mathcal{R}_i[\min(x_i)] == j\}| \leq c_j$ by Lemmas 1 and 3. Let $1 \leq i \leq n_R, j = \min(x_i), k \in [1, j[, h = \mathcal{R}_i[k]$. We show that $\sum_{m=1}^{n_R}(\mathcal{R}_m[\min(x_m)] == h) = c_h \wedge \forall l \underset{\mathcal{H}_h}{\succ\succ} i, \mathcal{R}_l[\min(x_l)] \neq h$. Note first that $h \in HFull$ (Lemma 2). Therefore, by using Lemma 3 we obtain $\sum_{m=1}^{n_R}(\mathcal{R}_m[\min(x_m)] == h) = c_h$. Recall that $\min(x_i) > \mathcal{R}_i^{-1}[h]$. Therefore, Constraint 2 of Γ enforces $\max(y_{h,k}) < \mathcal{H}_h^{-1}[i]$ for all $k \in [1, c_h]$. Thus, Constraint 5 removes $\mathcal{R}_l^{-1}[h]$ from $\mathcal{D}(x_l)$ for all $l \underset{\mathcal{H}_h}{\succ\succ} i$.   □

Lemma 6. *If Γ is AC then assigning all variables to their maximum value is solution.*

Proof. Here again we use Theorem 2 to prove the result. First observe that $\forall j \in [1, n_H], |\{i \mid \mathcal{R}_i[\max(x_i)] == j\}| \leq c_j$ by Lemmas 1 and 4. Let $1 \leq i \leq n_R, j = \max(x_i), k < j$, and $h = \mathcal{R}_i[k]$, we show that $\sum_{m=1}^{n_R}(\mathcal{R}_m[\max(x_m)] == h) = c_h \wedge \forall l \underset{\mathcal{H}_h}{\succ\succ} i, \mathcal{R}_l[\max(x_l)] \neq h$. Observe first that $h \in HFull$ (Lemma 2). Therefore, by using Lemma 4 we have $\sum_{m=1}^{n_R}(\mathcal{R}_m[\max(x_m)] == h) = c_h$. Next, we show that for any l such that $\mathcal{R}_l[\max(x_l)] = h$, then $\exists k \in [1, c_h]$ such that $\min(y_{h,k}) = \mathcal{H}_h^{-1}[l]$. In fact, if it's not the case, then by the proof of Lemma 4, we know that if $\min(y_{h,k}) = \mathcal{H}_h^{-1}[a]$ (for any $a \in [1, n_R]$), then $\mathcal{R}_a[\max(x_a)] = h$ and in this case we have $\sum_{m=1}^{n_R}(\mathcal{R}_m[\max(x_m)] == h) > c_h$ which contradicts Lemma 4 because $h \in HFull$.

Suppose now by contradiction that $\exists l \underset{\mathcal{H}_h}{\succ\succ} i, \mathcal{R}_l[\max(x_l)] = h$ and consider k such that $\mathcal{H}_h^{-1}[\min(y_{h,k})] = l$. Since $l \underset{\mathcal{H}_h}{\succ\succ} i$ then $\min(y_{h,k}) \geq \mathcal{H}_h^{-1}[i]$. Therefore, Constraint 2 enforces $\max(x_i) \leq \mathcal{R}_i^{-1}[h]$ which is false since $h \underset{\mathcal{R}_i}{\prec\prec} \mathcal{R}_i[\max(x_i)]$. Therefore we have $\forall l \underset{\mathcal{H}_h}{\succ\succ} i, \mathcal{R}_l[\max(x_l)] \neq h$. □

The following Theorem is an immediate consequence of Lemmas 5 and 6.

Theorem 3. *Enforcing AC on Γ makes 2-SIDEDSTABILITY$(\mathcal{X}, \mathcal{A}, \mathcal{B}, \mathcal{C})$ Bound $(\mathcal{D})$ consistent.*

4 Revisiting Bound($\mathcal{D}$) Consistency

We assume that a preprocessing step is performed where the GS-lists are computed and that the domain is updated accordingly. We suppose without loss of generality that $\exists n \in [1, n_R]$ such that $\forall i \in [1, n]$, $i \in RFree$ and $|RFree| = n$. Note that $\forall i > n$, $|\mathcal{D}(x_i)| = 1$ after the preprocessing step. Therefore, we shall assume that this holds for the rest of the section and we will focus only on $[x_1, \ldots, x_n]$. We show that BC(D) on 2-SIDEDSTABILITY$(\mathcal{X}, \mathcal{A}, \mathcal{B}, \mathcal{C})$ can be implemented with $O(L)$ time complexity down a branch of the search tree which improves the previous complexity $O(c \times L)$ (where $c = \max\{c_j | j \in [1, n_H]\}$). Next we show an adaptation of the filtering to achieve AC on this constraint.

4.1 Bound($\mathcal{D}$) Consistency

Our revision of BC(D) for this constraint is based essentially on Theorem 4.

Theorem 4. *2-SIDEDSTABILITY$(\mathcal{X}, \mathcal{A}, \mathcal{B}, \mathcal{C})$ is BC(D) iff assigning every variable to its maximum is a solution and assigning every variable to its minimum is a solution.*

Proof. ($\Rightarrow$) Let $\mathcal{D}$ be a domain where 2-SIDEDSTABILITY$(\mathcal{X}, \mathcal{A}, \mathcal{B}, \mathcal{C})$ is BC(D). Consider the encoding Γ on a domain $\mathcal{D}^*$ where $\mathcal{D}^*(x_i) = \mathcal{D}(x_i)$ for all $x_i \in \mathcal{X}$ and $\mathcal{D}^*(y_{j,k})$ equals to the initial domain detailed in Sect. 2.3 for all $j \in [1, n_H]$ and $k \in [0, c_j]$. Let $\mathcal{D}'$ be the domain obtained after enforcing AC on Γ. We know that $\mathcal{D}'$ cannot be empty (i.e., failure) since otherwise it will contradict the fact that 2-SIDEDSTABILITY$(\mathcal{X}, \mathcal{A}, \mathcal{B}, \mathcal{C})$ is BC(D). Moreover, no lower/upper bound can change after AC on Γ since every lower/upper bound has a support. Therefore, by using Lemmas 5 and 6, we know that assigning every variable to its minimum (respectively maximum) is a solution.

($\Leftarrow$) Straightforward. □

Theorem 4 shows that in order to maintain BC(D), it is sufficient to make sure that two specific solutions exist: one by assigning all variables to their minimum, and the other to their maximum. In the following, we show that one can maintain this property in $O(L)$ time down a branch of the search tree using an incremental algorithm.

Algorithm 1. BC(D)

---

1 while $\exists < i, j > \in New_{lb}$ **do**

    | UpdateLB$(i, j, \min(x_i) - 1, New_{lb})$;

    | Apply(i, New_{lb}) ;

2 while $\exists < i, j > \in New_{ub}$ **do**

    | UpdateUB(i, j, New_{ub}) ;

Given a domain $\mathcal{D}$, we define for each hospital h in $HFull$ the following:

- $MIN_h = \{k \mid \mathcal{R}_{\mathcal{H}_h[k]}[\min(x_{\mathcal{H}_h[k]})] = h\}$
- $MAX_h = \{k \mid \mathcal{R}_{\mathcal{H}_h[k]}[\max(x_{\mathcal{H}_h[k]})] = h\}$
- $maxofMAX_h = \max\{l \mid l \in MAX_h\}$.

For any $k \in MIN_h$ (respectively $k \in MAX_h$), if r is the correspondent resident (i.e., $r = \mathcal{H}_h[k]$), then h is the hospital of index $\min(x_r)$ (respectively $\max(x_r)$) in $\mathcal{R}_r$.

We also define $lastLeft_h$ for every hospital $h \in HFull$ as follows: $lastLeft_h = \max\{k \mid k \in [1, len^h{}_h] \wedge \mathcal{R}_{\mathcal{H}_h[k]}{}^{-1}[h] \in \mathcal{D}(x_{\mathcal{H}_h[k]})\}$. That is $lastLeft_h$ is the last index in the list of $\mathcal{H}_h$ where the corresponding resident still has the rank of h in its domain.

We suppose that MIN_h, MAX_h, $maxofMAX_h$, and $lastLeft_h$ are implemented as "reversible" data structures (i.e. their values are restored whenever the solver backtracks).

Let $\mathcal{D}_{BC(D)}$ be a domain that is BC(D) for 2-SIDEDSTABILITY$(\mathcal{X}, \mathcal{A}, \mathcal{B}, \mathcal{C})$. Let New_{lb} (respectively New_{ub}) be a set of pairs such that $< i, j > \in New_{lb}$ (respectively $< i, j > \in New_{ub}$) iff the lower (respectively upper) bound j has been removed from $\mathcal{D}_{BC(D)}(x_i)$. We show that Algorithm 1 maintains BC(D).

Take the case where only one variable x_i has a new lower bound $\min(x_i)$ and j was the previous lower bound. To maintain BC(D), we need to compute the new domain where assigning all variables to their minimum/maximum value is a solution. We therefore need to maintain $\forall 1 \leq j \leq n_H$, $\sum_{i=1}^{n_R}(\mathcal{R}_i[x_i] == j) \leq c_j$, and $\forall a \in [1, n], \forall h \underset{\mathcal{R}_a}{\prec} \mathcal{R}_a[\min(x_a)], \forall l \underset{\mathcal{H}_h}{\succ} a, \mathcal{R}_l[\min(x_l)] \neq h$. In other words, $\forall a \in [1, n], \forall v < \min(x_a), \forall k > idx, \min(x_r) \neq \mathcal{R}_r{}^{-1}[h]$ where $h = \mathcal{R}_i[v]$, $idx = \mathcal{H}_h{}^{-1}[i]$, and $r = \mathcal{H}_h[k]$.

Consider first the variable x_i. Since $\mathcal{D}_{BC(D)}$ is BC(D), then the property already holds for all $v < j$. Let $[a, b] = [j, \min(x_i) - 1]$. The property must be enforced for all $v \in [a, b]$. This is precisely what Algorithm 2 does. Let $h \leftarrow \mathcal{R}_i[v]$. Observe that $\forall k > lastLeft_h$, if $r = \mathcal{H}_h[k]$ then $\mathcal{R}_r{}^{-1}[h] \notin \mathcal{D}(x_r)$ (from the definition of $lastLeft_h$). Therefore, one needs only to enforce the property for $k \in [idx, lastLeft_h]$ (Line 1) where $idx = \mathcal{H}_h{}^{-1}[i]$ and perform the filtering in Line 5. Note that this value removal might change the lower (respectively upper) bound for a given x_r. In this case, the pair $\langle r, \min(x_r) \rangle$ (respectively $\langle r, \max(x_r) \rangle$) is added to New_{lb} (respectively New_{ub}) in Line 3 (respectively Line 4). The case where $k \in MIN_h$ is handled at Line 2 by removing k from MIN_h.

Algorithm 2. UpdateLB(i, a, b, New_{lb})

---

   for $v \in [a, b]$ do

      $h \leftarrow \mathcal{R}_i[v]$;

      $idx \leftarrow \mathcal{H}_h^{-1}[i]$;

1   for $k \in [idx, lastLeft_h]$ do

         $r \leftarrow \mathcal{H}_h[k]$;

         if $k \in MIN_h$ then

2          $MIN_h \leftarrow MIN_h \setminus \{k\}$;

         if $\mathcal{R}_r^{-1}[h] == \min(x_r)$ then

3          $New_{lb} \leftarrow New_{lb} \cup \{\langle r, \min(x_r)\rangle\}$;

         if $\mathcal{R}_r^{-1}[h] == \max(x_r)$ then

4          $New_{ub} \leftarrow New_{ub} \cup \{\langle r, \max(x_r)\rangle\}$;

5         $\mathcal{D}(x_r) \leftarrow \mathcal{D}(x_r) \setminus \{\mathcal{R}_r^{-1}[h]\}$;

      if $lastLeft_h > idx - 1$ then

6        $lastLeft_h \leftarrow idx - 1$;

Now once the call to UpdateLB $(i, j, \min(x_i) - 1, New_{lb})$ ends, we have to make sure that it is actually possible to assign x_i to its new minimum. This is performed by calling Algorithm 3 Apply(i, New_{lb}). More precisely, let $h = \mathcal{R}_i[\min(x_i)]$. Obviously if $\mathcal{H}_h^{-1}[i] > lastLeft_h$ then x_i cannot be assigned to its minimum (Lines 9 and 10). Otherwise, $\mathcal{H}_h^{-1}[i]$ is added to MIN_h (Line 1). Suppose now that MIN_h has more than c_h elements. We can easily show that this happens only if $|MIN_h| = c_h + 1$. In this case, we can see that the resident associated with the maximum index in MIN_h cannot be assigned to hospital h anymore. The maximum is computed by looking for the first index less than or equal to MIN_h (Line 3). Lines 4, 5, 6, 7, and 8 handle the fact that the corresponding resident cannot be assigned to hospital h.

At the end of the first loop in Algorithm 1, we may argue that $|MIN_h| = c_h$. Therefore $\forall\ 1 \leq j \leq n_H$, $\sum_{i=1}^{n_R}(\mathcal{R}_i[x_i] == j) \leq c_j$, and $\forall a \in [1, n], \forall h \underset{\mathcal{R}_a}{\prec\!\!\prec} \mathcal{R}_a[\min(x_a)], \forall l \underset{\mathcal{H}_h}{\succ\!\!\succ} a, \mathcal{R}_l[\min(x_l)] \neq h$. Thus assigning all variable to their minimum is a solution.

Consider now the case where only one variable x_i has a new upper bound $\max(x_i)$ and j was the previous upper bound. We use the following lemma to to compute the new domain.

Lemma 7. *Assigning all variables to their maximum is a solution iff $\forall\ h \in HFull, |MAX_h| = c_h$, and $\forall\ i \leq maxofMAX_h$, let $r = \mathcal{H}_h[i]$, and $l = \mathcal{R}_r^{-1}[h]$, then $i \notin MAX_h \implies \max(x_r) < l$.*

Proof. ($\Rightarrow$) Let M be the matching where each resident is assigned to the hospital corresponding to the maximum value in its domain. Let $h \in HFull$. We know that $|MAX_h| = c_h$ since M is stable. Consider now $i \leq maxofMAX_h$, such that $i \notin MAX_h$. Let $r = \mathcal{H}_h[i]$, and $l = \mathcal{R}_r^{-1}[h]$. Observe first that $\max(x_r) \neq l$

Algorithm 3. Apply(i, New_{lb})

$j \leftarrow \min(x_i)$;
$h \leftarrow \mathcal{R}_i[j]$;
if $\mathcal{H}_h^{-1}[i] \leq lastLeft_h$ **then**

  1 $MIN_h \leftarrow MIN_h \cup \{\mathcal{H}_h^{-1}[i]\}$;

     if $|MIN_h| = c_h + 1$ **then**

  2 $max \leftarrow lastLeft_h$;

         $max\_found = false$;

  3   **while** $not(max\_found)$ **do**

            if $max \in MIN_h$ **then**

               $max\_found = true$;

            else

               $max \leftarrow max - 1$

  4 $MIN_h \leftarrow MIN_h \setminus \{max\}$;

  5 $l = \mathcal{H}_h[max]$;

  6 $\mathcal{D}(x_l) \leftarrow \mathcal{D}(x_l) \setminus \{\mathcal{H}_l^{-1}[h]\}$;

  7 $New_{lb} \leftarrow New_{lb} \cup \{\langle l, \min(x_l)\rangle\}$;

  8 $lastLeft_h \leftarrow max$

else

  9 $\mathcal{D}(x_i) \leftarrow \mathcal{D}(x_i) \setminus \{\min(x_i)\}$;

 10 $New_{lb} \leftarrow New_{lb} \cup \langle i, \min(x_i)\rangle$;

(otherwise $i \in MAX_h$). Next, one can easily show that if $\max(x_r) > l$ then the pair (r, h) blocks M. Therefore, $\max(x_r) < l$.

($\Leftarrow$) We use Theorem 2 to show that assigning all variables to their maximum is a solution. We already have $\forall\, 1 \leq j \leq n_H$, $\sum_{i=1}^{n_R}(\mathcal{R}_i[\max(x_i)] == j) \leq c_j$. We show that $\forall\, 1 \leq i \leq n_R$, $\forall\, 1 \leq j \leq n_H + 1$, $\max(x_i) = j \implies \forall k \in [1, j[$, if $h = \mathcal{R}_i[k]$, then $\sum_{m=1}^{n_R}(\mathcal{R}_m[\max(x_m)] == h) = c_h \wedge \forall l \underset{\mathcal{H}_h}{\succ\!\succ} i, \mathcal{R}_l[\max(x_l)] \neq h$. Note first that the property is true for all $i \in [n+1, n_R]$. Let $i \in [1, n]$ $j = \max(x_i)$, and $h \underset{\mathcal{R}_i}{\prec\!\prec} \mathcal{R}_i[j]$ (note that $j \neq n_H + 1$). We have necessarily $h \in HFull$ (Lemma 2), and therefore $\sum_{m=1}^{n_R}(\mathcal{R}_m[\max(x_m)] == h) = c_h$ since $|MAX_h| = c_h$. Let $l \underset{\mathcal{H}_h}{\succ\!\succ} i$. We show that $\mathcal{R}_l[\max(x_l)] \neq h$. Suppose by contradiction that $\mathcal{R}_l[\max(x_l)] = h$, and let $m = \mathcal{H}_h^{-1}[l]$. We have $m \in MAX_h$, hence $m \leq \max\{MAX_h\}$. Therefore, $\mathcal{H}_h^{-1}[i] < \max\{MAX_h\}$ because $l \underset{\mathcal{H}_h}{\succ\!\succ} i$. Since $\mathcal{H}_h^{-1}[i] \notin MAX_h$, then $\max(x_i) < \mathcal{R}_i^{-1}[h]$ which contradicts the hypothesis that $h \underset{\mathcal{R}_i}{\prec\!\prec} \mathcal{R}_i[j]$. □

We maintain the property in Lemma 7 by calling Algorithm 4, UpdateUB (i, j, New_{ub}). Recall that, in this case, the current upper bound of x_i is strictly less than j.

Let $h = \mathcal{R}_i[j]$ and $idx = \mathcal{H}_h^{-1}[i]$. There are two cases to consider. First if $idx \in MAX_h$, we know that MAX_h has to be of cardinality c_h. Therefore, a new index needs to be added to MAX_h. From Lemma 7 we can argue that the

Algorithm 4. UpdateUB(i, j, New_{ub})

$h \leftarrow \mathcal{R}_i[j]$;
$idx \leftarrow \mathcal{H}_h^{-1}[i]$;
if $idx \in MAX_h$ **then**

1     $MAX_h \leftarrow MAX_h \setminus \{idx\}$;
     $new\_resident \leftarrow false$;
     $max = maxofMAX_h$;

2   **do**
        $max \leftarrow max + 1$;
        $r \leftarrow \mathcal{H}_h[max]$;
        if $\mathcal{R}_r^{-1}[h] \le \max(x_r)$ **then**
           $new\_resident \leftarrow true$;

    while *not new_resident*;

3     $MAX_h \leftarrow MAX_h \cup \{max\}$;

4     $maxofMAX_h \leftarrow max$;
     $rankOfh = \mathcal{R}_r^{-1}[h]$;
     if $rankOfh < \max(x_r)$ **then**

5        $New_{ub} \leftarrow New_{ub} \cup \{\langle r, \max(x_r)\rangle\}$;
        % make a new upper bound for x_r ;

6        $\mathcal{D}(x_r) \leftarrow \mathcal{D}(x_r) \cap \,] - \infty, rankOfh]$;

else
    if $j - 1 > \max(x_i)$ **then**
       $New_{ub} \leftarrow New_{ub} \cup \{\langle i, j - 1\rangle\}$;

---

new index cannot correspond to a value less than or equal to $maxofMAX_h$. Therefore, we start looking for a new index in the main loop (Line 2) starting from $maxofMAX_h + 1$. The loop ends when we find a replacement for r_i. The new index corresponds to a resident r such that $\mathcal{R}_r^{-1}[h] \le \max(x_r)$. The set MAX_h is updated accordingly in Lines 1 and 3. The upper bound of x_r is changed if h is better than the hospital corresponding to $\max(x_r)$ (Line 6). In this case, $\langle r, h \rangle$ is added to New_{ub} in Line 5.

Second, in the case where $idx \notin MAX_h$, we just need to make sure that if $j - 1$ is not the current upper bound of $\max(x_i)$, then we need to simulate the case where $j - 1$ was an upper bound for x_r and has been removed.

To maintain BC(D) in Algorithm 1, we loop over all the lower/upper bound changes and call the appropriate algorithms. The new domain is BC(D) as an immediate consequence of Theorems 4 and 2, and Lemma 7.

Complexity. We discuss now the complexity of this incremental algorithm. We assume that all set operations are implemented in constant time.

Observe that down a branch of the search tree the value of $lastLeft_h$ can always decrement (Line 6 in Algorithm 2 and Line 8 in Algorithm 3). Now consider one iteration of the main loop in Algorithm 2. The number of operations is bounded by $O(|idx - lastLeft_h|)$. Similarly, one call to Algorithm 3 is bounded by $O(|m - m'|)$ where m (respectively m') is the value of $lastLeft_h$ at Line 2 (respectively 8). Since $lastLeft_h$ can only decrement, then the time complexity of all lower bound operations down a branch is $O(len^h{}_h)$ for any hospital h. Therefore the complexity for lower bound operations is $O(L)$.

Consider now upper bound operations. Each call to Algorithm 4 is bounded by $O(|m' - m|)$ where m and m' are the values of $maxofMAX_h$ at Lines 1 and 4 respectively. And since $maxofMAX_h$ can only increment down a branch of the search tree, then the number of upper bound operations for any hospital h is $O(len^h{}_h)$. Therefore the total upper bound operations is $O(L)$. Hence the complexity of enforcing BC(D) is $O(L)$ down a branch of the search tree.

4.2 Arc Consistency

Assume that 2-SidedStability$(\mathcal{X}, \mathcal{A}, \mathcal{B}, \mathcal{C})$ is BC(D). We use a straightforward way to enforce AC based on the BC(D) propagator. For every variable x_i, we remove its upper bound u then we enforce BC(D). Let u' the new upper bound for x_i. Clearly any value $v \in]u', u[$ cannot be part of any support, hence should be removed from $\mathcal{D}(x_i)$. Moreover, the new upper bound u' has a support on the constraint because BC(D) is maintained. By repeating the process until reaching the lower bound of x_i, we are guaranteed that all values without a support in $\mathcal{D}(x_i)$ are removed. Therefore AC is maintained by using this procedure for every variable x_i. The overall complexity is $O(n \times L)$ since for each variable it takes $O(L)$ to enforce BC(D) from $\max(x_i)$ to $\min(x_i)$. Note that the algorithm is not incremental and the cost of $O(n \times L)$ is for each call to the propagator.

5 On the Complexity of the Hospital/Resident Problem with Forced and Forbidden Pairs

The variant of the Hospital/Resident problem with forced and forbidden pairs ($HRFF$) seeks to find a stable matching that includes or excludes, respectively, a number of pairs $\langle r, h \rangle$ (r denotes a resident and h denotes a hospital). To the best of our knowledge no polynomial algorithm exists in the literature to solve this problem; this observation is also true even if there is no forced pairs [7]. We show that the problem is indeed polynomial and can be solved in $O(L)$ time.

Recall that Theorem 4 states that once Bound(D) consistency is established, then assigning every variable to its maximum is a solution and assigning every variable to its minimum is a solution. Therefore, it is sufficient to enforce Bound(D) consistency on the problem and then take the minimum value in the domain of each variable as the assignment of the correspondent resident. Since BC(D) takes $O(L)$ time, then the complexity of solving HRFF is $O(L)$.

Consider now the particular case of stable marriage with forced and forbidden pairs. There exist a number of polynomial algorithms for solving this problem. The best algorithm for solving this problem runs in $O(n^4)$ time ([2] and Sect. 2.10.1 in [8]) where n is number of men/women. Now since this problem is a particular case of HRFF, then the complexity for solving it using the above

Table 1. Summary of the experimental results

Set	BC(D)-min			AC-min			BC(D)-max		
	Time	Nodes	Opt	Time	Nodes	Opt	Time	Nodes	Opt
2k	2	16.56	**100**	19	13.33	**100**	8	256.38	**100**
4k	5	18.14	**100**	151	14.69	**100**	37	394.86	**100**
6k	9	18.08	**100**	393	14.89	93	86	648.82	**100**
8k	19	18.16	**100**	332	15.80	79	131	491.50	**100**
	AC-max			BC(D)-rand			AC-rand		
	Time	Nodes	Opt	Time	Nodes	Opt	Time	Nodes	Opt
2k	28	204.53	**100**	4	67.36	**100**	12	81.91	**100**
4k	202	320.33	**100**	12	81.91	**100**	160	65.61	**100**
6k	564	535.65	85	25	120.58	**100**	502	99.20	92
8k	530	432.87	69	42	98.88	**100**	475	88.11	77

approach is also $O(L)$. Recall that in the case of stable marriage L is bounded by n^2, therefore the worst case complexity is $O(n^2)$, hence it is optimal.

6   Experimental Results

We consider a variant of the HR problem where some couples can express their desire to be matched together, assuming that they have the same preference lists. The problem is to find a stable matching maximizing the number of such couples who are matched together.

We generated a set of random instances as follows. Each instance is described by a tuple $\langle r, h, c \rangle$ where: $r \in \{2000, 4000, 6000, 8000\}$ is the number of residents; $h \in \{100, 200, 300, 400, 500\}$ is the number of hospitals; and $c \in \{100 + 50 * k \mid k \in [0, 8]\}$ is the number of couples. We implemented the AC and BC(D) propagators in Mistral-2.0 [6]. We use Intel Xeon E5-2640 processors for the experiments running on Linux. The variable ordering is fixed to be lexicographical for all experiments. As for the value ordering, we use three different branching strategies: minimum value (static); maximum value (static); and random min/max. We also use a geometric restart. Note that for random min/max, we use five different seeds since the heuristic is randomized. The time limit is fixed to 20 min for each instance and configuration.

The results are shown in Table 1 and Fig. 1. The table is split into two parts. In each part, each column depicts one configuration a-b where a is the propagator (BC(D) or AC) and b is the value branching strategy. Each row represents the results for one set of instances by their size (i.e. number of residents). We report for each configuration the runtime (Time) in second, the number of nodes (Nodes), and the percentage of instances where optimality where proven (Opt). All the statistics are given as averages across all successful runs (i.e. when optimality is found). Bold faced values show the best results in terms of percentage

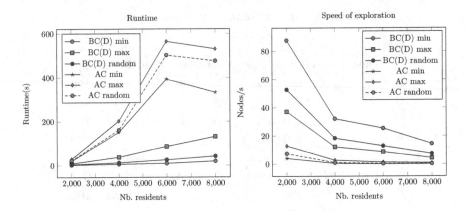

Fig. 1. Runtime and speed of exploration

of optimality. Figure 1 shows two plots corresponding to the runtime and the speed of exploration in terms of nodes explored per second.

There are a number of observations based on Table 1 and Fig. 1. First, clearly the models enforcing BC(D) outperform the AC models in terms of optimality. For instance, in the 8k set, with min value branching, BC(D) finds the optimal solution for all instances whereas the AC models finds optimality only for 79 % of the instances. Second, we observe that enforcing AC considerably slows down the exploration of the search tree. For instance, with set 2k and min value, BC(D) is 21 times faster at exploring the search space (see Fig. 1). This behaviour negatively affects the total runtime for AC. For instance, it takes AC 150 s to find optimal solutions for the 4k set with min value, whereas the BC(D) models needs about 5 s. Last, it should be noted that the search tree is slightly different between the two models. AC does prune some additional nodes as compared to BC(D), but the impact of the pruning does not pay off as it consumes an enormous amount of time. This behaviour is mainly due to the non-incrementality of the AC algorithm.

7   Conclusion

We addressed the filtering aspect of the 2-SIDEDSTABILITY constraint. We first showed that the filtering level of all previous approaches is BC(D). Then, we proposed an optimal BC(D) algorithm, as well as an AC algorithm for this constraint. Our experiments showed that, in practice, BC(D) completely outperforms AC on a variant of the HR problem involving couples. While the aim of this paper was to revisit the current filtering approaches for two-sided stability constraints, we found new theoretical results related to the complexity of the variant of HR with forced and forbidden pairs. We showed, for the first time, that this problem is polynomial and we improved the best known existing complexity for the particular case of stable marriage with forced and forbidden pairs by a factor of n^2.

References

1. Bessiere, C.: Constraint propagation. In: van Beek, P., Rossi, F., Walsh, T. (eds.) Handbook of Constraint Programming. Foundations of Artificial Intelligence, vol. 2, pp. 29–83. Elsevier, Amsterdam (2006)
2. Dias, V.M.F., da Fonseca, G.D., de Figueiredo, C.M.H., Szwarcfiter, J.L.: The stable marriage problem with restricted pairs. Theor. Comput. Sci. **306**(1–3), 391–405 (2003)
3. Gale, D., Shapley, L.S.: College admissions and the stability of marriage. Am. Math. Monthly **69**, 9–15 (1962)
4. Gent, I.P., Irving, R.W., Manlove, D., Prosser, P., Smith, B.M.: A constraint programming approach to the stable marriage problem. In: Walsh, T. (ed.) CP 2001. LNCS, vol. 2239, pp. 225–239. Springer, Heidelberg (2001)
5. Gusfield, D., Irving, R.W.: The Stable Marriage Problem - Structure and Algorithms. Foundations of computing series. MIT Press, Cambridge (1989)
6. Hebrard, E.: Mistral, a constraint satisfaction library. In: Proceedings of the CP-08 Third International CSP Solvers Competition, pp. 31–40 (2008)
7. Manlove, D.F., O'Malley, G., Prosser, P., Unsworth, C.: A constraint programming approach to the hospitals/residents problem. In: Van Hentenryck, P., Wolsey, L.A. (eds.) CPAIOR 2007. LNCS, vol. 4510, pp. 155–170. Springer, Heidelberg (2007)
8. Manlove, D.F.: Algorithmics of Matching Under Preferences. Series on Theoretical Computer Science, vol. 2. World Scientific, Singapore (2013)
9. Unsworth, C., Prosser, P.: A specialised binary constraint for the stable marriage problem. In: Zucker, J.-D., Saitta, L. (eds.) SARA 2005. LNCS (LNAI), vol. 3607, pp. 218–233. Springer, Heidelberg (2005)
10. Unsworth, C., Prosser, P.: An n-ary constraint for the stable marriage problem. CoRR, abs/1308.0183 (2013)

Optimal Flood Mitigation over Flood Propagation Approximations

Byron Tasseff[1]([⊠]), Russell Bent[1], and Pascal Van Hentenryck[2]

[1] Los Alamos National Laboratory, Los Alamos, NM 87545, USA
{btasseff,rbent}@lanl.gov
[2] University of Michigan, Ann Arbor, MI 48109, USA
pvanhent@umich.edu

Abstract. Globally, flooding is the most frequent among all natural disasters, commonly resulting in damage to infrastructure, economic catastrophe, and loss of life. Since the flow of water is influenced by the shape and height of topography, an effective mechanism for preventing and directing floods is to use structures that increase height, e.g., levees and sandbags. In this paper, we introduce the Optimal Flood Mitigation Problem (OFMP), which optimizes the positioning of barriers to protect critical assets with respect to a flood scenario. In its most accurate form, the OFMP is a challenging optimization problem that combines nonlinear partial differential equations with discrete barrier choices. The OFMP requires solutions that combine approaches from computational simulation and optimization. Herein, we derive linear approximations to the shallow water equations and embed them in the OFMP. Preliminary results demonstrate their effectiveness.

Keywords: Flood mitigation · Nonlinear programming · Mixed integer programming · Approximations

1 Introduction

Throughout human history, water-related natural disasters, e.g., the Johnstown Flood of 1889, the Great Mississippi Flood of 1927, and Hurricane Katrina in 2005, have caused immense human suffering and economic consequences. While the causes of such disasters (hurricanes, dam failures, excessive rainfall, etc.) vary, all are characterized by flooding, i.e., the flow of water into undesired areas. As a result, societies and governments have invested considerable resources into controlling and preventing the occurrence of floods. Despite these efforts, risks remain, and floods continue to be a subject of intense scrutiny [5,9,10].

One of the most influential factors in flooding is the shape of the ground surface (topography). As an example, under the influence of gravity, water naturally flows downhill and around areas of higher topographic elevation. Topography can be adjusted through construction of permanent structures, such as levees and berms, or temporary structures, such as sandbags. This paper introduces

© Springer International Publishing Switzerland 2016
C.-G. Quimper (Ed.): CPAIOR 2016, LNCS 9676, pp. 358–373, 2016.
DOI: 10.1007/978-3-319-33954-2_26

the Optimal Flood Mitigation Problem (OFMP), an optimization problem that aims to mitigate flooding by adjusting topographic elevation. Its goal is to select the position of barriers, e.g., sandbags or levees, to protect critical assets and/or enable the evacuation of threatened populations.

The OFMP is an inherently difficult optimization problem. Since these barriers divert flow, it is critical to accurately model the flood's propagation, traditionally captured by the two-dimensional (2D) shallow water equations. These nonlinear partial differential equations (PDEs) express flow conversation and momentum along two horizontal dimensions at every point in space and time. In practice, these equations are discretized over space and time, resulting in a set of nonlinear equations of high dimensionality. In addition, the OFMP aims to choose the position of barriers in space, introducing additional sources of nonconvexity and combinatorial explosion. However, unlike many control-related optimization problems, the OFMP optimizes only the initial conditions. Flood propagation is predetermined once initial conditions have been selected; there are limited opportunities to modify the flood behavior once the topography is adjusted. This observation provides the key intuition for our contribution: *the development of a principled approach for approximating the response of a flood to changes in topography that is tractable for current optimization technology.*

The main contributions of this paper can be summarized as follows:

- The formalization of the OFMP problem integrating simulation and optimization in the flood domain;
- The derivation of *linear* lower and upper space-time approximations to the PDEs describing flood propagation;
- The definition of optimization models for the OFMP based on these approximations;
- Preliminary empirical results that highlight the accuracy and tractability of the approximations and demonstrate the potential of optimization technology in this area.

The derivation of linear approximations to flood propagation is a critical step in bringing the OFMP into the realm of optimization technology. Our results show that these approximations can provide reasonable estimates of flood extent and water depth using the historical Taum Sauk dam failure as an example. The empirical results also demonstrate the potential of optimization technology on some preliminary case studies.

It is important to emphasize that the literature associated with optimizing the locations of barriers for flood mitigation is limited. To the best of our knowledge, the closest related work is [6]. They propose an interdiction model for flood mitigation and develop flood surrogates from simulation to serve as proxies for calculating flood response to mitigation efforts. However, these surrogates do not define strict relationships with the original PDEs. There are a number of papers focused on simulation-optimization approaches for flood mitigation, where the PDEs are treated as a black box. These papers are focused on controlling the release of water to prevent floods and do not attempt to exploit the structure of the PDEs themselves. Reference [2] is a recent example of this type of approach and contains an extensive literature review of these methods. The work in [4] has considered the PDEs, but their focus is on optimizing normal operations of an open-channel system. Hence their model only requires one spatial dimension, whereas the flooding application considered here has an inherent second spatial dimension and is thus significantly more difficult. Finally, the problem of optimizing dike heights with uncertain flood possibilities was considered in [1]. The PDEs for flood propagation are not considered, and probability models for maximum flood depths are used instead.

The rest of this paper is organized as follows: Sect. 2 discusses the background of flood modeling. Section 3 presents the linear flood relaxations. Section 4 introduces the OFMP and proposes a preliminary optimization model exploiting the linear flood relaxations. Section 5 describes empirical results, and Sect. 6 concludes the paper.

2   Background

The Two-Dimensional Shallow Water Equations. The 2D shallow water equations are a system of hyperbolic PDEs increasingly used to accurately model flooding phenomena. With recent advances in high-performance computing, numerical solutions to these equations have become tractable for large-scale simulation problems. They are especially useful in the context of urban flooding, where one-dimensional models fail due to increased topographic complexity. With bottom slope, bottom friction, and volumetric source terms, the 2D shallow water equations may be defined as

$$\begin{cases} \dfrac{\partial h}{\partial t} + \dfrac{\partial (hu)}{\partial x} + \dfrac{\partial (hv)}{\partial y} = R(x,y,t), \\[2mm] \dfrac{\partial (hu)}{\partial t} + \dfrac{\partial}{\partial x}\left(hu^2 + \dfrac{1}{2}gh^2\right) + \dfrac{\partial (huv)}{\partial y} = -gh\dfrac{\partial B}{\partial x} - \dfrac{\tau^x}{\rho}, \\[2mm] \dfrac{\partial (hv)}{\partial t} + \dfrac{\partial (huv)}{\partial x} + \dfrac{\partial}{\partial y}\left(hv^2 + \dfrac{1}{2}gh^2\right) = -gh\dfrac{\partial B}{\partial y} - \dfrac{\tau^y}{\rho}, \end{cases} \quad (1)$$

where h is the water depth, u and v are horizontal velocities, B is the bottom topography (or bathymetry), g is the acceleration due to gravity, τ^x and τ^y are horizontal components of the bottom friction, ρ is the water density, and R is a volumetric source term [3]. Although these equations represent the state of the art in flood modeling, even when discretized, they remain nonlinear and nonconvex, making them difficult to optimize over. It is thus beneficial to consider more tractable approximations.

A Hydrostatic Approximation. To obtain a more tractable approximation of flood propagation, we instead consider a simplified fluid model similar to that described by Mei et al. [8]. In this model, each cell (i, j) exchanges water content with adjacent cells using a set of virtual "pipes." For each time step, the model associates various information with each cell and pipe. In particular,

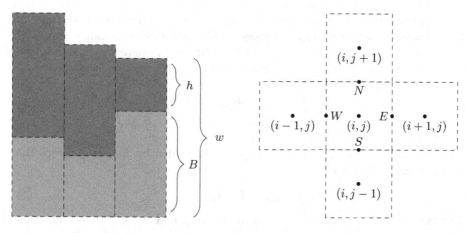

Fig. 1. The pipe flow model is discretized into columnar components, with h denoting the water depth, B the topographic elevation, and w the water surface.

Fig. 2. Two-dimensional representation of the pipe flow discretization containing cell (i, j), adjacent cells, and the four interfaces of (i, j).

h_{ijt} represents the depth of the water in cell (i, j) at time index t, w_{ijt} represents the water surface elevation, and B_{ij} represents the topographic elevation. Each cell (i, j) also has four connected pipes, one for each of its four neighboring cells, denoted by W(est), E(ast), N(orth), and S(outh). Each pipe is associated with an outgoing volumetric flux, f_{ijt}^k, which represents the flow of water from cell (i, j) to its neighbor in position k ($k \in W, E, N, S$) at time index t. For instance, f_{ijt}^W represents the outgoing flux from (i, j) to its "western" (left) neighbor (i.e., cell $(i - 1, j)$) at time index t.

In the model, the flux of a pipe is accelerated by the hydrostatic pressure difference between adjacent cells. The water volume V of a cell is integrated using the accumulated flux from all connected pipes. This corresponds to a change in the cell's depth and water surface elevation. These concepts are illustrated visually in Figs. 1 and 2.

For each cell, we first define the *estimated* flux vector $\tilde{\mathbf{f}}_{ijt} = (\tilde{f}_{ijt}^W, \tilde{f}_{ijt}^E, \tilde{f}_{ijt}^N, \tilde{f}_{ijt}^S)$ using the hydrostatically derived relation

$$\tilde{f}_{ijt}^k = \max\left(0, f_{ij,t-1}^k + \frac{Ag\Delta t}{\Delta s}\Delta w_{ij,t-1}^k\right), \tag{2}$$

where A is the cross-sectional area of the pipe, g is the acceleration due to gravity, Δs is the length of the virtual pipe (typically the grid cell spacing, e.g., Δx or Δy), Δt is the simulation time step, and Δw_{ijt}^k is the difference in water surface elevation between cell (i, j) and its k-neighbor at time index t, i.e.,

$$\Delta w_{ijt}^k = (B_{ij} + h_{ijt}) - (B_{ij}^k + h_{ijt}^k). \tag{3}$$

In this approximation, the estimated outgoing flux from a cell may exceed the available water content within that cell. This is obviously not desirable from a volume conservation standpoint. More importantly, if left uncorrected, this can lead to negative water depths and numerical instabilities. This may be resolved by scaling the outgoing flux with respect to the water content available in the cell. A scaling factor, K_{ijt}, for the outgoing flux vector may be defined as

$$K_{ijt} = \min\left(1, \frac{h_{ij,t-1}\Delta x \Delta y}{\left(\tilde{f}^W_{ijt} + \tilde{f}^E_{ijt} + \tilde{f}^N_{ijt} + \tilde{f}^S_{ijt}\right)\Delta t}\right). \tag{4}$$

The estimated outgoing flux vector $\tilde{\mathbf{f}}_{ijt}$ is then scaled by K_{ijt} to produce the actual outgoing flux vector $\mathbf{f}_{ijt}$, i.e.,

$$\mathbf{f}_{ijt} = K_{ijt}\tilde{\mathbf{f}}_{ijt}. \tag{5}$$

The change in water volume may then be computed using the accumulation of incoming flux, $\mathbf{f}^{in}$, and subtraction of outgoing flux, $\mathbf{f}^{out}$. For cell (i,j), the volumetric change in water is thus

$$\begin{aligned}
\Delta V_{ijt} &= \left(\sum f^{in}_{ijt} - \sum f^{out}_{ijt}\right)\Delta t \\
&= \left(f^E_{i-1,j,t} + f^W_{i+1,j,t} + f^N_{i,j-1,t} + f^S_{i,j+1,t} - \sum f^k_{ijt}\right)\Delta t.
\end{aligned} \tag{6}$$

Finally, the water depth in each cell may be integrated:

$$h_{ijt} = h_{ij,t-1} + \frac{\Delta V_{ijt}}{\Delta x \Delta y}. \tag{7}$$

For completeness, we also suggest the naive reflective boundary conditions

$$h_{ijt} = 0, \mathbf{f}_{ijt} = 0, \tilde{\mathbf{f}}_{ijt} = 0 \tag{8}$$

along the four boundaries of the domain.

3   Linear Approximations of the Pipe Flow Model

The pipe flow model described includes nonlinear terms, even when A, B, g, Δt, Δx, and Δy are treated as constants. Fortunately, these terms are only used for corrective measures, i.e., in Eq. 4. We now present two approximations to remove them. For convenience, we call them the lower and upper approximations because they underestimate and overestimate the water being sent from a cell to its neighbors instead of applying the scaling factor K.

Lower Approximation. The lower approximation is based on the following idea: if the estimated outgoing flux from a cell exceeds the available water content within that cell, the outgoing flux is approximated as zero, i.e., when

$$h_{ij,t-1}\Delta x \Delta y < \left(\tilde{f}^W_{ijt} + \tilde{f}^E_{ijt} + \tilde{f}^N_{ijt} + \tilde{f}^S_{ijt}\right)\Delta t, \tag{9}$$

$\mathbf{f}_{ijt}$ is approximated as zero. This bypasses the need for Eqs. (4) and (5). Intuitively, this means that, "when there is not enough water to be transferred," the water is held back within the cell.

Upper Approximation. The upper approximation implements another intuitive idea: if the estimated outgoing flux from a cell exceeds the available water content within that cell, the model assumes there is enough water, and no scaling occurs. This again bypasses the need for Eqs. (4) and (5).

However, it is important to note that, in the case of positive fluxes calculated as a result of differing dry topographies (and thus differing water surface elevations), Eqs. (4) and (5) provide an additional correction beside scaling. When the available water content within a cell is equal to zero, K_{ijt} is always forced to zero, and the resultant fluxes $\mathbf{f}_{ijt}$ are thus calculated as zero. Although a similar correction is achieved automatically by the lower approximation, it is necessary to impose the constraint $\mathbf{f}_{ijt} = 0$ when $h_{ij,t-1} = 0$ in the upper approximation.

4   The Optimal Flood Mitigation Problem

This section describes two optimization models based on the lower and upper approximations, respectively. Both optimization models aim to protect a set $\mathcal{A}$ of assets by minimizing maximum water depths at asset locations over time. To protect the assets, one or more barriers (e.g., sandbags or levees) can be placed on a cell to increase its elevation; a fixed number of barriers, n, are available for that purpose. The models are similar, differing only in the approximations used. We present them both to give a global view of the lower and upper approximations. Boundary conditions are omitted in the optimization models for simplicity.

Lower Approximation Optimization Model. The lower approximation optimization model is presented in Model 1. The objective function (10a) minimizes maximum water depths over the set $\mathcal{A}$ of assets. Constraints (10b) and (10c) limit the number of barriers, n_{ij}, that may be placed in each cell. This number must be no greater than M, the maximum allowable number of barriers per cell, as specified in (10b). The budget of barriers is limited by Constraint (10d). Constraint (10e) defines the water surface elevation as the sum of topographic elevation (i.e., base elevation and barrier additions, each with height ΔB) and water depth. Constraint (10f) defines estimated outgoing flux values, which must always be greater than or equal to zero. Constraints (10g) and (10h) define the outgoing flux values as prescribed by the lower approximation. Constraint (10i) provides a convenient definition for f^{in}, the sum of all incoming flux. Finally, the integration of water depth is defined using an Euler step in Constraint (10j).

Upper Approximation Optimization Model. The upper approximation optimization model is presented in Model 2. The model is clearly similar to the lower approximation optimization model. The only differences are in Constraints (11g) and (11h), which ensure that outgoing fluxes are nonzero only when the water depth within a cell is greater than zero, and in Constraint (11j), which ensures nonnegative depths: if the predicted net flux results in the transfer of water greater than what is contained within the cell, this depth is set to zero.

Model 1. Lower approximation optimization model

minimize	$\displaystyle\sum_{ij \in \mathcal{A}} \max_t h_{ijt}$	(10a)
subject to	$n_{ij} \in [0, M]$	(10b)
	$n_{ij} = 0; \; \forall (i, j) \in \mathcal{A}$	(10c)
	$\displaystyle\sum_{ij} n_{ij} = n$	(10d)
	$w_{ijt} = (B_{ij} + n_{ij}\Delta B) + h_{ijt}$	(10e)
	$\tilde{f}_{ijt}^k = \max\left(0, f_{ij,t-1}^k + \dfrac{Ag\Delta t}{\Delta s}(w_{ij,t-1} - w_{ij,t-1}^k)\right)$	(10f)
	$f_{ijt}^k = \tilde{f}_{ijt}^k \;$ if $\; h_{ij,t-1}\Delta x \Delta y \geq \Delta t \displaystyle\sum_k \tilde{f}_{ijt}^k$	(10g)
	$f_{ijt}^k = 0 \;$ if $\; h_{ij,t-1}\Delta x \Delta y < \Delta t \displaystyle\sum_k \tilde{f}_{ijt}^k$	(10h)
	$f_{ijt}^{in} = f_{i-1,j,t}^E + f_{i+1,j,t}^W + f_{i,j-1,t}^N + f_{i,j+1,t}^S$	(10i)
	$h_{ijt} = h_{ij,t-1} + \Delta t \dfrac{f_{ijt}^{in} - \sum_k f_{ijt}^k}{\Delta x \Delta y}$	(10j)

Model 2. Upper approximation optimization model

minimize	$\displaystyle\sum_{ij \in \mathcal{A}} \max_t h_{ijt}$	(11a)
subject to	$n_{ij} \in [0, M]$	(11b)
	$n_{ij} = 0; \; \forall (i, j) \in \mathcal{A}$	(11c)
	$\displaystyle\sum_{ij} n_{ij} = n$	(11d)
	$w_{ijt} = (B_{ij} + n_{ij}\Delta B) + h_{ijt}$	(11e)
	$\tilde{f}_{ijt}^k = \max\left(0, f_{ij,t-1}^k + \dfrac{Ag\Delta t}{\Delta s}(w_{ij,t-1} - w_{ij,t-1}^k)\right)$	(11f)
	$f_{ijt}^k = \tilde{f}_{ijt}^k \;$ if $\; h_{ij,t-1} > 0$	(11g)
	$f_{ijt}^k = 0 \;$ if $\; h_{ij,t-1} \leq 0$	(11h)
	$f_{ijt}^{in} = f_{i-1,j,t}^E + f_{i+1,j,t}^W + f_{i,j-1,t}^N + f_{i,j+1,t}^S$	(11i)
	$h_{ijt} = \max\left(0, h_{ij,t-1} + \Delta t \dfrac{f_{ijt}^{in} - \sum_k f_{ijt}^k}{\Delta x \Delta y}\right)$	(11j)

5   Empirical Results

This section reports some preliminary results regarding the proposed approximations and the associated optimization models.

5.1   Evaluation of the Flood Model Relaxations

This section compares differences among the discussed simulation models, i.e., the 2D shallow water equations, pipe flow, lower approximation, and upper approximation models. The comparison uses the historical Taum Sauk dam failure as an example scenario, with a thirty meter spatial resolution and a grid containing approximately 38,000 cells. In the models, a gravitational acceleration constant of $9.80665 \, \text{m/s}^2$ was used, and the dam failure was modeled as a time-dependent volumetric point source, using a hydrograph similar to a United States Geological Survey estimate [11]. In the shallow water equations model, a Manning's roughness coefficient of 0.035 was used, and time steps varied based on a Courant condition. In the remaining models, various constant cross-sectional pipe areas and time steps were used. Note that, in a simulation context, Mei et al. did not suggest using constant cross-sectional pipe areas or time steps; however, our intent was to simplify these models as much as possible.

For flood mitigation, we are primarily concerned with the accuracy of maximum depth estimates over a simulation's time extent.[1] Figure 3 compares images of maximum depth results from a 2D shallow water equations model (**SWE**) similar to [3], as well as pipe flow (**P**), lower approximation (**L**), and upper approximation (**U**) models which used various pipe areas and time steps.

The top row of Fig. 3 compares **SWE** with **P**, **L**, and **U** using a parameterization calibrated to minimize the root-mean-square error between **P** and **SWE**. **P** and **L** provided very reasonable estimates of **SWE**, but **U** greatly overestimated maximum depths. This is because, in **U**, the large pipe area of $500 \, \text{m}^2$ resulted in unrestricted large fluxes and thus poor volume conservation. In the second row, the pipe area was substantially decreased, and the pipe flow and lower approximation models overestimated **SWE**, although **U** behaved more reasonably. Finally, in the third row, as Δt was decreased, **U** began to converge upon **P** and **L**. Most model parameterizations provided reasonable simulated flood extents, similar to those found in the literature [7,11].

Finally, Fig. 4 reports volume conservation error for selected upper approximation parameterizations. As anticipated, the pipe flow and lower approximation models conserved volume well, with error on the order of machine epsilon. The upper approximation accumulated error more rapidly, although it displayed good convergence as the time step was decreased.

It is important to note a unique difference between pipe flow simulations and traditional two dimensional hydrodynamic simulations based on the shallow water equations. When using a full two-dimensional shallow water model, the

---

[1] This is different from evacuation settings, in which the flood arrival time at various locations is critical information.

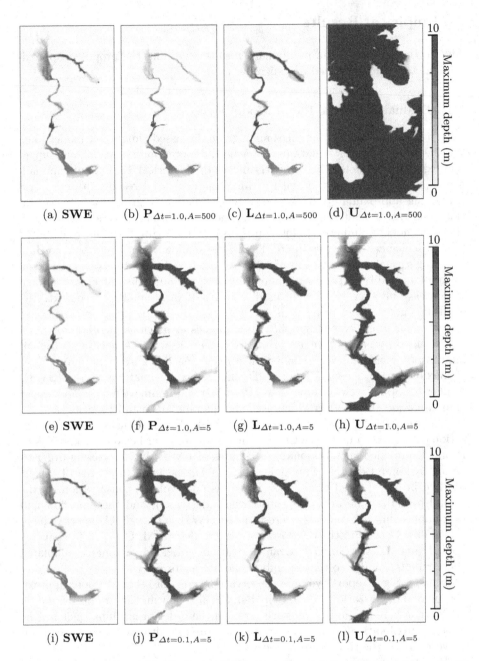

(a) **SWE**   (b) **P**$_{\Delta t=1.0, A=500}$ (c) **L**$_{\Delta t=1.0, A=500}$ (d) **U**$_{\Delta t=1.0, A=500}$

(e) **SWE**   (f) **P**$_{\Delta t=1.0, A=5}$ (g) **L**$_{\Delta t=1.0, A=5}$ (h) **U**$_{\Delta t=1.0, A=5}$

(i) **SWE**   (j) **P**$_{\Delta t=0.1, A=5}$ (k) **L**$_{\Delta t=0.1, A=5}$ (l) **U**$_{\Delta t=0.1, A=5}$

Fig. 3. Maximum flood depths for ten-hour simulations of the historical Taum Sauk dam failure using shallow water equations (**SWE**), pipe flow (**P**), lower approximation (**L**), and upper approximation (**U**) models. Pipe flow, lower approximation, and upper approximation models are compared using constant time steps (Δt) of 1.0 and 0.1 s and cross-sectional pipe areas (A) of 500 and 5 m^2.

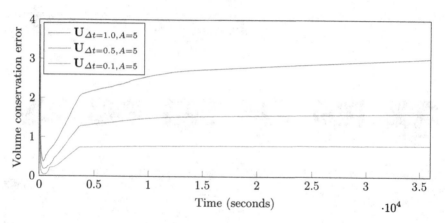

Fig. 4. Volume conservation error in upper approximation models (**U**) for ten-hour simulations of the Taum Sauk dam break using various time steps. Volume conservation error was computed as $(V_{computed} - V_{added})/V_{added}$.

Taum Sauk scenario can be almost fully simulated using a simulation time extent of three hours. In contrast, the pipe flow and approximated models allow for faster or slower propagation, depending on the model parameterization. As an example, the large pipe area used to produce simulation results in the top row of Fig. 3 resulted in fast propagation; the flood was fully propagated in less than an hour. The smaller pipe areas used in the second and third rows resulted in slower propagation; a time extent of roughly three hours was required. In general, as the cross-sectional pipe area decreased, a longer time extent was required for full propagation. Nonetheless, since flood mitigation is primarily concerned with protecting assets, and thus maximum water depth, we found differences in flood propagation speed acceptable for our current application.

5.2 The Potential of Optimization

This section describes some small case studies to highlight the potential and challenges of optimization for flood mitigation and, more generally, the integration of simulation and optimization.

Experimental Setting. The lower and upper optimization models were implemented using the C++ CPLEX interface and run on twenty Intel Xeon E5-2660 v3 cores at 2.60 GHz, with 128 GB of memory. Conditional expressions and min/max functions were eliminated using big-M transformations. No attempt was made to optimize the model or exploit problem structure.

A Simple Case Study. To validate the optimization model, an 8 × 8 scenario was constructed, with Δx and Δy equal to one meter. In this scenario, a topographic gradient was introduced, from the top to the bottom of the domain, with elevations linearly decreasing from 0.7 to zero meters in steps of 0.1 m. Four cells

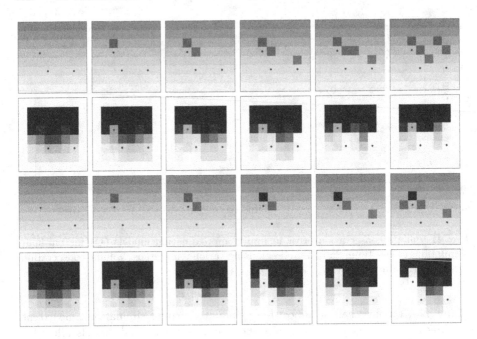

Fig. 5. Optimal elevation fields and maximum depths using the lower approximation, where the allowable number of barriers per cell is one (first two rows) or two (next two rows), and the total number of barriers ranges from zero to five. Darker oranges and blues correspond to larger topographic elevations and depths, respectively, and red circles correspond to assets that were protected. (Color figure online)

near the top of the domain were initialized to contain one meter of water depth. Under the influence of gravity and in the presence of the topographic gradient, the water was pushed down the domain over time. Three assets to protect were arbitrarily placed throughout the domain, and individual barrier heights (ΔB) of 0.5 m were employed. A constant time step of 0.1 s was used, and eight time steps were simulated. The optimization problem was varied to understand how solutions changed using various rules for resource allocation. In particular, the experiments studied limits on the total number of barriers and limits on the number of allowable barriers per cell.

Optimal Asset Protection. Figure 5 displays optimization results from the lower optimization models. Observe that, when only one barrier was allowed per cell, the optimization model tried to mitigate flooding in the asset regions almost one at a time, before placing more barriers in interesting places throughout the domain. When two barriers were allowed per cell, it clearly became preferential to protect the topmost asset, which received a large amount of water over the duration of the simulation. Figure 6 displays optimization results from the upper optimization models. These show similarly interesting outcomes. In the one barrier per cell case, the optimization decided to protect the topmost asset

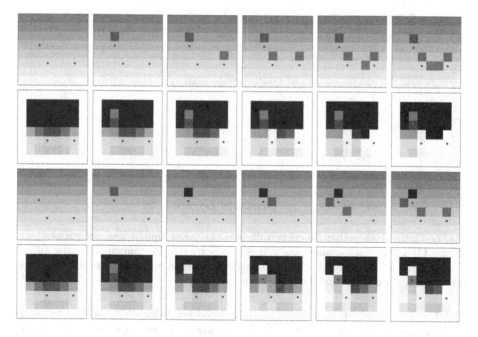

Fig. 6. Optimal elevation fields and maximum depths using the upper approximation (same setting as the lower approximation in Fig. 5). (Color figure online)

less in favor of protecting the bottom assets. When two barriers were allowed per cell and enough barriers were available, it was clearly beneficial to protect the topmost asset as much as possible from the water above it, which greatly reduced the objective value. It is also important to observe the non-monotonic behavior of the optimization results. Allowing more barriers often changed their optimal positions. This was the case when moving from four to five barriers in the top row and when moving from three to four in the bottom row. Since the barrier placements differed in both models, it was important to study how they behaved using the other model. These results are shown in the last two columns of Table 1. Column o_{opt} gives the optimal solutions, and the last column describes the objective value obtained when the optimal solution of the upper model was used in the lower model and vice-versa. They are particularly interesting, as they sometimes show significant differences in objective values. This indicates the need to apply robust optimization techniques. In practice, of course, solutions could be evaluated using full hydrodynamic simulations for various scenarios.

Evolution of the Objective Value. Figure 7 depicts the value of the objective function as the number of available barriers increased, for cases where the models allowed one or two barriers per cell. The critical information is the importance of using multiple barriers at a specific location, since it brings significant benefits as the number of barriers increases. We anticipate similar behavior when the number of allowable barriers per cell is increased to three or four. Note also

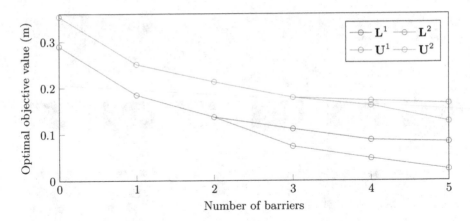

Fig. 7. Optimal objective function values using lower and upper approximations (**L** and **U**) and increasing allowable numbers of barriers. Also compared are the differences among objective values when one barrier is allowed per cell ($\mathbf{L}^1$ and $\mathbf{U}^1$) versus two barriers per cell ($\mathbf{L}^2$ and $\mathbf{U}^2$). (Color figure online)

that the lower and upper approximations behaved comparably as the number of maximum barriers was increased and, as expected, the upper objective value was always greater than the lower objective value.

Computational Results. Finally, Table 1 presents preliminary computational results. The first column describes the instance in terms of lower (**L**) or upper (**U**) approximations. The superscript represents the maximum number of barriers per cell, and the subscript represents the total number of barriers.

Table 1. General statistics and analysis of selected optimization models.

Model	t_{CPU} (s)	n_{nodes}	n_{var}	n_{con}	n_{bin}	o_{opt} (m)	o_{com} (m)
$\mathbf{L}_3^1$	49.65	37808	2280	4410	840	0.111607	0.134935
$\mathbf{L}_4^1$	83.66	71528	2280	4410	840	0.0878997	0.123499
$\mathbf{L}_5^1$	91.53	66485	2280	4410	840	0.0837807	0.112588
$\mathbf{L}_3^2$	82.74	86964	2646	4998	975	0.0739754	0.134935
$\mathbf{L}_4^2$	134.58	121477	2651	5004	978	0.0484597	0.155655
$\mathbf{L}_5^2$	80.61	56586	2651	5004	978	0.0248889	0.13016
$\mathbf{U}_3^1$	55.56	41487	2027	3840	840	0.178971	0.239858
$\mathbf{U}_4^1$	46.17	25037	2027	3840	840	0.17244	0.234095
$\mathbf{U}_5^1$	123.50	83850	2027	3840	840	0.166794	0.225759
$\mathbf{U}_3^2$	77.29	55398	2379	4292	1018	0.178971	0.197646
$\mathbf{U}_4^2$	263.72	203168	2382	4297	1019	0.161937	0.163313
$\mathbf{U}_5^2$	234.85	108909	2382	4297	1019	0.127602	0.157531

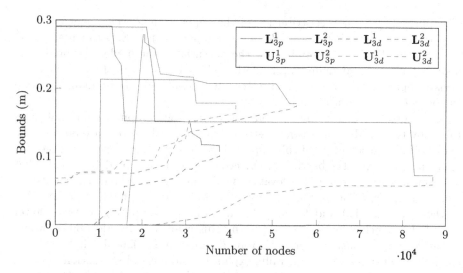

Fig. 8. Performance of CPLEX on problems where $n = 3$; p and d indicate the primal and dual, respectively. (Color figure online)

The second column denotes (wall clock) execution time, in seconds. The third column shows the number of nodes in the search tree. The fourth, fifth, and sixth columns describe the number of variables, constraints, and binary variables after presolve. Column o_{opt} describes the optimal objective value in meters of flood depth. The last column describes the objective value obtained when the optimal solution of the upper model was used for the lower model and vice-versa.

As mentioned earlier, no attempt was made to optimize the model or to exploit the problem structure. The instances have about 2,000 (mostly binary) variables and 4,000 constraints, and they can typically be solved in a few minutes. In general, CPLEX was not able to find high-quality solutions quickly, which substantially increased computation times. Integrating good primal heuristics should improve performance significantly. This is illustrated in Fig. 8, where CPLEX spent significant time improving the primal bound.

6   Conclusion

Each year, flood-related disasters cause billions of dollars in damage, loss of life, and significant human suffering. Resources such as levees and berms are constructed and utilized to mitigate the consequences of such events. The deployment of these mitigation efforts is often ad hoc and relies on subject matter expertise, as computational methods are immature due to the complexity of embedding flood models in modern optimization technologies. The goal of this paper was to establish the foundations for a more principled approach to flood mitigation. It introduced the Optimal Flood Mitigation Problem (OFMP), which aims at integrating simulation and optimization tightly by including flood simulation equations as part of the optimization model. To ensure the tractability of the approach, the main contribution of the paper is the development of linear, physics-based approximations of flood propagation models. Experimental results on the Taum Sauk dam failure show the potential of the models for predicting flood extent and maximum water depths. The integration of these approximations in optimization models was tested on a small case study, demonstrating the potential of optimization in this context.

Our current work is focused on addressing the computational challenges raised by the OFMP. Surprisingly, state of the art MIP solvers are not capable of exploiting the structure of this application. In particular, they do not seem to recognize that, once the barriers are placed, the problem is predetermined. That is, given a fixed topographic elevation field, only the deterministic simulation step remains. A combination of constraint programming (for fast propagation of the water depths) and linear programming (for computing a strong lower bound) has much potential in addressing this challenge. In addition, it would be interesting to consider whether strong dominance relationships hold between candidate solutions. More generally, exploiting the natural separation between mitigation decisions and flood propagation variables will be key when scaling to realistic problems.

Acknowledgments. We gratefully acknowledge our early discussions on flood modeling and mitigation with Feng Pan and David Judi at Pacific Northwest National Laboratory [6]. The work at LANL was carried out under the auspices of the National Nuclear Security Administration of the U.S. Department of Energy at Los Alamos National Laboratory under Contract No. DE-AC52-06NA25396.

References

1. Brekelmans, R., den Hertog, D., Roos, K., Eijgenraam, C.: Safe dike heights at minimal costs: the nonhomogeneous case. Oper. Res. **60**(6), 1342–1355 (2012). http://pubsonline.informs.org/doi/abs/10.1287/opre.1110.1028
2. Che, D., Mays, L.W.: Development of an optimization/simulation model for real-time flood-control operation of river-reservoirs systems. Water Resour. Manage. **29**(11), 3987–4005 (2015). http://link.springer.com/10.1007/s11269-015-1041-8
3. Chertock, A., Cui, S., Kurganov, A., Wu, T.: Well-balanced positivity preserving central-upwind scheme for the shallow water system with friction terms. Int. J. Numer. Methods Fluids **78**(6), 355–383 (2015). http://dx.doi.org/10.1002/fld.4023
4. Colombo, R.M., Guerra, G., Herty, M., Schleper, V.: Optimal control in networks of pipes and canals. SIAM J. Control Optim. **48**(3), 2032–2050 (2009). http://epubs.siam.org/doi/abs/10.1137/080716372
5. Downton, M.W., Miller, J.Z.B., Pielke, R.A.: Reanalysis of U.S. National Weather Service flood loss database. Nat. Hazards Rev. **6**(1), 13–22 (2005). http://ascelibrary.org/doi/10.1061/(ASCE)1527-6988(2005)6%3A1(13)
6. Judi, D., Tasseff, B., Bent, R., Pan, F.: LA-UR 14–21247: topography-based flood planning and optimization capability development report. Technical report, Los Alamos National Laboratory (2014)
7. Kalyanapu, A.J., Shankar, S., Pardyjak, E.R., Judi, D.R., Burian, S.J.: Assessment of GPU computational enhancement to a 2D flood model. Environ. Model. Softw. **26**(8), 1009–1016 (2011). http://dx.doi.org/10.1016/j.envsoft.2011.02.014

8. Mei, X., Decaudin, P., Hu, B.G.: Fast hydraulic erosion simulation and visualization on GPU. In: 15th Pacific Conference on Computer Graphics and Applications, Maui, Hawaii, pp. 47–56 (2007)

9. Pielke, R.A., Downton, M.W.: Precipitation and damaging floods: trends in the United States, 1932–97. J. Clim. **13**(20), 3625–3637 (2000). http://journals. ametsoc.org/doi/abs/10.1175/1520-0442(2000)013%3C3625%3APADFTI%3E2.0. CO%3B2

10. Proverbs, D., Mambretti, S., Brebbia, C., Wrachien, D.: Flood Recovery Innovation and Response III. WIT Press, Southampton (2012)

11. Rydlund, P.H.: Peak discharge, flood profile, flood inundation, and debris movement accompanying the failure of the upper reservoir at the Taum Sauk pump storage facility near Lesterville, Missouri. Technical report (2006)

A Bit-Vector Solver
with Word-Level Propagation

Wenxi Wang[✉], Harald Søndergaard, and Peter J. Stuckey

Department of Computing and Information Systems,
The University of Melbourne, Melbourne, VIC 3010, Australia
wenxiw2@student.unimelb.edu.au

Abstract. Reasoning with bit-vectors arises in a variety of applications in verification and cryptography. Michel and Van Hentenryck have proposed an interesting approach to bit-vector constraint propagation on the word level. Each of the operations except comparison are constant time, assuming the bit-vector fits in a machine word. In contrast, bit-vector SMT solvers usually solve bit-vector problems by bit-blasting, that is, mapping the resulting operations to conjunctive normal form clauses, and using SAT technology to solve them. This also means generating intermediate variables which can be an advantage, as these can be searched on and learnt about. Since each approach has advantages it is important to see whether we can benefit from these advantages by using a word-level propagation approach with learning. In this paper we describe an approach to bit-vector solving using word-level propagation with learning. We provide alternative word-level propagators to Michel and Van Hentenryck, and give the first empirical evaluation of their approach that we are aware of. We show that, with careful engineering, a word-level propagation based approach can compete with (or complement) bit-blasting.

1 Introduction

Since most time-critical and safety-critical software is built on fixed-width integers, it is vital to reason about fixed-width integers correctly and accurately in a software verification context. We consider the problem of how to support this reasoning with modern constraint solving techniques.

SAT Modulo Theory (SMT) solvers are the most common tools in this area, and almost all the modern SMT solvers ultimately rely on bit-blasting [4,5,9, 13,17] to solve bit-vector constraints, that is, translating constraints to propositional logic form. But bit-blasting tends to cause two problems. First, it may result in very large propositional formulas that even the most powerful current SAT solvers struggle to handle. Second, it disperses important word level information during the encoding—much is obscured in translation. Here we investigate alternatives to bit-blasting, replacing it with word-level propagation entirely to produce a pure word-level bit-vector SMT solver.

ⓒ Springer International Publishing Switzerland 2016
C.-G. Quimper (Ed.): CPAIOR 2016, LNCS 9676, pp. 374–391, 2016.
DOI: 10.1007/978-3-319-33954-2_27

Using word-level propagation was suggested by Michel and Van Hentenryck [18] who viewed the problem as a Constraint Satisfaction Problem (CSP). Each variable is associated with a "bit-vector domain" which is progressively tightened using word-level constraint propagation rules (we will make these clear shortly). The idea is appealing, as the propagation rules can be made to run in constant time (as long as the bit width of the bit-vector is less than or equal to the size of machine registers). An additional rule to check if a tightened domain remains valid also runs in constant time. However, we are not aware of any experimental evaluation of the method. Moreover, there is no "learning" mechanism in Michel and Van Hentenryck's proposal. We show that real improvement relies on the communication of explanations for the propagated bits. We also propose alternative ("decomposed") word-level propagators for some operations, based on insights in Warren's compendium [22] and we investigate the relative strengths and weaknesses of decomposed and composed propagators.

Additionally we use our solver to investigate different algorithmic possibilities. In a learning solver we can generate explanations in a "forward" manner, as propagation progresses, as is done in a SAT solver, or we can generate them in a "backward" manner during conflict analysis, as in an SMT solver. Forward explanation is simpler to implement, while backward explanation may require less explanation work overall. In our experiments we compare the two approaches.

Another potential benefit of word-level propagation is deeper conflict analysis. Normally, using bit-blasting, conflict analysis starts as soon as the first conflict is found. In the word-level solver, we could do the same, to find the first conflict clause and backtrack to the level indicated by this conflict (we call this "standard backjumping"). With word-level propagation, since we can discover several conflicts at once, we may find several learnt clauses at once, corresponding to several backtrack levels. We choose the smallest level from them in order to backtrack to the highest level of the search tree and add all the learnt clauses along the way to prevent all the conflicts from happening again (we call this "multi-conflict backjumping"). We also offer a comparison of these two approaches.

To construct the solver we have extended MiniSAT [6] so that it can keep track of opportunities for word-level propagation and intersperse this kind of propagation with unit propagation. Our word-level propagators contribute to MiniSAT's powerful search and learning mechanism by providing clauses as explanations for word-level propagated bits. In this way, the word level propagators become lazy clause generators [20] for a SAT solver extended with constraint programming technology [21]. In summary the main contributions of this paper are:

- a word-level propagating (but bit-level explaining) constraint solver;
- algorithms for generation of explanations for word-level propagators;
- an investigation of the algorithmic design space in building word-level propagation with explanation;

- the first (as far as we know) empirical evaluation of the proposal by Michel and Van Hentenryck [18], and comparison with the standard bit-blasting approach to these problems;
- results suggesting that, with careful engineering, a word-level propagation approach can be competitive with, or a useful supplement to, bit blasting.

The remainder of the paper is arranged as follows. Section 2 introduces bit-vector constraints and notation. Section 3 outlines the architecture of MiniSAT extended with word-level propagation. Section 4 introduces the propagators used in our solver. Section 5 explores several options for the design of the word-level solver and Sect. 6 evaluates these options and also compares to pure bit-blasting through experiments using standard benchmarks. Section 7 outlines related work and Sect. 8 concludes. The reader is assumed to have a basic understanding of modern SAT-solving technology.

2   Bit-Vector Constraints

In the following we shall need to distinguish word-level logical operations from Boolean operations carefully. As bit-wise operations we use $\sim$, $\&$, $|$, and $\oplus$ for bit complement, conjunction, disjunction, and exclusive or, respectively. As Boolean connectives, we use $\neg$, $\wedge$ and $\vee$ for negation, conjunction, and disjunction, respectively.

A *bit-vector* $x_{[w]}$ is a sequence of w binary digits (bits) and x_i denotes the ith bit in this sequence. The elements of the sequence are indexed from right to left, starting with index $0 : x = x_{w-1}...x_1 x_0$. Here we take Boolean variables as bit-vectors of length 1. A "trit-vector" (for bit-width w) is a sequence of w elements taken from $\{0, 1, *\}$. Here the $*$ represents an undetermined bit, so a trit-vector x corresponds to the cube $(\bigwedge_{i \in I_0} \neg x_i) \wedge (\bigwedge_{i \in I_1} x_i)$, where I_0 is the set of index positions that hold a 0, and I_1 is the set of index positions that hold a 1.

In an implementation, the trit-vector can be represented by a pair of bit-vectors: $\langle \mathsf{lo}(x), \mathsf{hi}(x) \rangle$, where $\mathsf{lo}(x)$ and $\mathsf{hi}(x)$ are bit-vectors representing the lower and upper bound of x respectively, with

$$\mathsf{lo}(x)_i = \begin{cases} 0 & \text{if } x_i = * \\ x_i & \text{otherwise} \end{cases} \qquad \mathsf{hi}(x)_i = \begin{cases} 1 & \text{if } x_i = * \\ x_i & \text{otherwise} \end{cases}$$

For example, trit-vector $z = 011*0*11$ is written $\langle 01100011, 01110111 \rangle$ in this "lo-hi" form. The advantage of this form is that, as long as the bit width of a trit-vector x is less than or equal to the size of machine registers, $\mathsf{lo}(x)$ and $\mathsf{hi}(x)$ can be treated as unsigned integers, that is, z is $\langle 99, 119 \rangle$. Supported by an implementation language (such as C) that can utilise word-level operations, we can rephrase bit-propagation on a trit-vector as word-level operations on its bounds. For an example, consider $y = *1110***$ corresponding to $\langle 01110000, 11110111 \rangle$, and the constraint $y = z$. We can utilize the word-level rule: $\mathsf{lo}(y) = \mathsf{lo}(z) = \mathsf{lo}(y) \mid \mathsf{lo}(z)$, $\mathsf{hi}(y) = \mathsf{hi}(z) = \mathsf{hi}(y) \& \mathsf{hi}(z)$ to obtain the new lo-hi form of y (and z): $\langle 01110011, 01110111 \rangle$ representing 01110*11. Instead of propagating the bits one by one, we effectively fix the bits y_7, y_1, y_0 and z_4 simultaneously with the word-level operations on the bounds.

The lo-hi form allows for invalid representations of trit-vectors. That happens when, for some x, a bit in $\mathsf{lo}(x)$ is 1 while the corresponding bit in $\mathsf{hi}(x)$ is 0. As will be seen, propagation can produce such invalid forms, but this happens when, and only when, an inconsistency is present in the current set of constraints. The validity checking rule is simple:

$$\mathsf{valid}(x) = \mathsf{lo}(x) \mid \mathsf{hi}(x) \tag{1}$$

The result for a valid bit-vector lo-hi form should be a bit-vector of all 1 bits with the same bit width of the bit-vector variable; otherwise it is invalid, and the 0 bits in the result are the bits that cause the invalidity.

The following predicates on trit-vectors will prove useful:

$$\mathsf{fixed}(x) \equiv \mathsf{lo}(x) = \mathsf{hi}(x) \qquad\qquad \mathsf{pos}(x) = \{\mathsf{lit}(x_i) \mid \mathsf{lo}(x_i) = 1\}$$
$$\mathsf{msb}(x_{[w]}) = x_{w-1} \qquad\qquad\qquad \mathsf{neg}(x) = \{\mathsf{lit}(x_i) \mid \mathsf{hi}(x_i) = 0\}$$
$$\mathsf{lit}(b) = \begin{cases} \ulcorner b \urcorner & \text{if } \mathsf{lo}(b) = 1 \\ \ulcorner \neg b \urcorner & \text{if } \mathsf{hi}(b) = 0 \end{cases} \qquad \mathsf{lits}(x) = \mathsf{pos}(x) \cup \mathsf{neg}(x)$$

We use $\mathsf{fixed}(x)$ to return a Boolean value indicating whether every bit in bit-vector x is fixed. We use $\mathsf{msb}(x_w)$ to denote the most significant bit of bit-vector x. We use $\mathsf{lit}(b)$ to return the *literal* corresponding to the Boolean bit b under the condition that b is fixed (hence the use of Quine corners). We use $\mathsf{pos}(x)$ ($\mathsf{neg}(x)$) to return the set of literals fixed to 1 (resp. 0) in bit-vector x, and $\mathsf{lits}(x)$ to return the set of fixed literals in x. In the later algorithms, we take the set of literals to mean the conjunction of the literals.

Our solver handles all operations in the QF_BV category of SMT-LIB2 except for multiplication, division, modulus and remainder. The operations have the usual semantics [15]. We summarize the most important constraints:

Logical Constraints. Logical constraints include bitwise equality $x = y$, bitwise negation $x = {\sim} y$, bitwise conjunction $z = x \mathbin{\&} y$, bitwise disjunction $z = x \mid y$, bitwise exclusive-or $z = x \oplus y$, bitwise nand, bitwise nor, reified equality $b \Leftrightarrow x = y$, and if-then-else operation $ite(b, x, y) = z$ where b is Boolean. The semantics of $ite(b, x, y) = z$ is $(b \wedge (z = x)) \mid (\neg b \wedge (z = y))$.

Arithmetic Constraints. Arithmetic constraints include (fixed-width) addition $x + y = z$, two's complement unary minus $y = -x$ which is equivalent to $y = {\sim} x + 1$, subtraction $z = x - y$ which is equivalent to $z = x + ({\sim} y + 1)$, unsigned inequality $b \Leftrightarrow x \leq_u y$, $b \Leftrightarrow x <_u y$, $b \Leftrightarrow x \geq_u y$, $b \Leftrightarrow x >_u y$, and the corresponding signed inequality constraints. Signed inequality constraints can be translated into unsigned inequality constraints. For instance, $b \Leftrightarrow x \leq_s y$ is equivalent to $b \Leftrightarrow (x \leq_u y) \oplus x_{w-1} \oplus y_{w-1}$.

Structural Constraints. Structural constraints include left shift ($\ll$), unsigned and signed right shift ($\gg_u$, $\gg_s$), left and right rotate (*rotl*, *rotr*), concatenation (::), extraction ($extract(x, n, m) = y$) where y is the extraction of bits

---

Algorithm 1. General algorithm for MiniSAT and word-level solver

---

add the input into the system   ▷ initialization; CNF or word-level formulas
if PROPAGATE() $\neq$ *true* **then**   ▷ unit/word-level propagation; top level conflict
    return UNSAT
while true **do**
    if PROPAGATE() $=$ *true* **then**   ▷ no conflict
        if all variables are assigned **then**
            return SAT
        else
            decide()
    else   ▷ conflict happens
        if top-level conflict found **then**
            return UNSAT
        else
            learnt_clause := *conflict_analyze*()
            backjump(*learnt_clause*)

---

n down to m from x, signed and unsigned extension (ext_u, ext_s), and repeat ($repeat(x, n) = y$) where y is the concatenation of n copies of x.

3   Extending MiniSAT

MiniSAT [6] is a small, complete, and efficient SAT solver which was designed with domain specific extension in mind. The general algorithm for both the MiniSAT and word-level SAT based solver is suggested in Algorithm 1 [6,15], based on the architecture shown in Fig. 1.

3.1   The Architecture and SAT Solving Process in MiniSAT

The input to MiniSAT is a CNF formula, that is, the conjunction of clauses. Each clause is the disjunction of literals, that is, Boolean variables or their negation. The output is either the assignment of all the variables that satisfies the input CNF formula, or "UNSAT" if the formula is unsatisfiable.

First, a literal ℓ is dequeued from the propagation queue, to see if any new literal can be propagated based on this literal, by looking up its Boolean watch list ($BWatch(\ell)$) and sending the related clauses to do the unit propagation. The unit propagation is the only propagation method applied in MiniSAT which finds clause C where all literals except for one literal ℓ' have been made false, then propagates ℓ' to *true*. After each round of unit propagation, either a new literal may be propagated in which case this literal will be enqueued into the Boolean propagation queue ($BPQueue$), and the clause C will be added to the explanation database as the reason for this variable b ($Reasons(b)$); or a conflict happens in which case the clause C becomes the conflict clause. If clause C is at the top-level then it means the whole problem is unsatisfiable; otherwise the conflict clause is analyzed based on the explanations of the fixed literals and

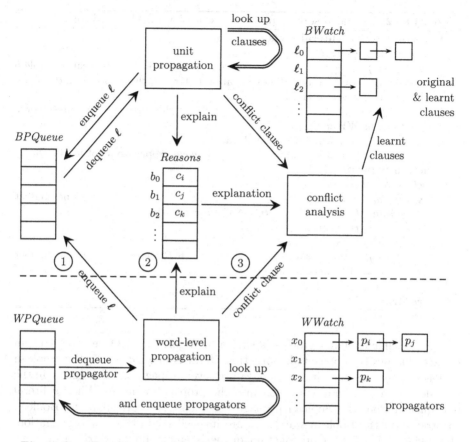

Fig. 1. Overall architecture: MiniSAT (top) and word-level mechanism (bottom)

a learnt clause is synthesized to direct "back-jumping". In addition, the learnt clause is added into the clause database to avoid the same conflict from occurring in the future, which is known as "no-good learning".

3.2   Architecture and Solving Process in Word-Level Solver

The extended architecture for our word-level SAT based solver is shown in the bottom part of Fig. 1. The input of the word-level solver is both the word-level formulas which are for bit-vector operations, and the CNF formulas which are for Boolean operations. A separate static watch list ($WWatch(x)$) for the word-level propagators of each related bit-vector is added to MiniSAT. Correspondingly, a separate word-level propagator queue ($WPQueue$) for the word-level propagators is added (it will have a lower priority than the Boolean propagation queue). Note that at the beginning, we put all the propagators into the propagator queue and run them to the fix-point.

Algorithm 2. Extended solving process in word-level SAT based solver

---

   function ENQUEUE(**literal** ℓ, **clause** C)
     $BPQueue.enqueue(\ell)$
     $b := var(\ell)$ ▷ get the corresponding boolean variable b
     $Reasons[b] := C$ ▷ add the explanation C for b to the explanation database
     if ℓ is in a bit-vector **then**
       $x := word(b)$ ▷ get the corresponding bit-vector x
       for p in $WWatch(x)$ **do**
         if p is not in $WPQueue$ **then**
            $WPQueue.enqueue(p)$ ▷ enqueue propagators not in $WPQueue$
   function PROPAGATE( )
     clause $confl := true$
     while $confl = true$ **do**   ▷ no conflict
       while $\neg BPQueue.isEmpty() \wedge confl = true$ **do**
       $\ell := BPQueue.dequeue()$
       $confl := unit\_prop(\ell)$
       if $confl = true$ **then**   ▷ $BPQueue$ is empty, no conflict
       $p := WPQueue.dequeue()$
       $confl := word\_prop(p)$
   return $confl$

The extended solving process is shown in Algorithm 2. Once a bit ℓ of an integer x is newly propagated, both this literal is enqueued into the Boolean propagation queue, and all the related word-level propagators of integer x in the word-level watch list are enqueued into the propagator queue. When a literal is dequeued from the Boolean propagation queue, the corresponding Boolean constraints in the Boolean watch list are invoked to do the unit propagation. Only when the Boolean propagation queue is empty do we start to dequeue propagators from the propagator queue. We thus favour unit propagation since it is faster but weaker, while the word level propagation is more powerful but slower. In addition, since the previously learnt clauses are in the priority queue, previous conflicts can be avoided earlier.

As can be seen from Fig. 1, the interactions[1] between MiniSAT (top part) and the word-level mechanism (bottom part) are the propagated literals ①, explanations ②, and the conflict clauses ③. The word-level propagators are required to return explanations for the literals they propagate and return conflict clauses when they detect conflict. Without these capabilities, word-level propagators cannot benefit from the learning capabilities of the SAT solver, including back-jumping and powerful autonomous search.

4   Word-Level Propagators with Bit-Level Explanation

Bit-blasting is the most common approach to bit-vector constraint solving. Bit-blasting rewrites all the word level formulas into large number of low-level propositional formulas although many of them may be redundant and never used in

---

[1] The algorithms below point out where/when the interactions ①, ②, ③ occur.

the solving process. Instead of doing bit-blasting, we use word-level propagation. The propagators perform propagation, and they also generate explanations in the form of clauses, for literals fixed by propagation. They can be seen as lazy clause generators, generating clauses only as these are needed.

The input of the word level propagators are all bit-vectors. Inside the propagator, we utilize and also extend the propagation rules introduced in [18] to do the propagation on the word level. At the same time we explain every propagated bit at the bit level. After each round of propagation for the bit-vector interval, validity checking (1) is applied on the new intervals. After the checking, either some bits are propagated, or a conflict happens which means a conflict clause (or several conflict clauses) should be returned. The explanation for the propagated bit is a set of literals which are the reason for making the propagated bit fixed. Note that the explanation for each fixed bit can also explain the conflict which happens because of this bit.

Logical Constraints. The detailed word-level propagation rules for the logical constraints can be found in [18]. The explanations for the basic logical constraints are as following. We take the bitwise equality $(x = y)$ as an example: when the ith bit of integer x_i is fixed to 1, we know that the reason is that y_i is already fixed to 1. So the clause $c_2 : \neg y_i \vee x_i$ is the explanation that could explain why x_i is fixed to 1. The explanation for reified equality constraint $b \leftrightarrow x = y$ is introduced in Sect. 5.1.

- Bitwise Equality $(x = y)$:

$$c_1 : \neg x_i \vee y_i \qquad c_2 : \neg y_i \vee x_i$$

- Bitwise Conjunction $(z = x \wedge y)$:

$$c_1 : \neg z_i \vee x_i \qquad c_2 : \neg z_i \vee y_i \qquad c_3 : \neg x_i \vee \neg y_i \vee z_i$$

- Bitwise Negation $(x = {\sim} y)$:

$$c_1 : \neg x_i \vee \neg y_i \qquad c_2 : x_i \vee y_i$$

- Bitwise Disjunction $(z = x \vee y)$:

$$c_1 : \neg x_i \vee z_i \qquad c_2 : \neg y_i \vee z_i \qquad c_3 : x_i \vee y_i \vee \neg z_i$$

- Bitwise Exclusive Or $(z = x \oplus y)$:

$$c_1 : x_i \vee y_i \vee \neg z_i \qquad c_2 : x_i \vee \neg y_i \vee z_i \qquad c_3 : \neg x_i \vee y_i \vee z_i$$
$$c_4 : \neg x_i \vee \neg y_i \vee \neg z_i$$

- Bitwise Conditional $(ite(b, x, y) = z)$:

$$c_1 : \neg b \vee \neg x_i \vee z_i \qquad c_2 : \neg b \vee x_i \vee \neg z_i \qquad c_3 : b \vee \neg y_i \vee z_i$$
$$c_4 : b \vee y_i \vee \neg z_i \qquad c_5 : \neg x_i \vee \neg y_i \vee z_i \qquad c_6 : x_i \vee y_i \vee \neg z_i$$

Arithmetic Constraints. A constraint $z = x + y$ is translated into constraints that introduce two new variables: c for the sequence of carry-ins, and d for carry-outs. As pointed out by Michel and Van Hentenryck [18], a full adder can then be captured with the constraints

$$z = x \oplus y \oplus c$$
$$d = (x \,\&\, y) \mid (c \,\&\, (x \oplus y))$$
$$c = d \ll 1$$

where the last constraint connects the carry-in bit-vector c and carry-out bit-vector d. By adding the intermediate variables into the addition constraint, the propagator for addition can be divided into several *decomposed* propagators which solve the basic constraints individually. The explanations for the addition constraint simply combines the explanations of these basic constraints. The propagation rules and explanations for inequality constraints will be introduced in Sect. 5.1.

Structural Constraints. We have extended the propagation rules mentioned in [18] to solve all the structural constraints in the QF_BV category of SMT-LIB2. We can take the structural constraints as the variants of the bitwise equality constraints for bit manipulation. Therefore, the propagation rules for structural constraints are based on the propagation rules of the bitwise equality constraints but with different "masks" designed to fix the particular bits to be 1 or 0. The explanations for the structural constraints are also similar to the bitwise equality constraints but with some bit shift ($\ll, \gg_u, \gg_s, rotl, rotr$), or fixing some bits value ($\ll, \gg_u, \gg_s, ext_u, ext_s$).

5   Word-Level Propagation Solving

5.1   Propagators: Composed vs Decomposed

To solve a complicated constraint, one way we can proceed is to create a single "composed" propagator. This propagator may be complex to implement, and may end up finding long explanations. In many cases it can be worth splitting the complicated constraint into several smaller constraints thus decomposing it. Not only are the decomposed components easier to implement, but more importantly, in a learning solver, the intermediate variables introduced may be useful for both search as well as making explanations shorter. Of course the end line for this approach is effectively full bit-blasting. On the other hand, the composed propagators are compact, while the decomposed propagators need the communication among the components. We propose both single propagators and decompositions to implement the reified equality constraint $b \Leftrightarrow x = y$ and the reified inequality constraint $b \Leftrightarrow x \leq_u y$.

Composed Propagators. Propagators return a conflict clause (*"true"* indicates no conflict) and enqueue the propagated literals together with their explanations.

Algorithm 3. Propagator for $b \Leftrightarrow x = y$

> **function** PROP_REIFEQ(**bit** b, **bit-vec** x, y)
>> **if** $\mathsf{lo}(b) = 1$ **then**
>>> **return** PROP_EQ(x, y) ③
>>
>> **else if** $\mathsf{hi}(b) = 0$ **then**
>>> **return** PROP_DISEQ(x, y) ③
>>
>> **else**
>>> **if** $\mathsf{fixed}(x) \wedge \mathsf{fixed}(y) \wedge \mathsf{lo}(x) = \mathsf{lo}(y)$ **then**   ▷ $x = y$
>>>> $Explanation := \mathsf{lits}(x) \wedge \mathsf{lits}(y) \to b$
>>>> ENQUEUE$(b, Explanation)$ ① ②
>>>
>>> **else**
>>>> $z := \mathsf{lo}(x) \ \& \sim \mathsf{hi}(y) \ | \sim \mathsf{hi}(x) \ \& \ \mathsf{lo}(y)$
>>>> **if** $z \neq 0$ **then**   ▷ $x \neq y$
>>>>> choose i with $z_i = 1$
>>>>> $Explanation := \mathsf{lit}(x_i) \wedge \mathsf{lit}(y_i) \to \neg b$
>>>>> ENQUEUE$(\neg b, Explanation)$ ① ②
>>
>> **return** *true*

We can implement a propagator for the reified equality constraint: $b \Leftrightarrow x = y$ as shown in Algorithm 3. The propagator reuses the implementation of the propagators for $x = y$ and $x \neq y$, or checks that $x = y$ in the current domain in which case it explains b, or that $x \neq y$ in the current domain, in which case it explains $\neg b$.

Algorithm 4. Propagator for $x \neq y$

> **function** DISEQ(**bit-vec** x, y)
>> **if** $\mathsf{fixed}(x) \wedge (\mathsf{lo}(x) = \mathsf{lo}(y) \vee \mathsf{lo}(x) = \mathsf{hi}(y))$ **then**   ▷ x fixed; y possibly not fixed
>>> $f := \mathsf{lo}(y) \oplus \mathsf{hi}(y)$
>>> **if** unique 1 bit in f **then**   ▷ only one bit of y is unknown
>>>> find i with $f_i = 1$
>>>> **if** $\mathsf{lit}(x_i) = x_i$ **then** $\ell := \neg y_i$ **else** $\ell := y_i$
>>>> $Explanation := \mathsf{lits}(x) \wedge \mathsf{lits}(y) \to \ell$
>>>> ENQUEUE$(\ell, Explanation)$ ① ②
>>
>> **return** *true*
>
> **function** PROP_DISEQ(**bit-vec** x, y)
>> **if** $\mathsf{fixed}(x) \wedge \mathsf{lo}(x) = \mathsf{lo}(y)$ **then**   ▷ $x = y$
>>> **return** $\mathsf{lits}(x) \wedge \mathsf{lits}(y) \to$ *false* ③
>>
>> **if** $\mathsf{lo}(x) \ \& \sim \mathsf{hi}(y) \ | \sim \mathsf{hi}(x) \ \& \ \mathsf{lo}(y)$ **then**   ▷ $x \neq y$
>>> **return** *true*
>>
>> DISEQ(x, y)
>> DISEQ(y, x)
>> **return** *true*

The propagator for $x \neq y$ (Algorithm 4) first checks whether x and y are known to be equal and if so, returns a failure explanation. If x and y are known to differ, it simply returns $true$. Otherwise if there is at most one unfixed bit, and they are otherwise equal it explains why the unfixed bit should be set to the opposite value of the corresponding fixed bit in the other variable. For example, if $x = 11010$, $y = 110 * 0$, we propagate bit $y_1 = 0$, the explanation is $x_4 \wedge x_3 \wedge \neg x_2 \wedge x_1 \wedge \neg x_0 \wedge y_4 \wedge y_3 \wedge \neg y_2 \wedge \neg y_0 \rightarrow \neg y_1$.

Similar to the way of solving the equality constraint, for the inequality constraint $b \Leftrightarrow x \leq_u y$, we also need two propagators, one for the constraint $x \leq_u y$ and one for $x >_u y$. Since the propagation rules and the way of generating the explanations of these two constraints are similar, we show only the propagator for $x \leq_u y$, as Algorithm 5. First, we still need to check if there is a conflict, that is, if the lower bound of x is (unsigned) greater than the upper bound of y, in which case a conflict clause needs to be returned. To generate the conflict clause, we go through every bit of x and y bit by bit from the most significant bits to find the first "bit pair" of 1 bit in x and 0 bit in y, and add all the 1 bits in x and 0 bits in y before the "bit pair" (included) to the conflict clause. For example, if $x = 10010**$, $y = 1000*11$ then the conflict clause is $x_6 \wedge x_3 \wedge \neg y_5 \wedge \neg y_4 \rightarrow false$.

After the conflict checking, we start the propagation which utilizes the propagation rules introduced in [18]. We take the propagation for the bits in variable x as an example. In the propagation of constraint $x \leq_u y$, we can only fix x to 0. But we pretend to fix the first free bit (from left) of x to 1 to see if there is a conflict in which case we know that this free bit must be fixed to 0; otherwise we cannot propagate anything. The way of generating the explanation for this fixed bit is similar to how the conflict clause was generated, but with the pretend lower bound of x. For example, if $x = 1100*1*$ and $y = 11000**$ then $xl = 1100110$, and the explanation is $x_6 \wedge x_5 \wedge \neg y_4 \wedge \neg y_3 \wedge \neg y_2 \rightarrow \neg x_2$.

The explanations generated by the composed propagators are often large, especially when the bit width of the involved bit-vectors is large. In comparison, each explanation for a basic constraint introduced in Sect. 4 contains at most three literals.

Decomposed Propagators. The decomposed propagator for equality constraint $b \Leftrightarrow x = y$ is based on this observation [22]:

$$b = \mathsf{msb}(\sim((x - y) \mid (y - x)))$$

We add intermediate variables to split this constraint into several basic constraints which can be processed by the word level propagators already introduced. Note that the $m_1 = x - y$ constraint will be further split as the arithmetic constraint introduced in Sects. 2 and 4. The explanation for the reified equality constraint $b \Leftrightarrow x = y$ is made up by those of the basic constraints—several small explanations with the intermediate literals involved.

$$m_1 = x - y; \quad m_2 = -m_1; \quad m_3 = m_1 \mid m_2; \quad m_4 = \sim m_3; \quad b = \mathsf{msb}(m_4)$$

Algorithm 5. Propagator $x \leq_u y$

---

if $\mathsf{lo}(x) >_u \mathsf{hi}(y)$ **then**
    $f := \mathsf{lo}(x) \, \& \sim \mathsf{hi}(y)$
    $i :=$ first 1 bit position in f ▷ find first bit pair: 1 bit of x and 0 bit of y
    return $\mathsf{pos}(x) \setminus \{x_j \mid j < i\} \wedge \mathsf{neg}(y) \setminus \{\neg y_j \mid j < i\} \to \textit{false}$ ③
for $i := w - 1$ **downto** 0 **do**   ▷ propagate the bits in x
    if $\neg\mathsf{fixed}(x_i)$ **then**
        $xl := \mathsf{lo}(x) \mid (1 \ll i)$ ▷ pretend ith bit is fixed to 1
        if $xl >_u \mathsf{hi}(y)$ **then**
            $\ell := \neg x_i$ ▷ fix x_i to 0
            $f := \mathsf{lo}(xl) \, \& \sim \mathsf{hi}(y)$
            $i :=$ first 1 bit position in f
            $\textit{Explanation} := \mathsf{pos}(x) \setminus \{x_j \mid j < i\} \wedge \mathsf{neg}(y) \setminus \{\neg y_j \mid j < i\} \to \ell$
            ENQUEUE$(\ell, \textit{Explanation})$ ① ②
        else
            break
/* the similar algorithm to propagate the bits in y */
return $\textit{true}$

The decomposed propagator for inequality constraint $b \Leftrightarrow x \leq_u y$ is based on this observation:

$$b = \mathsf{msb}((\sim x \mid y) \, \& \, ((x \oplus y) \mid \sim(y - x)))$$

The way to solve an inequality constraint with decomposed propagators is the same as for the equality constraint.

It is worth pointing out that the two kinds of propagator do not lead to identical search trees. The presence of intermediate variables introduced by the decomposition makes a considerable difference to activity based search, since there are new variables to search on and different initial activities.

5.2   Explanation: Forward vs Backward

Normally in a SAT solver, for every fixed Boolean literal, a reason why it became *true* is required for conflict analysis. Therefore, normally when we fix a Boolean literal in our word-level propagator, we return an explanation for it eagerly, so-called "forward explanation." Another approach, standard for SMT theory solvers [19] and discussed by Gent et al. [10], is to generate the explanation only during conflict analysis where the reason for a propagated literal is required. Compared to the forward explanation method, this has the advantage that explanations are only generated as needed. Furthermore, the "backward explanation" is especially good for our word-level propagator. Our propagators have two parts: one is the propagation part, the other is the explanation generation part which is the more time consuming. Therefore, backward explanation makes propagation faster, but possibly makes conflict analysis slower.

5.3   Conflict Analysis: First vs Highest Level

As already mentioned we can detect conflicts in many bit positions simultaneously. But to choose which one to do the conflict analysis on remains a question.

With bit-blasting, as soon as the first conflict is found, conflict analysis is started, returning a learnt clause of the form $C \vee \ell$, where ℓ is the unique literal (UIP) at the current decision level, and the maximum decision level in the remainder of the clause C determines the level to backjump to. One way to manage conflict analysis for word-level propagation is to choose the first conflict to do the conflict analysis as usual for SAT. We call this "standard backjumping".

An alternative approach is to generate a conflict clause for each bit position that is in conflict. We can then add all the learnt clauses generated to the clause database and then jump to the highest decision level indicated by one of them. This has the advantage of generating more information from the failure, and potentially higher backjumps. We call this "multi-conflict backjumping".

6   Experimental Evaluation

For the experimental data, we pick the folders from the QF_BV category of SMT-LIB2 benchmarks which do not make use of multiplication, division, modulus and remainder, and the bit width for the bit-vector operations is no greater than 64 (the size of our machine register). In total there are more than 12000 test cases. We split them into two categories: easy and difficult, according to the per-problem solve time of the bit-blaster baseline solver. We use a time limit of 500 s. In Tables 1 and 2, "time" means the total time in seconds for all the successful test cases in the folder; "TO" is the number of cases that timed out; "Total" is the total time of all successful test cases; "Overall time" is "Total" plus 500 s penalty for each unsuccessful case, which gives an overall "score" similar to what is used in SMT competitions. All the experiments were performed on a commodity computer with a Core-i7 CPU (2.7 GHz) and 5 GB RAM.

The first experiment compares forward explanation (F) versus backward explanation (B), as well as standard backjumping (S) versus multi-conflict backjumping (M). We implemented three variants of the word-level bit-vector solvers which all use the decomposed word-level propagators for equality and inequality constraints. The reason we only look at three variants is that the two parameters (F/B, S/M) do not interact with each other. Table 1 shows that, first, backward explanation outperforms the forward explanation significantly, especially when the test cases are easy. Second, the multi-conflict backjumping outperforms the standard backjumping considerably in both categories.

The second experiment compares bit-blasting with word-level bit-vector solving using composed and decomposed propagators. We implemented a vanilla bit-blaster as a baseline to compare against, which uses the decomposition of equality and inequality applied in the decomposed word-level solver Deq + Dle. Since Table 1 suggests the B+M combination has merit, all word-level bit-vector solvers listed in Table 2 use backward explanation and multi-conflict backjumping. However, they use different combinations of composed propagators (C) and

Table 1. Forward explanation vs backward explanation and standard backjumping vs multi-conflict backjumping (times are in seconds)

| Problem | | F + S | | B + S | | B + M | |
Name	Number	Time	TO	Time	TO	Time	TO
sage: app1	1176	727	0	432	0	**416**	0
sage: app2	475	8	0	**5**	0	**5**	0
sage: app5	990	44	0	**27**	0	29	0
sage: app6	245	0	0	0	0	0	0
sage: app7	339	4	0	**3**	0	**3**	0
sage: app8	1760	662	1	447	1	542	**0**
sage: app9	2096	370	1	732	**0**	587	**0**
sage: app12	4905	2118	0	**1226**	0	1262	0
stp_samples	424	18	0	**13**	0	**13**	0
bench_ab	284	0	0	0	0	0	0
Total	12694	3951	2	2885	1	**2857**	**0**
Overall time		4951		3385		**2857**	

brummayerbiere3	42	495	33	310	33	**271**	33
spear: cvs_v1.11.22	5	0	4	0	4	0	4
spear: openldap_v2.3.35	6	0	6	0	6	0	6
spear: samba_v3.0.24	4	0	4	0	4	494	3
rubik	7	524	2	308	2	407	1
uclid_contrib_smtcomp09	7	149	6	90	6	**72**	6
Total	71	1168	55	708	55	1244	53
Overall time		28668		28208		**27744**	

decomposed propagators (D) for equality (eq) and inequality constraints (le). Table 2 shows the resource consumption including the running time and average memory usage (mem) in MB. The results show that the Deq + Dle word-level propagator is typically faster than bit-blasting on the easy cases, using less memory. For the difficult cases, the Ceq + Cle word-level propagator outperforms the bit-blasting in some cases and also uses much less memory. But in general the bit-blasting method is more robust.

Table 3 shows the average number of conflicts per second (fail/sec), and the average number of inspections[2] in thousand per second (insp(k)/sec) that occur during the search. We compare bit-blasting only against the best word-level solver as identified above, that is, Deq + Dle for easy cases and Ceq + Cle for difficult cases. Note that the bit-blasting often finds fewer conflicts during the

---

[2] A call to a unit or word-level propagator (which may or may not result in fixing new bits).

Table 2. Resource consumption: bit-blaster vs word-level bit-vector solver and composed propagators vs decomposed propagators (memory is in MB)

Problem	bit-blaster			Deq + Dle			Ceq + Dle			Deq + Cle			Ceq + Cle		
Name	Time	TO	mem	Time	TO	mem	Time	TO	mem	Time	TO	mem	Time	TO	mem
app1	393	0	27	416	0	25	381	0	23	1985	32	16	905	32	**12**
app2	9	0	**7**	5	0	11	60	1	11	252	18	10	6311	1	8
app5	49	0	23	29	0	20	271	1	15	1505	15	16	1025	17	**10**
app6	0	0	**7**	0	0	8	0	0	8	0	0	8	0	0	8
app7	3	0	**7**	3	0	8	3	0	8	1	14	8	1	14	8
app8	1127	0	16	542	0	16	828	1	14	2062	2	13	1585	2	**10**
app9	863	0	15	587	0	14	303	2	13	2122	1	12	1387	3	**9**
app12	1052	0	19	1262	0	20	994	2	16	972	6	17	595	8	**11**
stp_sam	37	0	39	13	0	31	8	0	23	13	0	30	7	0	**20**
bench_ab	1	0	**7**	0	0	8	0	0	8	0	0	8	0	0	8
Total	12694	0	167	2857	0	161	2848	7	139	8912	88	138	11816	77	**104**
Overall time	3534			**2857**			6348			52912			50316		
brumm3	402	31	33	271	33	41	422	33	31	228	32	27	208	32	**16**
cvs	688	2	9	0	4	10	0	4	9	0	5	10	0	5	**8**
openldap	176	5	353	0	6	240	0	6	**46**	0	6	238	0	6	47
samba	0	4	1005	494	3	676	24	0	124	0	4	751	5	0	**87**
rubik	87	2	**7**	407	1	16	838	1	11	589	1	13	56	2	10
uclid	0	7	**7**	72	6	109	710	3	25	0	7	138	393	4	25
Total	1353	51	1414	1244	53	1092	1994	47	246	817	55	1177	662	49	**193**
Overall time	26853			27744			25494			28317			**25162**		

search with more propagation, while the word-level solvers often find more conflicts with less propagation. That is because propagating and checking the conflicts at word-level is parallel in some sense, resulting in a higher rate of conflict-finding as well as the reduction in inspections.

7 Related Work

Word-level reasoning on bit-vector logic is NEXPTIME-complete [14]. In spite of this, the problem has received much attention recently, albeit with limited progress. Current related work falls into one of or the combination [1] of three categories:

Word-Level Reasoning Based on Lazy SMT Techniques: Hadarean et al. [12] propose two word-level solvers an equality solver and inequality solver as the theory solver in their lazy bit-vector solver. But they cannot express the conflict at the bit-level which significantly affects the efficiency of the method as they showed in [12].

Word-Level Reasoning Based on Constraint Programming: Bardin et al. [2] propose two word-level propagators based on the Constraint Logic Programming framework. One is called Is/C which is to solve linear arithmetic constraints, and the other is the BL (Bit-List) propagator which runs in linear

Table 3. Conflicts and inspections during the search (a '-' indicates a time of 0)

Problem	Bit-blaster		Deq + Dle	
Name	fail/sec	insp(k)/sec	fail/sec	insp(k)/sec
app1	2.6	14.4	3.1	12.6
app2	4.8	20.3	9.2	24.0
app5	1.0	6.3	2.1	9.1
app6	-	-	-	-
app7	76.3	66.0	61.3	36.7
app8	2.6	21.3	4.5	11.7
app9	2.2	16.0	3.8	10.4
app12	1.5	4.2	1.4	3.2
stp_sam	0.3	10.1	1.1	10.4
bench_ab	2.0	23.0	-	-

Problem	Bit-blaster		Ceq + Cle	
Name	fail/sec	insp(k)/sec	fail/sec	insp(k)/sec
brumm3	41.5	507.0	109.9	297.0
cvs	1507.7	6041.5	11134.8	6645.2
openldap	211.5	2404.3	930.7	1833.7
samba	84.0	2675.5	200.0	1800.0
rubik	280.6	4155.5	875.0	2141.1
uclid	198.9	335.1	3506.5	7354.8

time to solve the linear bitwise constraints. Constraint propagators for modular arithmetic constraints have been proposed by Gotlieb et al. [11] who utilize so-called clockwise intervals in a linear fragment of modular integer constraints. None of these CP approaches support learning, or compare with bit-blasting.

Word-Level Reasoning Based on Linear Programming: This approach is to transform the problem into linear programming constraints [3,23]. For RTL verification, the performance of LP solvers are often no better than SMT solvers as reported in [16].

8   Conclusion

We have extended word-level propagation algorithms of Michel and Van Hentenryck [18] to produce an explaining solver. We have introduced decomposed counterparts to the proposed propagators, as these were not constant time. We also utilize a concept of multi-conflict backjumping, capitalizing on the fact that word-level propagation can detect multiple failures simultaneously. We have given an empirical comparison of word-level propagation versus bit-blasting, the standard approach to these problems. Our solver is a prototype, still to be tuned. Nevertheless it shows that, with careful engineering, a word-level propagation solver can compete with bit-blasting, particularly on easier problems.

For future work, it may be advantageous to apply some word-level simplification as done with the linear solver in STP [8,9]. We also need to deal with non-linear arithmetic operations, one way or other. Finally, an interesting line of research would be to combine word-level propagation with word-level search, especially stochastic local search as recently suggested by Fröhlich *et al.* [7].

Acknowledgment. This work is supported by the Australian Research Council under ARC grant DP140102194.

References

1. Achterberg, T., Berthold, T., Koch, T., Wolter, K.: Constraint integer programming: a new approach to integrate CP and MIP. In: Trick, M.A. (ed.) CPAIOR 2008. LNCS, vol. 5015, pp. 6–20. Springer, Heidelberg (2008)
2. Bardin, S., Herrmann, P., Perroud, F.: An alternative to SAT-based approaches for bit-vectors. In: Esparza, J., Majumdar, R. (eds.) TACAS 2010. LNCS, vol. 6015, pp. 84–98. Springer, Heidelberg (2010)
3. Brinkmann, R., Drechsler, R.: RTL-datapath verification using integer linear programming. In: VLSI Design, pp. 741–746. IEEE Computer Society (2002)
4. Brummayer, R., Biere, A.: Boolector: an efficient SMT solver for bit-vectors and arrays. In: Kowalewski, S., Philippou, A. (eds.) TACAS 2009. LNCS, vol. 5505, pp. 174–177. Springer, Heidelberg (2009)
5. Cook, B., Kroning, D., Sharygina, N.: Cogent: accurate theorem proving for program verification. In: Etessami, K., Rajamani, S.K. (eds.) CAV 2005. LNCS, vol. 3576, pp. 296–300. Springer, Heidelberg (2005)
6. Eén, N., Sörensson, N.: An extensible SAT-solver. In: Giunchiglia, E., Tacchella, A. (eds.) SAT 2003. LNCS, vol. 2919, pp. 502–518. Springer, Heidelberg (2004)
7. Fröhlich, A., Biere, A., Wintersteiger, C.M., Hamadi, Y.: Stochastic local search for satisfiability modulo theories. In: Giunchiglia, E., Tacchella, A. (eds.) Proceedings of the 29th AAAI Conference on Artificial Intelligence, pp. 1136–1143. AAAI Press (2015)
8. Ganesh, V.: Decision Procedures for Bit-Vectors, Arrays and Integers. Ph.D. thesis, Stanford University (2007)
9. Ganesh, V., Dill, D.L.: A decision procedure for bit-vectors and arrays. In: Damm, W., Hermanns, H. (eds.) CAV 2007. LNCS, vol. 4590, pp. 519–531. Springer, Heidelberg (2007)
10. Gent, I.P., Miguel, I., Moore, N.C.A.: Lazy explanations for constraint propagators. In: Carro, M., Peña, R. (eds.) PADL 2010. LNCS, vol. 5937, pp. 217–233. Springer, Heidelberg (2010)
11. Gotlieb, A., Leconte, M., Marre, B.: Constraint solving on modular integers. In: Proceedings of the Ninth International Workshop on Constraint Modelling and Reformulation (ModRef 2010) (2010)
12. Hadarean, L., Bansal, K., Jovanović, D., Barrett, C., Tinelli, C.: A tale of two solvers: eager and lazy approaches to bit-vectors. In: Biere, A., Bloem, R. (eds.) CAV 2014. LNCS, vol. 8559, pp. 680–695. Springer, Heidelberg (2014)

13. Hutter, F., Babic, D., Hoos, H.H., Hu, A.J.: Boosting verification by automatic tuning of decision procedures. In: Formal Methods in Computer Aided Design (FMCAD 2007), pp. 27–34. IEEE Comp. Soc. (2007)
14. Kovásznai, G., Fröhlich, A., Biere, A.: On the complexity of fixed-size bit-vector logics with binary encoded bit-width. In: Proceedings of SMT 2012, pp. 44–55 (2012)
15. Kroening, D., Strichman, O.: Decision Procedures: An Algorithmic Point of View. Springer, Heidelberg (2008)
16. Kunapareddy, S., Turaga, S.D., Sajjan, S.S.T.M.: Comparison between LPSAT and SMT for RTL verification. In: Proceedings of the International Conference on Circuit, Power and Computing Technologies, pp. 1–5. IEEE Computer Society (2015)
17. Limaye, R.S., Seshia, S.A.: Beaver: An SMT solver for quantifierfreebit-vector logic. Master's thesis, EECS Department, University of California, Berkeley, May 2010
18. Michel, L.D., Van Hentenryck, P.: Constraint satisfaction over bit-vectors. In: Milano, M. (ed.) CP 2012. LNCS, vol. 7514, pp. 527–543. Springer, Heidelberg (2012)
19. Nieuwenhuis, R., Oliveras, A., Tinelli, C.: Solving SAT and SAT modulo theories: from an abstract Davis-Putnam-Logemann-Loveland procedure to DPLL(T). J. ACM **53**(6), 937–977 (2006)
20. Ohrimenko, O., Stuckey, P.J., Codish, M.: Propagation via lazy clause generation. Constraints **14**(3), 357–391 (2009)
21. Schulte, C., Stuckey, P.J.: Efficient constraint propagation engines. ACM Trans. Program. Lang. Syst. **31**(1), 2:1–2:43 (2008)
22. Warren Jr., H.S.: Hacker's Delight. Addison Wesley, Boston (2003)
23. Zeng, Z., Kalla, P., Ciesielski, M.: LPSAT: a unified approach to RTL satisfiability. In: Design, Automation and Test in Europe (DATE 2001), pp. 398–402. IEEE Press (2001)

A New Solver for the Minimum Weighted Vertex Cover Problem

Hong Xu$^{(\boxtimes)}$, T.K. Satish Kumar, and Sven Koenig

University of Southern California, Los Angeles, CA 90089, USA
{hongx,skoenig}@usc.edu, tkskwork@gmail.com

Abstract. Given a vertex-weighted graph $G = \langle V, E \rangle$, the minimum weighted vertex cover (MWVC) problem is to choose a subset of vertices with minimum total weight such that every edge in the graph has at least one of its endpoints chosen. While there are good solvers for the unweighted version of this NP-hard problem, the weighted version—i.e., the MWVC problem—remains understudied despite its common occurrence in many areas of AI—like combinatorial auctions, weighted constraint satisfaction, and probabilistic reasoning. In this paper, we present a new solver for the MWVC problem based on a novel reformulation to a series of SAT instances using a primal-dual approximation algorithm as a starting point. We show that our SAT-based MWVC solver (SBMS) significantly outperforms other methods.

1 Introduction

Given a directed or undirected graph $G = \langle V, E \rangle$, a *vertex cover* of G is defined as a collection of vertices $S \subseteq V$ such that every edge in E has at least one of its endpoint vertices in S. A *minimum vertex cover* (MVC) of G is a vertex cover of minimum cardinality. When G is vertex-weighted—i.e., each vertex $v_i \in V$ has a non-negative weight w_i associated with it—the *minimum weighted vertex cover* (MWVC) for it is defined as a vertex cover of minimum total weight.

Two important combinatorial problems equivalent to the MVC problem are the *maximum independent set* (MIS) problem and the *maximum clique* (MC) problem [8]. The MVC problem and its equivalent MIS and MC problems have numerous real-world applications such as in AI scheduling, logistics and operations management, and VLSI design. More recent applications have also been discovered in information retrieval, signal processing, and sequence alignment in computational genomics [14].

Since the MVC problem is a special case of the MWVC problem, the latter not only captures all of the real-world combinatorial problems that the MVC problem can model but also captures a wide range of other combinatorial problems central to AI. For example, consider a simple combinatorial auction problem. We are given a set of items with bids placed on subsets of the items. Each bid has a valuation. The goal is to pick a set of winning bids that maximizes the total valuation—i.e., revenue of the auctioneer—and allocates each item to

© Springer International Publishing Switzerland 2016
C.-G. Quimper (Ed.): CPAIOR 2016, LNCS 9676, pp. 392–405, 2016.
DOI: 10.1007/978-3-319-33954-2_28

at most one winning bid. This can be modeled as a *maximum weighted independent set* (MWIS) problem—equivalent to the MWVC problem—as follows. We create a vertex for each bid such that the weight of the vertex is equal to the valuation of that bid. Two vertices are connected by an edge if and only if their corresponding bids have a non-empty intersection. It is easy to see that the winning bids correspond to the vertices in the MWIS for the graph.

In [19, 20], the MWVC problem has also been identified as being fundamental to solving *weighted constraint satisfaction problems* (WCSPs). Any combinatorial problem posed as a WCSP is equivalent to the MWVC problem for its associated *constraint composite graph* [19, 20]. An efficient solver for the MWVC problem, therefore, has important implications on how well we can solve the plethora of real-world problems that can be modeled as WCSPs. Examples include—but are not limited to—representing and reasoning about user preferences [3], over-subscription planning with goal preferences [10], and various resource allocation problems. Quite importantly, WCSPs also arise as *energy minimization problems* (EMPs) in probabilistic settings. In computer vision applications, for example, tasks such as image restoration, total variation minimization, and panoramic image stitching can be formulated as EMPs derived in the context of *markov random fields* [17, 18].

The MVC problem has received a lot of recent attention in response to the DIMACS Implementation Challenge [14]. There are both exact and heuristic algorithms for solving the MVC problem. Exact algorithms mainly use branch-and-bound techniques [21, 28]. While they guarantee optimality, they may not scale efficiently to be able to solve large problem instances. Heuristic and local search methods, on the other hand, can provide near-optimal solutions to larger and harder problem instances [6, 26]. As a matter of fact, the NuMVC solver [5] integrates many interesting local search techniques for the MVC problem and performs very well in practice.

While there are reasonably good solvers for the MVC problem, the MWVC problem remains understudied. Clearly, the MWVC problem is a generalization of the MVC problem and is harder to solve efficiently. Exact algorithms based on the branch-and-bound technique are not expected to do well for large instances of the MWVC problem simply because they do not scale well even for large instances of the MVC problem. Moreover, the local search techniques used in the best solvers for the MVC problem are also not expected to generalize well to the MWVC problem. This is because the MVC problem is fixed-parameter tractable while the MWVC problem is not [7]. The local search solvers for the MVC problem [5, 26] heavily rely on this property as they solve the fixed-parameter vertex cover problem in their inner loops.

In this paper, we present a new solver for the MWVC problem based on a novel reformulation to a series of SAT instances using a primal-dual approximation algorithm as a starting point. Our SAT-based MWVC solver (SBMS) implements an anytime algorithm that trades off running time with the quality of the produced solution. Moreover, SBMS also reports on how good the produced solution is guaranteed to be with respect to the optimal solution. In many cases,

SBMS converges to the optimal solution in a few iterations and reports it within the allocated amount of time. Empirical results show that SBMS significantly outperforms other methods.

2   Background

The MVC problem is a well known NP-hard problem [8]. There exists a simple factor-2 approximation algorithm for it that runs in polynomial time [29].[1] There are also polynomial-time algorithms that yield slightly better approximation factors but are more involved [15]. However, the MVC problem is also known to be APX-complete. It cannot be approximated arbitrarily well unless P = NP [29]. Furthermore, PCP theorems yield inapproximability results for designing polynomial-time algorithms with approximation factors better than 1.36 [9].[2]

The MWVC problem is harder than the MVC problem since it is a generalization of the latter. The negative results associated with the MVC problem therefore carry over to the MWVC problem. Fortunately, the MWVC problem is still amenable to a fairly simple polynomial-time factor-2 approximation algorithm based on the idea of *linear programming duality* [29]. However, unlike the MVC problem, the MWVC problem is not fixed-parameter tractable [7]. The MVC problem is in fact studied as a central problem in parameterized complexity theory and can be formulated as a half-integral linear programming problem whose dual yields a maximum matching in the corresponding graph [29].

The good solvers for the MVC problem are based on local search [5,6,26]. They implicitly exploit the fixed-parameter tractability of the MVC problem in their inner loops. In order to solve the k-vertex cover problem—i.e., find a vertex cover of size k—in their inner loops, they maintain a current set of vertices of size k and iteratively exchange two vertices—one inside and one outside of this set—until it becomes a valid vertex cover. The state-of-the-art solver for the MVC problem, NuMVC [5], also exploits the fixed-parameter tractability of the MVC problem but with added optimizations.

The NuMVC solver mainly introduces two new techniques not present in its predecessors [5]. The first optimization decomposes the exchange process into two stages—one stage for removing a vertex from the set and the other for adding a vertex to the set. This decomposition leads to linear-time subroutines for each stage instead of the original quadratic-time subroutine that deliberates all pairs of vertices for a possible exchange. The second optimization involves weighting the edges across different iterations while simultaneously employing a mechanism to forget weighting decisions made too far in the past [5].

---

[1] While the MVC is approximable within a constant factor, this has no implications on the MIS problem. In fact, the MIS problem is one of the hardest combinatorial problems and has no polynomial-time constant-factor approximation algorithm unless P = NP [29].

[2] This inapproximability result is tighter under the *unique games conjecture* [16].

3 Reformulations of the MWVC Problem

Given that there are no standard solvers for the understudied MWVC problem, we develop a solver based on reformulating it to a series of SAT instances. We study the usefulness of this reformulation in comparison to modeling the MWVC problem as an Integer Linear Program (ILP), a Pseudo-Boolean Optimization (PBO) problem, a MAX-SAT problem, or an Answer Set Program (ASP).

3.1 Reformulation as an ILP or a PBO Problem

For a given vertex-weighted undirected (or directed) graph $G = \langle V, E \rangle$, the MWVC problem can be formulated as an ILP as follows. We simply associate a 0/1 variable X_i with each vertex $v_i \in V$. X_i indicates the presence of v_i in the MWVC. Here, w_i is the non-negative weight associated with vertex v_i.

$$\text{Minimize} \quad \sum_{i=1}^{|V|} w_i X_i \tag{1}$$
$$\forall \, v_i \in V : \quad X_i \in \{0, 1\}$$
$$\forall \, (v_i, v_j) \in E : \quad X_i + X_j \geq 1$$

To reformulate the MWVC problem as a PBO problem, we simply change the "type" of each variable X_i in the ILP formulation from a 0/1 integer to a Boolean variable.

3.2 Reformulation as a MAX-SAT Problem

The MWVC problem can also be formulated as a weighted MAX-SAT problem—simply referred to as the "MAX-SAT problem" here. In a MAX-SAT problem, we are given a set of clauses on Boolean variables. Each clause has a reward associated with satisfying it. The goal is to find a complete assignment of Boolean values to all variables so as to maximize the sum of the rewards associated with the satisfied clauses. The MAX-SAT problem is a well known NP-hard problem [8].

The reformulation of the MWVC problem to the MAX-SAT problem is easy to understand by first modeling the complement of the MWVC problem—i.e., the MWIS problem—as a MAX-SAT problem. Once again, we associate a Boolean variable X_i with each vertex $v_i \in V$ of weight w_i. For each edge $(v_i, v_j) \in E$, we create the clause $(\overline{X_i} \vee \overline{X_j})$ with a very high reward so that there is no incentive to violate it.[3] These clauses represent an independent set in the graph. For each vertex $v_i \in V$, we also add the singleton clause X_i with an associated reward of w_i. It is easy to see that solving the MAX-SAT problem over all these clauses with their associated rewards solves the MWIS problem on the given graph.

---

[3] It suffices for this reward to be greater than the sum of the weights of all vertices in the graph.

3.3   Reformulation as an ASP

To formulate the MWVC problem as an ASP, we use a constant to represent
each vertex. We define a predicate "edge" to represent the edges in the graph. We
also define a predicate "picked" to represent whether a vertex is in the MWVC.
We define a function "cost" to denote the cost of picking a vertex. Equation 2
captures the nature of undirected edges and vertex cover constraints. The goal
is to minimize the sum of the costs of all picked vertices.

$$
\begin{aligned}
edge(X, Y) &\leftarrow edge(Y, X) \\
picked(X) \vee picked(Y) &\leftarrow edge(X, Y)
\end{aligned}
\tag{2}
$$

3.4   Reformulation as a Series of SAT Instances

An instance of the MWVC problem can be reformulated as a series of SAT
instances with each SAT instance answering the question: "Is there a vertex cover
of weight less than a given test weight w_{test}?" Solving these SAT instances iteratively converges to a solution of the MWVC problem since we can conduct binary
search for the cost of the optimal solution within the interval $[0, \sum_{i=1}^{|V|} w_i]$.[4]

Formulating Each SAT Instance: Consider associating a Boolean variable
X_i with each vertex $v_i \in V$ of weight w_i. X_i indicates the presence of v_i in the
MWVC. Each SAT instance is intended to search for a vertex cover of weight
less than a test weight w_{test}. The clauses in the SAT instance should therefore
encode two properties: (a) the validity of the vertex cover; and (b) the weight of
the vertex cover being less than w_{test}.

The validity of the vertex cover is enforced simply by having a clause $(X_i \vee X_j)$
for each $(v_i, v_j) \in E$. The weight of the vertex cover being less than w_{test} is
enforced by converting the arithmetic operations involved into Boolean operations just like in a digital circuit.

Figure 1 illustrates how to make use of a digital circuit to enforce that the
weight of a vertex cover is less than a test weight. In other words, it enforces
the condition $\sum_{i=1}^{|V|} w_i X_i - w_{test} < 0$. For simplicity of exposition, assume that
all weights are non-negative integers. Each given weight w_i is first converted
to its 2's complement representation. For example, w_1 is converted to '0101'
in the figure. Replacing the '1's in this binary representation by X_i represents
the term $w_i X_i$. To represent $-w_{test}$ on the left side of the condition, we simply
use its 2's complement. For example, $-w_t = -3$ in the figure is represented
as '1101'. A hierarchy of adder circuits adds all these terms—two numbers at a
time as shown in Fig. 1—and produces a final output that represents the quantity
$\sum_{i=1}^{|V|} w_i X_i - w_{test}$. Since we require it to be negative, we simply enforce that
the final sign-bit s is '1'.

---

[4] We can compute much more informed lower and upper bounds as explained later.

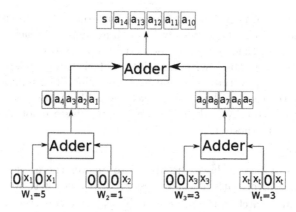

Fig. 1. Shows how to use a digital circuit to enforce the weight of a vertex cover to be less than a given test weight. Assume that there are 3 vertices, v_1, v_2, and v_3, with associated Boolean variables X_1, X_2, and X_3, respectively. The corresponding weights are $w_1 = 5$, $w_2 = 1$, and $w_3 = 3$. The test weight is $w_t = 3$. w_1 is converted to its binary representation '0101' and the '1's are replaced by X_1 to represent the term $w_1 X_1$. w_2 and w_3 are converted in a similar way. For w_t, however, the binary representation '0011' is converted to its negative '1101' in 2's complement representation to represent $-w_t$ (X_t is set to '1'). The final output of the adder circuits represents the quantity $w_1 X_1 + w_2 X_2 + w_3 X_3 - w_t$. The internal variables of the hierarchy of adder circuits are added to the SAT encoding. The constraints dictated by the gates of the digital circuit are added as clauses to the SAT encoding. The final sign-bit s is set to '1' in the SAT encoding to enforce that the result is negative as required.

Once we have a digital circuit, we can convert it into a CNF Boolean formula—i.e., a SAT instance—with Tseitin transformation. The internal variables of the hierarchy of adder circuits are added to the SAT encoding. The constraints dictated by the gates of the digital circuit are also added as clauses to the SAT encoding.[5] Each integer is represented using a non-redundant number of bits. When we add two integers with the longer of the two having k bits, the result is allocated $k + 1$ bits. All operations are done consistently with the 2's complement representation of integers. This reformulation is similar to [30] in the context of translating CSPs into SAT, to [11] in the context of translating pseudo-Boolean constraints into SAT, to [24] in the context of solving disjunctive temporal reasoning problems efficiently, and to [4,27] in the context of solving planning problems.

Several issues need to be addressed in this reformulation of the MWVC problem. Some of them are: (a) the number of auxiliary variables in the SAT instances; (b) the number of clauses in the SAT instances; and (c) the precision of the numbers used to specify the weights. However, these issues have already been addressed in [4,11,24,27] for SAT encodings of other combinatorial problems. The arithmetic operations that we encode using the digital circuit are very

---

[5] We skip a detailed discussion of this transformation since it is similar to the works of various authors mentioned later.

simple: addition ('+'), negation ('-'), and comparison ('<'). This makes the circuit representation compact with only logarithmic depth. If each weight has an L-bit representation, then there are about $|V|$ numbers with L bits each in the bottom level, $|V|/2$ numbers with $L+1$ bits each in the next level, and so on. This leads to $O(L|V|)$ variables in the SAT encoding. The number of internal gates is thus of the same order. This makes the SAT encoding small enough to be solvable by powerful SAT solvers.[6] When the weights are not integral, scaling techniques similar to those in [24] can be used.

Optimizations: Once we have the ability to answer the question of whether there is a vertex cover with a weight less than a given test weight w_{test}, we can employ binary search in the interval $[0, \sum_{i=1}^{|V|} w_i]$ to converge to the MWVC. However, this naive strategy is not very effective without the following optimizations that significantly reduce the number of iterations—i.e., the number of SAT instances to be solved.

The first optimization, *quasi binary search*, is based on the following observation. Suppose, in some iteration, the binary search is in the interval $[L, U]$, the test weight is $w_{test} = (L+U)/2$, and the SAT solver determines that there exists a vertex cover with a weight less than the given test weight w_{test}. Then, the SAT solver is also able to produce a candidate solution with weight $w' < w_{test}$. In the next iteration, therefore, the interval for the binary search can be reduced to $[L, w']$ instead of $[L, w_{test}]$. This can reduce the number of iterations significantly whenever we find a "good" solution, i.e., a small w'.

The second optimization is to make use of an approximation algorithm to produce tighter lower and upper bounds for use in the very first iteration instead of the conservative interval $[0, \sum_{i=1}^{|V|} w_i]$. Clarkson's primal-dual factor-2 approximation algorithm can be used to do so [29]. This algorithm is motivated by a linear programming perspective on the MWVC problem. Using a simple greedy strategy, it constructs integral primal and integral dual solutions simultaneously with the cost of the primal solution being at most twice the cost of the dual solution. The cost of the optimal solution should be in between; and, therefore, the greedily constructed primal solution serves as a factor-2 approximation. If the cost of such an approximate solution is S, then we can set $[S/2, S]$ as the binary search interval in the very first iteration. It is unlikely that we can do better since finding a $2 - \epsilon$ approximation for the MVC or MWVC problem is UG-hard [16].[7]

The third optimization is to run an MVC solver by ignoring all the weights before the first iteration. The cost of the MVC solution produced can then be evaluated to serve as an upper bound for the first iteration of the binary search. However, this method is not guaranteed to be effective since it is completely oblivious to the weights. Nonetheless, it could often produce something useful.

---

[6] In fact, this approach is employed by CircuitTSAT, a state-of-the-art solver for disjunctive temporal reasoning problems [24].

[7] UG-hard means "Unique Games-hard", i.e., hard under the unique games conjecture.

4   Empirical Evaluation

We now compare the ILP, MAX-SAT, PBO, ASP and SAT-based approaches on a variety of MWVC problem instances. We also make important observations about the behaviors of these solvers. For the ILP-based solver, we use Gurobi [13], a state-of-the-art solver for mathematical programming, and lp_solve [1], a popular open source mixed integer linear programming solver. For the MAX-SAT-based solver, we use EvaSolver [23], a state-of-the-art MAX-SAT solver. For the PBO-based solver, we use WBO [22]. For the ASP-based solver, we use clingo from Potassco—the Potsdam Answer Set Solving Collection [12]. Because the MWVC problem is equivalent to the MWIS problem, we can also use a

Table 1. Shows the performances of SBMS, Gurobi and cliquer on unweighted BHOSLIB benchmark problem instances. The column "Iteration" indicates the number of iterations needed to produce the optimal solution or reach the running time limit of 2 h. The column "Initial Bounds" indicates the bounds generated by Clarkson's algorithm. The column "Running Time" indicates the running time in seconds. When the running time exceeds the running time limit, the upper bound in column "Bounds" indicates the cost of the current candidate solution and the lower bound indicates that there cannot be a solution of lower cost. If either the lower bound or upper bound is not specified, it is marked with a '-'. When a problem instance is solved within the running time limit, the cost of the produced solution matches the entry in column "MVC" and the column "Bounds" is marked with a '-' in such a case.

| Graph | | | SBMS | | | | Gurobi | | cliquer | |
Instance	Vertices	MVC	Running Time	Iteration	Bounds	Initial Bounds	Running Time	Bounds	Running Time	Bounds
frb30-15-1	450	420	49.83	8	-	[218, 437]	22.80	-	15.29	-
frb30-15-2	450	420	40.84	8	-	[219, 438]	11.76	-	30.26	-
frb30-15-3	450	420	36.22	8	-	[218, 437]	34.05	-	120.33	-
frb30-15-4	450	420	40.84	8	-	[219, 439]	29.10	-	0.99	-
frb30-15-5	450	420	34.84	8	-	[219, 438]	10.38	-	0.15	-
frb35-17-1	595	560	65.73	8	-	[292, 584]	84.87	-	14.20	-
frb35-17-2	595	560	84.39	8	-	[292, 584]	>7200	[560, 561]	53.66	-
frb35-17-3	595	560	66.97	8	-	[291, 582]	>7200	[560, 561]	>7200	[-, 582]
frb35-17-4	595	560	55.37	8	-	[292, 584]	>7200	[560, 561]	5189.27	-
frb35-17-5	595	560	54.70	8	-	[290, 581]	>7200	[560, 561]	98.84	-
frb40-19-1	760	720	90.76	8	-	[371, 743]	>7200	[720, 722]	>7200	[-, 736]
frb40-19-2	760	720	131.52	9	-	[372, 745]	>7200	[720, 722]	>7200	[-, 733]
frb40-19-3	760	720	127.73	9	-	[372, 744]	>7200	[720, 721]	273.22	-
frb40-19-4	760	720	243.98	9	-	[372, 744]	>7200	[720, 722]	1555.14	-
frb40-19-5	760	720	198.27	9	-	[372, 745]	>7200	[720, 722]	42.77	-
frb45-21-1	945	900	2955.26	9	-	[465, 930]	>7200	[900, 904]	>7200	[-, 917]
frb45-21-2	945	900	235.59	9	-	[465, 930]	>7200	[900, 903]	>7200	[-, 917]
frb45-21-3	945	900	2036.46	9	-	[465, 930]	>7200	[900, 902]	>7200	[-, 913]
frb45-21-4	945	900	884.90	9	-	[465, 931]	>7200	[900, 902]	>7200	[-, 914]
frb45-21-5	945	900	1958.17	9	-	[465, 931]	>7200	[900, 903]	>7200	[-, 922]
frb50-23-1	1150	1100	3208.50	10	-	[556, 1133]	>7200	[1100, 1104]	>7200	[-, 1102]
frb50-23-2	1150	1100	>7200	9	[1100, 1101]	[567, 1135]	>7200	[1100, 1103]	>7200	[-, 1113]
frb50-23-3	1150	1100	111.09	10	-	[567, 1135]	>7200	[1100, 1105]	>7200	[-, 1112]
frb50-23-4	1150	1100	113.10	10	-	[567, 1135]	>7200	[1100, 1104]	1868.10	-
frb50-23-5	1150	1100	113.68	10	-	[568, 1137]	>7200	[1100, 1104]	>7200	[-, 1129]
frb53-24-1	1272	1219	>7200	8	[1219, 1221]	[625, 1250]	>7200	[1219, 1225]	>7200	[-, 1232]
frb53-24-2	1272	1219	114.87	10	-	[625, 1251]	>7200	[1219, 1224]	>7200	[-, 1239]
frb53-24-3	1272	1219	>7200	9	[1219, 1220]	[628, 1256]	>7200	[1219, 1224]	>7200	[-, 1237]
frb53-24-4	1272	1219	>7200	9	[1219, 1220]	[628, 1257]	>7200	[1219, 1224]	>7200	[-, 1228]
frb53-24-5	1272	1219	120.37	10	-	[627, 1255]	>7200	[1219, 1226]	>7200	[-, 1247]
frb56-25-1	1400	1344	>7200	9	[1344, 1345]	[692, 1384]	>7200	[1344, 1350]	>7200	[-, 1365]
frb56-25-2	1400	1344	>7200	9	[1344, 1345]	[691, 1383]	>7200	[1344, 1352]	>7200	[-, 1371]
frb56-25-3	1400	1344	6717.57	10	-	[692, 1384]	>7200	[1344, 1348]	>7200	[-, 1377]
frb56-25-4	1400	1344	>7200	9	[1344, 1345]	[692, 1385]	>7200	[1344, 1350]	>7200	[-, 1348]
frb56-25-5	1400	1344	120.31	10	-	[690, 1381]	>7200	[1344, 1350]	>7200	[-, 1379]
frb59-26-1	1534	1475	>7200	9	[1475, 1476]	[757, 1514]	>7200	[1475, 1482]	>7200	[-, 1493]
frb59-26-2	1534	1475	>7200	9	[1475, 1476]	[757, 1515]	>7200	[1475, 1481]	>7200	[-, 1513]
frb59-26-3	1534	1475	>7200	9	[1475, 1476]	[757, 1514]	>7200	[1475, 1482]	>7200	[-, 1509]
frb59-26-4	1534	1475	>7200	8	[1475, 1477]	[756, 1513]	>7200	[1475, 1481]	>7200	[-, 1516]
frb59-26-5	1534	1475	131.04	10	-	[759, 1519]	>7200	[1475, 1481]	>7200	[-, 1496]

Table 2. Shows the performance of SBMS on a subset of the weighted BHOSLIB benchmark problem instances. "Q" refers to enabling quasi binary search. "C" refers to enabling Clarkson's algorithm. "N" refers to enabling the use of an initial upper bound derived from running NuMVC for 30 s. "None" refers to disabling all optimizations. All running times include the time to perform optimizations as well as the time to perform the actual search. The running times in the "Q+C+N" column and the "C+N" column are almost identical because the evolution of the bounds for these benchmark instances is not affected much by quasi binary search.

Graph			Running time of SBMS (mins)							
Instance	Vertices	MWVC	Q+C+N	C+N	Q+C	Q+N	Q	C	N	None
frb30-15-1	450	825	38.33	38.32	37.68	60.00	35.10	37.49	29.99	35.23
frb30-15-2	450	825	59.97	59.98	58.98	75.12	74.87	59.00	75.00	74.80
frb30-15-3	450	790	0.84	0.84	36.43	0.87	36.84	36.32	0.86	36.73
frb30-15-4	450	825	16.92	16.84	14.47	18.79	18.33	14.39	18.80	18.71
frb30-15-5	450	827	28.28	28.34	47.80	27.73	43.13	47.77	27.75	44.35

clique-based solver that searches for the maximum weighted clique in the edge-complement graph. We therefore additionally use one such state-of-the-art solver in our experiments. In particular, we use cliquer [25] for this purpose. For the SAT-based solver, we use SBMS, which makes use of Lingeling [2], a state-of-the-art complete SAT solver. For SBMS, we also use Clarkson's primal-dual factor-2 approximation algorithm for the MWVC problem [29] to generate the initial lower and upper bounds for the quasi binary search. For the BHOSLIB and DIMACS benchmark problems described below, SBMS also runs NuMVC [5] for 30 s in order to yield a possibly tighter upper bound. Except for Gurobi and Eva-Solver for which we used prebuilt binaries, all solvers were implemented in C++, were compiled by gcc 4.9.2 with the -O3 option, and were run on a GNU/Linux workstation with Intel Xeon Processor E3-1240 v3 (8 MB Cache, 3.4 GHz) and 16 GB RAM.

Since the MWVC problem has not received much attention, there do not exist any benchmark instances for it. However, benchmark instances for the MVC problem do exist, such as the BHOSLIB and DIMACS suites used in [5]. We created MWVC versions of these instances by arbitrarily assigning a weight of $i \bmod 3 + 1$ to a vertex with index i to achieve repeatability of the experiments. As argued before, the number of variables in the SAT encoding increases linearly with the size of the bit representations of the weights and thus only logarithmically with their values (scaled to be non-negative integers).

Clearly, any good MWVC solver should also perform well on regular MVC problem instances. Our first experiment, therefore, used the unweighted version of the BHOSLIB instances. In essence, we solved hard benchmark instances of the MVC problem using a complete solver. Table 1 shows our performance results. We solved more than 50 % of these benchmark instances quite comfortably. Even in the cases that were not solved within the running time limit, SBMS returned solutions with the guarantee that they were no more than a cost of 1 away from

Table 3. Shows the performances of Gurobi, lp_solve, EvaSolver, WBO, clingo, cliquer and SBMS on weighted DIMACS benchmark problem instances. When the running time exceeds the running time limit of 1 h, the corresponding entry is marked with ">3600".

Graph			Running times (secs)						
Instance	Vertices	Edges	Gurobi	lp_solve	EvaSolver	WBO	Clingo	Cliquer	SBMS
brock200_2	200	10024	32.41	168.44	83.86	>3600	23.28	<0.01	212.34
brock200_4	200	6811	52.64	1078.76	491.10	>3600	558.61	0.08	1438.92
brock400_2	400	20014	>3600	>3600	>3600	>3600	>3600	226.74	>3600
brock400_4	400	20035	>3600	>3600	>3600	>3600	>3600	208.74	>3600
brock800_2	800	111434	>3600	>3600	>3600	>3600	>3600	3220.27	>3600
brock800_4	800	111957	>3600	>3600	>3600	>3600	>3600	2826.08	>3600
C1000.9	1000	49421	>3600	>3600	>3600	>3600	>3600	>3600	>3600
C125.9	125	787	0.72	28.64	1649.65	>3600	>3600	3.37	>3600
C2000.5	2000	999164	>3600	>3600	>3600	>3600	>3600	>3600	>3600
C2000.9	2000	199468	>3600	>3600	>3600	>3600	>3600	>3600	>3600
C250.9	250	3141	2058.10	>3600	>3600	>3600	>3600	>3600	>3600
C4000.5	4000	3997732	>3600	>3600	>3600	>3600	>3600	>3600	>3600
C500.9	500	12418	>3600	>3600	>3600	>3600	>3600	>3600	>3600
DSJC1000.5	1000	249674	>3600	>3600	>3600	>3600	>3600	43.42	>3600
DSJC500.5	500	62126	>3600	>3600	>3600	>3600	>3600	0.46	>3600
gen200_p0.9_44	200	1990	3.38	>3600	>3600	>3600	>3600	1722.84	>3600
gen200_p0.9_55	200	1990	0.10	2921.30	872.60	>3600	>3600	43.05	>3600
gen400_p0.9_55	400	7980	>3600	>3600	>3600	>3600	>3600	>3600	>3600
gen400_p0.9_65	400	7980	>3600	>3600	>3600	>3600	>3600	>3600	>3600
gen400_p0.9_75	400	7980	381.13	>3600	>3600	>3600	>3600	>3600	>3600
hamming10-4	1024	89600	>3600	>3600	>3600	>3600	>3600	>3600	>3600
hamming8-4	256	11776	1.34	>3600	800.69	>3600	>3600	0.97	71.42
keller4	171	5100	0.63	475.42	55.65	>3600	132.62	0.03	69.69
keller5	776	74710	1510.35	>3600	>3600	>3600	>3600	>3600	>3600
keller6	3361	1026582	>3600	>3600	>3600	>3600	>3600	>3600	>3600
MANN_a27	378	702	<0.01	0.11	0.12	0.16	>3600	>3600	43.66
MANN_a45	1035	1980	0.01	1.30	0.82	1.26	>3600	>3600	242.38
MANN_a81	3321	6480	0.03	19.54	10.12	18.85	>3600	>3600	1783.03
p_hat1500-1	1500	839327	>3600	>3600	>3600	>3600	>3600	0.98	>3600
p_hat1500-2	1500	555290	>3600	>3600	>3600	>3600	>3600	>3600	>3600
p_hat1500-3	1500	277006	>3600	>3600	>3600	>3600	>3600	>3600	>3600
p_hat300-1	300	33917	56.33	309.62	17.88	>3600	3.01	<0.01	131.43
p_hat300-2	300	22922	94.83	>3600	>3600	>3600	1512.84	0.13	>3600
p_hat300-3	300	11460	1764.59	>3600	>3600	>3600	>3600	47.11	>3600
p_hat700-1	700	183651	>3600	>3600	2887.61	>3600	1548.55	0.04	3532.76
p_hat700-2	700	122922	>3600	>3600	>3600	>3600	>3600	1247.51	>3600
p_hat700-3	700	61640	>3600	>3600	>3600	>3600	>3600	>3600	>3600

the optimal ones.[8] The running times of lp_solve, EvaSolver, WBO and clingo are not listed in Table 1 since none of them could solve any of the benchmark instances within 2 h. It is also easy to see that SBMS significantly outperforms

---

[8] frb53-24-1 and frb59-26-4 are the only two exceptions with a gap of 2.

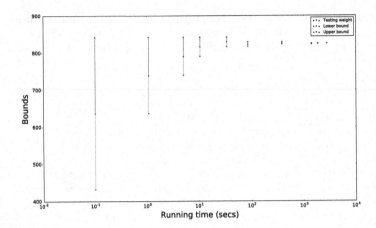

Fig. 2. Shows the evolution of the lower and upper bounds with the running time of SBMS on the weighted BHOSLIB instance frb30-15-1. The mid-point of the bounds is used as the testing weight for the SAT instance posed at that time.

even Gurobi and cliquer both in terms of the number of problem instances solved as well as the quality of the bounds produced.

As expected, the running times of NuMVC are smaller on these benchmark instances compared to the running times of SBMS [5]. However, NuMVC also fails to find an optimal solution on a few of these benchmark instances. In addition, NuMVC solves only MVC problem instances and, furthermore, is an incomplete solver that cannot prove the optimality of the produced solution nor provide optimality bounds. Some other state-of-the-art complete solvers, like MaxCLQdyn+EFL+SCR [21], were not included in the evaluation of NuMVC in [5] since their performance was poor.[9] SBMS, therefore, is a state-of-the-art complete solver for MVC instances.

Our second experiment used the weighted BHOSLIB instances. None of lp_solve, EvaSolver, WBO, clingo or cliquer could solve any of these instances in less than 2 h. Table 2 shows the performance of SBMS—and its variants with various optimization features enabled or disabled—on the first five weighted benchmark instances that it could solve.[10] For the instances that it could not solve, SBMS still produced useful bounds. For generality, our third experiment used a different set of benchmark instances—the weighted DIMACS instances. Table 3 shows the performances of Gurobi, lp_solve, EvaSolver, WBO, clingo, cliquer and SBMS on these instances. Once again, SBMS produces useful bounds when it cannot solve a problem instance.

To understand the anytime property of SBMS, we also ran experiments to observe patterns in its behavior. Figures 2 and 3 show the typical behavior of

---

[9] See the second paragraph on page 18 of [5] that states "... MaxCLQdyn+EFL+SCR is not evaluated on BHOSLIB benchmark which is much harder and requires more effective technologies for exact algorithms ...".

[10] Gurobi was competitive with SBMS on these five instances.

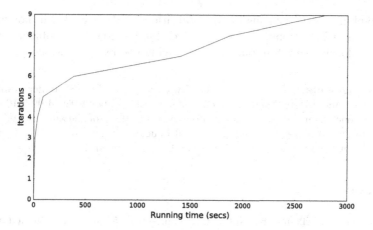

Fig. 3. Shows the iteration number as a function of the running time of SBMS on the weighted BHOSLIB instance frb30-15-1.

SBMS on a fixed benchmark instance. Figure 2 illustrates that the intervals between the optimality bounds typically decrease very quickly, and the solver thus finds a good solution fast. SBMS spends most of the time in trying to improve a good solution to the optimal solution. This "diminishing returns" property which is so pronounced that it is apparent despite a log scale used in the figure is very desirable for an anytime algorithm. Figure 3 reinforces this observation by showing that the SAT instances in the early iterations are much easier to solve than in later iterations (which have smaller intervals between optimality bounds). Thus, by the time the SAT instances get hard to solve, the solver has already found a good solution and is only trying to improve it further.

5 Conclusions and Future Work

In this paper, we presented a SAT-based solver for the MWVC problem. We first argued that, because the MWVC problem is not fixed-parameter tractable, none of the state-of-the-art methods for the MVC problem can be easily modified to tackle the MWVC problem. We compared several solvers based on ILP, MAX-SAT, PBO, ASP and SAT reformulations. Our reformulation of the MWVC problem as a series of SAT instances yields an anytime algorithm that exhibits the "diminishing returns" property and quickly converges to a good solution. In most cases, SBMS significantly outperforms the other methods. SBMS uses quasi binary search in the inner loop and a primal-dual approximation for the MWVC to provide a good starting point.

While SBMS appears to provide an alternative to the few competitive solvers that currently exist for the MWVC problem, we presented it here mostly as a strawman solver for the purpose of gaining interest among AI researchers to study this combinatorial problem more closely. Recent results—like in [19,20]—have demonstrated the importance of the MWVC problem for a wide range of other combinatorial problems in AI applications. Future work will not only be directed toward developing a better solver for the MWVC problem but also toward exploring the full implications of having good solvers available.

Acknowledgments. The research at USC was supported by NSF under grant numbers 1409987 and 1319966 and a MURI under grant number N00014-09-1-1031. The views and conclusions contained in this document are those of the authors and should not be interpreted as representing the official policies, either expressed or implied, of the sponsoring organizations, agencies or the U.S. government.

References

1. Berkelaar, M., Eikland, K., Notebaert, P.: lp_solve 5.5 open source (mixed integer) linear programming software (2004). http://lpsolve.sourceforge.net/5.5/
2. Biere, A.: Lingeling, plingeling and treengeling entering the SAT competition 2013. In: Proceedings of the SAT Competition 2013. Department of Computer Science Series of Publications B, vol. B-2013-1, pp. 51–52 (2013)
3. Boutilier, C., Brafman, R.I., Domshlak, C., Hoos, H.H., Poole, D.: CP-nets: a tool for representing and reasoning with conditional ceteris paribus preference statements. J. Artif. Intell. Res. **21**, 135–191 (2004)
4. Büttner, M., Rintanen, J.: Satisfiability planning with constraints on the number of actions. In: The Proceedings of the International Conference on Automated Planning and Scheduling, pp. 292–299 (2005)
5. Cai, S., Su, K., Luo, C., Sattar, A.: NuMVC: an efficient local search algorithm for minimum vertex cover. J. Artif. Intell. Res. **46**(1), 687–716 (2013)
6. Cai, S., Su, K., Sattar, A.: Local search with edge weighting and configuration checking heuristics for minimum vertex cover. Artif. Intell. **175**(9–10), 1672–1696 (2011)
7. Chen, J., Kanj, I.A., Xia, G.: Improved parameterized upper bounds for vertex cover. In: Královič, R., Urzyczyn, P. (eds.) MFCS 2006. LNCS, vol. 4162, pp. 238–249. Springer, Heidelberg (2006)
8. Cormen, T.H., Leiserson, C.E., Rivest, R.L., Stein, C.: Introduction to Algorithms, 3rd edn. MIT Press, Cambridge (2009)
9. Dinur, I., Safra, S.: On the hardness of approximating minimum vertex cover. Ann. Math. **162**(1), 439–485 (2005)
10. Do, M.B., Benton, J., Briel, M.V.D., Kambhampati, S.: Planning with goal utility dependencies. In: Proceedings of the International Joint Conference on Artificial Intelligence, pp. 1872–1878 (2007)
11. Eén, N., Sörensson, N.: Translating pseudo-boolean constraints into SAT. J. Satisfiability Boolean Model. Comput. **2**, 1–26 (2006)
12. Gebser, M., Kaufmann, B., Kaminski, R., Ostrowski, M., Schaub, T., Schneider, M.: Potassco: the Potsdam answer set solving collection. AI Commun. **24**(2), 107–124 (2011)
13. Gurobi Optimization, I.: Gurobi optimizer reference manual (2015). http://www.gurobi.com
14. Johnson, D.J., Trick, M.A. (eds.): Cliques, Coloring, and Satisfiability: Second DIMACS Implementation Challenge. American Mathematical Society, Boston (1996)

15. Karakostas, G.: A better approximation ratio for the vertex cover problem. ACM Trans. Algorithms **5**(4), 1–8 (2009)
16. Khot, S., Regev, O.: Vertex cover might be hard to approximate to within 2-ε. J. Comput. Syst. Sci. **74**(3), 335–349 (2008)
17. Kolmogorov, V., Zabih, R.: What energy functions can be minimized via graph cuts? IEEE Trans. Pattern Anal. Mach. Intell. **26**(2), 147–159 (2004)
18. Kolmogorov, V.: Primal-dual algorithm for convex Markov random fields. Technical report MSR-TR-2005-117, Microsoft Research (2005)
19. Kumar, T.K.S.: A framework for hybrid tractability results in boolean weighted constraint satisfaction problems. In: Stuckey, P.J. (ed.) CP 2008. LNCS, vol. 5202, pp. 282–297. Springer, Heidelberg (2008)
20. Kumar, T.K.S.: Lifting techniques for weighted constraint satisfaction problems. In: Proceedings of the International Symposium on Artificial Intelligence and Mathematics (2008)
21. Li, C.M., Quan, Z.: Combining graph structure exploitation and propositional reasoning for the maximum clique problem. In: Proceedings of the IEEE International Conference on Tools with Artificial Intelligence, pp. 344–351 (2010)
22. Manquinho, V., Marques-Silva, J., Planes, J.: Algorithms for weighted boolean optimization. In: Proceedings of the International Conference on Theory and Applications of Satisfiability Testing, pp. 495–508 (2009)
23. Narodytska, N., Bacchus, F.: Maximum satisfiability using core-guided MaxSAT resolution. In: Proceedings of the AAAI Conference on Artificial Intelligence, pp. 2717–2723 (2014)
24. Nelson, B., Kumar, T.K.S.: CircuitTSAT: a solver for large instances of the disjunctive temporal problem. In: Proceedings of the International Conference on Automated Planning and Scheduling, pp. 232–239 (2008)
25. Niskanen, S., Östergård, P.R.J.: Cliquer user's guide, version 1.0. Technical report T48, Communications Laboratory, Helsinki University of Technology, Espoo, Finland (2003)
26. Richter, S., Helmert, M., Gretton, C.: A stochastic local search approach to vertex cover. In: Hertzberg, J., Beetz, M., Englert, R. (eds.) KI 2007. LNCS (LNAI), vol. 4667, pp. 412–426. Springer, Heidelberg (2007)
27. Rosa, E.D., Giunchiglia, E., Maratea, M.: Solving satisfiability problems with preferences. Constraints **15**(4), 485–515 (2010)
28. Tomita, E., Kameda, T.: An efficient branch-and-bound algorithm for finding a maximum clique with computational experiments. J. Global Optim. **37**(1), 95–111 (2007)
29. Vazirani, V.V.: Approximation Algorithms. Springer, Heidelberg (2003)
30. Walsh, T.: SAT v CSP. In: Dechter, R. (ed.) CP 2000. LNCS, vol. 1894, pp. 441–456. Springer, Heidelberg (2000)

Optimal Upgrading Schemes for Effective Shortest Paths in Networks

Eduardo Álvarez-Miranda[1(✉)], Martin Luipersbeck[2], and Markus Sinnl[2]

[1] Department of Industrial Engineering, Universidad de Talca, Curicó, Chile
ealvarez@utalca.cl
[2] Department Statistics and Operations Research,
University of Vienna, Vienna, Austria
{martin.luipersbeck,markus.sinnl}@univie.ac.at

Abstract. In this paper, a generalization of a recently proposed optimal path problem concerning decisions for improving connectivity is considered [see 6]. Each node in the given network is associated with a connection delay which can be reduced by implementing upgrading actions. For each upgrading action a cost must be paid, and the sum must satisfy a budget constraint. Given a fixed budget, the goal is to choose a set of upgrading actions such that the total delay of establishing paths among predefined node pairs is minimized. This model has applications in areas like multicast communication planning and wildlife reserve design.

A novel formulation is provided along with an ad-hoc branch-and-cut and a stabilized Benders decomposition algorithm. These strategies exploit connections of the considered problem with other well-known network design problems. Computational results on a large set of instances show the efficacy of the proposed preprocessing methods and optimization algorithms with respect to existing alternatives for the problem. Complementary, the scalability of the models and the corresponding algorithms is investigated with the aim of answering questions raised by [6].

1 Introduction and Motivation

Finding optimal paths in networks is a fundamental task in a plethora of decision making contexts involving traffic in some form. The basic variant consists of finding a minimum distance path between two predefined points (nodes), source and target. The decision on which connections (or edges) must be chosen to connect source and target is typically taken by considering the sum of the lengths or weights of these edges (which is generally minimized), while respecting some other set of topological and/or operative requirements (the reader is referred to [10,16] and the references therein, which propose models and algorithms for different optimal path problems).

Although edge-weighted networks offer a broad range of modeling possibilities in several applications, there exist problems in which decisions must be taken based on the set of nodes traversed by a given path. Hence, the performance of a path depends on node weights rather than edge weights. For instance, in

© Springer International Publishing Switzerland 2016
C.-G. Quimper (Ed.): CPAIOR 2016, LNCS 9676, pp. 406–420, 2016.
DOI: 10.1007/978-3-319-33954-2_29

a multicast communication setting a backbone server broadcasts a signal to many subscribers; the layout of such communication network should be such that delays (which express when the signal traverses a node) between the server and all subscribers must be minimal or bounded. In such a problem, the decision maker seeks for an arrangement of nodes, technologies, and connections such that a positive function of node delays is minimized or it fulfills a Quality-of-Service (QoS) requirement.

The Upgrading Shortest Path problem (USP), originally proposed by [6], fits within the above mentioned context. In this problem, an *upgrading* action can be taken, at a certain cost, in order to decrease the delay induced by nodes. The optimization problem corresponds to (i) finding an upgrade strategy that induces a minimum network delay while respecting a total upgrade cost budget, or (ii) finding an upgrade strategy that yields a minimum total upgrade cost while ensuring that the overall delay is not greater than a given QoS bound. Moreover, and as stated in [6], the USP can be regarded as a decision aid tool for the design of wildlife reserves (see, e.g., [5]).

The problem of finding optimal upgrading schemes for improving network effectiveness has been addressed before. In the seminal work [14] several variants of network upgrading problems are proposed. For these variants delays can be caused both along edges and across nodes, so the upgrade decisions involve both components. The complexity of these problems is provided, showing that they range from polynomially solvable problems up to NP-hard problems. Later on, in [12], a problem closely related to the USP is proposed. If a node is upgraded at a given cost, the weight of all incident edges is decreased. The goal is to find an upgrade strategy so as to reduce the total weight of a corresponding minimum spanning tree in the graph. Only complexity results are provided. In [3] a set of arc-based upgrading problems is proposed; in all cases the aim is to find an upgrading strategy so that a min-max type of objective is optimized. Complexity results are provided as well as heuristic approaches.

From an algorithmic point of view, [6] first prove that the USP is NP-hard. Additionally, the authors provided a mixed integer linear programming (MIP) formulation for the USP and designed two greedy algorithms. The performance of these algorithms is contrasted with results obtained by solving the formulation using a stand-alone MIP solver. The MIP model and the heuristics are able to tackle synthetic grid instances with up to 20×20 nodes and a medium size real-world instance taken from a wildlife planning application. Although interesting, the obtained results reveal the need of developing more sophisticated exact tools able to solve larger instances while still providing reasonable guarantees of optimality.

Contribution and Paper Outline. The aim of this paper is to provide different exact algorithmic tools to solve the budget constrained variant of the generalized counterpart of the USP. Generalized means that one can choose among several upgrading actions at each node. This generalization is mentioned in [6] as an interesting topic for further work. Experimental results on a large set of benchmark instances show that the proposed methods are capable of outperforming the results provided by the compact formulation presented in [6] and, moreover, are capable of solving larger instances.

The paper is organized as follows. A formal definition of the problem and a formulation based on node separators, along with a corresponding exact algorithm, are presented in Sect. 2. A decomposable formulation along with a Benders decomposition scheme is provided in Sect. 3. Computational results on different data sets are reported in Sect. 4. Finally, concluding remarks are drawn in Sect. 5.

2   Cut-Based Formulation

In this Section a formal definition of the problem is first presented. Afterwards, a formulation based on connectivity cuts is given along with a B&C scheme for tackling it.

Problem Definition. Let $G = (V, E)$ be an undirected graph, where V is the set of nodes and E is the set of edges. Set $P \subseteq V \times V$ corresponds to the set of node pairs, say $p = (s, t)$, that must be connected by paths. Let $d_v \geq 0$, be the delay of node v, $v \in V$; likewise, for each upgrading level $l \in L$ and node $v \in V$, let $d_v^l \geq 0$ be the *reduced* delay of node v, if the node is upgraded to level l. Complementary, $c_v^l \geq 0$ corresponds to the cost of *upgrading* node $v \in V$ to level $l \in L$. Finally, let $B \geq 0$ be the total cost budget.

An *upgrading scheme* S is a partition $V^0 \cup V^1 \cup \ldots \cup V^l$ of the node set V, with the meaning that a node $i \in V^l$ is updated to level l, a node $i \in V^0$ is not updated. An upgrading scheme S is feasible, if the cost of the upgrading actions induced by S do not exceed B. Let $\mathcal{S}$ denote the family of all upgrading schemes. Let $D_p(S)$ be the delay of the shortest path connecting $p = \{s, t\}$ under upgrading scheme S. Using this notation, the problem can be formulated as follows

$$\min \left\{ \frac{1}{|P|} \sum_{p \in P} D_p(S) \mid \sum_{l \in L} \sum_{v \in V^l} c_v^l \leq B, S \in \mathcal{S} \right\}.$$

In other words, we look for a feasible upgrading scheme that induces a minimum average path delay. This definition corresponds to the budget-constrained variant of the general USP. Note that even in the case where $|L| = 1$, the problem has been proven to be NP-hard [6]. In the following, the constant term $\frac{1}{|P|}$ will be neglected for ease of exposition.

2.1   Node Separators and MIP Formulation

Let $\mathbf{x} \in \{0, 1\}^{|V| \times |L|}$ be a vector of binary variables such that $x_v^l = 1$ if node $v \in V$ is upgraded to level $l \in L$, and $x_v^l = 0$ otherwise. Likewise, let $\mathbf{y} \in \{0, 1\}^{|V| \times |P|}$ be a vector of binary variables such that $y_{pv} = 1$ if node $v \in V$ is part of the path connecting the pair $p = (s, t) \in P$, and $y_{pv} = 0$ otherwise. Complementary, let $\mathbf{z} \in \{0, 1\}^{|V| \times |P| \times |L|}$ be a vector of binary variables such that $z_{pv}^l = 1$ if the node $v \in V$, upgraded to level $l \in L$, is part of the path connecting the pair $p = (s, t) \in P$, and $z_{pv}^l = 0$ otherwise. For an arbitrary set of nodes, say $S \subseteq V$, for any pair $p \in P$ and for a given $l \in L$, the notation $\mathbf{y}_p(S) = \sum_{v \in S} y_{vp}$ and $\mathbf{z}_p^l(S) = \sum_{v \in S} z_{vp}^l$ will be used. The following definition is required.

Definition 1 (Node separator). *For a given pair $(s,t) \in P$, a subset of nodes $N \subseteq V\backslash\{s,t\}$ is called (s,t) node separator if and only if after eliminating N from V there is no (s,t) path in G. A separator N is minimal if $N\backslash\{i\}$ is not a (s,t) separator, for any $i \in N$. Let $\mathcal{N}(s,t)$ denote the family of all (s,t) separators.*

With these elements, a feasible set of paths along with an upgrading scheme must fulfill the following set of constraints,

$$\mathbf{y}_p(N) + \sum_{l \in L} \mathbf{z}_p^l(N) \geq 1, \qquad \forall N \in \mathcal{N}(s,t), p = (s,t) \in P \qquad \text{(C.1)}$$

$$z_{pv}^l \leq x_v^l, \qquad \forall l \in L, \ \forall v \in V\backslash\{s,t\}, \ \forall p = (s,t) \in P \qquad \text{(C.2)}$$

$$\sum_{l \in L} \sum_{v \in V} c_v^l x_v^l \leq B. \qquad \text{(C.3)}$$

Constraints (C.1) ensure that for every pair $p = (s,t)$ in P, there is a path comprised by a combination of normal nodes or upgraded nodes. Constraint (C.2) imposes that an upgraded node can be used ($z_{pv}^l = 1$), if and only if it has been actually upgraded ($x_v^l = 1$). Finally, constraint (C.3) imposes that any feasible upgrading scheme must meet the budget limitation. Hence, one can formulate the budget constrained USP as follows,

$$\text{(NODE)} \qquad \min \sum_{(s,t)=p \in P} \sum_{v \in V|_{v \neq s,t}} \left(d_v y_{pv} + \sum_{l \in L} d_v^l z_{pv}^l \right) \qquad (1)$$

$$\text{s.t.} \qquad \text{(C.1)--(C.3)} \qquad (2)$$

$$(\mathbf{x}, \mathbf{y}, \mathbf{z}) \in \{0,1\}^{|V| \times |L| + |V| \times |P| + |V| \times |P| \times |L|}. \qquad (3)$$

Note that although this formulation contains an exponential number of constraints (C.1), it can be solved efficiently by a branch-and-cut (B&C) algorithm in which these constraints are added on-the-fly.

2.2 Branch-and-Cut Algorithm

The main ingredient of the B&C approach is its separation scheme, in which violated constraints of type (C.1) are identified during the exploration of the branch-and-bound tree. Moreover, two primal heuristic procedures are also discussed in this Section.

Separation Schemes. Each time (NODE) is solved, the current LP solution $(\tilde{\mathbf{x}}, \tilde{\mathbf{y}}, \tilde{\mathbf{z}})$ is used to compute a set of violated inequalities (C.1). To perform separation, the following transformation of G into a bi-directed graph $G_A' = (V', A')$

is needed (see also [1,7] for similar transformations). This graph is obtained by first bi-directing G (i.e., each edge is replaced by two anti-parallel arcs), and then splitting each node $i \in V$ into an arc (i_1, i_2). In other words, a graph $G'_A = (V', A')$ is created such that $V' = \{i_1 \mid i \in V\} \cup \{i_2 \mid i \in V\}$, $A' = \{(i_2, j_1) \mid (i, j) \in A\} \cup \{(i_1, i_2) \mid i \in V\}$. To separate inequalities (C.1) for a path $(s, t) = p \in P$, arc capacities in G'_A are defined as follows:

$$cap_{uv} = \begin{cases} \tilde{y}_{pi} + \sum_{l \in L} \tilde{z}^l_{pi}, & \text{if } u = i_1, v = i_2, i \in V, i \neq s, t \\ \infty, & \text{otherwise.} \end{cases}$$

Next, the maximum flow/minimum cut between s_2 and t_1 in G'_A is calculated. Note that due to the choice of arc capacities, a minimum (s_2, t_1) cut in G'_A solely contains split-arcs, and thus corresponds to an (s, t) node separator in G. If the computed maximum flow is smaller than one, the associated inequality of type (C.1) is violated, and subsequently added to (NODE).

Alternatively, if the current LP solution $(\tilde{x}, \tilde{y}, \tilde{z})$ is integer at a given node of the search tree, the following more efficient separation scheme runs in linear time for each $p \in P$:

Let $\tilde{G}^p = (\tilde{V}^p, \tilde{E}^p)$ be the subgraph induced by $\tilde{V}^p = \{v \in V \mid \tilde{y}_{pv} + \sum_{l \in L} \tilde{z}^l_{pv} = 1\}$. If $\tilde{G}^p$ contains a path between s and t, no violated inequality exists. Otherwise, $\tilde{G}^p$ contains at least two disconnected components H^p_s and H^p_t, such that $s \in H^p_s$ and $t \in H^p_t$.

Let $\bar{H}^p_t$ be the set of neighboring nodes of H^p_t in G, i.e., $\bar{H}^p_t = \{v \in V \backslash H^p_t \mid \exists \{u, v\} \in E \text{ and } u \in H^p_t\}$. A minimal separator between s and t can be found as follows: (i) delete from G all edges induced by $H^p_t \cup \bar{H}^p_t$; (ii) apply a BFS from s, and let $R(s)$ be the set of all the reached nodes; finally, (iii) the set $\mathcal{N}_{s,t} = R(s) \cap \bar{H}^p_t$ defines a minimal (s, t) node separator, and the corresponding cut of type (C.1) is added to the model.

Primal Heuristic. In order to accelerate the convergence of the method, a simple, but effective, LP-based procedure has been designed with the aim of using the current LP values for the construction of feasible (and eventually incumbent) solutions.

Let $(\tilde{x}, \tilde{y}, \tilde{z})$ be the current LP solution; the primal heuristic works as follows,

Step 1: For every $v \in V$, find $\ell_v = \arg\max_{l \in L}(\tilde{x}^l_v)$, and calculate $\tilde{d}_v = (1 - \tilde{x}^{\ell_v}_v)d_v + \tilde{x}^{\ell_v}_v d^{\ell_v}_v$.

Step 2: Compute the shortest path (SP) for every $(s, t) = p \in P$ using the delay values calculated in **Step 1**; let $\tilde{Y} = \{\tilde{Y}^1, \tilde{Y}^2, \ldots, \tilde{Y}^{|P|}\}$ be such paths.

Step 3: For every $v \in V$ and $l \in L$, compute $\gamma^l_v = \frac{d_v - d^l_v}{c^l_v} \sum_{p \in P} |\tilde{Y}^p \cap v|$. Afterwards, use a knapsack-like heuristic to pack as many upgrades into the paths as possible (which defines **x**), considering the order given by the values γ.

In this procedure, the values of the **x** variables are set in Step 3; hence, the values of the $(\mathbf{y}, \mathbf{z})$ variables can be straightforwardly calculated from the paths obtained in Step 2.

Local Branching. Along with the above mentioned construction heuristic, a state-of-the-art procedure for generating primal solutions using an MIP-solver as black-box, known as Local Branching is implemented (see [8]; the technique is also known as Limited Discrepancy Search [11] within the constraint programming community).

Roughly speaking, for a given (feasible) upgrade scheme $\tilde{\mathbf{x}}$, Let $S = \{(l, v) \mid \tilde{x}_v^l = 1, \forall v \in V, \forall l \in L\}$ be a set of pairs such that each element denotes if a node is upgraded using a certain upgrade type in the current incumbent solution. The goal is to find an improved *neighboring* solution containing at least $|S| - k$ upgrades from the current incumbent solution. This is achieved by solving the current model via branch-and-bound (B&B) after adding the following so-called asymmetric local branching constraint $\sum_{(l,v) \in S}(1 - x_v^l) \leq k$.

Initially, $k := 10$ and a B&B node limit of 10000 and time limit of 10 s are imposed. If within the current neighborhood no improving solution is found for the given node and time limit, k is increased by 5. The procedure is repeated as long as $k \leq 20$. As soon as an improving solution has been found, the B&B is restarted, and k is reset to its initial value. In each B&B node the proposed primal heuristic is executed. Only integer solutions are cut off, i.e., inequalities are only separated when the current LP solution is integral. The cuts separated are gathered in a cut pool and added to the model for subsequent iterations.

At the beginning of the resolution process, the previously described primal heuristic is used to produce a starting solution, using the original delay values.

3   Benders-Based Formulation

3.1   Decomposable Formulation

The USP problem embodies the typical structure of a two-stage like problem; in a first stage one would decide over the upgrading scheme, and on a second stage one would define the corresponding shortest paths. Therefore, the USP becomes a natural candidate to be solved via Benders Decomposition: the master problem decides over the values of vector **x** (upgrading decisions); this solution is then used as parameter when solving the corresponding slave problems (shortest paths) whose solutions are mapped back in the master in the form of so-called (optimality) *Benders* cuts.

Considering the definition of variables presented before, the master problem is given by

$$(\text{BF}) \; \min \left\{ \sum_{p \in P} \theta_p \mid \boldsymbol{\theta} \geq \boldsymbol{\Phi}(\mathbf{x}, P), \; (\text{C.3}) \; \mathbf{x} \in \{0, 1\}^{|V| \times |L|} \text{ and } \boldsymbol{\theta} \in \mathbb{R}_{\geq 0}^{|P|} \right\}, \quad (\text{MP})$$

where $\boldsymbol{\theta} \in \mathbb{R}_{\geq 0}^{|P|}$ corresponds to a set of $|P|$ auxiliary variables, where each of them serves as a surrogate of the lower-bound, given by $\boldsymbol{\Phi}(\mathbf{x}, p)$, of corresponding p-th path.

Recall the graph transformation described before for the separation of connectivity cuts for the linear case. For an optimal solution, say $\mathbf{x}^*$, of (MP), and a given path $(s, t) = p \in P$, the underlying slave problem corresponds to

$$\text{(BF-Sub)} \quad \varPhi(\mathbf{x}^*, p) = \min \sum_{v \in V | v \neq s, t} \left(d_v \bar{y}_v + \sum_{l \in L} d_v^l \bar{z}_v^l \right) \tag{SP.1}$$

$$\text{s.t} \qquad \bar{z}_v^l \leq x_v^{l*} \qquad\qquad\qquad \forall l \in L, \forall v \in V \tag{SP.2}$$

$$\sum_{e \in \delta^-(s^-)} f_{pe} = 0 \text{ and } \sum_{e \in \delta^+(s^+)} f_{pe} = 1 \tag{SP.3}$$

$$\sum_{e \in \delta^-(t^-)} f_{pe} = 1 \text{ and } \sum_{e \in \delta^+(t^+)} f_{pe} = 0 \tag{SP.4}$$

$$\sum_{e \in \delta^-(v^-)} f_{pe} = \bar{y}_v + \sum_{l \in L} \bar{z}_v^l \text{ and } \sum_{e \in \delta^+(v^+)} f_{pe} = \bar{y}_v + \sum_{l \in L} \bar{z}_v^l, \forall v \in V \backslash \{s, t\} \tag{SP.5}$$

$$(\bar{\mathbf{y}}, \bar{\mathbf{z}}) \in \{0, 1\}^{|V| + |V| \times |L|} \text{ and } \mathbf{f} \in [0, 1]^{|A'|}, \tag{SP.6}$$

where $\mathbf{f} \in [0, 1]^{|A'|}$ is a set of flow variables that enable to model an s, t-path on G' and associate the corresponding delay values in the objective function.

The algorithmic scheme designed on the basis of this decomposable formulation will be outlined in detail in Sect. 3.2.

3.2 Benders Decomposition

In the following, the generation of Benders cuts and the details of the implemented stabilization procedure are described. Note that in this paper the Benders decomposition has been implemented within a B&C framework.

Benders Cuts: Fractional and Integer Case. Due to the structure of the above presented formulation, it holds that the slave problem is always feasible for any master solution; therefore the generated cuts are then regarded as *optimality* cuts.

The separation of Benders cuts depends on whether the current master solution $\tilde{\mathbf{x}}$ is integer or not. If $\tilde{\mathbf{x}}$ is fractional, the corresponding slave problem (SP.1)–(SP.6) is solved as a linear problem and the dual multipliers are then used to build the cut.

Note that this decomposition scheme falls within the general scheme for solving fixed-charged (uncapacitated) network design problems (see [4,13], for further details). In particular, for integer $\tilde{\mathbf{x}}$, the subproblem for a $p \in P$ reduces to a shortest-path problem in the graph induced by the upgrades selected in $\tilde{\mathbf{x}}$. Let $SP_p(\tilde{\mathbf{x}})$ be the value of a shortest-path in this graph and for each $l \in L$, let $S^l = \{v \in V \mid \tilde{x}_v^l = 1\}$. If for a given $p \in P$ it holds that $\tilde{\theta}_p < SP_p(\tilde{\mathbf{x}}, p)$, the following inequality cuts off the current integer point,

$$\theta_p \geq SP_p(\tilde{\mathbf{x}}, p) - \sum_{l \in L} \sum_{v \notin S^l} (d_v - d_v^l) x_v^l. \tag{CC}$$

The validity of (CC) can be explained as follows. Clearly, removing any node from S^l cannot improve the value of the shortest path. Moreover, adding a node v to some S^l may improve the value $SP_p(\tilde{\mathbf{x}}, p)$ obtained with the currently selected updates, but the improvement is bounded by $(d_v - d_v^l)$.

Stabilization. Benders decomposition frequently exhibits a strong tailing-off effect, i.e., cutting planes get significantly less effective as the lower bound increases. A possible strategy to address this issue is to include some form of stabilization into the performed separation scheme. In the proposed implementation, a simple stabilization procedure similar to the in-out-method (see [2,9]) is applied at the root node. Instead of performing separation for the (optimal) master LP solution $\tilde{\mathbf{x}}$, a separation point $\mathbf{x}_{sep}$ is computed as linear combination between a stabilization point $\bar{\mathbf{x}}$ and the optimal LP solution $\tilde{\mathbf{x}}$.

For $\bar{\mathbf{x}}$, the vector $\mathbf{1}^n$ is used. The separation point $\mathbf{x}_{sep}$ is computed as $\mathbf{x}_{sep} := \gamma \, \bar{\mathbf{x}} + (1 - \gamma) \, \tilde{\mathbf{x}}$, for some $\gamma \in (0.1, 1]$. In each cutting plane iteration, separation points are iteratively generated until the generated point is violated. The parameter γ is chosen in the form of a binary-search, approaching $\tilde{\mathbf{x}}$ with each iteration. If no violated point is found within five iterations, the stabilization procedure is terminated and separation is performed for $\tilde{\mathbf{x}}$.

When performing separation based on $\tilde{\mathbf{x}}$, adding a small ε to $\tilde{\mathbf{x}}$ improves the strength of cuts. However, during the final cutting plane iterations, this approach may lead to numerical difficulties, so the ε is removed once the lower bound increase between iterations is below a fixed threshold. If the removal of ε does not decrease tailing-off, the cut-loop is terminated and branching is performed. After the root node, separation of optimality cuts is performed without stabilization and the number of cutting plane iterations is limited to five per B&B node.

Primal Heuristic. As for the B&C approach, a scheme for generating primal (master) feasible solutions is embedded into the Benders decomposition. This scheme is basically equivalent to the one designed for the B&C: the master (optimal) solution $\tilde{\mathbf{x}}$ is used for computing a vector of delays $\tilde{\mathbf{d}}$ (Step 1), which are then used to compute a new feasible vector $\check{\mathbf{x}}$ along with values $\check{\theta}^p$ (the value of the corresponding p-th shortest path). The pair $(\check{\mathbf{x}}, \check{\boldsymbol{\theta}})$ is therefore a candidate of a new incumbent solution.

4   Computational Results

Experimental Setting. The algorithmic schemes described in Sects. 2 and 3 have been implemented in C++ using the CPLEX 12.6 Concert framework. All experiments have been performed on an Intel Xeon CPU with 2.5 GHz and 20 cores (only one core is used per run). A fixed memory limit of 16 GB and a time limit of 1800 s have been imposed. The budget B has been set to βB_{max}, where B_{max} corresponds to the total budget necessary to achieve the shortest delay possible. Three different budget configurations have been tested, i.e., $\beta \in \{0.1, 0.25, 0.5\}$.

Benchmark Instances. Two types of instances are considered. The first type corresponds to $N \times N$ grid graphs (see [6]). The second type are random instances generated by the following scheme: Given the number of nodes, arcs are placed randomly between nodes until a specified density $\alpha = |E|/|V|$ is reached and the graph is connected. For each type, set P is defined by randomly selecting $|P|$ pairs of nodes.

For both types of instances, the upgrade costs **c** have been chosen uniformly at random from the range $[50, 1000]$. For defining the delay of upgraded nodes, the following schemes have been considered [6]: (i) **Scaled** – each upgraded delay value is set to $d'_v = cd_v$, $\forall v \in V$, where $c \in [0, 1]$ *scales* d_v. For experiments, c has been set to 0.1, 0.5 and 0.9. (ii) **Constant** – each upgraded delay value $d'_v = 50$. (iii) **Tiered** – for $d_v \in (500, 1000]$, $d'_v = 500$, for $d_v \in (100, 500]$, $d'_v = 75$, and for $d_v \in [50, 100]$, $d'_v = 50$. Thus in total five upgrading schemes are considered: *Scaled = 0.1, Scaled = 0.5, Scaled = 0.9, Constant, Tiered.*

One grid instance has been generated for every combination between upgrading schemes, graph size $N \in \{20, 30\}$ and number of paths $|P| \in \{5, 10, 20, 40\}$ (40 grid instances). Similarly, one dense instance has been generated for every combination between upgrading schemes, number of nodes $|V| \in \{1000, 2000\}$ and graph density $\alpha \in \{4, 8, 6, 32\}$, with a fixed number of paths $|P| = 20$ (40 dense instances).

4.1 Algorithmic Performance

First, experiments are reported which measure the average effect of all implemented algorithmic components separately, i.e., the stabilization procedure, primal heuristics and preprocessing. Afterwards, a detailed comparison of the algorithmic strategies is given. These strategies include B&C algorithms based on the proposed Benders formulation (BF) and cut formulation based on node separators (NODE). As a third strategy the multi-commodity flow formulation (MCF) proposed in [6] is considered.

Table 1 shows the average influence of the implemented stabilization procedure. For each considered budget slack β, all instances have been run once with and without stabilization. Columns $t'_R(s)$ and $t_R(s)$ display the average root relaxation solution time for formulation (BF), with and without stabilization, respectively. The column $g_R(\%)$ lists the average root gap to the best known solution. Results show that the speedup increases with the value of β, reaching

Table 1. Comparison of the average root relaxation solution time for formulation (BF) with (t_R) and without stabilization (t'_R).

β	$t_R(s)$	$t'_R(s)$	$g_R(\%)$
0.1	21.49	19.88	1.59
0.25	58.19	44.62	3.64
0.5	573.48	23.43	0.69

one order of magnitude for $\beta = 0.5$. The gap of the root relaxation is on average already close to the optimum, suggesting that both (MCF) and its Benders reformulation (BF) achieve high-quality bounds.

Table 2 compares the influence of primal heuristics. For each considered budget slack β, all instances have been run once with and without primal heuristics. The columns compare running time and gap for each configuration. Average results are reported only for formulations (NODE) and (BF). For formulation (MCF), the implemented primal heuristics did not manage to outperform the default CPLEX heuristics, and were thus switched off for (MCF) in all subsequent runs. The results show that the primal heuristics play a crucial role for formulations (BF) and (NODE), where less information is available for the LP solver to exploit than for (MCF). Note that since the implemented local branching heuristic is potentially very time-consuming, as the exploration of neighborhoods involves the solution of LPs, in our implementation it is only applied once after solving the root relaxation.

Table 2. Average influence of primal heuristics on running time and gap.

	$\beta = 0.1$		$\beta = 0.25$		$\beta = 0.5$	
	t(s)	g(%)	t(s)	g(%)	t(s)	g(%)
W/O HEUR	535	5.51	600	8.93	470	4.47
HEUR	391	2.68	451	4.70	365	2.04

As a preprocessing step the same procedure as proposed in [6] is implemented, which can be directly incorporated into (BF) and (NODE). On average, the percentage of fixed node variables is 2.82 % (constant), 4.64 % (scaled = 0.1), 25.28 % (scaled = 0.5), 57.24 % (scaled = 0.9) and 24.55 % (tiered). The results show that the preprocessing is not very effective for the delay types which were established as difficult in [6].

Tables 3 and 4 compare the algorithms' performance for $|L| = 1$. For each setting both the running time (in seconds, columns "t(s)") and optimality gap (in percent, columns "g(%)") are reported. All results are partitioned based on budget slack β and delay structure (scaled, tiered, constant). If a run exceeds its time limit, the corresponding time column contains TL. If for a run the formulation's root relaxation could not be solved within the time limit, the gap column contains "–". For each configuration the best results are marked in bold.

The results in Table 3 compare scalability with respect to the number of paths on grid graphs 20×20 and 30×30. For $\beta = 0.1$, the performance of (BF) is best for delay structures scaled = 0.5, scaled = 0.9 and tiered, where (MCF) is outperformed even for small values of $|P|$. For delay structures scaled = 0.1 and constant, the performance is more erratic, and (MCF) frequently achieves comparable or better performance even for $|P| = 40$.

For higher budgets, (BF) also tends to perform worse in general. As already observed by [6], delay structures constant and scaled = 0.1 are more difficult for (MCF), and this also holds for (BF). The worse performance of (BF) for the aforementioned configurations can be explained by the fact that in these cases

Table 3. Test results on grid graphs.

	β = 0.1						β = 0.25						β = 0.5					
	MCF		BF		NODE		MCF		BF		NODE		MCF		BF		NODE	
|P|	t(s)	g(%)	t(s)	g(%)	t(s)	g(%)	t(s)	g(%)	t(s)	g(%)	t(s)	g(%)	t(s)	g(%)	t(s)	g(%)	t(s)	g(%)
N = 20 × 20, Constant																		
5	4	0.0	4	0.0	26	0.0	16	0.0	159	0.0	143	0.0	6	0.0	229	0.0	28	0.0
10	5	0.0	5	0.0	21	0.0	30	0.0	76	0.0	63	0.0	1	0.0	7	0.0	12	0.0
20	271	0.0	266	0.0	TL	0.1	741	0.0	TL	5.8	TL	17.1	146	0.0	254	0.0	549	0.0
40	TL	10.6	TL	18.5	TL	24.0	TL	15.0	TL	22.3	TL	41.3	693	0.0	TL	0.8	TL	5.5
N = 20 × 20, Scaled = 0.1																		
5	2	0.0	4	0.0	20	0.0	14	0.0	72	0.0	187	0.0	10	0.0	61	0.0	44	0.0
10	10	0.0	12	0.0	59	0.0	52	0.0	178	0.0	1363	0.0	20	0.0	90	0.0	78	0.0
20	184	0.0	112	0.0	1279	0.0	1662	0.0	TL	0.5	TL	22.1	197	0.0	530	0.0	1692	0.0
40	TL	5.7	TL	4.1	TL	24.2	TL	18.1	TL	15.8	TL	43.7	TL	1.6	TL	6.1	TL	31.3
N = 20 × 20, Scaled = 0.5																		
5	2	0.0	2	0.0	9	0.0	3	0.0	2	0.0	16	0.0	2	0.0	2	0.0	7	0.0
10	2	0.0	1	0.0	18	0.0	2	0.0	3	0.0	22	0.0	1	0.0	1	0.0	15	0.0
20	9	0.0	5	0.0	93	0.0	6	0.0	6	0.0	74	0.0	29	0.0	28	0.0	117	0.0
40	310	0.0	42	0.0	TL	0.0	328	0.0	124	0.0	1736	0.0	215	0.0	197	0.0	1485	0.0
N = 20 × 20, Scaled = 0.9																		
5	1	0.0	1	0.0	1	0.0	1	0.0	1	0.0	2	0.0	0	0.0	0	0.0	2	0.0
10	1	0.0	1	0.0	3	0.0	1	0.0	1	0.0	2	0.0	1	0.0	1	0.0	2	0.0
20	2	0.0	1	0.0	10	0.0	2	0.0	1	0.0	10	0.0	3	0.0	3	0.0	8	0.0
40	5	0.0	2	0.0	53	0.0	6	0.0	3	0.0	45	0.0	5	0.0	4	0.0	31	0.0
N = 20 × 20, Tiered																		
5	1	0.0	7	0.0	2	0.0	2	0.0	10	0.0	2	0.0	2	0.0	2	0.0	10	0.0
10	2	0.0	7	0.0	1	0.0	1	0.0	7	0.0	1	0.0	1	0.0	1	0.0	13	0.0
20	86	0.0	27	0.0	417	0.0	20	0.0	12	0.0	154	0.0	26	0.0	17	0.0	106	0.0
40	333	0.0	72	0.0	TL	1.1	372	0.0	54	0.0	TL	0.1	29	0.0	20	0.0	438	0.0
N = 30 × 30, Constant																		
5	16	0.0	24	0.0	59	0.0	33	0.0	203	0.0	74	0.0	62	0.0	TL	3.2	138	0.0
10	663	0.0	TL	8.6	TL	17.0	1492	0.0	TL	33.5	TL	36.1	192	0.0	TL	17.1	857	0.0
20	TL	1.8	TL	11.1	TL	24.2	TL	66.1	TL	31.1	TL	46.2	TL	0.2	TL	5.1	TL	12.8
40	TL	–	TL	33.0	TL	48.2	TL	75.0	TL	41.1	TL	60.8	TL	5.4	TL	10.8	TL	20.4
N = 30 × 30, Scaled = 0.1																		
5	5	0.0	8	0.0	10	0.0	7	0.0	13	0.0	27	0.0	5	0.0	19	0.0	23	0.0
10	55	0.0	86	0.0	1778	0.0	1089	0.0	TL	1.9	TL	6.6	648	0.0	TL	6.0	TL	37.2
20	TL	1.9	TL	3.0	TL	15.8	TL	66.6	TL	34.5	TL	44.9	TL	20.8	TL	25.8	TL	48.0
40	TL	–	TL	29.2	TL	44.3	TL	74.8	TL	49.7	TL	62.1	TL	64.4	TL	28.5	TL	50.4
N = 30 × 30, Scaled = 0.5																		
5	4	0.0	5	0.0	49	0.0	12	0.0	17	0.0	55	0.0	14	0.0	46	0.0	56	0.0
10	5	0.0	3	0.0	24	0.0	13	0.0	6	0.0	42	0.0	6	0.0	6	0.0	24	0.0
20	381	0.0	36	0.0	1632	0.0	1687	0.0	316	0.0	TL	0.3	TL	0.1	1209	0.0	TL	0.6
40	TL	0.0	270	0.0	TL	5.4	TL	0.2	546	0.0	TL	9.9	TL	0.0	TL	0.0	TL	0.4
N = 30 × 30, Scaled = 0.9																		
5	1	0.0	1	0.0	1	0.0	1	0.0	1	0.0	2	0.0	1	0.0	1	0.0	1	0.0
10	5	0.0	5	0.0	70	0.0	9	0.0	4	0.0	84	0.0	9	0.0	4	0.0	68	0.0
20	8	0.0	7	0.0	95	0.0	7	0.0	4	0.0	96	0.0	4	0.0	10	0.0	80	0.0
40	69	0.0	53	0.0	TL	0.0	90	0.0	32	0.0	720	0.0	113	0.0	27	0.0	936	0.0
N = 30 × 30, Tiered																		
5	2	0.0	1	0.0	1	0.0	1	0.0	1	0.0	3	0.0	1	0.0	1	0.0	3	0.0
10	28	0.0	19	0.0	256	0.0	13	0.0	21	0.0	129	0.0	7	0.0	8	0.0	99	0.0
20	73	0.0	32	0.0	TL	0.5	228	0.0	80	0.0	TL	0.0	70	0.0	79	0.0	1635	0.0
40	378	0.0	65	0.0	TL	8.8	727	0.0	1034	0.0	TL	2.3	144	0.0	62	0.0	TL	1.6

Table 4. Test results on dense graphs.

	β = 0.1						β = 0.25						β = 0.5					
	MCF		BF		NODE		MCF		BF		NODE		MCF		BF		NODE	
|E|/|V|	t(s)	g(%)	t(s)	g(%)	t(s)	g(%)	t(s)	g(%)	t(s)	g(%)	t(s)	g(%)	t(s)	g(%)	t(s)	g(%)	t(s)	g(%)
|V| = 1000, |P| = 20, Constant																		
4	255	0.0	372	0.0	1277	0.0	1205	0.0	544	0.0	710	0.0	65	0.0	36	0.0	46	0.0
8	TL	1.3	537	0.0	366	0.0	173	0.0	444	0.0	112	0.0	112	0.0	1057	0.0	37	0.0
16	TL	0.4	855	0.0	307	0.0	815	0.0	404	0.0	108	0.0	81	0.0	63	0.0	12	0.0
32	797	0.0	811	0.0	42	0.0	460	0.0	415	0.0	25	0.0	118	0.0	87	0.0	12	0.0
|V| = 1000, |P| = 20, Scaled = 0.1																		
4	185	0.0	84	0.0	225	0.0	514	0.0	670	0.0	TL	1.2	379	0.0	320	0.0	328	0.0
8	102	0.0	63	0.0	33	0.0	225	0.0	168	0.0	86	0.0	381	0.0	330	0.0	83	0.0
16	62	0.0	73	0.0	9	0.0	101	0.0	103	0.0	11	0.0	83	0.0	138	0.0	14	0.0
32	51	0.0	108	0.0	4	0.0	123	0.0	104	0.0	8	0.0	118	0.0	81	0.0	6	0.0
|V| = 1000, |P| = 20, Scaled = 0.5																		
4	6	0.0	2	0.0	10	0.0	20	0.0	6	0.0	17	0.0	39	0.0	13	0.0	18	0.0
8	23	0.0	8	0.0	10	0.0	22	0.0	7	0.0	13	0.0	13	0.0	3	0.0	11	0.0
16	14	0.0	6	0.0	5	0.0	51	0.0	8	0.0	12	0.0	46	0.0	9	0.0	16	0.0
32	38	0.0	20	0.0	5	0.0	146	0.0	31	0.0	8	0.0	241	0.0	53	0.0	10	0.0
|V| = 1000, |P| = 20, Scaled = 0.9																		
4	10	0.0	2	0.0	13	0.0	11	0.0	3	0.0	13	0.0	6	0.0	1	0.0	6	0.0
8	15	0.0	3	0.0	10	0.0	20	0.0	10	0.0	8	0.0	24	0.0	8	0.0	8	0.0
16	21	0.0	3	0.0	5	0.0	25	0.0	4	0.0	5	0.0	20	0.0	4	0.0	4	0.0
32	34	0.0	8	0.0	4	0.0	42	0.0	7	0.0	4	0.0	43	0.0	8	0.0	4	0.0
|V| = 1000, |P| = 20, Tiered																		
4	23	0.0	11	0.0	40	0.0	63	0.0	21	0.0	31	0.0	26	0.0	27	0.0	20	0.0
8	29	0.0	22	0.0	23	0.0	39	0.0	22	0.0	22	0.0	76	0.0	32	0.0	25	0.0
16	13	0.0	9	0.0	6	0.0	38	0.0	10	0.0	6	0.0	17	0.0	7	0.0	5	0.0
32	70	0.0	26	0.0	10	0.0	53	0.0	29	0.0	6	0.0	71	0.0	8	0.0	4	0.0
|V| = 2000, |P| = 20, Constant																		
4	977	0.0	TL	4.1	TL	1.8	TL	60.9	TL	16.0	TL	33.7	983	0.0	1001	0.0	1466	0.0
8	TL	7.9	TL	7.3	TL	5.0	TL	3.6	TL	7.3	TL	26.7	494	0.0	439	0.0	83	0.0
16	TL	–	TL	17.2	TL	29.2	TL	64.5	TL	8.2	1792	0.0	291	0.0	510	0.0	42	0.0
32	TL	–	TL	19.3	776	0.0	TL	56.7	TL	9.4	273	0.0	1486	0.0	1466	0.0	96	0.0
|V| = 2000, |P| = 20, Scaled = 0.1																		
4	60	0.0	127	0.0	69	0.0	276	0.0	518	0.0	303	0.0	447	0.0	TL	0.2	210	0.0
8	85	0.0	139	0.0	30	0.0	222	0.0	383	0.0	69	0.0	484	0.0	973	0.0	133	0.0
16	560	0.0	912	0.0	125	0.0	563	0.0	520	0.0	86	0.0	380	0.0	542	0.0	38	0.0
32	98	0.0	261	0.0	14	0.0	385	0.0	403	0.0	31	0.0	893	0.0	1047	0.0	102	0.0
|V| = 2000, |P| = 20, Scaled = 0.5																		
4	12	0.0	5	0.0	35	0.0	28	0.0	7	0.0	36	0.0	31	0.0	13	0.0	47	0.0
8	86	0.0	21	0.0	60	0.0	64	0.0	30	0.0	67	0.0	72	0.0	24	0.0	58	0.0
16	81	0.0	29	0.0	23	0.0	48	0.0	20	0.0	19	0.0	77	0.0	26	0.0	25	0.0
32	70	0.0	42	0.0	19	0.0	324	0.0	148	0.0	76	0.0	298	0.0	63	0.0	17	0.0
|V| = 2000, |P| = 20, Scaled = 0.9																		
4	24	0.0	6	0.0	15	0.0	10	0.0	2	0.0	21	0.0	33	0.0	14	0.0	18	0.0
8	26	0.0	5	0.0	30	0.0	49	0.0	12	0.0	21	0.0	24	0.0	7	0.0	27	0.0
16	45	0.0	7	0.0	15	0.0	21	0.0	6	0.0	12	0.0	53	0.0	11	0.0	22	0.0
32	70	0.0	15	0.0	28	0.0	96	0.0	16	0.0	13	0.0	108	0.0	27	0.0	22	0.0
|V| = 2000, |P| = 20, Tiered																		
4	43	0.0	41	0.0	149	0.0	131	0.0	115	0.0	239	0.0	41	0.0	58	0.0	84	0.0
8	110	0.0	42	0.0	62	0.0	142	0.0	109	0.0	82	0.0	569	0.0	412	0.0	193	0.0
16	62	0.0	32	0.0	20	0.0	76	0.0	43	0.0	24	0.0	246	0.0	72	0.0	46	0.0
32	55	0.0	38	0.0	11	0.0	118	0.0	47	0.0	12	0.0	63	0.0	37	0.0	7	0.0

Table 5. Test results on grid graphs with three upgrade types per node.

	$\beta = 0.1$						$\beta = 0.25$						$\beta = 0.5$					
	MCF		BF		NODE		MCF		BF		NODE		MCF		BF		NODE	
$|P|$	t(s)	g(%)	t(s)	g(%)	t(s)	g(%)	t(s)	g(%)	t(s)	g(%)	t(s)	g(%)	t(s)	g(%)	t(s)	g(%)	t(s)	g(%)
$N = 20, |P| = 20, |L| = 3, Scaled = \{0.1, 0.5, 0.9\}$																		
5	**21**	**0.0**	237	0.0	173	0.0	**81**	**0.0**	TL	5.3	804	0.0	**23**	**0.0**	TL	3.3	152	0.0
10	**15**	**0.0**	16	0.0	192	0.0	78	0.0	**52**	**0.0**	84	0.0	**12**	**0.0**	46	0.0	44	0.0
20	TL	5.2	TL	7.8	TL	16.8	TL	12.2	TL	27.5	TL	44.4	**176**	**0.0**	1035	0.0	TL	0.1
40	TL	54.7	**TL**	**5.9**	TL	31.8	TL	9.9	TL	35.6	TL	51.6	**397**	**0.0**	540	0.0	TL	0.1
$N = 30, |P| = 20, |L| = 3, Scaled = \{0.1, 0.5, 0.9\}$																		
5	**14**	**0.0**	35	0.0	197	0.0	**34**	**0.0**	459	0.0	496	0.0	**27**	**0.0**	TL	13.4	201	0.0
10	TL	7.5	TL	17.9	TL	23.8	**TL**	**16.0**	TL	49.0	TL	53.4	**250**	**0.0**	TL	2.2	1365	0.0
20	TL	13.4	TL	14.6	TL	29.6	**TL**	**24.8**	TL	40.9	TL	42.5	**431**	**0.0**	TL	5.8	TL	3.0
40	TL	–	**TL**	**41.0**	TL	63.9	TL	–	**TL**	**53.9**	TL	67.5	**TL**	**6.0**	TL	13.7	TL	32.4

Table 6. Test results on dense graphs with three upgrade types per node.

	$\beta = 0.1$						$\beta = 0.25$						$\beta = 0.5$					
	MCF		BF		NODE		MCF		BF		NODE		MCF		BF		NODE	
$|E|/|V|$	t(s)	g(%)	t(s)	g(%)	t(s)	g(%)	t(s)	g(%)	t(s)	g(%)	t(s)	g(%)	t(s)	g(%)	t(s)	g(%)	t(s)	g(%)
$|V| = 1000, |P| = 20, |L| = 3, Scaled = \{0.1, 0.5, 0.9\}$																		
4	526	0.0	**405**	**0.0**	726	0.0	**94**	**0.0**	235	0.0	208	0.0	**608**	**0.0**	1100	0.0	1471	0.0
8	1346	0.0	809	0.0	**578**	**0.0**	1359	0.0	920	0.0	**805**	**0.0**	212	0.0	94	0.0	**38**	**0.0**
16	TL	4.5	TL	4.3	**1528**	**0.0**	TL	4.6	1558	0.0	**1275**	**0.0**	279	0.0	159	0.0	**31**	**0.0**
32	TL	23.6	1369	0.0	**290**	**0.0**	693	0.0	594	0.0	**77**	**0.0**	95	0.0	120	0.0	**21**	**0.0**
$|V| = 2000, |P| = 20, |L| = 3, Scaled = \{0.1, 0.5, 0.9\}$																		
4	1164	0.0	966	0.0	**398**	**0.0**	TL	14.7	**TL**	**13.5**	TL	36.4	952	0.0	1106	0.0	**709**	**0.0**
8	TL	–	**TL**	**21.7**	TL	23.1	TL	–	**TL**	**22.5**	TL	40.4	868	0.0	1648	0.0	**510**	**0.0**
17	TL	–	TL	4.5	**878**	**0.0**	TL	–	TL	28.3	**TL**	**26.3**	TL	0.3	1450	0.0	**336**	**0.0**
32	TL	–	TL	7.3	**1519**	**0.0**	TL	–	TL	31.2	**TL**	**9.8**	756	0.0	843	0.0	**105**	**0.0**

far more optimality cuts are generated, which in turn slow down the solution of the master problem. For (NODE), which clearly performs worst in most observed cases, a similar problem occurs. Here a large number of cuts is required to enforce connectivity on extremely sparse grid graphs. The high difficulty of this instance type for algorithms based on branch-and-cut is also known for similar problems, e.g., the Steiner tree problem, where large-scale grid graphs remain challenging even for state-of-the-art approaches (see, e.g., [15]).

In Table 4, results on dense graphs with varying values for $|E|/|V|$ are reported. Here (MCF) only manages to outperform other approaches on instances with constant delay structure which are relatively sparse. For all higher densities, (MCF) quickly becomes less practical, and is outperformed both by (BF) and (NODE). Here (NODE) clearly performs best, and is less affected by different delay structures and budget slacks. The performance of (BF) is similar to (NODE) except for delay structures constant and scaled = 0.1.

4.2   Multiple Upgrades

In this section the case $|L| > 1$ is explored. For this purpose grid and dense graphs with three upgrade levels have been constructed based on the scaled delay structure, i.e., for each node there exist three possible upgrades using 0.1,

0.5 and 0.9 as scaling factor. Again costs are computed randomly in the range of $[50, 1000]$, but are assigned to upgrades per node such that the upgrade with the lowest delay is assigned the highest cost.

Table 5 reports results for grid graphs with varying values of $|P|$. The results show that on the grid graphs (MCF) performs best for the considered weights. Only for low budgets and a high number of paths becomes (BF) more competitive. Table 6 reports results for dense graphs with varying values of $|E|/|V|$. As for $|L| = 1$, (NODE) performs best on average, only being outperformed by (MCF) for graphs of lowest density. For higher densities, the root relaxation of (MCF) cannot be solved within the time limit.

5   Conclusions and Future Work

In this paper, algorithmic expedients along with computational results are presented for the Upgrading Shortest Path Problem (USP). The USP is a recently proposed network optimization problem that enables to model a variety of decision making problems where the goal is to optimize the effectiveness of the sought network while respecting a design budget.

The proposed algorithms show to be effective for quite large instances, being able to reach rather small optimality gaps, within reasonable computing times, for instances with medium to large sizes. Moreover, these tailored strategies outperform the use of a compact formulation even if medium size instances are considered.

Acknowledgements. E. Álvarez-Miranda is supported by the Chilean Council of Scientific and Technological Research through the grant FONDECYT N.11140060 and through the Complex Engineering Systems Institute (ICM:P-05-004-F, CONICYT:FBO16). M. Sinnl is supported by the Austrian Research Fund (FWF, Project P 26755-N19). M. Luipersbeck acknowledges the support of the University of Vienna through the uni:docs fellowship programme.

References

1. Álvarez-Miranda, E., Ljubić, I., Mutzel, P.: The rooted maximum node-weight connected subgraph problem. In: Gomes, C., Sellmann, M. (eds.) CPAIOR 2013. LNCS, vol. 7874, pp. 300–315. Springer, Heidelberg (2013)
2. Ben-Ameur, W., Neto, J.: Acceleration of cutting-plane and column generation algorithms: applications to network design. Networks **49**(1), 3–17 (2007)
3. Campbell, A., Lowe, T., Zhang, L.: Upgrading arcs to minimize the maximum travel time in a network. Networks **47**(2), 72–80 (2006)
4. Costa, A.: A survey on benders decomposition applied to fixed-charge network design problems. Comput. OR **32**(6), 1429–1450 (2005)
5. Dilkina, B., Gomes, C.P.: Solving connected subgraph problems in wildlife conservation. In: Lodi, A., Milano, M., Toth, P. (eds.) CPAIOR 2010. LNCS, vol. 6140, pp. 102–116. Springer, Heidelberg (2010)

6. Dilkina, B., Lai, K.J., Gomes, C.P.: Upgrading shortest paths in networks. In: Achterberg, T., Beck, J.C. (eds.) CPAIOR 2011. LNCS, vol. 6697, pp. 76–91. Springer, Heidelberg (2011)
7. Fischetti, M., Leitner, M., Ljubić, I., Luipersbeck, M., Monaci, M., Resch, M., Salvagnin, D., Sinnl, M.: Thinning out steiner trees: a node-based model for uniform edge costs. In: Workshop of the 11th DIMACS Implementation Challenge (2014)
8. Fischetti, M., Lodi, A.: Local branching. Math. Program. **98**(1–3), 23–47 (2003)
9. Fischetti, M., Salvagnin, D.: An in-out approach to disjunctive optimization. In: Lodi, A., Milano, M., Toth, P. (eds.) CPAIOR 2010. LNCS, vol. 6140, pp. 136–140. Springer, Heidelberg (2010)
10. Fournier, J.: Optimal paths, pp. 119–147. ISTE (2010)
11. Harvey, W.D., Ginsberg, M.L.: Limited discrepancy search. In: IJCAI (1), pp. 607–615 (1995)
12. Krumke, S., Marathe, M., Noltemeier, H., Ravi, R., Ravi, S., Sundaram, R., Wirth, H.: Improving minimum cost spanning trees by upgrading nodes. J. Algorithms **33**(1), 92–111 (1999)
13. Magnanti, T., Mireault, P., Wong, R.: Tailoring benders decomposition for uncapacitated network design. In: Gallo, G., Sandi, C. (eds.) Netflow at Pisa, Mathematical Programming Studies, vol. 26, pp. 112–154. Springer, Heidelberg (1986)
14. Paik, D., Sahni, S.: Network upgrading problems. Networks **26**(1), 45–58 (1995)
15. Polzin, T.: Algorithms for the Steiner problem in networks. Ph.D. thesis, Universitätsbibliothek (2003)
16. Vassilevska, V.: Efficient algorithms for path problems in weighted graphs. Ph.D. thesis, Carnegie Mellon University, August 2008

Author Index

Printed in the United States
By Bookmasters

Printed in the United States
By Bookmasters